The Physics and Applications
of Amorphous Semiconductors

The Physics and Applications of Amorphous Semiconductors

Arun Madan
Glasstech Solar Inc.
Wheatridge, Colorado

Melvin P. Shaw
Department of Electrical and Computer Engineering
Wayne State University
Detroit, Michigan

ACADEMIC PRESS, INC.

Harcourt Brace Jovanovich, Publishers

Boston   San Diego
New York   London
Sydney   Tokyo   Toronto

Copyright © 1988 by Academic Press, Inc.
All rights reserved.
No part of this publication may be reproduced or
transmitted in any form or by any means, electronic
or mechanical, including photocopy, recording, or
any information storage and retrieval system, without
permission in writing from the publisher.

ACADEMIC PRESS, INC.
1250 Sixth Avenue, San Diego, CA 92101

United Kingdom Edition published by
ACADEMIC PRESS, INC. (LONDON) LTD.
24–28 Oval Road, London NW1 7DX

Library of Congress Cataloging-in-Publication Data

Madan, A. (Arun)
    The physics and applications of amorphous semiconductors/Arun
Madan, Melvin P. Shaw.
    p.   cm.
   Bibliography: p.
   Includes index.
   ISBN 0-12-464960-2
    1. Amorphous semiconductors.   I. Shaw, Melvin P. II. Title.
QC611.8.A5M33 1988
530.6′22—dc19

                                                              87-20881
                                                                 CIP

Printed in the United States of America
87 88 89 90   9 8 7 6 5 4 3 2 1

Dedicated to
Angie, Damian and Nadia Madan,
Bernetta Miller-Shaw,
and
The memory of David Adler

Contents

Preface ix

Chapter 1 Introduction

1.0 Glassy, Non-Crystalline and Amorphous Solids 1
1.1 Classification of Amorphous Materials 3
1.2 Basic Concepts 4
1.3 The Effect of Disorder 5

Chapter 2 Opto-Electronic Properties of Amorphous Silicon Based Materials

2.1 Growth Kinetics 12
2.2 Structure and Composition 27
2.3 Density of States (DOS) 35
2.4 Optical Absorption 62
2.5 Electrical Transport 72
2.6 Photoconductivity 97
2.7 Photoluminescence 106
2.8 Doping in Glow Discharge a-Si Based Alloys 115
2.9 Surface States on Amorphous Silicon 127
2.10 Hydrogenated Amorphous Silicon Using Techniques Other Than Glow Discharge Decomposition 134
2.11 Some Properties of Halogenated Amorphous Silicon Alloys 140
2.12 Narrow and Wide Band Gap Alloys 149

Chapter 3 Opto-Electronic Applications of Amorphous Silicon Based Materials

3.1 Solar Cells 163
3.2 Thin Film Transistors for Display Applications 276
3.3 Miscellaneous Applications 309

Chapter 4 Characterization and Properties of Amorphous Chalcogenide Alloys

4.1	Introduction	318
4.2	Optical Properties of Amorphous Chalcogenide Films	322
4.3	Electrical Transport Properties	331
4.4	Summary and Conclusions	351

Chapter 5 Electrical Switching and Memory Devices Employing Films of Amorphous Chalcogenide Alloys

5.1	Introduction	356
5.2	The Thermistor	359
5.3	Thermally Induced NDC	370
5.4	Thin Chalcogenide Films	382
5.5	Thermophonic Studies of Thick Chalcogenide Films	409
5.6	Electronic Models for Threshold Switching	433
5.7	Summary	454

Chapter 6 Electrophotography

6.1	Introduction	470
6.2	The Physics of Electrophotography	476
6.3	The Physics of the Development Process	489
6.4	Charging of the Toner Particles	491
6.5	Material Requirements for Photoreceptors	491
6.6	Electrophotographic Applications of Amorphous Hydrogenated Silicon	497

Chapter 7 Optical Memories

7.1	Introduction	501
7.2	General Requirements for the Optical Data Disc	505
7.3	Optical Massmemories Based on Crystalline/Amorphous Phase Transitions in Chalcogenide Glasses	507
7.4	Optical Massmemories Based on the Thermal Creation of Holes	512
7.5	Recent Technological Developments	516

References 517
Subject Index 539

Preface

During the past twenty years the subject of amorphous semiconductors has progressed from almost total obscurity to the attainment of device structures such as solar cells, flat panel displays and optical memories that are now commercially available. The subject was brought to the attention of the scientific and business community in 1968 by S. R. Ovshinsky, who advocated the use of chalcogenide materials for electronic switching, and promised a revolution in the electronics industry. However, these materials have so far proven to be useful only for optical applications. The present impetus has resulted from breakthroughs in thin film amorphous silicon alloy films made by W. E. Spear's group at the University of Dundee in Scotland. Their results led to the demonstration of viable solar cell devices by D. E. Carlson and C. R. Wronski of the RCA Laboratories.

The book is aimed at the semiconductor device physicist, engineer and student; a knowledge of solid state physics on a level with the text *Solid State Physics*, by Kittel (Wiley, New York, 1976) is assumed. Furthermore, it is hoped that the book will be useful to device engineers and designers wishing to apply these devices in novel and creative ways. To this end, we emphasize analytical and numerical treatments of both the material and device structures. The details of the physical principles involved in understanding amorphous solids in general are not emphasized in the text, but can be found in books such as *The Physics of Amorphous Solids*, by R. Zallen (Wiley, N.Y., 1983) and *Electronic Processes in Non-Crystalline Materials*, by N. Mott and E. A. Davis (Clarendon, Oxford, 1979). Our introductory chapter is used to classify glassy, non-crystalline and amorphous materials, and emphasize basic concepts and the effects of disorder. The next two chapters stress the opto-electronic properties and applications of amorphous silicon based materials. We treat important features such as film preparation, the density of states, transport and photoconductivity. Devices such as solar cells and thin film transistors are emphasized. Image scanners, optical recorders, charge coupled devices and gas sensors are also discussed.

Chapter four characterizes the amorphous chalcogenide alloys in a manner similar to chapter one. The last three chapters then focus on devices that have relied heavily on the properties of these materials: electrical switching and memory devices; electrophotography; optical memories. It is our hope that the treatment we provide in this monograph will allow students and researchers alike to develop a firm foundation for the understanding and predictability of the optical, electrical and physical behavior or amorphous materials and devices in general.

We are grateful to many of our colleagues and friends, worldwide, for their interest in our research and in the development of this book. Among them, special thanks for their support and encouragement are due to the late David Adler, Pawan Bhat, Mike Cicak, John Deneufville, Edward Fagen, Richard Flasck, Helmut Fritzsche, Irving Gastman, George Cheroff, Harold Grubin, Heinz Henisch, Scott Holmberg, Sergey Kostylev, James Kotz, Peter LeComber, Harold McMaster, Simon Moss, Nevill Mott, Norman Nitschke, Stanford Ovshinsky, Kurt Petersen, Howard Rockstad, Marvin Silver, Peter Solomon, Walter Spear, David Strand, Bolko von Roedern, and Jianping Xi.

We are deeply grateful to Marilyn Wazny and Renie Ayers for typing many revisions of the manuscript.

Special thanks for support, past and present, is also due the University of Dundee, Wayne State University, Energy Conversion Devices, Inc., the Solar Energy Research Institute and Glasstech Solar, Inc. Our work at these laboratories provided the foundation for the writing of this book.

The Physics and Applications of Amorphous Semiconductors

1

Introduction

1.0 Glassy, Non-Crystalline and Amorphous Solids 1
1.1 Classification of Amorphous Materials 3
1.2 Basic Concepts 4
1.3 The Effect of Disorder 5

1.0 Glassy, Non-Crystalline and Amorphous Solids

Amorphous semiconductors have become increasingly important because of their significance in many diverse applications. The technological aspects of these materials can be roughly divided into two classes: (1) hydrogenated amorphous silicon (a-Si:H) type alloys, whose applications include inexpensive solar cells, thin film transistors, image scanners, electrophotography, optical recording and gas sensors; (2) amorphous chalcogenides, whose applications include electrophotography, switching and memory elements. We shall treat a-Si:H alloys in the first part of the book (Chaps. 2 and 3) and chalcogenide materials in the second (Chaps. 4 through 7).

The term glassy and/or non-crystalline material has been used synonymously with the term amorphous material. It is important to understand that amorphous semiconductors are metastable thin solid film materials, whereas a glass is a viscous liquid that is normally formed by the continuous hardening of a cooled liquid. The latter is to be distinguished from a solid that is formed either by a discontinuous solidification in which a solid mass appears and grows in the liquid, the solidification occuring at the liquid-solid interface, or, as will be prevalent in this book, a thin film that condenses onto a substrate from the gaseous phase.

An ideal glass or an amorphous material is a solid in internal equilibrium in which there is, just as in crystalline solids, a definite set of equilibrium positions about which the atoms oscillate. However, in contrast to crystalline solids, neither the ideal glass nor the amorphous material exhibits

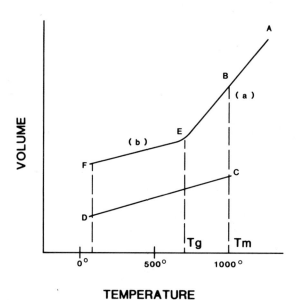

Fig. 1.1. Changes in glass volume on cooling. The line ABCD represents a substance that crystallizes at B; the line ABEF, the glassy form of the same material.

translational symmetry. The difference between a glass and the corresponding liquid or supercooled liquid may be conveniently demonstrated by the Volume-Temperature (V-T) relationship shown in Fig. 1.1. Upon slowly cooling a liquid from any temperature, T, to the melting point of the crystal, T_m, a path "a" will be followed. However, when crystallization is inhibited by rapid cooling, the V-T relationship follows path "b", to produce a supercooled liquid. Upon further cooling to the glass transition temperature T_g, when the viscosity of the supercooled liquid is approximately 10^{13}–10^{14} poise, a break in the curve occurs at a point indicated by E. There is an arbitrary choice of the value of the viscosity that characterizes a glass; it is commonly taken as 10^{15} poise. Below T_g, a non-crystalline material is formed that exhibits a high viscosity. Here the atomic motions are retarded and the relaxation times exceed the time generally taken for performing quantitative measurements.

Present views of the structure of glasses and amorphous materials are derived largely from the postulates of Zachreisen (1932), who argued that since the mechanical strength of glass is on the same order as the strength of crystals, then the atoms in glassy materials are linked by forces that are essentially the same as those in crystals; the atoms must oscillate about definite equilibrium positions. As in crystals, the atoms in a glass form an extended three dimensional network, but as we shall discuss later (Section 2.2), there is an absence of a sharp X-ray diffraction pattern beyond the second nearest neighbor distance; the glass network is not periodic on an extended scale—hence the term "amorphous" or "disorder" is most descriptive. From the condition that the energy content should be comparable with that of the corresponding crystal, it follows that the coordination number, z, of an atom in the non-crystalline case must be approximately the same as in the corresponding crystal.

1.1 Classification of Amorphous Materials

As in crystals, the classification of amorphous semiconductors is determined by the type of chemical bonding that is primarily responsible for the cohesive energy of the material, i.e. on the first coordination number (Adler, 1971). In crystalline solids the five major classes are: (a) ionic; (b) covalent; (c) metallic; (d) Van der Waals; (e) hydrogen-bonding. Materials of type (d) and (e) generally have low cohesive energies and thus low melting points; little effort has been devoted to these classes of amorphous materials. Amorphous materials of type (c) have been studied extensively and their properties form a subject outside the scope of this book. The interested reader is referred to the books by Lubovsky (1983) and Hasigawa (1983). The ionic materials, (a), have been studied in great detail, notably the halides, oxide glasses and the transition metal glasses such as NiO. The reader is referred to the book by Uhlmann and Kreidl (1980) for further discussion of these materials.

In this book we shall concentrate on type (b) materials, that can be subdivided, further, as follows. The first subdivision includes the tetrahedrally bonded materials such as Si and Ge, in which there is only positional disorder. As we shall discuss in later sections, these materials are of little technological use because of their inherently high density of localized states. It is only with the addition of hydrogen and/or halogen atoms that the density of states is reduced to a point where device possibilities emerge. Therefore, in this broad category we include alloys of the type a-Si:H, a-Si:F:H and a-Si:Cl:H, where we expect positional as well as some compositional disorder. Other materials such as a-Si:Ge, a-Si:C, a-Si:Sn and a-Si:N based alloys are also included, where we also expect compositional disorder.

The second subdivision of covalently bonded materials includes S, Se and Te, which are two fold coordinated; here we expect positional disorder. Multicomponent glasses, such as As_2S_3, As_2Se_3, SiTeAsGe, GeTe, will possess compositional as well as positional disorder. They are the subject of the second part of this book.

1.2 Basic Concepts

The major success of solid state physics has been the theoretical prediction of the behavior of metals, insulators and semiconductors. The success of energy band theory has relied on the derivation of the E-k (energy-wave vector) relationship, which involves the consideration of a periodic array of atoms and the derivation of Bloch wave functions, from which the classes of energy states available to electrons and holes can be predicted. This leads to the concept of a forbidden energy gap. Implicit in these treatments is that the concept of the periodicity of atoms is vital. However, the resistivity of most materials is observed not to change at the melting point, where the periodicity is lost, but where the density of the material changes markedly. Since there is no change in resistivity at the melting point, then the long range periodicity is apparently not of great importance, because the general electrical characteristics of the material remain unaltered (see, e.g., Adler, 1971). As Adler (1971) has pointed out, the emphasis on periodicity is in reality a theoretical device to make the problem tractable from a quantitative viewpoint. Periodicity dictates that only the primitive unit cell need be considered; by using translational symmetry the opto-electronic properties of the material can be predicted. From a conceptual viewpoint, short range order is important in determining the opto-electronic properties. For example, from X-ray and electron diffraction experiments it is found that the nearest neighbor environments in a-Ge and a-Si are approximately the same as those found in their crystalline counterparts (see Section 2.2). Further, there is evidence from photoemission data (Ley et al., 1972) suggesting that the density of states in a liquid or non-crystalline materials does not differ greatly from the corresponding crystalline state. As shown in Fig. 1.2, the density of states spectra of amorphous solids do not exhibit the sharp structure due to singularities that are associated with the long range order of the corresponding crystal. Nevertheless, as shown in Fig. 1.2, there are three main bands in crystalline Si, whereas for a-Si, only two bands are found. Calculations have indicated that this is primarily due to changes in the short range order, rather than the loss of long-range order, because in the amorphous case there are odd numbered rings (five fold and seven fold), whereas in the crystalline case only sixfold rings are present (Joannopoulos and Cohen, 1973; Alben et al., 1972; Ortenburger et al., 1972). The topological nature of the network is apparently more important than the long range order. The concept of the density of states, $g(E)$, the number of one electron states in the energy range $E, E + dE$, can thus be retained in the description of amorphous solids. Therefore, the average occupancy of states using Fermi–Dirac statistics remains a meaningful measure.

THE EFFECT OF DISORDER

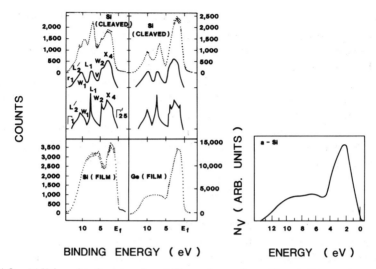

Fig. 1.2. (a) Valence band spectra of crystalline and amorphous silicon and germanium excited with monochromatized Al K_α X-rays. The spectra have been corrected for a background of inelastically scattered electrons. Also shown are theoretical densities of states, unbroadened (lower solid lines) and broadened, in order to facilitate comparison with the experimental spectra. (b) Experimental valence density of states of a-Si obtained from spectrum of (a) after correction for variations in photoelectric cross sections (Ley *et al.*, 1972).

Another concept of considerable value is the mobility, μ (velocity per unit applied electric field). The mobility depends on the entire state of the system and in general cannot be associated with a given one-electron state. For a perfectly periodic crystal at absolute zero all states are delocalized and the mobility of a carrier is infinite. In real crystals the value of μ is limited by scattering arising from deviations of atoms from perfect periodicity. In highly disordered systems, intense scattering due to large fluctuations in the one-electron potential can reduce the mobility to small values, despite the delocalized nature of the states.

Once $g(E)$ and the corresponding $\mu(E)$ are known, the electrical properties of the materials, such as the conductivity and thermoelectric power, can be determined (see Section 2.5). The same information enables us to predict optical properties such as absorption, reflectivity, photoemission and photoconductivity (see, e.g., Section 2.4).

1.3 The Effect of Disorder

As expected, the disorder inherent in amorphous semiconductors has a profound influence on their electronic and optical properties, providing for distinct differences between them and their crystalline counterparts. To understand this, let us begin by recalling that the free electron model developed for a periodic lattice has proven to be very successful in describing the behavior of metals. Indeed, by considering the interaction of electrons with a periodic lattice, considerable insight has been gained regarding the nature of the difference between insulators and conductors. One commonly used approach is the tight binding approximation; here the wave function of an electron is developed from a consideration of the interaction of the particles with the ion cores of the lattice. The wave function is normally constructed from a linear combination of atomic orbitals:

$$\psi_{kt}(r) = N \sum_{R_i} \exp(i\underline{k} \cdot \underline{R_i}) \Phi_t(\underline{r} - \underline{R_i}), \qquad (1.1)$$

where $\Phi_t(\underline{r} - \underline{R_i})$ represents the atomic orbital for a free atom with its center situated at R_i. The exponential term is simply a phase factor and N is a normalizing term. (From Eq. 1.1 it is evident that the atomic orbital dominates when $\underline{R_i} \to 0$.) The solution of Schrodinger's equation for this wave function leads to an eigenvalue, $\langle E \rangle$, given by,

$$\langle E \rangle = E_t + \langle V'(\underline{r} - \underline{R_j}) \rangle + \frac{1}{N} \sum_{R_i \neq R_j} \exp\{i\underline{k} \cdot (\underline{R_j} - \underline{R_i})\} \gamma \qquad (1.2)$$

where the overlap integral, γ, is given by

$$\gamma = \int \Phi_t^*(\underline{r} - \underline{R_i}) V'(\underline{r} - \underline{R_j}) \Phi_t(\underline{r} - \underline{R_j}) \qquad (1.3)$$

In Eq. 1.2, $\langle E \rangle$ is the energy of the electron and differs from that of the electron in the free atom, E_t, by a term $\langle V' \rangle$ and a term that is a function of the wave vector, k. It is the latter term that transforms the discrete atomic orbitals into an energy band in the solid. For a simple cubic lattice it can be shown that the width of the band, $J = 2z\gamma$ which increases with increasing overlap.

Next, consider the effect on the band width of a non-periodic potential. This can result from either, (a) the displacement of each atom by a random amount, (b) the addition of a random potential energy, $V_0/2$, to each well in such a way that the energy of an electron, E_t, is changed to $E_t + V_0/2$; (c) both (a) and (b). It has been shown (e.g., Mott and Davis, 1979) that the mean free path, L, of an electron in a lattice having a random potential (case b above) can be described by

$$L = 16\pi a \left(\frac{\gamma}{V_0}\right)^2 \qquad (1.4)$$

THE EFFECT OF DISORDER

As the disorder increases, L decreases; ultimately, for $V_0/\gamma \cong 7$, $L \cong a$ (Anderson, 1958), in which case the sign of the wave function will vary in a random fashion from well to well, as shown in Fig. 1.3. At this point, the band width will increase and is given by,

$$J \sim \frac{2zV_0}{\sqrt{16\pi}} \qquad (1.5)$$

Hence, the introduction of disorder results in scattering that leads to a finite mean free path of the carriers, even at $T = 0\,\text{K}$. If L is large, a wave function constructed as above would be valid between the scattering events. However, when a moderate amount of disorder makes for $L \sim a$, the wave

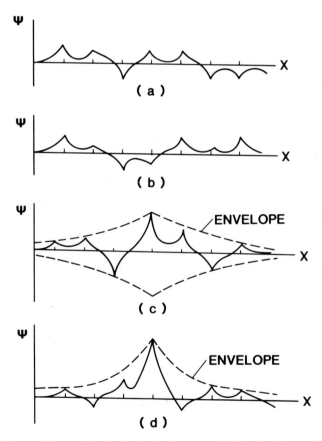

Fig. 1.3. Form of the wavefunction in the Anderson model: (a) when $L \sim a$; (b) when states are just non-localized ($E \geq E_c$); (c) when states are just localized ($E \leq E_c$); (d) strong localization (Mott and Davis, 1979).

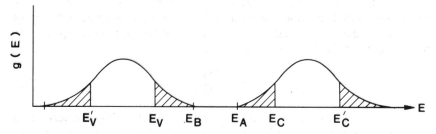

Fig. 1.4. Density of states in the Anderson model when states are non localized in the center of the band. Localized states are shown shaded, E_c, E_c' separate the ranges of energy where states are localized and non-localized (Mott and Davis, 1979).

function will fluctuate randomly, at which point k is no longer a good quantum number (Ioffe and Regel, 1960). With further increase in disorder the atomic orbitals given by Eq. 1.1 dominate the constructed wave functions. Now the wave function decays exponentially at a rate related to the magnitude of disorder potential V_0, also shown in Fig. 1.3. From the above argument, when $L \cong a$, $V_0/\gamma \simeq 7$ represents the demarcation between localized and non-localized states, which are now separated by a critical energy, E_c, that is a function of V_0. Hence, the energy band picture can be visualized as shown in Fig. 1.4, where E_v and E_c define critical energies for the valence and conduction bands, respectively. It is immediately apparent that in amorphous semiconductors, the notion of a forbidden energy gap, as in the crystalline counterpart, is not valid.

The position of E_c (to be discussed below) and the extent of the band tails has received considerable attention (Economou and Cohen, 1970; Schönhammer, 1971; Abou-Chacra and Thouless, 1974). For weak disorder the band tail width (the separation of E_c from the bottom of the band at E_A) is given by

$$E_c - E_A \approx \frac{1}{4} \frac{V_0^2}{K\gamma} \qquad (1.6)$$

where $K = z - 1$. The band tail width evidently increases with the disorder potential.

The behavior of the wave function for energies near E_c has also been the subject of considerable study (see, e.g., Mott and Davis, 1979). The form of the wave function for $E < E_c$ is shown in Fig. 1.3c and 1.3d, and can be described by

$$\Psi \alpha \sum \phi(\underline{r} - \underline{R}_i)e^{-\alpha_L r} \qquad (1.7)$$

where $e^{-\alpha_L r}$ represents the envelope. As E approached E_c, α tends to zero and the wave function fluctuates randomly: the states are called "extended". For $E < E_c$, the decay factor, α_L, can be described by

$$\alpha_L \approx \alpha_0 \left[\frac{E_c - E}{E_c} \right]^s \qquad (1.8)$$

where $s = 0.6$ and α_0 is the value for α_L far from the critical energy, E_c. Therefore, for $E < E_c$ the states become localized and hopping transport between the states becomes possible (see Section 2.5).

The changes in the form of the wave function below and above E_c has an effect on the basic transport mechanism. The conductivity (σ) in the two regions can be expressed as

$$\langle \sigma_E \rangle = 0 \quad \text{for } E < E_c, \text{ excluding "hopping" transport,} \quad (1.9a)$$

$$\langle \sigma_E \rangle > 0 \quad \text{for } E > E_c. \qquad (1.9b)$$

As previously discussed, the wave functions are randomized for $E \approx E_c$, with the consequence that the mean free path is on the order of the interatomic distance. Cohen (1970) has pointed out that here the motion of the carriers can be likened to a diffusive Brownian motion. From a random phase model (Mott, 1970; Hindley, 1970), in which the wave functions are described by a linear combination of atomic wave functions, the extended state mobility is given by

$$\mu_{\text{ext}} = \frac{2\pi}{3} \cdot \frac{ea^5}{\hbar} z\gamma g(E_c) \qquad (1.10)$$

where $g(E_c)$ is the effective density of states. Estimates of μ_{ext} range from 0.1 to greater than 10 cm² V⁻¹ s⁻¹ for conduction at $E \geq E_c$ (see Section 2.5.1).

Conduction for $E \leq E_c$ occurs via phonon assisted hopping within the localized states. It can be shown that in this case the hopping mobility, μ_{hop}, is given by,

$$\mu_{\text{hop}} \sim \frac{1}{6} \nu_{ph} \left(\frac{er^2}{kT} \right) \exp - \left(2\alpha_L r + \frac{W}{kT} \right), \qquad (1.11)$$

where the factor $\frac{1}{6}$ arises from an averaging procedure, ν_{ph} ($\sim 10^{12}$ s⁻¹) is the phonon frequency, r is the hopping distance and W is the energy difference of the two hopping sites. The prediction of μ_{hop} lead to values less than 10^{-2} cm² s⁻¹ V⁻¹ (see Section 2.5.3).

From Eqs. 1.10 and 1.11 $\mu_{\text{ext}}/\mu_{\text{hop}} > 10^3$; the concept of a mobility edge at E_c, and correspondingly at E_v, is realized. The energetic range $E_v < E < E_c$ is defined as the mobility gap or the pseudogap. It is this aspect that leads to the semiconducting behavior of certain amorphous materials; its counterpart is the forbidden energy gap in crystalline semiconductors. In a pure crystal the optical and electronic energy band gaps are equal. In an amorphous material the electronic properties are described by a mobility gap and, as we shall see, the optical properties can be related to an optical band gap that is generally greater than the mobility gap.

2

Opto-Electronic Properties of Amorphous Silicon Based Materials

2.1	Growth Kinetics	12
	2.1.1 Introduction	12
	2.1.2 Glow Discharge in Silane Gas	13
	2.1.3 Glow Discharge Using Disilane Gas	25
	2.1.4 Glow Discharge Involving Halogens	26
2.2	Structure and Composition	27
2.3	Density of States (DOS)	35
	2.3.1 Some Models for the Density of States	35
	2.3.2 Experimental Determination of the DOS	44
2.4	Optical Absorption	62
2.5	Electrical Transport	72
	2.5.1 Conductivity in Extended States	73
	2.5.2 Conductivity in Tail States	74
	2.5.3 Hopping Conductivity	75
	2.5.4 Seebeck Coefficient in Extended and Defect States	76
	2.5.5 Conduction Processes in Amorphous Silicon	77
	2.5.6 Drift Mobility	84
2.6	Photoconductivity	97
2.7	Photoluminescence	106
2.8	Doping in Glow Discharge a-Si Based Alloys	115
	2.8.1 The Effect of Phosphorous Doping	117
	2.8.2 The Effect of Boron Doping	122
	2.8.3 The Effect of Compensation	125
2.9	Surface States on Amorphous Silicon	127
2.10	Hydrogenated Amorphous Silicon Using Techniques Other Than Glow Discharge Decomposition	134
	2.10.1 Chemical Vapor Deposition (CVD)	134
	2.10.2 Reactive Sputtering Techniques	135
	2.10.3 Miscellaneous Techniques	139
2.11	Some Properties of Halogenated Amorphous Silicon Alloys	140
	2.11.1 Fluorinated Amorphous Silicon	140
	2.11.2 Chlorinated Amorphous Silicon Films	148

2.12 Narrow and Wide Band Gap Alloys.. 149
    2.12.1 Amorphous Silicon–Germanium Alloys.. 149
    2.12.2 Amorphous Silicon–Tin Alloys.. 153
    2.12.3 Amorphous Silicon–Carbon Alloys.. 155
    2.12.4 Amorphous Silicon–Nitrogen Alloys.. 159

In this chapter we review some of the characterization techniques and the opto-electronic properties of amorphous silicon, with an emphasis on a-Si : H alloys prepared by the radio frequency glow discharge technique. We begin with the growth kinetics, follow with the structural properties and then discuss in detail some of the opto-electronic properties of the material. Finally, we will review results for both a-Si : H films deposited using different deposition techniques and alloys other than a-Si : H.

2.1 Growth Kinetics

2.1.1 Introduction

Amorphous Silicon films have been prepared using numerous deposition techniques such as: glow discharge (GD or Plasma Enhanced Chemical Vapor Deposition (PECVD)); CVD; reactive sputtering; reactive evaporation. Although it had been demonstrated that in principle a continuous random network (CRN) of a-Si (devoid of dangling bonds or traps) could be produced, results for evaporated Si or sputtered films revealed that the density of localized states (DOS) in the mobility gap was far too high to be of any use in device applications. It was the accidental discovery by researchers at Standard Telephone Laboratories and the University of Dundee in the U.K. in the early 1970s that revealed that the GD or PECVD a-Si films prepared from SiH_4 gas possessed a low DOS. It was subsequently discovered that these films contained approximately 10 atomic percent Hydrogen and hence were appropriately referred to as a-Si : H type alloys. GD is presently the most important deposition technique and is widely used in the industry.

    A glow discharge of a gas can be created by using either a dc or a rf electric field. In a dc glow discharge electrons are created by the impact of positive ions that collide with the cathode. These electrons are accelerated back into the glow discharge and collide with and ionize molecules. The energy gained by the electrons comes from the applied electric field. However, in a rf type discharge the electrons gain energy from the time varying fields at almost any plasma-surface boundary, in the bulk plasma, or from the cathode electric field. Consequently, rf discharges can be

GROWTH KINETICS

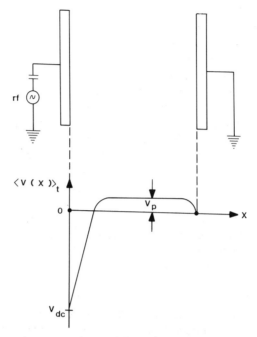

Fig. 2.1. Approximate time-averaged potential versus distance for a capacitively coupled planar rf glow discharge system with wall area much greater than cathode electrode area.

operated at lower pressures than dc discharges. GD a-Si:H films have been prepared using inductively (Chittick *et al.*, 1969; LeComber *et al.*, 1970) and capacitively coupled planar type systems. The latter approach is now predominantly used, and Fig. 2.1 shows the time-average potential distribution between the rf plate (cathode) and the grounded plate (anode). Note that the system is operated such that the discharge is created with the anode sufficiently close to the cathode so that the positive column is eliminated. Further, the plasma potential, V_p, is more positive than the potential of any electrode in contact with the plasma, provided that the negative charge carriers are electrons and not ions. This results because of the high mobility of the electrons that tend to deplete the glow discharge of negative ions near the positively biased electrode and thereby cause the plasma potential to increase.

2.1.2 Glow Discharge in Silane Gas

The opto-electronic properties of the GD a-Si:H films depend upon many deposition parameters such as the: pressure of the gas; flow rate; substrate temperature; power dissipation in the plasma; excitation frequency.

Table 2.1a shows an example of the deposition conditions that are generally employed to produce high quality opto-electronic material. (The deposition parameters are to some extent system-dependent.) In Table 2.1b some of the present state-of-the-art properties are summarized. Further improvements in the material are expected to result from a better understanding of the relationship between the processing conditions and the specific chemical reactions taking place in the plasma that promote film growth. To understand and formulate such reactions, various techniques such as optical emission spectroscopy (OES), mass spectroscopy (MS) and Coherent Anti-Stokes Raman Spectroscopy (CARS) have been utilized. In the following we consider some of the current ideas concerned with growth kinetics and make an attempt to relate them to the influence they have on the opto-electronic properties of the materials.

It has been shown (Spear *et al.*, 1972; Madan, 1973; Madan *et al.*, 1976) that the density of localized states (DOS) is critically controlled by the deposition temperature, T_s, of the substrate. For example, from their field effect data (see Section 2.3.2.1), the DOS was seen to decrease by orders of magnitude when T_s was raised from 300 K to 570 K (see Fig. 2.27). This vast change in the DOS has been attributed to basic structural changes occurring within the film. It has been shown by Lucovsky *et al.* (1979) that films produced at low T_s generally possess H concentrations in excess of 30%, in which case the materials can be described as polymeric type structures containing an excess of (SiH_2) and (SiH_3) chains. Although, as shown in Fig. 2.2, the H/Si content ratio within the film, and the consequent optical band gap, E_g, is reduced when T_s is increased (Perrin *et al.*, 1979), it is the mode of incorporation of H into the Si matrix that appears to have a major effect on the electronic behavior. It is found that for the high deposition temperature samples (T_s = 600 K), the infra-red peaks (see Section 2.2) occur at 2000 cm^{-1} ($\equiv$ Si–H stretch mode of vibration) and at 640 cm^{-1} ($\equiv$ Si–H wag mode),

Table 2.1a. Typical Conditions Employed in the GD of SIH$_4$ Gas in a Research Type System

Parameters	Comments	Typical values
Gas Composition		100% SiH$_4$
Flow rate	High	>40 sccm
Pressure	Low	200–600 m torr
Deposition Temperature		250°C
R.F. Power	Low	<25 mW cm^{-2}
Anode–Cathode Distance		~2–4 cm

Table 2.1b. Typical Opto-Electronic Parameters Obtained for GD A-Si:H(F) Alloys

Undoped

Hydrogen Content		~10%
Dark Conductivity at 300°K	σ_D	~10^{-10} (Ω cm)$^{-1}$
Activation Energy	ΔE or E_σ	~0.8 – 0.9 eV
Pre-exponent Conductivity Factor	σ_0	>10^3 (Ω cm)$^{-1}$
Optical Band Gap at 300°K	E_g	1.7 – 1.8 eV
E_g Variation with Temperature	$E_g(T)$	2-4.10^{-4} eV K^{-1}
Density of States at the Minimum	g_{min} or $g(E_f)$	>10^{15} – 10^{17} cm^{-3} eV^{-1}
Density of States at the Conduction Band Edge	$g(E_c)$	~10^{21} cm^{-3} eV^{-1}
ESR Spin Density	N_s	~10^{15} cm^{-3}
Infra-red Spectra		2000/640 cm^{-1}
Photoluminescence Peak at 77 K		~1.25 eV
Extended State Mobility—Electrons	μ_n or μ_e	>10 cm^2 s^{-1} V^{-1}
—Holes	μ_p or μ_h	~1 cm^2 s^{-1} V^{-1}
Drift Mobility—Electrons	μ_n or μ_e	~1 cm^2 s^{-1} V^{-1}
—Holes	μ_p or μ_h	~10^{-2} cm^2 s^{-1} V^{-1}
Conduction Band Tail Slope		25 meV
Valence Band Tail Slope		40 meV
Hole Diffusion Length		~1 μm

Doped—Amorphous

n-type—~1% addition of PH$_3$ to gas phase	σ_D ~ 10^{-2} (Ω cm)$^{-1}$
	ΔE ~ 0.2 eV
p-type—~1% addition of B$_2$H$_6$ to gas phase	σ_D ~ 10^{-3} (Ω cm)$^{-1}$
	ΔE ~ 0.3 eV

Doped—Microcrystalline

n-type—~1% PH$_3$ added to dilute SiH$_4$/H$_2$ or 500 vppm, PH$_3$ added to SiF$_4$/H$_2$ (8/1) gas mixtures. Relatively high powers are involved.	$\sigma_D \geq 1$ (Ω cm)$^{-1}$ $\Delta E \leq 0.05$ eV
p-type—~1% B$_2$H$_6$ added to dilute SiH$_4$/H$_2$ gas mixtures.	$\sigma_D \geq 1$ (Ω cm)$^{-1}$ $\Delta E \leq 0.05$ eV

whereas the films deposited at low T_s (e.g., room temperature) exhibit, in addition to the above, peaks centered at 2100 cm^{-1} ($\equiv$ Si-H stretch mode in the SiH$_2$ mode) and a corresponding scissors mode of vibration at 900 cm^{-1} (Brodsky et al., 1977; Lucovsky et al., 1979). It is the fundamental change from a polymeric type structure for samples deposited at low temperatures, to a monohydride at higher T_s, that has a major effect on the reduction of the DOS.

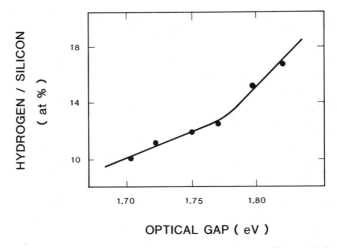

Fig. 2.2. Correlation between the optical gap E_g and the hydrogen–silicon ratio for constant electronic and gas conditions: rf power = 70 W, SiH_4–H_2 gas ratio of 17%. The experimental points correspond to films deposited at different temperatures T_s in the range 140–400°C (Perrin *et al.*, 1979).

The pressure of the gases during deposition affects the properties of the film, since gas phase polymerization is encouraged at high pressures (Brodsky, 1977; Turban *et al.*, 1979), with the consequence that the SiH_2 and SiH_3 grouping within the film becomes more pronouced. Brodsky (1977) has pointed out that the different bonding arrangements within the film can be understood from Paschens Law, which states that the voltage required to sustain a glow discharge in a plasma between two electrodes is a two branched function of the product of plasma pressure (p) and the distance between the two electrodes (d) (Thomson and Thompson, 1933). This is shown representatively in Fig. 2.3. In region I (as p is lowered or as d is decreased) the electron energy is limited by collisions with the electrodes. When p or d are made to decrease, more voltage is required to make up for the energy lost to the walls of the system. In region II (increase of p or d), the electrons are more likely to collide with the plasma constituents than with the electrodes. The voltage required to sustain the discharge also increases, this time to make up for the energy lost to collisions within the plasma, which in turn promotes polymerization and can lead to the inclusion of $(SiH_2)_n$ chains within the film.

Matsuda *et al.* (1982) have identified the relative concentration of ionic and neutral species contained within the plasma by mass spectroscopy. Figure 2.4 shows the comparison of mass spectra data with the ionizer on and off. It is clearly evident that for low power the density of the ionic species is lower by $\sim 10^4$ compared to the density of the neutral species. Ions such as SiH_3^+,

GROWTH KINETICS

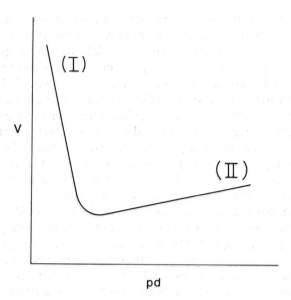

Fig. 2.3. A schematic illustration of Paschens law for the voltage, V, needed to sustain a glow discharge in plasma of pressure, p, between electrodes separated by a discharge, d (Brodsky, 1977).

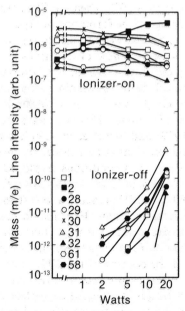

Fig. 2.4. Line intensities of the observed mass numbers (m/e) from a SiH$_4$ plasma as functions of rf power (5 sccm, 50 m torr) for ionizer-on and off operation (Matsuda *et al.*, 1982).

SiH_2^+, SiH^+, H^+, $Si_2H_2^+$ increase by more than two orders of magnitude as the rf power is increased from 2 W to 20 W. (Unequivocal interpretation of the species in the plasma is not always possible using mass spectrometry, since there is an experimental difficulty in transferring a sample species from the relatively high pressure reaction system to the ion source of a low pressure mass spectrometer without losing radicals through collisions both in the gas phase or due to recombination at the surface if the path is too long. Further, energetic short-lived species may not be detected since it is not always possible to transport them quickly enough. It should also be recognized that since the threshold energy of the SiH_4 gas dissociation (~ 8 eV) is lower than the electron energy for ionization (> 11.9 eV) then the ionization as well as decomposition of SiH_4 molecules into neutral species can take place within the ionization cell such that the mass spectra observed could be somewhat different from that of the glow discharge of the original SiH_4 gas plasma.) As seen in Fig. 2.4, a rapid rise in the population of ions with power implies an increase in the electron density. Further, the predominant ionic species appears to be SiH^{3+}, in agreement with the results of Drevillon et al. (1981). Note also that as the power is increased there is a considerable increase in the density of H_2 molecules; we shall return to this point later.

The results of Turban et al. (1980) shown in Fig. 2.5 illustrate that for low power levels (e.g., ~ 5 W) up to 50% of the SiH_4 remains undissociated and this percentage decreases to ~ 20% when the films are prepared at a power above 50 W. Figure 2.5c shows the formation of H_2 due to the decomposition of SiH_4 gas and Fig. 2.5d shows that the decomposition proceeds more readily with lower flow rates. These observations, coupled with infra-red measurements on samples, suggest that for conditions under which the silane is not entirely decomposed the films contain a majority of SiH units. Those films deposited under conditions in which silane is strongly dissociated contain a majority of SiH_2 units. This is in agreement with the work of Knights and Lujan (1979) and Street et al. (1978a), who have obtained a correlation between the spin density (a measure of the defect density) of the a-Si:H film and the deposition conditions employed. Similar conclusions were also arrived at by Hirose et al. (1981) using OES. By studying the emission lines of Si (299 nm), SiH (414 nm) and H (656 nm) it was noted that the emission corresponding to atomic H increased faster than that for SiH when the flow rate was decreased. The importance of the reactive SiH and H species is shown by the correlation between the concentration ratio of SiH to H in the plasma and the content of the SiH_2 units in the resulting film, as shown in Fig. 2.6. The incorporation of SiH_2 units is significant at low flow rates where the SiH_4 gas is expected to deplete rapidly.

The asymmetry of the reactions taking place within the capacitive plates have been demonstrated by the use of the CARS technique, in which the

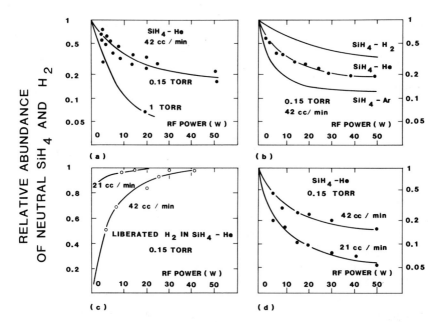

Fig. 2.5. Relative concentration of neutral molecules in a silane discharge as a function of power: (a), (b) and (d) show the concentration of SiH_4: (c) concentration of H_2 released by silane decomposition (Turban et al., 1980).

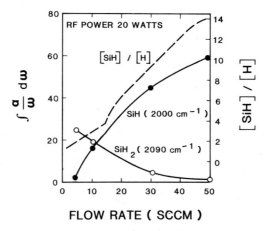

Fig. 2.6. Integrated absorption of SiH and SiH_2 stretching modes as a function of silane flow rate. The dashed line refers to the concentration ratio of the reactive SiH to H [SiH]/[H] (Hirose et al., 1981).

20 OPTO-ELECTRONIC PROPERTIES OF AMORPHOUS SILICON MATERIALS

signal is a measure of the number density of the sample molecule. Hata *et al.* (1983), showed that SiH_4 dissociated at the highest rate in the plasma near the cathode; this may occur because the electric field is largest in the vicinity of the cathode (see Fig. 2.1) where a large number of electrons with higher kinetic energies excite and dissociate SiH_4 molecules. This affects the opto-electronic properties via the incorporation of a H/Si ratio that can vary because of the superimposed substrate bias, as shown in Fig. 2.7 (Perrin *et al.*, 1979).

The conditions under which the films are deposited has also been related to the stress (Harbison *et al.*, 1984). The change in the film stress as a function of the H_2 to SiH_4 dilution is shown in Fig. 2.8a and the corresponding change in the growth rate is shown in Fig. 2.8b. Note that there is a considerable drop in stress for films fabricated with a 3% SiH_4 gas mixture. It is reported that this is accompanied by the onset of columnar growth within the film, but there is a suprising absence of polysilanes within the film. It was concluded that the large compressive stress is indicative of the incorporation of large amounts of impurity atoms within the film.

The frequency of the discharge also has an affect on the internal stress, as reported by Matsuda *et al.* (1984). They noted that as the excitation frequency is increased from 10 KHz to 1 MHz the stress increases, but for larger frequencies the stress decreases by nearly a factor of 5. The stress was

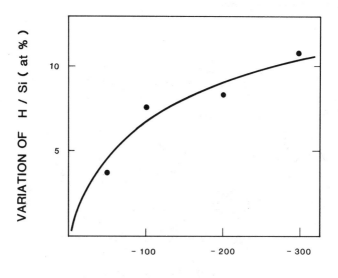

Fig. 2.7. Variation of the hydrogen concentration with applied substrate bias at a constant target voltage $V = -400$ V (Perrin *et al.*, 1979)

GROWTH KINETICS

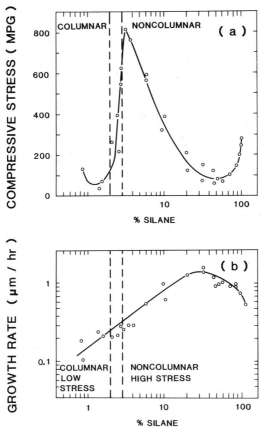

Fig. 2.8. (a) Compressive stress measured in films grown on crystalline silicon substrates as a function of silane dilution. (b) Growth rate as a function of silane dilution (Harbison et al., 1984).

noted to be compressive for all frequencies investigated, whereas the hydrogen incorporation showed a weak dependence on frequency (about 20 at.% for a film prepared at 10 KHz and 15 at.% for the 13.56 MHz film).

In general, the hydrogen content, C_H, of a-Si based alloys is approximately in the range $0.1 - 0.3$, which is much smaller than the ratio H/Si = 4 for the case of a pure SiH_4 plasma. Therefore, a model for growth of the film needs to incorporate the hydrogen elimination process, and also be able to explain the increase in the dihydride formation (polysilanes, SiH_2 and SiH_3) that seems to occur for a large dissociation of silane within the plasma. A qualitative attempt to model the growth kinetics has been made by Kampas and Griffith (1981) and Kampas (1982). They assumed that SiH_2-free radicals react with the Si:H bond on the growing surface and produce an activated complex that can eliminate H$_2$ molecules. The amount of hydrogen remaining in the a-Si:H film is determined by a competition between H$_2$ elimination and deactivation of the complex. Further, it was concluded that changes of the deposition parameters that control the deposition rate also increase the atomic hydrogen flux on the growing surface, which results in a higher hydrogen content in the film.

The evidence that SiH$_2$ could be the predominant species involved in the growth is provided by photolysis of SiH$_4$ by 9.4 eV photons, which reveal the following reactions (Perkins et al., 1979)

$$h\nu + \text{SiH}_4 \rightarrow \text{SiH}_3 + \text{H} \qquad (2.1)$$
$$\rightarrow \text{SiH}_2 + 2\text{H},$$

with quantum yields of 0.17 and 0.83 respectively. Also, the source of H, in a rf plasma, can be envisaged (Kampas, 1982) using the following reactions, with the energies required indicated in the parentheses:

$$(4.04 \text{ eV}) \quad e^- + \text{SiH}_4 \rightarrow \text{SiH}_3 + \text{H} + e^- \qquad (2.2)$$

$$(2.16 \text{ eV}) \quad e^- + \text{SiH}_4 \rightarrow \text{SiH}_2 + \text{H}_2 + e^- \qquad (2.3)$$

$$(4.53 \text{ eV}) \quad e^- + \text{H}_2 \rightarrow 2\text{H} + e^- \qquad (2.4)$$

The electron energy required to produce SiH$_2$ units evidently is less than that required for SiH$_3$ units. Therefore, using SiH$_2$ as the units impinging onto the growing film, Kampas (1982) proposed the following reaction to explain the H elimination process during film growth:

$$\text{SiH}_2(g) + \text{Si}_x\text{H}_y(s) \xrightarrow{r_d} \text{Si}_{x+1}\text{H}^*_{y+2}(s)$$

$$\xrightarrow{r_1} \text{Si}_{x+1}\text{H}_{y+2}(s) \xrightarrow{r_2} \text{Si}_{x+1}\text{H}_y(s) + \text{H}_2(g) \qquad (2.5)$$

where r_d, r_1, r_2 are the deposition, deactivation and hydrogen elimination rates, g and s (in parentheses) represent the gaseous and the solid phases. It is speculated that a plot of log (r_2/r_1) versus $10^3/T$ should give a straight line whose slope gives the difference in the activation energies of r_2 and r_1. A plot of this kind is shown in Fig. 2.9 for GD a-Si:H films prepared under different conditions: (a) low power (2 W), pure silane (2 S); (b) higher power (18 W) with 5% silane in Ar gas mixture (18 A); (c) high power (25 W), 5% silane in Ar gas mixture (25 A). Note that as the conditions change from (a) to (c), the slope changes significantly and that the concentration of H within the films increases. The preparation conditions corresponding to (b) and (c) are generally associated with a defective material having a large C_H. Kampas (1982) concluded that the smaller slope is correlated with an increased

GROWTH KINETICS 23

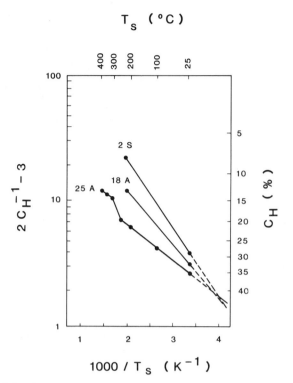

Fig. 2.9. Log($C_H^{-1} - 3$) vs. 1000/T, for films deposited under three sets of conditions: 2 W, pure silane (2 S); 18 W, 5% silane in argon (18 A); and 25 W, 5% silane in argon (25 A) (Kampas, 1982).

number of SiH$_2$ units within the film. Further, any deposition condition in which SiH$_4$ gas is heavily depleted will have more defects than films produced where the SiH$_4$ is not as heavily depleted. To some extent, support for this is shown in Fig. 2.5, where it is noted that at large rf powers, the concentration of H$_2$ within the plasma is similar to that of the SiH$_4$ within the plasma. Figure 2.4 also shows that there is a considerable increase in H$_2$ neutrals with increasing power, accompanied by a rapid increase in the electron density. Therefore, it may be that the reactions given in Eqs. (2.3) and (2.4) will increase with increasing power, which in turn leads to an increase in the H flux at the growing surface resulting in the breakage of weak Si–Si bonds. The model put forward by Kampas et al. (1981) involving SiH$_2$ addition, H$_2$ elimination and cross linking is shown in Fig. 2.10. It corresponds to any deposition condition that leads to a low depletion of SiH$_4$, such as low power or high flow rate. On the other hand, when the deposition conditions are altered such that the SiH$_4$ gas is heavily depleted (e.g., high

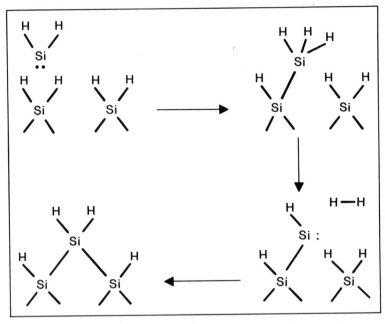

Fig. 2.10. Schematic of the SiH$_2$ addition, H$_2$ elimination, and cross-linking steps that compose the proposed model for film growth and hydrogen elimination at the surface (Kampas and Griffith, 1981).

power, low flow rate) then as discussed above, the H flux at the growing surface is increased so that the weak Si–Si bonds can be broken. This is most likely to occur between two growing islands, since cross linking must occur between atoms which are not part of the same network. The breakage of these bonds leads to an increase in the number of dangling bonds and seems to be consistent with the observation of an increased number of spins (Street *et al.*, 1978a) and the occurrence of columnar structure (Knights and Lujan, 1979) within the film.

A further confirmation of the above model is provided by effects observed due to the presence of a magnetic field during growth. Taniguchi *et al.* (1980) noted via OES measurements that the emission intensity of the H$_\alpha$, SiH and H$_2$ lines decrease with increasing magnetic field. Since a magnetic field applied to a plasma in general lowers the electron gas temperature, it would therefore lead to a reduction of the reaction given by Eqs. (2.2) and (2.4). This seems to be consistent with the fact that films prepared at low power and zero magnetic field (5 W, 0 KG) are similar to films prepared at high power, but with a superimposed magnetic field (20 W, 0.42 KG).

GROWTH KINETICS

There is strong evidence that rather than SiH_2, it is the SiH_3 species that could be more important during the film growth. Robertson *et al.* (1983) have measured the flux of free radicals directly from 10 m-Torr silane-150 m Torr Argon discharges using mass spectrometry. They concluded that SiH_3 is the dominant species, followed by SiH, Si and SiH_2. Further support of this is provided by Longeway *et al.* (1984), who found that the presence of NO in the plasma, which scavenges SiH_3, prevented film deposition.

2.1.3 Glow Discharge Using Disilane Gas

Good opto-electronic properties of a-Si:H from Silane gas are generally obtained at relatively low deposition rates (1-5 Å s^{-1}). Much higher deposition rates are readily obtained by the use of disilane. For example, Scott *et al.* (1980) have shown that films prepared in this fashion are compositionally similar to those produced from monosilane gas except that the deposition rate exceeded 14 Å s^{-1}. The reason for the high deposition rate is conjectured to be the removal of the rate determining step involving the formation of the silane diradical (SiH_2) step in the reaction $SiH_4 \rightarrow SiH_2 + 2H_2$. This is overcome in the disilane decomposition since the reaction could proceed via $Si_2H_6 \rightarrow SiH_2 + SiH_4$, which is considerably faster. Another reason cited for the higher deposition rate is that there is a considerable H by-product involving SiH_4, but not when Si_2H_6 is used, leading to a minimization of back etching on the growing film (Webb *et al.*, 1979).

Although deposition rates of 60 Å s^{-1} have been obtained (Ogawa *et al.*, 1981), it is found that the opto-electronic properties of the resultant films degrade with deposition rate (Matsuda *et al.*, 1983). They have also noted that while all the H is bonded to Si in monosilane type films, there appears to be large amounts of H not bonded to Si for the films prepared using Si_2H_6. This tendency seems to be enhanced in the higher rf power ranges. Further, it seems that the growth rate using Si_2H_6 gas does not always exceed that compared to the use of SiH_4. Kuboi *et al.* (1984) show that this may be true for low powers, but the deposition rates using the two gases approach one another as the power is increased. Ohnishi *et al.* (1984) have shown that as the rf power is increased the H/SiH emission lines from the plasma increase, and concomitantly, the ratio of the Si-H to Si-H_2 vibrational modes from their infra-red data also increases, which is an indication of improved film quality. This type of correlation seems to be opposite to that obtained for films prepared using SiH_4 gas. It appears that the quality of the a-Si:H alloy obtained using Si_2H_6 is not as good as that obtained from SiH_4.

2.1.4 Glow Discharge Involving Halogens

One of the major reasons that a-Si:H based alloys possess good opto-electronic properties is that many of the dangling bonds are satiated by H. Since F is more electronegative than H, then good materials should in principle also result for a-Si:F alloys. (A more complete discussion on halogenated a-Si materials is given in Section 2.11). The first production of a device quality material of this type (a-Si:F:H) was by Madan *et al.* (1979), who employed the GD in SiF_4 and H_2 gas mixtures. Using OES, Matsuda *et al.* (1982) have identified the following emissions lines in a SiF_4 plasma: SiF (4393 Å), F_2(5731 Å) and F(6855 Å). With the addition of H to the SiF_4 plasma (SiF_4/H_2 or SiF_4/SiH_4 gas mixtures), other additional emission lines are observed, such as SiH, H_2, and H. It was recognized that amorphous Si films could not be produced in the absence of excited species such as H, H_2 and SiH. This can be understood if we consider the dissociation energy inequalities for the possible bonding arrangements given by

$$H-F > Si-F > H-H > Si-Si > Si-H > F-F \qquad (2.6)$$

The reaction of the film surface between molecules and the excited species in the gas phase can be qualitatively determined by the difference in the dissociation energies of the relevant chemical bond. In the first step of the film deposition, [SiF] or [SiH] molecules are adsorbed on the surface of the substrate. In the second step, for the case of a pure SiF_4 glow-discharge, the first monolayer involving Si free bonds is completely passivated by F atoms and further reactions do not take place since the Si-F bond is the strongest among the chemical bonds associated with the SiF_4 plasma. On the other hand, for the glow-discharge plasma in SiF_4, diluted by SiH_4 (or H_2), [H], [SiH] and [H_2] radicals could play an essential role in the surface reaction process, i.e., the second and third Si layers can be formed successively by the following exothermic reaction

$$\underset{\text{(adsorbed)}}{Si-F} + \underset{\text{(gas)}}{[SiH]} \to \underset{\text{(solid)}}{Si-Si} + \underset{\text{(gas)}}{H-F}. \qquad (2.7)$$

In an earlier discussion (Section 2.2.2) a view was put forth that a large flux of H arriving at the growing surface was detrimental to the properties of the films since weak Si-Si bonds can be broken. Hence, the OES intensity ratio (H/SiH) within the plasma could correlate with the film properties. In Fig. 2.11 (Matsuda *et al.*, 1982), the H/SiH ratio is plotted against material parameters, such as the dark and light conductivities, for different ratios of SiF_4 to SiH_4. A lower H/SiH ratio apparently corresponds to a lower flux of H atoms and an improved photoconductivity of the material (see Section 2.6).

STRUCTURE AND COMPOSITION

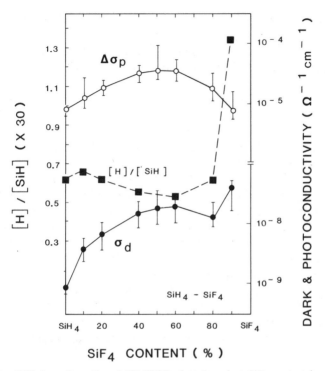

Fig. 2.11. OES intensity ratio of [H]/[SiH] plotted against SiF$_4$ content in a SiH$_4$-SiF$_4$ mixture. Photoconductivity ($\Delta\sigma_p$) and dark conductivity (σ_d) are also shown (Matsuda et al., 1982).

2.2 Structure and Composition

Although the long-range periodicity is lost in a-Si based materials, a high degree of short-range order is retained because of the chemical bonding. Hence, the nearest neighbor separation and bond lengths are nearly equal, as in the corresponding crystalline Si. Generally, the structure of amorphous solids can be classified into (Zallen, 1983): (a) a continuous random network (CRN) appropriate to the structure of covalent type materials; (b) a random close packing appropriate to the structure of simple metallic glasses; (c) a random coil model appropriate to polymeric organic glasses. The common feature is that of a statistical distribution of the position of the atoms. In this section we will concern ourselves primarily with materials of type (a).

It is widely believed that the bulk atomic structure of a-Si is such that on the average a Si atom has four others around it at the same distance (as in the crystal) forming a regular tetrahedron. Further, each of these atoms in turn has three more neighbors, similarly arranged, giving rise to a well defined second shell of 12 atoms, none of which is a neighbor to any other. The structural data show that the resemblance of a-Si to crystal Si ceases beyond the second shell; the commonly held view is the structure can be described as a CRN. This was first proposed by Zachareisen (1932) who applied it to oxide type glasses. Although the number of atoms associated with the nearest neighbors are the same, there is nevertheless a variation in the interbond angles that rapidly leads to a loss of local order and ultimately to the absence of long-range order.

For tetrahedrally bonded amorphous semiconductors, such as a-Si, a model built by Polk (1971), first demonstrated the possibility of building up an expanded CRN with the coordination number $z = 4$ without developing undue bond length strain, but allowing for a spread in bond angles ($\pm 10°$), as shown in Fig. 2.12. Polk (1971) has shown that allowing a small spread in bond lengths ($\sim 1\%$) can lead to a smaller bond angle distortion ($\pm 7°$), since a small increase in bond length distortion energy is compensated for by a larger decrease in bond angle distortion energy.

A higher order angular characteristic of a tetrahedrally coordinated covalent material is the distribution of dihedral angles that describe the relative orientation of the adjacent tetrahedra. (The dihedral angle is the average angle separating a bond projection belonging to one atom from the nearest bond projection to the other atom.) In the eclipsed configuration the

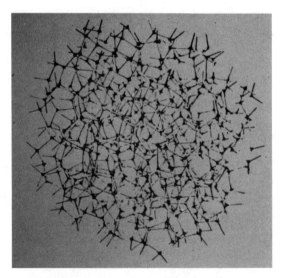

Fig. 2.12. Cluster composed of snap-together units of the type used by Polk (1971) in constructing his four-coordinated continuous random network for amorphous silicon and germanium.

projections line up and hence the dihedral angle is zero, whereas in the staggered configuration the dihedral angle is close to its maximum value of 60°. In the CRN model it is found that there is a continuous distribution of dihedral angles and, as can be seen in Fig. 2.12, the structure possesses a substantial number of 5-, 6- and 7-membered rings, with a density within 1% of that of a diamond-structure crystal having the same bond length.

Various techniques such as X-ray and electron diffraction (Barna et al., 1977; Mosserei et al., 1979, neutron diffraction (Postol et al., 1980), EXAFS (Knights et al., 1977a), ^{29}Si NMR (Reimer et al., 1981; Lamotte et al., 1981; Jeffrey et al., 1981) and Raman scattering (Brodsky et al., 1977) have been used to study the structure of a-Si and a-Si:H type materials.

The radial distribution function (rdf) is widely used to characterize these materials. It is derived via Fourier transforms of the diffraction data. In these experiments a beam incident on the sample consists of monoenergetic electrons, X-rays, or neutrons at a selected energy, E. E is chosen so that the de Broglie wavelength ($\lambda = h\sqrt{2mE}$) of the incident energetic particles of mass m (in the electron or neutron experiments) is on the order of 10^{-8} cm, which is comparable to the interatomic spacings. The measured quantity is the scattering interference function $I(k)$, where the scattering vector, k, is related to the observed scattering angle, θ, by $k = (4\pi/\lambda) \sin \theta$. The radial distribution function, $\rho(r)$ is obtained as the real-space transform of $I(k)$. Figure 2.13 shows the electron diffraction pattern of non-hydrogenated a-Si before and after partial crystallization (Moss and Graczyk, 1979). The width

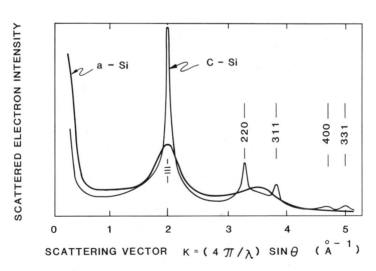

Fig. 2.13. Electron diffraction pattern (Moss and Graczyk, 1969) of amorphous silicon and of the same film after partial crystallization.

of the first peak in the amorphous solid is evidently very similar to that for the crystalline case. (The peak width corresponds to the spread in the nearest neighbor distances, and has a finite width in crystals because of vibrational motion). Since the first peak widths of the two types of solids are identical, then the bonding interactions responsible must be nearly identical and the short range order is retained in the a-Si case. In both cases, crystalline and amorphous, the coordination number, $z = 4$, is determined by integrating the area under the first peak. Further, the second nearest neighbor coordination is retained for the amorphous case but the absence of long-range order is exemplified by the fact that there is no discernible peak beyond the second nearest neighbor. It should be stressed that although the rdf data provides crucial information about the short range order (and hence the nature of the chemical bonding), it is not able to provide a unique structural model because a wide range of topologically distinct models, having the same rdf-defined short-range order, can fit the experimentally observed data.

In Fig. 2.14 we show the reduced density function for GD a-Si:H deposited under various deposition conditions (Graczyk, 1979). (The dotted lines are values calculated using the CRN model.) Samples (a) and (c) (deposited at room temperature) use conditions that normally lead to a high defect state density and curve (b) corresponds to samples deposited at a $T_s = 250°C$, which yield low defect state densities (see Section 2.3 and 2.5) Graczyk (1979) and Barna *et al.* (1977) both concluded that samples of type (c) could be described structurally by the CRN model, which was modified to include 20% of a staggered configuration of tetrahedral units to account for some additional structure observed at $r = 4.55$ Å. Upon annealing the samples to 580°C, crystallization appears as shown in Fig. (2.13), where additional structure becomes evident at positions corresponding to the third and fourth nearest neighbor distances. Both Graczyk (1979) and Mosserei *et al.* (1979) concluded that material deposited on low temperature substrate exhibited significant deviations from the CRN configuration. Graczyk (1979) noted that there was an additional peak at 4.29 Å that could be due to densely packed polymeric chains (SiH_2) and concluded that such samples might therefore consist of a mixture of polymeric and CRN regions.

Infrared (ir) spectroscopy has also proven to be useful in the study of local bonding configurations. Specifically, the SiH environments are characterized by a bond stretching mode at 2000 cm^{-1} and a bond bending mode at 640 cm^{-1}, whereas sites with more than one H-atom exhibit additional features in the bond bending frequency range, $800-950 \text{ cm}^{-1}$, as well as bond stretching modes between 2050 and 2150 cm^{-1}. Figure 2.15 shows several anode type GD samples fabricated with SiH_4 gas diluted by A_r (5%) as a function of RF power with the substrate temperature held at $T_s = 230°C$ (Lucovsky *et al.*, 1979). The low power sample is dominated by

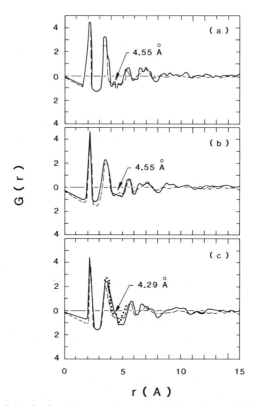

Fig. 2.14. Reduced density function vs. r from transmission electron diffraction for a-Si:H plasma-deposited onto cleaved NaCl; (a) plasma pressure (p) = 0.05 torr, substrate temperature (T_s) = 25°C, (b) p = 0.3 torr, T_s = 250°C, (c) p = 0.3 torr, T_s = 25°C. (———) experimental data; (– – –) calculated function for a continuous random network (Graczyk, 1979).

two strong absorption bands, one at 2000 cm^{-1} and the other at 630 cm^{-1}. With increasing power an additional band at 2090 cm^{-1} appears and is accompanied by a strong absorption band in the 850–900 cm^{-1} range. The absorption spectra for anode films deposited onto room temperature substrates are displayed in Fig. 2.16. These spectra show the bond stretching (1900–2200 cm^{-1}) and bond bending (880–900 cm^{-1}) regimes, with two sets of doublets in the 800–900 cm^{-1} region.

The absorption peaks can be interpreted by recognizing that there can be either 1, 2 or 3 H atoms situated at a Si-site. Since the mass of H is small compared to that of Si, then the vibrational modes can be described by considering only the displacement of the H-atoms. For example, the modes of SiH$_2$ group connected to the Si network in a tetrahedral environment are

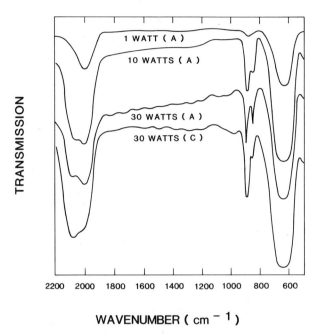

Fig. 2.15. Infra-red transmission for a-Si:H samples prepared at different power with T_s held at 230°C (Lucovsky et al., 1979).

obtained from the cluster, Si_2–SiH_2 (Knights et al., 1978). The molecular clusters relevant to the three local bond configurations, as well as a schematic representations of the vibrational modes involving H-motions, are shown in Fig. 2.17 (Lucovsky et al., 1979). Table 2.2 summarizes the assignments made in terms of four local atomic environments: SiH; SiH_2;$(SiH_2)_n$; SiH_3.

Two points are worth noting: the H concentration for low T_s samples is much greater than for the high T_s samples (see also Fig. 2.2); the low T_s samples exhibit a peak at 845 cm^{-1}. The latter observation indicates that chains of SiH_2 groups [$(SiH_2)_n$ or polysilane] exist within the material. This seems to be supported by the appearance of columnar morphology as observed via Scanning Electron Microscopy (SEM) (Knights and Lujan, 1979); it becomes more pronounced with an increase in the intensity of the 845 cm^{-1} band. It can be concluded that polysilane like chains exist in interstitial regions between columns of less-hydrogenated material. Further evidence that low temperature films are structurally different from high temperature films is also provided by H effusion measurements (see, e.g., Woodyard et al., 1985). In this type of experiment a sample is heated in an evacuated chamber; at specific temperatures there is a sudden increase in pressure due to the effusion of H (Fritzche et al., 1978). A sample deposited

STRUCTURE AND COMPOSITION

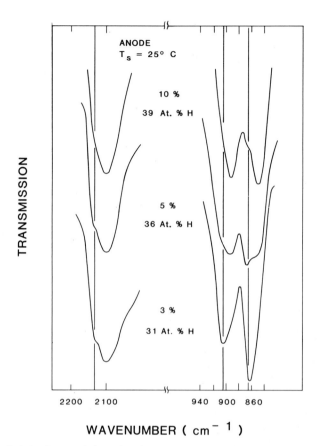

Fig. 2.16. Infra-red transmission of a-Si:H for $T_s = 25°C$. The relative dilution of SiH$_4$ in Ar is given as the atomic percent of hydrogen (Lucovsky et al., 1979).

at low T_s produces two distinct pressure changes, one at $T = 350°C$ and the other at $T = 550°C$, whereas a high T_s sample only shows the latter effect. This confirms the above points: the H is bonded in different configurations which in turn is related to the deposition parameters; the $T = 350°C$ effusion is associated with the H bonded in the SiH$_2$ (dihydride) configuration; the H effusion at $T = 550°C$ is related to the H bonded in a Si–H (monohydride) configuration.

As discussed above, the H concentration can vary with the deposition temperature T_s. However, in any GD a-Si:H or a GD a-Si:F:H type film there are also other species, primarily contaminants such as O, C and N. These affect the opto-electronic properties of the material and consequently the device performance; they will be discussed in Section 3.1.4.5. Various

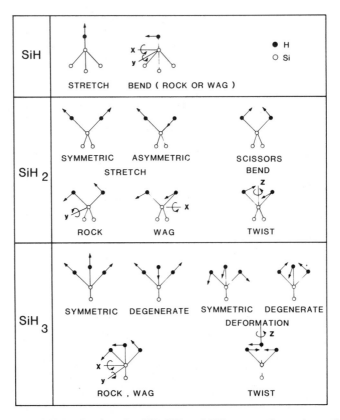

Fig. 2.17. Local Si–H vibrations for SiH, SiH$_2$ and SiH$_3$ groups (Lucovsky *et al.*, 1979).

techniques have been used to determine the concentration of H, F, O and C in the films. The most commonly used technique is Secondary Ion Mass Spectrometry (SIMS), which is able to profile many species, an example of which is given in Fig. 3.18. The other techniques for measuring primarily the H content of the films are hydrogen evolution (Fritzche *et al.*, 1978; Fritzche *et al.*, 1979a), which we have discussed above, and nuclear reaction analysis, in which a high energy ion beam is incident on the sample and the products of the reaction of the ion beam with the atomic constituents of the sample, such as γ-rays, are analyzed. For H content evaluation, reactions of the type $H + {}^{15}N \rightarrow {}^{4}He + {}^{12}C + \gamma$ and $H + {}^{19}F \rightarrow {}^{4}He + {}^{16}O + \gamma$ are generally used (Lanford *et al.*, 1976). An estimate of the H concentration in a-Si:H films can also be made from the infra-red data (Brodsky *et al.*, 1977). Here, the area under the peak corresponding to the Si–H wag mode (640 cm^{-1}) is determined and is related to the number of such bonds per unit volume.

DENSITY OF STATES

Table 2.2. Assignments for the Principal IR and Raman Features in a-Si:H as a Function of the Local Structural Groups: SiH; SiH_2; $(SiH_2)_n$; SiH_3 (Lucovsky et al., 1979).

Group	Frequency (cm^{-1})	Assignment
SiH	2000	Stretch
	630	Bend
SiH_2	2090	Stretch
	880	Bend-scissors
	630	Rock
$(SiH_2)_n$	2090–2100	Stretch
	890	Bend-scissors
	845	Wag
	630	Rock
SiH_3	2140	Stretch
	907	Symmetric Deformation
	862	Degenerate Deformation
	630	Rock

The H concentration is then determined by assuming an oscillator strength corresponding to the Si–H active vibrational mode. This can lead to an erroneous result since there is a controversy as to whether a-Si is homogeneous or whether there are H-rich regions embedded in an a-Si matrix.

It is found that the H concentration in a-Si:H can be as high as 40–50 at.%, depending upon the deposition conditions. However, "device quality" a-Si:H materials generally have a H concentration of less than 10%. For a-Si:F:H type material the H concentration is between 0.5% to 4% (Madan et al., 1979). In GD a-Si:H, the H concentration appears also to be influenced by doping (Müller et al., 1980). As shown in Fig. 2.18, (Demond et al., 1981) heavily B doped films undergo a large reduction in the H concentration, which could explain in part the decrease in the band gap of this type of material, the implications of which will be discussed in Sections 2.8.2 and 3.1.4.6a.

2.3 Density of States

2.3.1 Some Models for the Density of States

The determination of the complete density of states (DOS) structure of a given material generally requires knowledge of the equilibrium position of the atoms (atomic structure), their normal modes of vibration (phonon

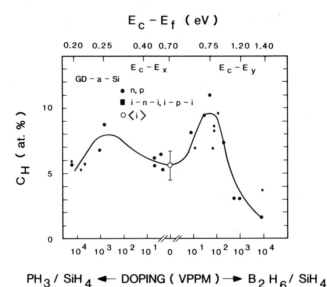

Fig. 2.18. Hydrogen concentration versus doping ratio of GD a-Si samples. n, p means bulk material of the respective type; i–n–i, i–p–i refers to a three-layer structure in which a central zone of n or p-type material is sandwiched between two intrinsic layers. Fermi-level positions corresponding to the doping ratios are given with reference to the conduction band edge (Demond et al., 1981).

structure) and the derivation of the excited electronic structure. Although this can be achieved to a limited extent in crystalline materials, where the periodicity aids in solving the problem by approximations, it is more difficult to achieve in amorphous semiconductors. One reason for this is that amorphous semiconductors (such as amorphous Silicon) are prepared from a non-equilibrium situation and are therefore metastable. Consequently, their DOS depends upon the details of the preparation conditions. Several factors that determine the DOS for a given material consist of, among others, the coordination number, the interatomic distance and the type of chemical bonding.

As discussed in Section 1.3, the lack of long-range order inherent in amorphous solids results in conduction and valence band tails, the extent of which depends upon the amount of inherent disorder. Virtually every model for DOS spectra embodies this. The early models of Cohen–Fritzche–Ovshinsky (1969) shown in Fig. 2.19a and by Mott and Davis (1968) in Fig. 2.19d recognized this. These models were appropriate to multicomponent chalcogenide glasses that have positional disorder in addition to the compositional disorder, with the result that there is a large DOS in the mid-gap region. This leads to a pinning of the Fermi level, E_f, in this region. In the former model the pinning is achieved by the overlap of the tails in the mid-gap

NON – CRYSTALLINE SEMICONDUCTORS

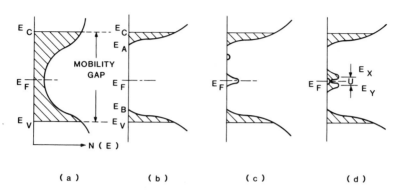

Fig. 2.19. Various forms proposed for the density of states in amorphous semiconductors. Localized states are shown shaded. (a) Overlapping conducting and valence band tails as proposed by Cohen et al. (1969), the CFO model; (b) a real gap in the density of states, suggested here as being appropriate for a continuous random network without defects; (c) the same as (b) but with partially compensated band of defect levels; (d) the same as (b) but with overlapping bands of donor (E_Y) and acceptor (E_X) levels arising from the same defect (Mott and Davis, 1979).

region, which provides for a transfer of electrons from the top of the valence band tail to the bottom of the conduction band tail. In the latter model (Fig. 2.19d) a band of localized states is assumed to exist near mid gap, which in turn ensures the pinning of E_f. More recent models for chalcogenide glasses involving valence alternation pairs that are based on lone pair bonding are discussed in Chapter 4.

In a covalent bonded material, such as a-$Si_{1-x}H_x$, the strong local forces determine the coordination number z, the bond lengths of the Si–Si and Si–H bonds, and the bond angles associated with Si. Phillips (1979a) has shown that the constraints within the a-$Si_{1-x}H_x$ solid can be satisfied if $z = 2.45$. Since $z > 3$ for GD a-Si:H films, then considerable amounts of strain will be exhibited by these materials. However, some of the strain can be relieved by allowing for bond angle and bond length distortions, as shown by Polk (1971) and Graczyk (1979) and discussed in Section 2.2. Adler (1984) argues that since tetrahedral coordination is the maximum possible using s and p electrons only, over coordinated Si defects are not likely to occur [although Pantelidis (1986, 1987) argues otherwise (Fedders and Carlsson, 1987)]. The simplest possible defect that might be present is the isolated dangling bond, shown in Fig. 2.20b, corresponding to a T_3^0 center. [In this nomenclature, T_z^q represents a tetrahedral (T) bond with a coordination of z and q is the local charge state (Adler, 1978).] The normal structural bonding configuration,

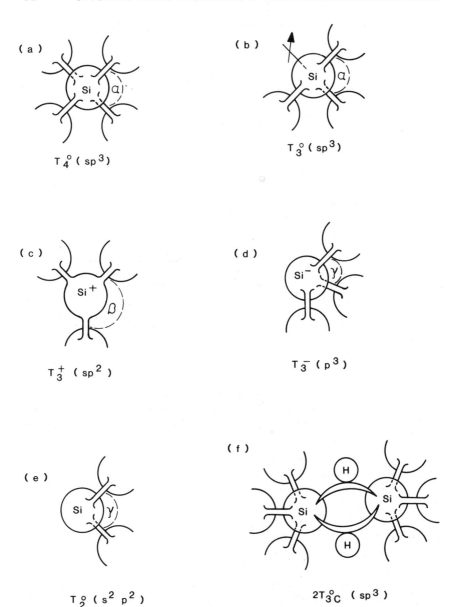

Fig. 2.20. Sketches of the optimal local coordinations of a Si atom in several different configurations: (a) ground state, T_4^0; (b) neutral dangling bond, T_3^0; (c) positively charged dangling bond, T_3^+; (d) negatively charged dangling bond, T_3^-; (e) two fold-coordinated Si atom, T_2^0; (f) complex consisting of two three-center bonds with bridging H atoms. The bond angles identified are $\alpha = 109.5°$, $\beta = 120°$, $\gamma = 95°$ (Adler, 1984).

T_4^0, is shown in Fig. 2.20a. Adler (1982) points out that the presence of localized states within the mobility gap leads to the breakdown of the two major approximations of the band theory of solids: the one electron approximation and the adiabatic approximation. These neglect the notion that two electrons can correlate their motion to minimize their mutual electrostatic repulsion. The repulsion between two electrons with opposite spins is called the correlation energy, U. U is negligible for extended states, but for localized states U is on the order of a few tenths of an eV. Tight binding estimates made by Adler (1978) suggest that T_3^0 defects yield two states in the gap, a lower filled state and an upper empty state separated by U. If an electron is removed from a T_3^0 center, it converts to T_3^+, shown in Fig. 2.20c. The removal of the electron can induce strong chemical forces leading to a distortion of the local environment. Adler (1982) speculates that T_3^+ optimally bonds in a planar sp^2 configuration with a 120° bond angle. On the other hand, the conversion of T_3^0 to T_3^- leads to a decrease of the bond angle to 95° (Fig. 2.20d). This implies that $U < 0$ (or more correctly an effective $U_{eff} < 0$), whereas the calculations of Allen and Jonanopolous (1980) indicate that $U_{eff} > 0$. The resulting DOS spectra is shown in Fig. 2.21 and the corresponding DOS diagram for $U_{eff} < 0$ is depicted in Fig. 2.22.

The evidence that $U_{eff} > 0$ is provided by the ESR data of Dersch et al. (1981), indicating the presence of a well defined unpaired spin with $g = 2.0055$ and a peak to peak linewidth of 7.5 G that is almost universally associated with a T_3^0 center. However Adler (1983) argues against this since it is believed that a-Si:H is an overconstrained network and further that NMR data (Reimer et al., 1980) suggest that even "device quality" a-Si:H films could contain regions with relatively low hydrogen concentrations interspersed in regions with considerably higher concentrations. Phillips (1979b) has suggested that internal edges exist in these type films. "Surface" reconstruction is therefore possible such that neighboring dangling bonds pair to eliminate any unpaired spins that are always observed in a-Si$_{1-x}$H$_x$ alloys. This model could account for the fact that much more hydrogen can be introduced into a-Si structure than the values of the spin density indicates. This is consistent with Adler's (1984) suggestion that isolated T_3^0 centers having $U_{eff} > 0$ can coexist with $T_3^+ - T_3^-$ pairs, where $U_{eff} < 0$. That is, a range of U_{eff} exists that extends through both positive and negative values (Silver et al., 1986). The DOS structure envisaged is shown in Fig. 2.23. (This model has been used to explain some of the degradation phenomena in solar cells and will be discussed in Section 3.1.6.) It is also possible that more complicated defects arise in a-SiH type alloys, such as T_2^0 centers, T_5^0 centers, (Pantelidas, 1986), defect complexes such as vacancy type defects (Hirose, 1981) and divacancies (Spear, 1973, Madan, 1973). The reader is referred to these articles for more details.

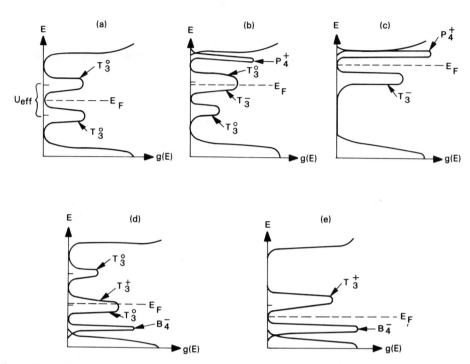

Fig. 2.21. Effective one-electron density of states for a tetrahedrally bonded amorphous semiconductor with dangling-bond defects characterized by $U_{eff} > 0$: (a) undoped sample; (b) moderately P-doped sample; (c) heavily P-doped sample; (d) moderately B-doped sample; (e) heavily B-doped sample. The defect bands are labeled by their state when filled if they are located below E_F and when empty if they are located above E_F (Adler, 1984).

One of the more remarkable aspects of a-Si : H type alloys is that they can be doped and made to exhibit variations in conductivity of up to 10 orders of magnitude (see Section 2.8). However, doping with Phosphorous (P) or Boron (B) does not lead to degeneracy, unlike the case of crystalline Si. Further, it is found that the doping efficiency is less than unity (Spear, 1977) and indicates that the dopant atoms introduce additional defect states that act as recombination centers (Magarino et al., 1982). The clearest evidence for this is given by the light induced ESR signals, where there is a rapid rise in the spin density of defects of the dangling bond type (Knights et al., 1977b) with doping. The proper interpretation of doping must therefore embody the above experimental facts.

Adler (1984) and Robertson (1985) point out that for Phosphorous doping, P_3^0 is the most stable configuration, with $P_4^+ - T_3^-$ as the next most stable configuration, as shown in Figs. 2.24 and 2.25. Adler (1984) indicates that the electronic structure arising from the presence of $P_4^+ - T_3^-$ pairs depends

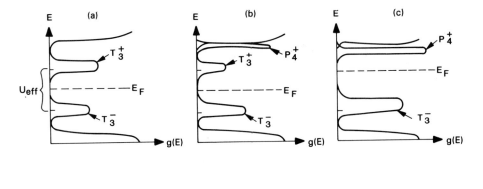

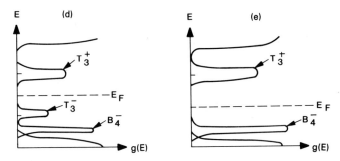

Fig. 2.22. Effective one-electron density of states for a tetrahedrally bonded amorphous semiconductor with well-separated dangling-bond defects characterized by $U_{\text{eff}} < 0$: (a) undoped sample; (b) moderately P-doped sample; (c) heavily P-doped sample; (d) moderately B-doped sample; (e) heavily B-doped sample (Adler, 1984).

critically on the sign of the correlation energy, as shown in Figs. 2.21 and 2.22. The doping results imply that U_{eff} may be negative in some regions of the film, since the ESR data (Magarino *et al.*, 1982) indicate that as P is introduced, the concentration of T_3^0 center decreases because when E_f approaches E_c, the T_3^0 sites are converted into spin-paired T_3^- centers. With further addition of P the signal corresponding to the T_3^0 spin is completely quenched and is replaced by a new signal with $g = 2.004$ and a linewidth of 10 G, which is speculated to arise from T_3^- centers. Street (1982) has suggested that additional Si dangling bonds are created to accommodate the extra electrons in T_3^- centers and these are consequently present in equal numbers. Doping now occurs because E_f is forced above the T_3^- levels, but the doping efficiency gradually declines as $N^{-1/2}$, where N is the total doping concentration.

The introduction of B lowers the position of E_f and the spin signal corresponding to the T_3^0 center is also quenched. It is therefore possible that

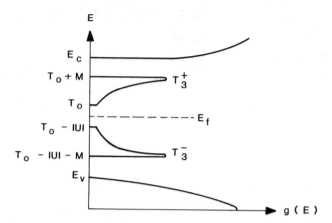

Fig. 2.23. Effective one-electron density of states for a tetrahedrally bonded amorphous semiconductor containing a distribution of spatially correlated dangling-bond defects with $U_{\text{eff}} < 0$. T_0 represents the energy of the nonbonded electron on a neutral dangling bond and M is the electrostatic attraction between a T_3^+ center and a T_3^- center located on nearest neighbor sites (Adler, 1984).

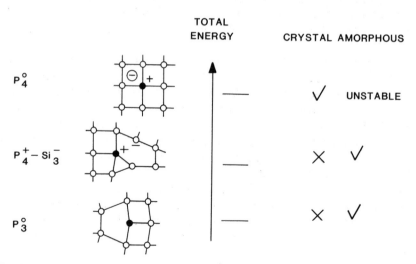

Fig. 2.24. Configurations, total energy, and stability in crystalline Si and a-Si:H of P_3^0, $P_4^+ - Si_3^-$ and P_4^0 sites (Robertson, 1985).

DENSITY OF STATES

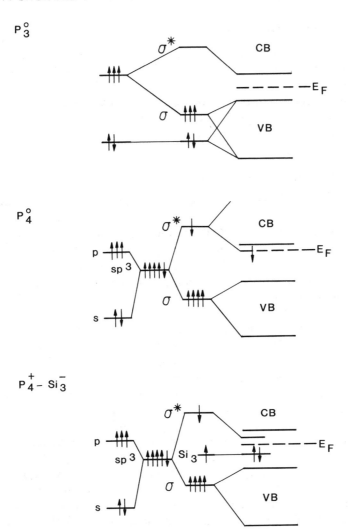

Fig. 2.25. Electronic configurations of P_3^0, P_4^0, $P_4^+ - Si_3^-$. VB and CB denote valence and conduction bands, respectively (Robertson, 1985).

the $T_3^+ - B_4^-$ complex will depend upon the sign of the correlation energy, as shown in Figs. 2.21 and 2.22. Because of the complexity of Boron chemistry, the situation is not totally analogous to P; there is evidence from NMR data (Carlos et al., 1982a and Carlos et al., 1982b) that the vast majority of the B is three-fold coordinated, which could lead to the presence of B_2^0 centers, thereby creating extra defect states.

2.3.2 Experimental Determination of the DOS

There has been a major effort to measure the DOS spectra of a-Si:H type films, with results ranging from 10^{15} cm^{-3} eV^{-1} to 10^{17} cm^{-3} eV^{-1} at $E = E_f$ for similarly produced films. Many techniques have been employed such as: field effect; space charge limited currents (SCLC); deep level transient spectroscopy (DLTS), capacitance–voltage (C–V). In the following, we shall review some of the more widely used techniques.

2.3.2.1 Field Effect Technique

Field effect measurements have been widely used to determine the number and distribution of surface states in crystalline semiconductors (Many et al., 1965; Egerton, 1971; Barbe, 1971). However, in amorphous semiconductors, in addition to the surface states there are significant amounts of localized bulk states. Generally, the effect of surface states has been neglected in the analysis of the field effect as applied to amorphous semiconductors; consequently, the bulk state densities have been overestimated (Spear and LeComber, 1972; Madan, 1973; Madan et al., 1976; Madan and LeComber, 1977; Madan, 1980). The specimen geometry used is shown in Fig. 2.26a, where S and D are the source and drain contacts separated by a channel length L and the contacts are of width W. Figure 2.26b shows that with the application of a positive voltage, V_G, to the gate electrode the induced negative charge causes the bands to bend downward with the result that the occupancy of the localized states around the Fermi level changes from empty to full. Conversely, the application of a negative V_G results in the bands bending in the opposite direction. The creation of a space charge therefore affects the number and density of mobile carriers above the band edge. This change is measured by the current, I_{sd}, flowing through the film when a constant potential difference, V_{sd}, is established between the source and drain electrodes. The key in the field effect analysis is to relate $I_{sd}(V_G)$ to the surface potential, its variation in the bulk [$V(x)$] and in turn relate that to the DOS spectra, $g(E)$. At any point in the semiconductor, the potential profile within the space charge region is given by the solution of the Poisson equation

$$\frac{d^2 V}{dx^2} = -\frac{\rho}{\varepsilon_s \varepsilon_0}, \qquad (2.8)$$

where ε_s is the relative permitivity of the semiconductor. If the charge density, ρ, is only a function of $V(x)$ then

$$\rho(V) = -e \int g(E)\{f(E - eV) - f(E)\} \, dE \qquad (2.9)$$

DENSITY OF STATES

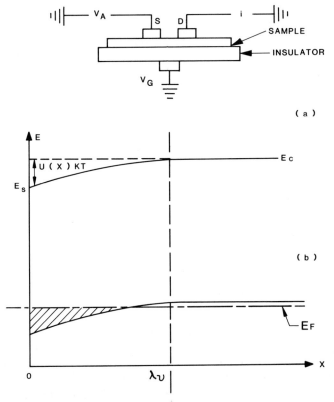

Fig. 2.26. (a) Specimen geometry used in the field effect technique; (b) electron energy E plotted as a function of distance from the specimen surface when a positive potential V_G is applied to the gate electrode (this alters the occupancy of states near the surface region, shown by the shaded area).

where $f(E)$ is the Fermi–Dirac function. Manipulation of Eqs. (2.8) and (2.9) leads to the field, $F(x)$, at any point given by,

$$F^2(x) = \frac{2}{\varepsilon_s \varepsilon_0} \int_0^V \rho(V)\, dV \qquad (2.10)$$

The conductance, G, is given by

$$G = \frac{W}{L} \left\{ \int_0^d n_0 e\mu_n \left[\exp\left(\frac{eV(x)}{kT}\right) - 1 \right] dx \right.$$
$$\left. + \int_0^d p_0 e\mu_p \left[\exp\left(-\frac{eV(x)}{kT}\right) - 1 \right] \right\} dx \qquad (2.11)$$

where d is the thickness of the semiconductor, μ_n and μ_p are the mobilities of electrons and holes and, n_0 and p_0 are the equilibrium electron and hole carrier densities, respectively. For flat band conductance, G_0, $V(x) = 0$ in Eq. (2.8). Hence, the change in conductance, $\Delta G = (G - G_0)/G_0$, is given by

$$\frac{\Delta G}{G_0} = \frac{e\mu_n n_0 \int_0^d \left\{\exp\left[\frac{eV(x)}{kT}\right] - 1\right\} dx + e\mu_p p_0 \int_0^d \left\{\exp\left[-\frac{eV(x)}{kT}\right] - 1\right\} dx}{d(e\mu_n n_0 + e\mu_p p_0)} \quad (2.12)$$

$\Delta G/G_0$ is an experimentally measured quantity; it is related to $V(x)$ via the solution of Eqs. (2.8) and (2.9). In the solution of the problem the boundary conditions are that at the surface ($x = 0$), $V(0) = V_s$, at $x = d$, $dV/dx = 0$, and that for a sufficiently thick dielectric material, the capacitance of the insulator is much less than that of the semiconductor. The induced charge, Q_{ind}, can be related to the surface field, F_s, which in turn can be related to the gate voltage via

$$Q_{\text{ind}} = C_i V_G = \frac{\varepsilon_i \varepsilon_0 V_G}{d_i} = \varepsilon_s \varepsilon_0 F_s, \quad (2.13)$$

where d_i is the thickness and ε_i is the permitivity of the insulator. By assuming a $g(E)$, $\rho(V)$ can be found from Eq. (2.9). From Eqs. (2.8) and (2.10), $V(x)$ (and hence V_s) and $F(x)$ can be found; hence Eq. (2.13) can be related to the experimentally defined gate potential, V_G. Using Eq. (2.12), $\Delta G/G_0$ can then be computed and compared with the experimental results. The $g(E)$ spectra is then changed until consistency is achieved. Similarly, $\Delta G/G_0$ for the depletion and inversion cases can also be generated (Madan, 1973; Madan, 1980).

The earliest results for the DOS were made by the use of this technique and gave an indication of a surprisingly low value (Spear and LeComber, 1972; Madan, 1973; Madan et al., 1976) for GD a-Si:H alloys. Apart from the neglect of surface states, the early analysis assumed a definite form of the band bending (parabolic or exponential), zero temperature statistics and abrupt junction formation. These assumptions produced inaccuracies in the DOS. Later work (Madan and LeComber, 1977) showed that removal of the abrupt approximation had underestimated the DOS by a factor of 3.

Figure 2.27 (Madan, 1973; Spear, 1973; Madan et al., 1976) shows that for GD a-Si:H the DOS decreases by orders of magnitude as T_s is increased. As discussed in Section 2.5, the transport properties changed from hole hopping behavior to electron extended state conduction as T_s was increased from 300 K to 570 K. The reason for the change in the DOS spectra is attributed to the change in the fundamental structure of the film, since the

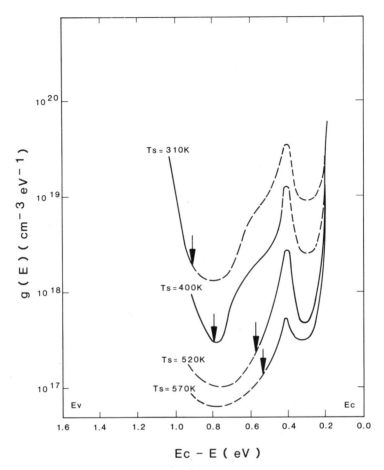

Fig. 2.27. Localized state density, $g(E)$, for glow discharge a-Si:H alloy. T_s refers to the substrate temperature. Full lines are obtained from the field effect data and the arrows indicate the Fermi level position of the samples (Madan, 1973; Spear, 1973; Madan *et al.*, 1976).

infrared spectra, discussed in Section 2.2, indicated that the films changed from a dihydride to a monohydride configuration over this temperature range.

Since the field effect response is ultimately limited by the dielectric strength of the insulator, then only small portions of the $g(E)$ spectra can be generated, as shown by the solid lines in Fig. 2.27. However, by doping the material with P and B, different portions of the DOS spectra can be obtained and a DOS distribution covering a larger range of energies can be obtained, as shown in Fig. 2.28. In the creation of this composite DOS curve it was implied that $g(E)$ did not change with doping, which is an incorrect assumption, since,

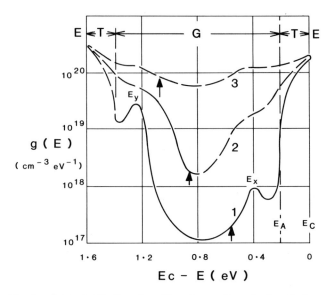

Fig. 2.28. Density of state distributions for a-Si specimens. Curve 1, glow discharge specimen, $T_s = 520$ K; curve 2, glow discharge specimen, $T_s = 350$ K; curve 3, evaporated specimen. The full lines indicate results obtained from field-effect experiments and the arrow on each curve shows the position of the Fermi level, E, extended states; T, tail states; G, gap states (Madan et al., 1976; Madan and LeComber, 1977).

as discussed in Sections 2.3.1 and 2.8, not all of the donor (e.g., P) or acceptor (e.g., B) atoms enter tetrahedral configurations; extra defect states are created.

Another notable feature from Figs. 2.27 and 2.28 is the presence of the two peaks situated at E_x and E_y, which have been used extensively in the interpretation of other experimental data (Spear, 1977). These peaks, together with the veracity of the field effect technique itself, have been questioned, since the various assumptions used in the analysis can lead to an erroneous $g(E)$ spectra. The validity of these assumptions have been investigated by Goodman et al. (1980) and Powell (1981). For example, Powell (1981) used the $g(E)$ curve shown in Fig. 2.29 and calculated the expected conductance shown in Fig. 2.30 using different assumptions in the analysis. The curve (a) in Fig. 2.30 corresponds to a parabolic profile with an abrupt approximation curve, (b) corresponds to an exponential profile with an abrupt approximation, (c) and (d) show the effects of removing the abrupt approximation, (e) is a more exact analysis using $T = 0$ K statistics, and (f) is an exact analysis using finite temperature statistics. It is evident that different field effect conductance curves can yield the same $g(E)$ spectra. Goodman et al., (1980) have questioned whether the field effect technique can resolve the structure

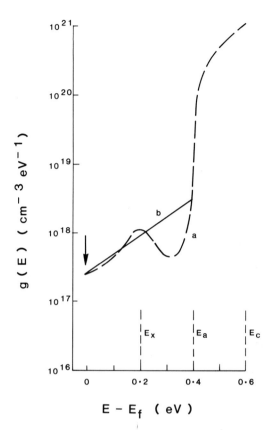

Fig. 2.29. Published $g(E)$ for glow discharge amorphous silicon [curve (a)] and a simple structureless model $g(E)$ [(curve (b)]. The arrow indicates the Fermi level position for the field-effect conductance calculation (Powell, 1981).

at energies E_x and E_y shown in Figs. 2.27 and 2.28. In their study the hypothetical $g(E)$ curves shown in Fig. 2.31, which differ from each other only slightly, were used to calculate the conductance curves shown in Fig. 2.32. From this, it is clear there seems to be a very little difference in the "data" from different $g(E)$ spectra shown in Fig. 2.31. This leads to the conclusion that only the broad features of the DOS can be determined; a unique $g(E)$ spectra may not be readily obtainable.

Other concerns regarding this technique have been the neglect of surface effects, the inability to define a flat band position (Weisfield et al., 1981), the preparation and hence the quality of the semiconductor-insulator interface (Goodman et al., 1980). In addition, it is assumed that the semiconductor is homogeneous and that no charges reside at the insulator–semiconductor

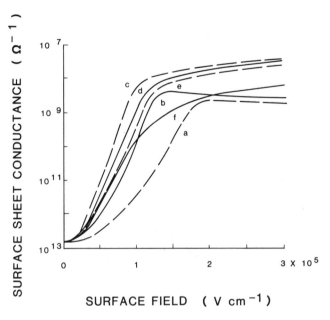

Fig. 2.30. Calculated field-effect conductance for the density of states in curve (a) of Fig. 2.29 according to different methods: (a) abrupt approximation, parabolic profile; (b) abrupt approximation, exponential profile; (c) non-abrupt approximation, parabolic profile; (d) non-abrupt approximation, exponential profile; (e) zero-temperature statistics; (f) finite-temperature statistics (Powell, 1981).

interface or within the insulator itself. Studies have also shown that there could be a large number of surface states and it is possible that the growth mechanism can alter, which would in effect change the properties of the film. Since with increased band bending the region adjacent to the insulator, where the majority of the current is carried, becomes very thin (< 100 Å), then any irregularity in that surface could lead to an error in the $g(E)$ spectra. Nevertheless, the technique, despite its limitations, is very important since the field effect is employed in the thin film transistor (TFT) device used to drive flat panel displays, which we will discuss in detail in Section 3.2.

2.3.2.2 Deep Level Transient Spectroscopy (DLTS)

The DLTS technique has been used widely in the characterization of deep trap levels in crystalline semiconductors. It was originally proposed by Lang (1974) as an alternative to the thermally stimulated capacitance (TSCAP) technique and other similar steady state methods. It is capable of measuring thermal emission rates, activation energies, trap concentration profiles and

DENSITY OF STATES 51

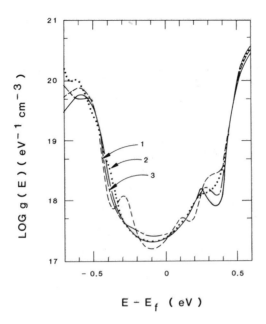

Fig. 2.31. Three density of states curves which fit the "data" shown in Fig. 2.32. The original $g(E)$ of Fig. 2.28 is shown by the solid curve (Goodman and Fritzche, 1980).

capture rates. The distinguishing features of DLTS is the concept of the "rate window". If a filling pulse is applied to a Schottky Barrier (SB) type device (see Section 3.1.3), a series of transient signals are produced. The time constant, τ, of the transients is a rapidly varying function of both the energy of the defect level and temperature. When the sample temperature is slowly varied, response peaks occur at the temperature where the emission rate for a certain trap level is approximately equal to the rate window. The resulting DLTS capacitance transient is then given by $C(t_1) - C(t_2)$ and has the form $e^{-t_1/\tau} - e^{-t_2/\tau}$, which when maximized with respect to τ yields.

$$\tau_{max} = \frac{t_1 - t_2}{l_n(t_1/t_2)} \qquad (2.14)$$

where t_1 and t_2 are the thermal evolution times. A trap level with emission rate $e_n(T) = 1/\tau_{max}$ will produce a peak in the DLTS spectrum, as shown in Fig. 2.33 (Lang, 1974; Cohen,, 1984). The energy depth of each level with respect to the closer band edge is proportional to the temperature according to

$$E = kT \ln\left\{\frac{vt}{\ln 2}\right\}, \qquad (2.15)$$

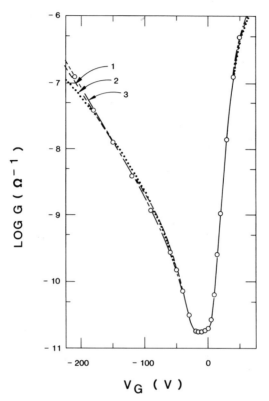

Fig. 2.32. Three computer fits to field effect "data" points generated from the $g(E)$ of Fig. 2.31 (Goodman and Fritzche, 1980).

where ν relates to carrier emission from traps. For the capacitance transient case the positive sign of the signal indicates hole emission to the valence band and a negative sign corresponds to electron emission to the conduction band. The magnitude of the signal is related to the number of states at each level. When the number of deep levels, N_T, is much smaller than the number of shallow traps, N_D (see, e.g., Cohen, 1984)

$$\frac{N_T}{N_D} \cong 2 \frac{\Delta C}{C_0}, \qquad (2.16)$$

where ΔC is the size of the DLTS capacitance and C_0 is the quiescent bias capacitance.

To observe traps below the midgap region a laser pulse can be used to produce a non-equilibrium population of both holes and electrons. This shifts the quasi-Fermi levels so that traps located below midgap are able to

DENSITY OF STATES

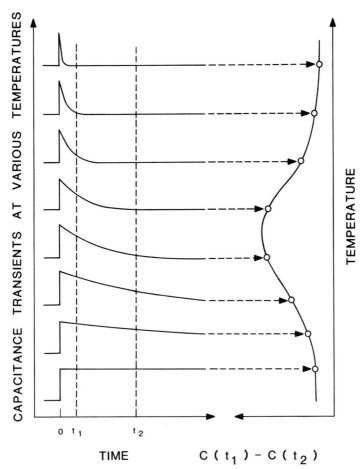

Fig. 2.33. Basic principle of the DLTS transient analysis technique as illustrated for emission from a discrete trap level (Cohen, 1984).

capture holes. After the laser pulse is removed, holes are thermally emitted to the valence band, which results in an increase of the depletion width. The capacitance transient will then exhibit an opposite sign to that due to electron emission, whereas the current transient will show the same sign for electrons and holes. If ΔC in Eq. (2.16) is due only to holes, then N_T, which is now associated with hole traps, can be found.

An important feature of DLTS is its ability to distinguish between bulk states and surface states. The DLTS signal produced by discrete surface states can be separated by varying the amplitude of the filling pulse. A bulk feature in the DLTS spectrum will exhibit a monotonic increase as the pulse amplitude is increased, but a feature due to surface states will respond at a specific threshold voltage, after which the DLTS signal rises abruptly and saturates (Cohen et al., 1980). When both surface and bulk states are present the shape of the DLTS spectrum changes with the pulse voltage. If the distribution of the surface states is continuous in the energy gap, a surface feature is more easily distinguished by the fact that it will not only decrease but also shift to a higher temperature (Yamasaki et al., 1979). In this respect, the capacitance transient approach offers an advantage over the current transient approach because of the difference in the spatial sensitivity of the two methods. Current transients are more sensitive to charge emission near the barrier interface, whereas in the capacitance transient technique charges emitted at the edge of the depletion region have the greatest effect on the capacitance.

For a-Si based alloys the DOS can be determined from the DLTS spectrum via the simplified analytical expression (Huyn et al., 1982)

$$g(E) = 3\frac{\Delta C}{C_0}\frac{N_{SD}}{\delta E}, \qquad (2.17)$$

where δE is the DLTS energy linewidth (Yamasaki et al., 1979) defined as

$$\delta E \cong kT \ln\left(\frac{t_2}{t_1} + 10\right), \qquad (2.18)$$

where t_2/t_1 is the ratio of the two sampling times and N_{SD} is the ionized shallow level concentration. Use of Eq. (2.17) leads to an accuracy within a factor of 3 in the $g(E)$ spectra.

Cohen (1984) has used a trial $g(E)$ to compute the DLTS spectra and compared that with the experimental data. The trial $g(E)$ spectra is then varied point by point until good agreement is obtained, as shown in Fig. 2.34. Figure 2.35 shows the $g(E)$ results for a Schottky Barrier device with the active layer doped with a small amount of P; it shows that there are peaks in the DOS spectra at 1.4 eV, at 1.0 eV, and a deep minimum centered at 0.5 eV, below E_c. The two peaks are attributed to the two charged states of the dangling bond (Lang et al., 1982). The major disagreement between DLTS and other techniques, such as field effect (Section 2.3.2.2) and SCLC, to be discussed in the next (Section 2.3.2.3), is the energetic location of some of the features in the $g(E)$ spectra. In the range of energy between 0.8 eV and 0.6 eV below E_c, $g(E)$ obtained using other techniques does not fall off rapidly with energy, rather, it rises with energy. To reconcile these differences Okushi et al. (1982), have studied the isothermal capacitance transient spectra (ICTS) of an a-Si:H Schottky Barrier (SB) type device and found that the capture cross sectional areas of the traps are energy dependent. This is physically plausible because the capture rate must be related to the number

DENSITY OF STATES

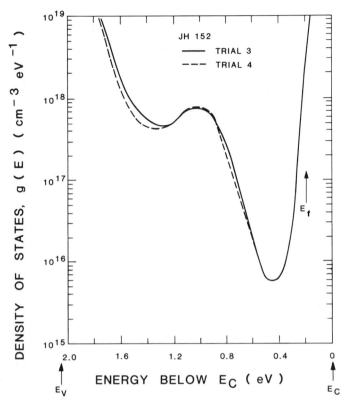

Fig. 2.34. Two very nearly identical densities of states employed as trial functions to fit the DLTS spectra (Cohen, 1984).

of phonons an electron must create in order to conserve energy during the capture process. With the inclusion of this effect, the $g(E)$ deduced from the ICTS results shows a peak, in contrast to the minimum in $g(E)$ obtained from DLTS measurements; it is shown in Fig. 2.36 (Weisfield, 1985). The peak in the ICTS is attributed to the negatively charged (T_3^-) dangling bond state. This figure also shows the $g(E)$ spectra taken by the field effect technique; it shows reasonable agreement with the ICTS measurement.

2.3.2.3 The Space Charge Limited Current (SCLC) Technique

In the SCLC technique voltage is applied across a diode, which causes an electrode (cathode) to inject a non-equilibrium density of electronic charge that populates the empty gap states above E_f in the vicinity of the cathode. When the applied voltage is large enough extra charge extends across the

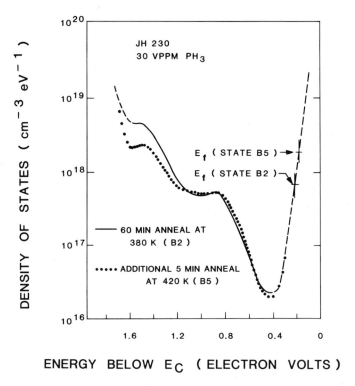

Fig. 2.35. Densities of states determined from DLTS measurements for a single n-type a-Si:H sample with different degrees of light exposure and dark annealing. These data indicate the ability of the DLTS method to discern very specific small changes in $g(E)$ (Lang et al., 1982).

entire sample thickness and the current collected by the counter electrode (anode) becomes space charge limited. Almost all the charge is trapped, but an exponentially small fraction is thermally promoted to the conduction band, increasing the current. It is this bulk enhancement mechanism that is analyzed to provide information about the density of electronic states above the equilibrium Fermi level.

The charge injected into the semiconductor is given by

$$Q_{\text{inj}} = \frac{\varepsilon_s \varepsilon_0}{L} V \tag{2.19}$$

where L is the sample thickness. The time taken to move across the semiconductor is the transit time of the carriers, t_t, given by

$$t_t = \frac{L^2}{\mu V}. \tag{2.20}$$

DENSITY OF STATES

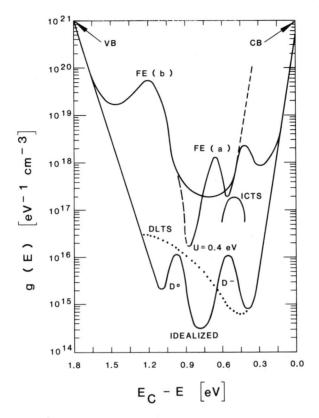

Fig. 2.36. Idealized g(E) in undoped GD a-Si:H, along with g(E) data from field effect (FE) in (a) sputtered a-Si:H from Harvard (Weisfield et al., 1981) and (b) glow discharge a-Si:H from Dundee (Fig. 2.28); DLTS (Lang et al., 1983) and ICTS (Okushi et al., 1982).

Hence, the space charge limited current density, J_{SCLC}, is simply given by

$$J_{SCLC} = \frac{Q_{inj}}{t_t} = \frac{\varepsilon_s \varepsilon_0 \mu V^2}{L^3}. \tag{2.21}$$

In real crystals the major portion of the injected charge resides in traps; when the injected charge is sufficiently large to fill the traps, the current will rise sharply. Further, if excess charge is placed into a semiconductor the free charge redistributes to compensate for the excess charge in a dielectric relaxation time given by $\tau_{rel} = \varepsilon_s \varepsilon_0/\sigma$, where σ is the conductivity. If $t_t < \tau_{rel}$, the injected charge can dominate the conductivity of the material, which is the case for a-Si:H alloys.

Dark I-V (Current-Voltage) measurements on a-Si:H SB diodes have been interpreted as evidence for SCLC, from which a quasi exponential DOS spectra was deduced (Ashok *et al.*, 1980). However, the diodes in that study were so poor in quality that any estimates from that particular work must be considered to be questionable. In its stead, we now discuss structures of the type n^+-i-n^+, where the holes are blocked and the current is primarily carried by electrons (single injection).

For a quasi-exponential distribution of states given by

$$g_t(E) = \frac{g_t}{kT_t}\exp\left(\frac{E - E_c}{kT}\right) \qquad (2.22)$$

where g_t and T_t characterizes the trap distributions (den Boer, 1981), it can be shown that the dark log J vs. log V characteristics of a n^+-i-n^+ structure should be linear, i.e., $J\alpha V^m$ with $m > 2$ and the current density, J, can be described by

$$J = g(E_c)e\mu_n\left(\frac{\varepsilon_s^l}{g_t(l+1)}\right)^l\left(\frac{2l+1}{l+1}\right)^{l+1}\frac{V^{l+1}}{L^{2l+1}} \qquad (2.23)$$

where $g(E_c)$ is the effective density of states at the conduction band edge and $l = T_t/T$.

However, in general, the log J vs. log V data are not linear, and in the presence of injection, E_f must be replaced by a quasi Fermi level. The shift in the Fermi level, ΔE_f, position is given by

$$\Delta E_f = kT \ln\frac{J_2 V_1}{J_1 V_2}. \qquad (2.24)$$

The average trap density within this interval can be written as,

$$\tilde{g}_t = \frac{2\varepsilon_s\varepsilon_0 \Delta V}{eL^2 \Delta E_f} \qquad (2.25)$$

Using this technique, den Boer (1981) has found that for GD a-Si:H samples deposited at $T_s \cong 300°C$, $g(E_f) \cong 2.10^{16} \text{ cm}^{-3} \text{ eV}^{-1}$.

Figure 2.37 shows log J vs. log V for samples prepared in the n^+-i-n^+ configuration and deposited at $T_s = 300°C$ (McKenzie *et al.*, 1982). For low voltages, the exponent m approaches unity and the current is therefore in the ohmic regime. Below 320 K, the J-V curves show a reasonably well defined transition from linear to V^2 behavior, which is representative of the expected SCLC flow. Further, the scaling law of $J/L\alpha(V/L^2)$ is also satisfied, at least in the thickness, L, range from 0.7 μm to 2.0 μm. McKenzie *et al.* (1982) have compared the step by step method, as proposed by den Boer (1981), to determine $g(E)$, as well as a more elaborate method proposed by Nespurek *et al.* (1980), which requires an accurate knowledge of the J-V curve as a

DENSITY OF STATES

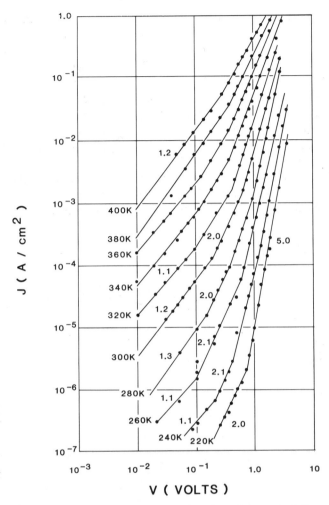

Fig. 2.37. Typical logarithmic J-V graphs for a n^+-i-n^+ specimen at the indicated temperatures. The numbers next to the curves denote the power b in the relation $j \sim V^b$ (McKenzie et al., 1982).

function of V, i.e., $m(V) = d \ln J/dV$. The $g(E)$ is then given by

$$g(E_{fn}) = \frac{X_1 \varepsilon_s \varepsilon_0}{eL^2 kT} \frac{V}{m(V) - 1} \tag{2.26}$$

and

$$E_c - E_{fn} = kT \ln\left(\frac{eg(E_c)\mu X_2}{L}\right) + kT \ln\left(\frac{V}{J}\right), \tag{2.27}$$

where X_1 and X_2 (≤ 1) are the correction factors that take into account the nonuniformity of the internal field and carrier density. The DOS thus deduced was a factor of about two lower than that deduced by the step by step method. The use of an exponential DOS distribution of the form given by Eq. (2.22) leads to $J\alpha V^2$. The $g(E)$ spectra derived in this way, from the data of Fig. 2.37, are shown in Fig. 2.38. They lie between the step by step method and the differential method of Nespurek et al. (1980).

Figure 2.39 shows a comparison of the $g(E)$ spectra derived from the field effect and the SCLC techniques for various types of samples (McKenzie et al, 1982). The solid line marked D.B. represents the average $g(E)$ determined by den Boer (1981); it fits well with the results reported by McKenzie et al. (1982). An upward trend in $g(E)$ in the energy range between 0.75 eV and 0.46 eV below the conduction band edge is to be noted for all samples.

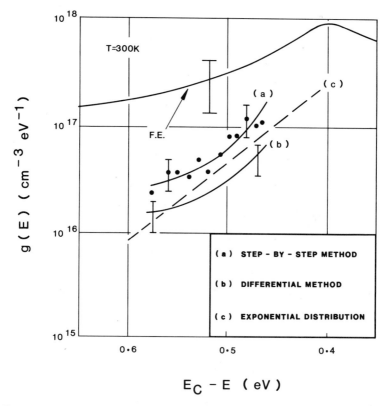

Fig. 2.38. Comparison of the density of states distribution $g(E)$ calculated from the same set of experimental results by the three methods of analysis. Curve F.E. shows typical field effect results (McKenzie et al., 1982)

DENSITY OF STATES

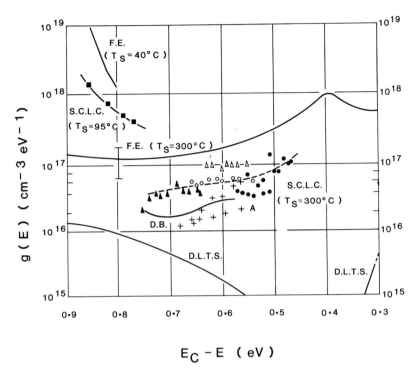

Fig. 2.39. Comparison of density of state distributions obtained from SCLC, field effect (F.E.) and DLTS measurements. T_s-deposition temperature; curve D.B., average values obtained by den Boer (1981); A, B and C refer to specimens deposited under different conditions (McKenzie et al., 1982)

It should also be emphasized that the general trend and the shape of the field effect agrees with the volume distribution of states deduced from the SCLC technique. The similarity between the SCLC and field effect techniques in terms of the DOS spectra have also been confirmed by Bhattacharya et al. (1982). Weisfield (1985) has extended the above analysis by calculating the energy and spatial distributions of electrons across a thin film for an arbitrary $g(E)$ at any level of current. This uniquely determines the dark J-V characteristic of a diode in terms of sample geometry and the $g(E)$ spectra. A representative set of DOS data, as measured by Weisfield (1985), are shown in Fig. 2.40.

Other techniques to measure $g(E)$ have also been used, such as tunnelling in MIS junctions (Balberg, 1980); the results were in substantial agreement with the field effect technique. Techniques involving capacitance variations such as C-V (see Section 3.1.3.6) or $C(T)$ have also been employed; the reader is referred to a review covering these by Cohen (1984).

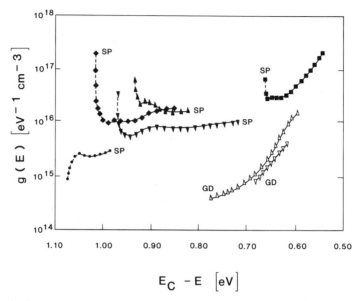

Fig. 2.40. $g(E)$ data from sputtered (SP) and GD a-Si:H samples (Weisfield, 1985).

2.4 Optical Absorption

The optical properties of a material are defined by the spectral dependence of its complex index of refraction, $\varepsilon_r = n(E) + ik(E)$ where n is the refractive index and k is the absorption index. The spectral region of major interest for a semiconductor is in the vicinity of the absorption edge, since it can provide information on the optical gap, E_g, as well as on the density of states within the gap. For semiconductors such as a-Si:H, n and k are usually obtained from transmission (T), and reflection (R) measurements (Heavens, 1970; Cody et al., 1980). n and k are derived using the equations outlined by Tomlin (1968). For a film of thickness d, which has a complex refractive index, $n_1 - ik_1$, and is on a substrate of complex refractive index $n_2 - ik_2$, the relevant equations are

$$(4n_0 n_2 A)^{-1}\{F[B\cosh(2\alpha_1) + 2D\sinh(2\alpha_1)]$$
$$+ G[C\cos(2\gamma_1) + 2E\sin(2\gamma_1)]\} - \frac{1-R}{T} = f_1(n_1, k_1) = 0; \quad (2.28a)$$

$$(2n_2 A)^{-1}\{n_1[B\sinh(2\alpha_1) + 2D\cosh(2\alpha_1)]$$
$$+ k_1[C\sin(2\gamma_1) - 2E\cos(2\gamma_1)]\} - \frac{1-R}{T} = f_2(n_1, k_1) = 0, \quad (2.28b)$$

where $A = n_1^2 + k_1^2$, $B = n_1^2 + n_2^2 + k_1^2 + k_2^2$, $C = n_1^2 - n_2^2 + k_1^2 - k_2^2$, $D = n_1 n_2 + k_1 k_2$, $E = n_1 k_2 - n_2 k_1$, $F = n_0^2 + n_1^2 + k_1^2$, $G = n_0^2 - n_1^2 - k_1^2$, $\alpha_1 = 2\pi k_1 d/\lambda$, $\gamma_1 = 2\pi n_1 d/\gamma$ and λ is the wavelength. The calculation is initiated by estimating the value of n_1 (for GD a-Si:H, n_1 is usually between 3 and 5). For each value of n_1, a value of k_1 is computed from Eq. (2.28a), i.e., $f_1(n_1, k_1)$. This value is then fed into Eq. (2.28b) to determine the condition $f_2(n_1, k_1) \cong 0$. The procedure is repeated until the best solution is found.

Using a procedure outlined by Kühl *et al.* (1974), the ratio Q, of the fringe amplitude as measured in reflection is

$$Q = \frac{R_{max\lambda 1}}{R_{min\lambda 1}}, \qquad (2.29)$$

Using Q, the absorption coefficient, α, can be calculated by

$$\exp(\alpha d) = \left\{\frac{r_{01}r_{12} - (r_{12}/r_{01})}{Q - 1}\right\}\left(\frac{Q+1}{2} + Q^{1/2}\right)$$
$$+ \left[\left(\frac{r_{01}r_{12} - (r_{12}/r_{01})}{Q-1}\right)^2\left(\frac{Q+1}{2} + Q^{1/2}\right)^2 + (r_{12})^2\right]^{1/2}, \qquad (2.30)$$

where $r_{01} = (n_0 - n_1)/(n_0 + n_1)$ and $r_{12} = (n_1 - n_2)/(n_1 + n_2)$.

Figure 2.41 shows α as a function of energy in the range 1.1 eV to 2.4 eV (Abeles *et al.*, 1980) for a GD a-Si:H alloy. α, as determined by transmission measurements, is shown by the open squares for photon energies between 1.64 eV to 2.40 eV. The optical absorption in the weakly absorbing regions was inferred from the transmission measurements by averaging the interference fringes. To extend the absorption measurements below 1.6 eV, one needs to use either very thick films or other photoelectric techniques, such as collection efficiency measurements. Here, the optical absorption is inferred from the collection efficiency, $Y(E)$ (see Section 3.1.4.7a) of a SB solar cell. The collection efficiency of a cell can be defined by

$$Y(E) \sim (1 - \exp - \alpha(E)X_c), \qquad (2.31)$$

where X_c is the collection width. Using Eq. (2.31) the solid circles in Fig. 2.41 show a measure of α down to approximately 1 cm^{-1}.

In Fig. 2.41 a comparison of low α values is also made by measuring the photoconductivity using a uniformly absorbed light, (a) in a coplanar configuration and (b) in a solar cell configuration. In the diode configuration the photoconductivity was obtained from the diode series resistance measured from the far forward bias condition (Wronski, 1978). The good agreement between the three techniques for measuring α over several orders of magnitude is significant in view of the major differences involved in the

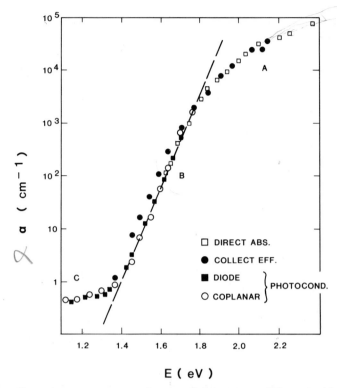

Fig. 2.41. Absorption constant, α, as a function of photon energy E determined from direct measurements of optical transmission, from collection efficiency and from photoconductivity in the diode and coplanar structures (Abeles *et al.*, 1980).

different experiments. In the photoconductivity measurements the transport depends on the majority carrier electrons, whereas the collection efficiency measurement involves the extraction of both types of carriers. The close agreement between the photoconductivity data and the collection efficiency data suggest that the generation efficiency remains independent of photon energy down to about 1.4 eV.

There are three major regions of interest, A, B and C, as indicated in Fig. 2.41; these are discussed below.

In considering region A we recall that in crystalline materials there are two types of optical transitions that can occur at the fundamental absorption edge, direct and indirect. Both involve the absorption of an electromagnetic wave by an electron in the valence band that is raised above the forbidden gap to the conduction band. However, indirect transitions involve the simultaneous interaction with phonons. Here the wavevector of the electron changes in the optical transition, the momentum change being taken or given

OPTICAL ABSORPTION

up by phonons. Neglecting exciton formation (the electron-hole interaction), $\alpha(h\nu)$ depends on the joint density of states for bands containing the initial and final states. For simple parabolic bands, $g(E) \alpha E^{1/2}$. For direct transitions, the absorption coefficient is given by

$$\alpha n_1 \hbar \omega = B(\hbar \omega - E_g)^n \qquad (2.32)$$

where $n = 2$ or 3, depending on whether the transition is quantum mechanically allowed or forbidden.

The most important features of optical absorption processes in amorphous semiconductors is the relaxation of certain selection rules (particularly that of k conservation) that apply to optically induced transition in crystalline-materials. Equation (2.32) is generally used for a-Si:H alloys with $n = 2$ and implicitly assumes parabolic bands and equal matrix elements for all optical transitions for photon energies in excess of the band gap (Cody et al., 1980; Abeles et al., 1980). It has also been reported that a better fit to the data is obtained when $n = 3$ is used (Klazes et al., 1982). Also, exception to the square root formula has been found, for example, in a group of multicomponent chalcogenide glasses (Fagen et al., 1970). Using $n = 2$ for a-Si:H material generally leads to band gaps in the range 1.7–1.8 eV. The prefactor B, in Eq. (2.32) for amorphous semiconductors is given by (Mott and Davis, 1979)

$$B = \frac{4\pi \sigma_{\min}}{n_1 c \, \Delta E_W}, \qquad (2.33)$$

where $\sigma_{\min}$ is the minimum metallic conductivity, c is the velocity of light and ΔE_W is the extent of the band tailing, $E_c - E_A$, or $E_v - E_B$ (Section 1.3). Equation (2.33) is derived, however, by assuming that the densities of states at the band edges are linear functions of energy. For GD a-Si:H alloys, B is on the order of 3 to 6×10^5 cm^{-1} eV^{-1} (Sakata et al., 1981), but it should be emphasized that similar values are also found for other a-semiconductors such as GeTe (Tsu et al., 1970) or As$_2$Te$_3$ (Weiser et al., 1970).

It has been experimentally shown that for GD a-Si:H alloys E_g, as defined by Eq. (2.32), correlates with the hydrogen content within the film, as shown in Fig. 2.42 (Hama et al., 1983) (see also Fig. 2.2; it should be noted that there are differences between the two results which should not be too surprising since it can be deposition system dependent but nevertheless the trends are similar). This result seems to be consistent with the theoretical predictions made by Papaconstanopolous et al. (1981) using tight binding calculations, It is also consistent with the recession of the top of the valence band observed in photoemission experiments, which shows that E_g depends on the H content (von Roedern et al., 1978).

In virtually all amorphous semiconductors an exponential region B, as shown in Fig. 2.41, is observed. This is a well known phenomenon in

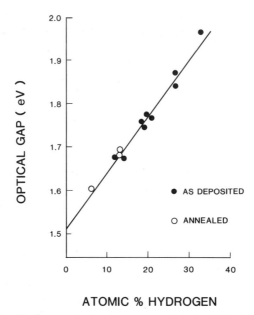

Fig. 2.42. Relationship between hydrogen content and the optical energy gap (Hama *et al.*, 1983).

crystalline materials, in particular in the alkali halides, CdS and trigonal Se. It has also been observed in chalcogenide glasses and the silicate glasses (Mott and Davis, 1979). The region is referred to as the "Urbach tail" (Urbach, 1953) and several mechanisms have been proposed to explain it, such as band gap variations due to density fluctuations, broadening of the band edge, or the presence of excitonic states induced by internal electric fields. Al Jalali *et al.* (1980), from electroabsorption studies of a-Si:H, suggest that the absorption edge reflects the tailing of states into the gap region due to fluctuations of bond lengths and bond angles. In addition, they propose that the transition probability changes rapidly in the absorption tail as the energy states become increasingly localized. It has also been suggested by many authors (Tauc, 1970; Langon, 1963) that this tail could arise from electronic transitions between (localized) states in the band-edge tails, the density of which is assumed to fall off exponentially with energy. Davis *et al.* (1970) point out that since the slopes of the observed exponential absorption edges are very much the same in a variety of materials, then it would seem unlikely that the tailing would be so similar. Instead, they suggest the existence of a correlation between the slope of the Urbach tail and the coordination number, or the valency. They speculate that materials with lower coordination (higher valency) more easily form an ideal amorphous network with fewer defects

OPTICAL ABSORPTION

and voids and perhaps exhibit less departure from optimum covalent bond angles and lengths than in those with higher coordination numbers.

The dashed line in Fig. 2.41 corresponds to an exponential edge with a characteristic energy, $E_0 = 0.05$ eV. This differs substantially from $E_0 = 0.09$ eV, which was deduced using the primary photocurrent technique (Crandall, 1980) for GD a-Si:H film. Two explanations for this vast difference have been put forward: (a) the primary photocurrent technique contains a substantial photoemission component at low photon energies; (b) the material investigated could have been of poor quality.

For a-Si:H type alloys, Cody *et al.* (1981a) have concluded that their data on the Urbach tail is consistent with the interpretation that both the width of the exponential edge and the optical gap are controlled by the amount of disorder, structural and thermal, in the network. Further, they conclude that hydrogen affects the band gap only indirectly. In order to compare the effects of structural and thermal disorder on the absorption edge they induced structural disorder intentionally in the films by introducing dangling bonds via thermal evolution of H. In Fig. 2.43, we show their optical measurements

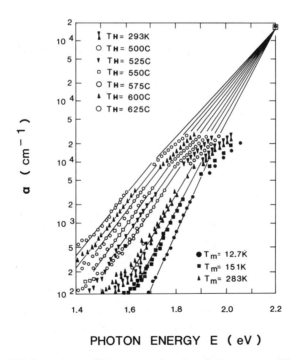

Fig. 2.43. Optical absorption coefficient, α, as a function of photon energy. The solid symbols refer to data obtained at different measurement temperatures T_m. The open symbols refer to a film that has been isochronally heated at temperature T_H (Cody *et al.*, 1981b).

(Cody et al., 1981b) as a function of photon energy, E, at $T = 12.7$, 151 and 293 K on as prepared films of composition $SiH_{0.13}$. The figure also shows data taken at $T = 293$ K on a similar film, from which hydrogen was evolved in a stepwise manner via isochronal heating in a vacuum at 25°C intervals from 400°C to 600°C for 30 minutes at a time. It should be noted that the absorption edge broadens and shifts to a lower energy either with increasing thermal disorder or with structural disorder due to isochronal annealing. The absorption coefficient $\alpha(E, T)$ can be expressed by the Urbach form given by

$$\alpha(E, T) = \alpha_0 \exp\left[\frac{(E - E_1)}{E_0(T, X)}\right] \qquad (2.34)$$

where $E_0(T, X)$ is the width of the exponential tail and X is a parameter that defines the structural disorder. α_0 and E_1 are experimentally determined parameters with values 1.3×10^6 cm^{-1} and 2.17 eV, respectively. By fitting the temperature dependence of the optical energy gap, E_g, the expression

$$[\alpha(E, T)E]^{1/2} = C[E - E_g(T)] \qquad (2.35)$$

results where C is a constant, 6.9 (eV μm)$^{-1/2}$. By using Eqs. (2.34) and (2.35), $E_g(T)$ and $E_0(T)$ can be deduced and correlated together as shown in Fig. 2.44. Note that in the figure, E_g and E_0 for films of differing amounts of H are also included.

Standard treatments of the Urbach edge in crystalline semiconductors conclude that the width of the absorption edge, E_0, is proportional to $\alpha \langle U^2 \rangle_T$, where U is the displacement of the atoms from their equilibrium positions (Tauc, 1976). The affect of structural disorder on E_0 is included by conjecturing that (Cody et al., 1981b)

$$E_0(T, X) \approx \{\langle U^2 \rangle_T + \langle U^2 \rangle_x\}, \qquad (2.35)$$

where $\langle U^2 \rangle_x$ is the contribution of the topological disorder to the mean square deviation of the atomic positions from a perfectly ordered configuration. They further show that E_g and E_0 can be related by

$$E_g(T, X) = E_g(0, 0) - \langle U^2 \rangle_0 D\left[\frac{E_0(T, X)}{E_0(0, 0)} - 1\right], \qquad (2.36)$$

which is in agreement with the results of Fig. 2.44. Note that this figure also includes two reactively sputtered films (with H concentrations of 8 and 22 at.%) and data that was inferred for a-Si films using the CVD approach. The experimental data suggests that the optical band gap, E_g, is a linear function of E_0 and through it, of temperature and structural disorder. Although the hydrogen content varies for the films shown in Fig. 2.44 from <1% to 20%, they conclude that the bonded H content itself is not the

OPTICAL ABSORPTION

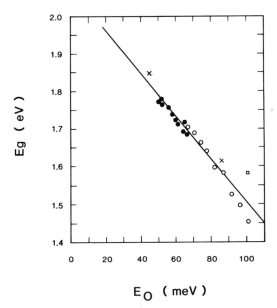

Fig. 2.44. Optical gap $E_g(T, X)$ as a function of $E_0(T, X)$ for the measurements at variable T and constant X (solid circles) ($H = 12$ at.%) and for measurements at constant T and variable X (open circles) ($H = 12$-1 at.%). The square is derived for CVD a-Si at $T = 300$ K. The crosses are obtained from measurements on reactively sputtered films ($H = 22$ at.%, $H = 8$ at.%) (Cody et al., 1981b).

fundamental controlling factor in the optical bandgap. Rather, it is the structural disorder, which is influenced by the H content. This result seems to have been confirmed by Ferraton et al. (1983) who further show that the affect of doping a-Si:H with P on the optical gap and the Urbach tail parameter of a-Si:H are negligible.

Region C of Fig. 2.41 can be related to intraband transitions and to the DOS. The procedure for determining the DOS has been to measure the absorption in excess of that identified with an extrapolated Urbach tail. As indicated above and shown in Fig. 2.41, photoelectric effects have been used to measure α down to 1 cm^{-1}. However, it has been argued that errors could arise if surface states are neglected and if the energy dependence of the $\mu\tau$ products are not taken into account (Yamasaki et al., 1981). Two alternative, highly sensitive techniques for measuring low values of α have been employed: Photothermal Deflection Spectroscopy (PDS) (Bocarra et al., 1980); Photoacoustic Spectroscopy (PAS) (Yamasaki et al., 1981). Using the PDS technique, absorptances as small as 10^{-5} cm^{-1} for a 1 μm film can be resolved. This technique owes its high sensitivity to the fact that the absorbed energy can be measured directly via the deflection of a laser beam that can detect a small temperature rise in the sample. In the latter technique a sample is placed inside a closed cell containing air and a sensitive microphone. The sample is then illuminated with modulated monochromatic light. The periodic heating caused by the non-radiative processes associated with optical absorption provides a periodic heat flow to the air and the resultant acoustic signal is detected by a microphone.

From the integrated excess absorption (region C) the number of defects, N_{sa}, can be calculated from (Dexter, 1956; Jackson et al., 1982)

$$N_{sa} = \frac{cnm}{2\pi^2 \hbar} \left[\frac{(1 + 2n)^2}{e^2 f_{0j} g n^2} \right] \int \alpha_{ex} \, dE, \tag{2.38}$$

where c is the speed of light, n (= 3.8) is the index of refraction of a-Si:H, f_{0j} is the oscillator strength of the absorption transition and m is the electron mass. α_{ex} is the excess absorption due to the sub-gap states, which can be computed from $\alpha_{ex} = \alpha - \alpha_0 \exp(\hbar w/E_0)$, where α_0 and E_0 are determined by a fit to the exponential region.

In Fig. 2.45, we show $\alpha(h\nu)$ for undoped GD a-Si:H material as a function of the rf power employed during the deposition process. The excess absorption increases with increasing power and there is a progressive decrease in the slope of the exponential edge. These results are consistent with increased

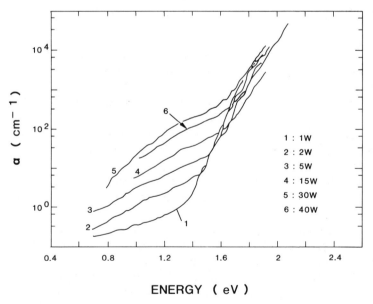

Fig. 2.45. Absorption coefficient vs. energy for undoped a-Si:H prepared at various rf powers with substrate temperature $T_s = 230°C$ (Jackson and Amer, 1982)

OPTICAL ABSORPTION

disorder and an increase in the dangling bond density, since a correlation has been reported between $N_{s_a}(\text{ABS})$ found using Eq. 2.38 and the spin density, N_s measured by the ESR technique (Jackson et al., 1982). Further, as discussed in Section 2.1.2, good quality a-Si:H is generally obtained with conditions that do not deplete SiH_4, such as low power levels, which seems to be consistent with the above results.

The affect of doping and compensation on sub-bandgap absorption is shown in Fig. 2.46 (Jackson et al., 1982a). The sub-bandgap absorption increases with the level of the dopant. This agrees with the results obtained

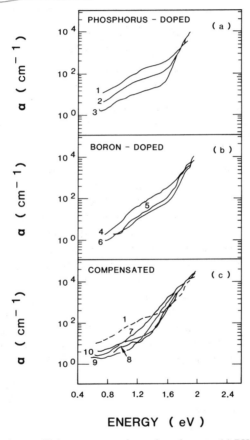

Fig. 2.46. Absorption coefficient vs. energy for various dopants, (a) PH_3 doping concentration is 1: 1×10^{-3}, 2: 3×10^{-4} and 3: 1×10^{-5}. (b) B_2H_6 doping concentration is 4: 10^{-3}, 5: 3×10^{-4} and 6: 10^{-4}. (c) compensated samples. All have 10^{-3} PH_3 and the B_2H_6 concentrations are 1: 0, 7: 2×10^{-4}, 8: 4×10^{-4}, 9: 2×10^{-3} and 10: 4×10^{-3}. All concentrations refer to the relative concentration of the dopant in the gas phase; $T = 230°C$ and rf power is 2 W (Jackson and Amer, 1982).

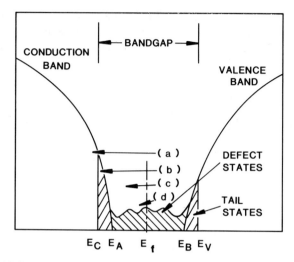

Fig. 2.47. DOS diagram for a-Si:H alloys showing the distinction between tail states and deep defect states (see also Fig. 1.4). (a), (b), (c) and (d) represent the expected conduction mechanisms for electrons.

from the PAS technique, which shows that doping with P introduces defects at ~1.2 eV below the conduction band (Yamasaki *et al.*, 1981). It is suggested that the energy and the cross sections of these defects are nearly identical to that of the dangling bonds (Jackson, 1982). The placement of the Si dangling bond at 1.2 eV below E_c is also consistent with the theoretical prediction made by Joanoppolous (1980). This suggestion is based on the fact that P doped a-Si:H alloys are not expected to have an ESR signal, but rather a LESR signal, which would arise from a doubly occupied dangling bond defect with one electron removed by optical excitation. The excess absorption for the P doped material, as shown in Fig. 2.46, was found to be proportional to the LESR signal. Further, it has been suggested that the shift of the spectra between undoped and P doped material would provide a value of the positive correlation energy which Jackson deduced to be about 0.35 eV (Jackson, 1982).

2.5 Electrical Transport

In this section we shall first review briefly the basic conduction processes that occur in amorphous semiconductors and then review the transport mechanisms found primarily in GD a-Si:H films.

We have already noted in Chapter 1 that the major distinguishing feature of an amorphous semiconductor is the continuum of localized states (within

ELECTRICAL TRANSPORT

the mobility gap) that can be roughly divided into two parts: (a) inherent states due to the disorder that leads to band tails; (b) point defects that lead to a mid-gap DOS. Consequently, the DOS can be depicted, as shown in Fig. 2.47. The critical energy E_c (conduction band edge) separates localized states from those that are extended. Hence, the transport processes can be divided into several categories (see, e.g., Mott and Davis, 1979) that we now discuss.

2.5.1 Conductivity in Extended States

Extended state transport occurs via carriers excited beyond the mobility edge into states at or above E_c (or below E_v), as shown in Fig. 2.47. The conductivity, σ, for a non-degenerate semiconductor can be written as

$$\sigma \cong e \int g(E)\mu(E)f(E)\, dE. \tag{2.39}$$

Since the conduction is expected to occur within about kT of the edge at $E = E_c$, then the conductivity for n-type conduction can be written as

$$\sigma = eg(E_c)\mu_{\text{ext}} kT \exp\left[-\frac{(E_c - E_f)_T}{kT}\right], \tag{2.40}$$

where $(E_c - E_f)_T = (E_c - E_f)_0 - \delta_F T$. Use of this leads to

$$\sigma = \sigma_{\min} \exp\left[-\frac{(E_c - E_f)_T}{kT}\right], \tag{2.41}$$

where $\sigma_{\min}$ is the minimum metallic conductivity, defined such that for a system at $T = 0$ K, where the electron states are occupied up to E_f, it is the smallest non-zero value of the conductivity at $T = 0$. Since the wave function of an electron varies in a random fashion at $E = E_c$ (Section 1.3), it can be shown that (Mott and Davis, 1979)

$$\sigma_{\min} = \frac{\pi}{4z} \cdot \frac{e^2}{\hbar a}\left(\frac{J}{V_0}\right)^2, \tag{2.42}$$

where z is the coordination number, a is the interatomic distance, J is the bandwidth (given by $2z\gamma$ where γ is the overlap integral) and V_0 is the disorder parameter. As discussed in Section 1.3, if $V_0/\gamma \gg 1$, the wavefunction for each isolated atom would be perturbed only slightly by the other atoms and hence the states become localized. When the mean free path of an electron is approximately equal to the interatomic distance, a, the wave function fluctuates at random. Anderson (1958) has shown that localization begins at $(V_0/J)_{\text{crit}} \cong 2$. Using this, the minimum metallic conductivity $\sigma_{min} \sim 200 \, (\Omega \, cm)^{-1}$ and the prefactor, $\sigma_0 = \sigma_{min} \exp \delta_F/k \approx 1 - 2 \times 10^3 \, (\Omega \, cm)^{-1}$. However, the idea of a minimum metallic conductivity has been challenged and the present situation is somewhat obscure (see, e.g., Mott, 1985). The concept itself may be inapplicable to many physical systems.

The value of μ_{ext} can be estimated by the use of Eqs. (2.40) and (2.41) and written as

$$\mu_{ext} = \frac{\sigma_{min}}{eg(E_c)kT}. \tag{2.43}$$

Knowledge of the effective DOS at $E_c (g(E_c) \approx 1.10^{21} \, cm^{-3} \, eV^{-1})$ provides an estimate of μ_{ext}, which is on the order of $10 \, cm^2 \, s^{-1} V^{-1}$ (Mott and Davis, 1979). However, care should be taken when the concept of a σ_{min} is employed; the validity of Eq. (2.43) is questionable.

2.5.2 Conductivity in the Tail States

The tail states shown in Fig. 2.47, generated by disorder, lie in the energy ranges $E_A < E < E_c$ and $E_v < E < E_B$, for the conduction and valance bands respectively. The conduction at an arbitrary energy $E = E_x$ in the tail states, shown by path (b), can be written as

$$\sigma = e\mu_{hop} \int_{E_A}^{\infty} g(E_x) \exp\left(-\frac{E_x - E_f}{kT}\right) dE, \tag{2.44}$$

where $(E_x - E_f)_T = (E_x - E_f)_0 - \delta T$. The hopping mobility, μ_{hop}, can be estimated from the diffusion coefficient, D (Cohen, 1970):

$$D = \tfrac{1}{6} v_{ph} R^2 e^{-W_1/kT}, \tag{2.45}$$

where the factor $\tfrac{1}{6}$ arises from an averaging procedure, v_{ph} is the phonon frequency, R is the hopping distance and W_1 is the energy difference between the states. Using the Einstein relation, $\mu = eD/kT$, the hopping mobility can be written as

$$\mu_{hop} = \frac{1}{6} v_{ph} \left(\frac{er^2}{kT}\right) e^{-W_1/kT}. \tag{2.46}$$

If the predominent conduction path is at E_x for n-type ($E_A < E_x < E_c$), then the conductivity in this regime is given by

$$\sigma = e\mu_{hop} g(E_x) kT \exp\left(-\frac{E_x - E_f + W_1}{kT}\right). \tag{2.47}$$

2.5.3 Hopping Conductivity

The hopping of carriers between localized states situated in the mid-gap region ($E_A < E < E_B$) occurs either at low T, or at high T in materials with a high defect state density; this is shown by path (c) for electrons in Fig. 2.47. The probability that a carrier jumps from one localized state to another state of higher energy will depend upon: (a) $\exp[-W_2/kT]$, where W_2 is the energy difference between the two states; (b) v_{ph}, the phonon spectra; (c) $e^{-2\alpha_L r}$, the overlap of the wavefunctions, where r is the distance separating the two localized states.

Considering only the electrons at E_f, for weak fields the conductivity is written as (Mott and Davis, 1979)

$$\sigma = 2e^2 r^2 v_{ph} g(E_f) e^{-2\alpha_L r} \exp(-W_2/kT). \tag{2.48}$$

For strong localization ($\alpha_L r \gg 1$), nearest neighbor hopping is to be expected. However, in the weak localization case, the electrons will have a wider choice of centers to hop to. Here the material exhibits variable range hopping (Mott, 1968). In this case the hopping distance r increases with decreasing temperature since the electrons will have a higher probability of jumping to a more distant site where W_2 is smaller. By maximizing this probability, it can be shown that the conductivity is given by

$$\sigma = e^2 g(E_f) r^2 v_{ph} \exp(-B/T^{1/4}), \tag{2.49}$$

where

$$r = \frac{3^{1/4}}{\{2\pi \alpha_L g(E_f) kT\}^{1/4}} \tag{2.50a}$$

and

$$B = 1.66 \left(\frac{\alpha_L^3}{k g(E_f)} \right)^{1/4}. \tag{2.50b}$$

Equation (2.49) is Mott's $T^{1/4}$ law, applicable for transport in the three dimensional case. Consideration of transport in the two dimensional case leads to $\sigma \alpha T^{-1/3}$ (Mott et al., 1975). A test of this theory is to look for a change from $T^{-1/3}$ to $T^{-1/4}$ behavior as the sample thickness is increased.

From the above, the electron conductivity, σ, as a function of T [Eqs. (2.40), (2.47), (2.48) and (2.49)] can be expressed as

$$\sigma = \sigma_{01} e^{\delta/k} \exp\left(-\frac{E_c - E_f}{kT}\right) + \sigma_{02} \exp\left(-\frac{E_x - E_f + W_1}{kT}\right)$$
$$+ \sigma_{03} \exp\left(-\frac{W_2}{kT}\right) + \sigma_{04} \exp\left(-\frac{B}{T^{1/4}}\right), \tag{2.51}$$

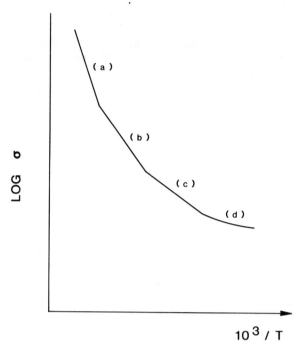

Fig. 2.48. Idealized $\ln \sigma$ vs. $1/T$ corresponding to the conduction mechanism depicted in Fig. 2.47.

and a plot of $\ln \sigma$ vs. $10^3/T$ for the various regions can be represented as shown in Fig. 2.48. Similar expressions can be derived for holes. The details of the DOS will determine which component and carrier type dominates the transport over a given temperature range.

2.5.4 Seebeck Coefficient in Extended and Defect States

A useful parameter in the study of basic electrical transport is the Seebeck coefficient, S. The expression for the Seebeck coefficient is

$$S\sigma = -\frac{k}{e} \int \frac{(E_f - E)}{kT} \sigma(E)\, dE, \qquad (2.52)$$

or

$$S\sigma = -\frac{k}{e} \int \frac{e^2 L^2}{\tau k T} \left(\frac{E - E_f}{kT} \right) f(1-f) g(E)\, dE, \qquad (2.53)$$

where the energy carried is measured relative to E_f, and the conductivity

ELECTRICAL TRANSPORT

(isotropic) in a band is given by (Cutler et al., 1969)

$$\sigma = \int \frac{e^2 L^2}{\tau k T} f(1-f) g(E)\, dE, \tag{2.54}$$

where L is the mean free path and τ is the scattering time.

The sign convention is that $S < 0$ for electrons at energies $E > E_f$ and vice versa. If τ and $g(E)$ can be defined by $K_2 E^{-s}$ and $K_1 E^l$ where K_1 and K_2 are constants, the generalized expression for the Seebeck coefficient (thermopower) becomes

$$S = -\frac{k}{e}\left[\left(\frac{E_c - E_f}{kT}\right) + (1 - s + l)\right] = -\frac{k}{e}\left[\frac{E_c - E_f}{kT} + A_c\right]. \tag{2.55}$$

If $g(E)$ and μ vary linearly with energy above E_c, then $A_c = 3$. Similarly, for holes (i.e., conduction below E_f), the Seebeck coefficient can be written as

$$S = \frac{k}{e}\left[\frac{E_f - E_v}{kT} + A_v\right]. \tag{2.56}$$

The above expressions are only valid for the non-degenerate cases such that $E_c - E_f \gg kT$. For the degenerate case (use $f(1-f) = -kT\, \partial f / \partial E$ and expand $g(E)\mu(E)$ in terms of Taylor series), S is given by,

$$S = -\frac{\pi^2}{3}\frac{k}{e} kT \frac{d \ln \sigma}{dE}\bigg|_{E=E_f} \tag{2.57}$$

and since the conductivity is due to hopping, then S is rewritten as,

$$S = -\frac{\pi^2}{3}\frac{k}{e}\left[kT \frac{d}{dE}\ln \sigma_0 - \frac{dW}{dE}\right]. \tag{2.58}$$

2.5.5 Conduction Processes in Amorphous Silicon

As discussed in Section 2.2 and 2.4 the structural and optical properties of GD a-Si:H films are linked to the deposition parameters. Low defect state densities are generally obtained when gas phase nucleation is suppressed; the optimum conditions are outlined in Table 2.1(a). Optimal quality films a-Si:H films exhibit a straight line over many orders of magnitude on a $\ln \sigma_D$ vs. $10^3/T$ plot (Fig. 2.49) and hence exhibit a well defined activation energy for conduction. (σ is now written as σ_D, the dark conductivity.) Films of this type have $S < 0$; the conduction mechanism is dominated by electrons. Since the conductivity and thermopower data of these films generally give the same activation energy (E_σ and E_s respectively), then it is concluded that conduction occurs via extended states. The conductivity can

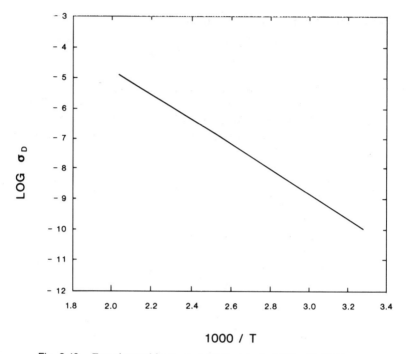

Fig. 2.49. Experimental ln σ_D vs. $1/T$ for a typical GD a-Si:H film.

be expressed via Eq. (2.40) with an experimental conductivity activation energy, $E_\sigma \geq 0.8$ eV with a room temperature dark conductivity, $\sigma_D \sim 10^{-10}$ $(\Omega\text{ cm})^{-1}$. Using the measured values of $\delta = 2.4 \times 10^{-4}$ eV K^{-1} (e.g., Mott and Davis, 1979), the prefactor is $\sigma_0 \cong 10^3$ $(\Omega\text{ cm})^{-1}$. Using Eq. (2.40) and assuming that $\mu_{\text{ext}} \cong 10$ cm^2 s^{-1} V^{-1}, the DOS at the conduction band edge, $g(E_c) \cong 10^{21}$ cm^{-3} eV^{-1} (LeComber et al., 1972).

As discussed in Section 2.2, the substrate temperature, T_s, critically dictates the structure and the DOS spectra. In Section 2.3.2.1, it was seen that $g(E)$ could be reduced by orders of magnitude when T_s is increased from room temperature to about 570 K. This reduction in the DOS is accompanied by a large increase in the recombination lifetime of the majority carriers (Loveland et al., 1974). The change has been ascribed to changes that occur in the fundamental structure of the films (see Section 2.2). These changes also are reflected in the dc conductivity behavior. In Fig. 2.50 we show the results, extracted from dc conductivity data, for σ_D, σ_0 and E_σ as functions of T_s (LeComber et al., 1972). The minimum in σ_D at $T_s \cong 350$ K is associated with a change from p-type hopping behavior to n-type behavior. Because of the discontinuity in σ_0 of 10^7 at $T_s = 350$ K, and since $\sigma_0 \cong 10^3$ $(\Omega\text{ cm})^{-1}$ for

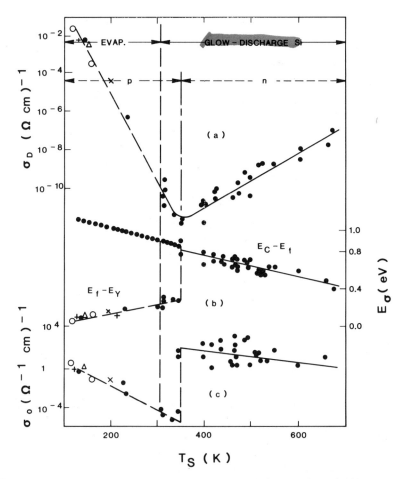

Fig. 2.50. Variation of (a) room temperature conductivity σ_D, (b) activation energy of conductivity (see text), (c) pre-exponential factor σ_0 for glow-discharge films of amorphous silicon, with deposition temperature T_s. Points on the left of this figure refer to evaporated films for which T_s has no significance (LeComber et al., 1972).

$T_s > 350$ K, then these samples are associated with n-type conduction with transport occurring within extended states. This interpretation is confirmed using various other techniques such as thermopower, field effect and drift mobility, which we discuss below (Spear, 1977). The low DOS in GD a-Si:H is evidently due to the crucial role played by H; this can perhaps best be illustrated by considering the results for sputtered Si deposited in a reactive H environment where the amount of H included within the film can be more readily controlled (Anderson and Paul, 1981). Figure 2.51 shows the effect

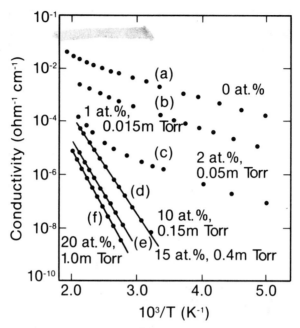

Fig. 2.51. Dark conductivity, σ_D, vs. $10^3/T$ for sputtered a-Si:H samples prepared at $T_s = 200°C$ using partial hydrogen pressure between 0 and 1 mTorr. Hydrogen content (at.%) and hydrogen partial pressure in 5 mTorr Ar are given on each curve (Anderson and Paul, 1981).

of increasing the H content in the film on the electronic properties of a-Si, with all the samples prepared at $T_s = 200°C$. In this figure, $\ln \sigma_D$ vs. $10^3/T$ is plotted for films prepared with differing hydrogen gas pressure in the sputtering system. (The hydrogen content is correlated to the hydrogen pressure.) For curve (a), an unhydrogenated sample, the value of σ_D is high and a unique activation energy cannot be defined. The conductivity is dominated by thermally assisted hopping, within the defect states, and is given by $\sigma_D = eg(E)kT\mu_{hop}$ where μ_{hop} is defined by Eq. (2.46). With the addition of 20 at.% H to the films, curve (f) results in an activated behavior over several orders of magnitude in conductivity, corresponding to extended state conduction above E_c, since the samples are now found to be slightly n-type. This change in the conduction behavior is a direct consequence of a reduction in the DOS within the mobility gap such that the predominant path for transport occurs at the conduction band edge. However, the inclusion of 10 to 20 at.% H in the a-Si network has one negative aspect as far as solar cells are concerned: the optical band gap increases to 1.7–1.8 eV, which is beyond the optimal point for solar energy conversion. We discuss this further in Section 3.1.

The evidence that the addition of H leads to the removal of dangling bonds is also provided by various indirect measurements. For example, the intensity and the peak position of the photoluminescence (PL) spectra (see Section 2.7) are determined by how efficiently hydrogen removes the dangling bonds, which are likely to determine the competing non-radiative recombination processes. The peak in the luminescence spectra will shift to higher energies as the DOS is reduced. It is found that for optimal GD a-Si:H the PL peak occurs at ~1.35 eV while unhydrogenated material shows a much reduced luminescence peak at a lower energy.

The evidence for the ln $\sigma_D \alpha T^{-1/4}$ law, is provided by the use of a highly defective material such as elemental evaporated Si. An example is shown in Fig. 2.52 for differently annealed states of the sample (Bahl and Bhagat, 1975). Using Eq. (2.50b), $g(E_f)$ is found to be in the 10^{18} cm^{-3} eV^{-1} range assuming the decay constant, $\alpha_L = 10^7$ cm^{-1}. However, using the same parameter in Eq. (2.50a) in the preexponential part of the $T^{1/4}$ law does not lead to the same value of $g(E_f)$ which casts some doubt as to the validity of this law. On the other hand, in situ measurements of the electrical conductivity in the hopping regime for a-Si films of varying thicknesses deposited in ultra high vacuum have been made by Knotek (1975). A transition from $T^{-1/4}$ to $T^{-1/3}$ behavior was observed as the film thickness was reduced to values comparable with the calculated hopping length. The fit of 3-dimensional and 2-dimensional variable range hopping theory to these results yields $\alpha_L^{-1} = 3$ Å for the radius of the localized wavefunction and 3×10^{19} cm^{-3}eV^{-1} for the DOS at E_f, which is consistent with what is usually found for unhydrogenated samples.

In an early work (LeComber et al., 1972), a change in conduction mechanism from extended states to activated hopping conduction within the band tail states was reported. Evidence for this was provided by a kink occurring simultaneously in the log σ_D vs. $1/T$ and log μ vs. $1/T$ data at $T \cong 250$ K, as shown in Fig. 2.53. In region (1), the conductivity is given by Eq. (2.40) with E_σ given by $E_c - E_f$; a kink occurred at 250 K with region (2) showing a reduced activation energy corresponding to $(E_x - E_f + W_1)$ [from Eq. (2.47)]. The difference in the two activation energies was found to be about 0.10 eV. A similar kink was observed at 250 K in then log μ vs. $1/T$ data (see Section 2.5.6). Here, region (1) in the mobility data was interpreted in terms of extended state conduction with the activation energy $(E_c - E_a \approx 0.2 \text{ eV})$ arising from trapping in shallow localized states and region (2) from hopping within these states (see Section 2.5.6) with an activation energy W, of about 0.1 eV. Hence, the predominant conduction for $T < 250$ K was concluded to be at $E_x \approx E_A$, the edge of the band tail states. Further confirmation of this self consistent approach was also provided by the field effect and thermopower data (e.g., Spear, 1977). However, subsequent reports from other

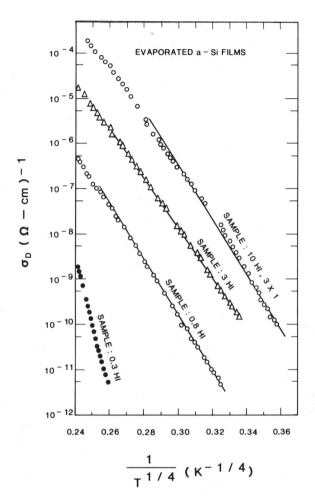

Fig. 2.52. Dark conductivity, σ_D plotted as a function of $(1/T)^{-1/4}$ (Bahl et al., 1975).

researchers did not reveal those kinks. Also, the recognition of dispersive transport behavior in the interpretation of the μ vs. $1/T$ data also casts some doubt on this model (to be discussed in Section 2.5.6).

We have shown above the strong evidence that H is responsible for the reduction in the DOS. As mentioned in Section 2.1.4, "device quality" halogenated amorphous Silicon alloys can also be produced by the use of rf glow discharge in SiF_4 and H_2 gas mixtures (Madan et al., 1979). They noted that by choosing different gas ratios, $(r = SiF_4/H_2)$, the conduction mechanism could be altered from hopping conduction to an activated

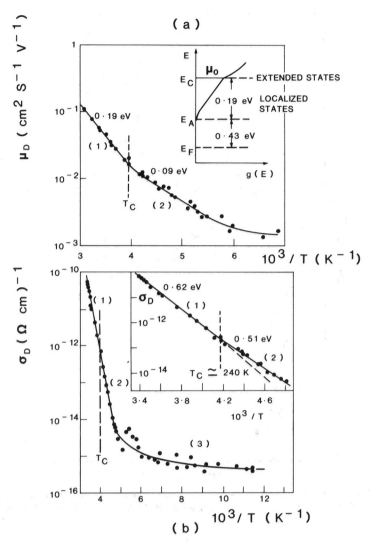

Fig. 2.53. Temperature dependence of (a) electron drift mobility μ_D. (b) conductivity σ_D in a glow-discharge film of silicon deposited at 500 K (LeComber and Spear, 1970).

behavior corresponding to conduction in extended states, as shown in Fig. 2.54. The change in the conduction mechanism was attributed to the reduction in the DOS and was brought about by an increase in H and F incorporation within the growing a-Si network. A more complete discussion regarding the role of Fluorine is given in Section 2.11.1.

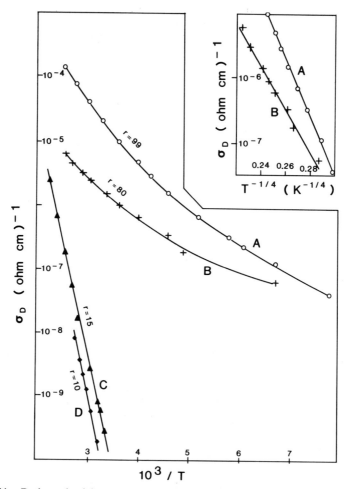

Fig. 2.54. Dark conductivity σ_D vs. $10/T$ plotted for samples fabricated with different gas ratios r. The inset shows σ_D plotted as function of $10^3/T$ for large values of r (Madan *et al.*, 1979).

2.5.6 Drift Mobility

In this type of experiment, depicted in Fig. 2.55, carriers are injected at one point in the sample and their transit time, t_t, to another point at a distance, d, is measured under the influence of an electric field, F_0. The drift mobility of the carriers is defined by

$$\mu_D = \frac{d}{F_0 t_t}. \tag{2.59}$$

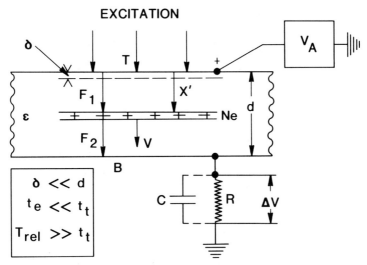

Fig. 2.55. Illustration of the principle underlying drift mobility measurements on resistive specimen t_e is the excitation time of the pulse (Spear, 1969).

Because of the high resistivities of amorphous semiconductors, the dielectric relaxation time, τ_{rel} ($= \varepsilon_s \varepsilon_0 / \sigma$), is much longer than the transit time of the carriers. Therefore, unlike the crystalline case, the excess carriers are not screened by the majority carriers and hence the measurement of the mobility of both types of carriers can thus be made by simply reversing the polarity of the applied field. Figure 2.55 illustrates that at $t = 0$, N_0 electron-hole pairs per unit area are created at a depth δ ($\ll d$). In the example, we suppose that N_0 carriers escape recombination and the field is arranged such that positive carriers are drawn into the sample, which perturbs the applied field F_0 ($= V_0/d$) into regions F_1 and F_2 as shown. By solving the Poisson equation and using the boundary conditions at $x \equiv 0$ and $V = 0$, it can be shown that the fields F_1 and F_2 are

$$F_1(x') = F_0 - \frac{eN_0}{\varepsilon_s \varepsilon_0}\left(1 - \frac{x'}{d}\right) \tag{2.60}$$

and

$$F_2(x') = F_0 + \frac{N_0 x'}{\varepsilon_s \varepsilon_0 d} \tag{2.61}$$

where the quantity $eN_0/\varepsilon_s \varepsilon_0$ is the self field which must be $\ll 1$ for the assumption of $F_0 = V_0/d$ to remain valid in Eq. (2.59). Referring to Fig. 2.55, the charge collected as a function of time in the absence of trapping and thermal release is given by (Spear, 1969)

$$q_T(t) = \frac{eN_0 x(t)}{d}. \tag{2.62}$$

In the presence of trapping, but not thermal release

$$q_T(t) = \frac{eN_0 \mu F_0 \tau}{d} \{1 - e^{-t/\tau}\}, \tag{2.63}$$

where τ defines the recombination lifetime. If trapping is not important then the mean free path of the carrier, $l \gg \mu F_0 t$, and hence,

$$q(t) = \frac{eN_0 \mu F_0}{d} \quad \text{for } t < t_t, \tag{2.64}$$

and

$$q(t) = eN_0 \quad \text{for } t > t_t. \tag{2.65}$$

This leads to a trace of type (a) shown in Fig. 2.56 with t_t as defined in the figure. With trapping Eq. (2.64) yields a type (b) response and the transit time is still identifiable.

In addition to trapping there is thermal release. Consider trapping centers of density N_t, situated at an energy E_t below E_c. If the charge distributions approach thermal equilibrium, then the free carrier concentration n is given by

$$n = N_c \exp\left(-\frac{E_c - E_f}{kT}\right), \tag{2.66}$$

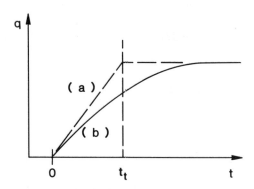

Fig. 2.56. Pulse shape observed in drift mobility when the carriers are generated close to the top electrode ($\delta \ll d$ of Fig. 2.55). The effect on the signal with (b) and without (a) deep trapping effects (Spear, 1969).

where $N_c = g(E_c)kT$ and the trapping concentration is given by

$$n_t = N_t \exp\left(-\frac{E_t - E_f}{kT}\right). \tag{2.67}$$

The lifetime of carriers with respect to these traps is given by τ and the thermal release times by τ_r

$$\tau_r = (N_c v_{th} C)^{-1} \exp\left(\frac{E_c - E_t}{kT}\right), \tag{2.68}$$

where C is the capture cross sectional area. If $\tau \ll \tau_{\text{rel}}$ and $\tau_r \ll \tau_{\text{rel}}$, the carriers are continually trapped and released and thermal equilibrium is achieved. In the presence of traps the drift mobility, μ_d, is defined as

$$\mu_D = \mu_0 \left\{1 + \frac{n_t}{n}\right\}^{-1}, \tag{2.69}$$

which, using Eq. (2.66) and (2.67) yields at low T

$$\mu_D \cong \mu_0 \frac{N_c}{N_t} \exp\left(-\frac{E_c - E_t}{kT}\right), \tag{2.70}$$

whereas at high T

$$\mu_D \cong \mu_0, \quad \text{since } \tau_r \ll \tau, \tag{2.71}$$

in which case the effect of traps is negligible. Note that as the temperature is lowered an increased fraction of the measured transit time will be due to the localization of carriers within the traps. Since a-Si based alloys possess a continuum of states, the above analysis can be extended to include different forms of localized state distributions (e.g., Mott and Davis, 1979). Note that μ_0 (or μ_{ext}) is the extended state mobility.

The charge collection as a function of time, $q(t)$, requires that the time constant $RC \gg t_t$. However, the transit pulse can also be displayed in the current mode, in which case $RC \ll t_t$. The voltage pulse is given by

$$\Delta V = \frac{RN_0}{v} d, \quad \text{for } 0 < t < t_t$$
$$= 0, \quad \text{for } t > t_t. \tag{2.72}$$

The pulse shapes with and without trapping are displayed in Fig. 2.57 [shown as (2) and (1) respectively]. Instead of traces of type (1) or (2), quite often traces of type (3) and (4) are observed, which show a high level of dispersion. Several attempts have been made to account for the dispersion (such as analyses involving various aspects of trap limited band transport under conditions where relatively few trapping and release events are experienced

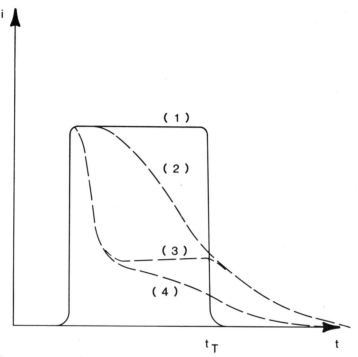

Fig. 2.57. Current pulse shapes observed with (2) and without (1) trapping effects in the absence of dispersion. In the presence of dispersion, pulse shapes of type (3) and (4) are quite often observed.

by the average carrier during transit), but without much success (Silver et al., 1971; Marshall and Owen, 1971; Fox and Locklar, 1972). A successful interpretation of dispersive transport with respect to both the time dependence and the universality of the pulse form has been provided by Scher and Montroll (1975).

In the non-dispersive case carrier transport is described by Gaussian statistics. Here the time development of the injected carriers can be described in terms of the mean displacement, l from the illuminated surface, as shown in Fig. 2.58 (Pfister and Scher, 1978). The dispersion parameter, β, characterizes the spread of the charge sheet about the mean. In the absence of trapping the transient current remains constant, as in trace (1) of Fig. 2.57, independent of β about the mean. In the presence of trapping and thermal release the carrier spreads according to Gaussian statistics, $(\beta/l) \alpha t_t^{-1/2}$.

In the dispersive (non-Gaussian) case, the carrier packet is not expected to grow symmetrically about its mean position, as shown in Fig. 2.59. Immediately after the generation of the carriers produced by the light

ELECTRICAL TRANSPORT

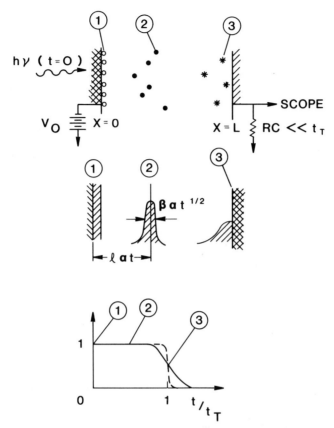

Fig. 2.58. Schematic representation of carrier propagation under Gaussian conditions. Top: Position of representative carriers in the sample bulk at $t = 0$ (○), $t < t_T$ (●) and $t \sim t_T$ (∗). Middle: Charge distribution in sample bulk at $t = 0$, $t < t_T$ and $t \sim t_T$. Bottom: Current pulse in external circuit induced by charge displacement. Units normalized to t_t and $i_T = i(t_t)$. Dashed line represents transient current for lower applied bias field, i.e., longer transit time (Pfister and Scher, 1978).

flash, some carriers will rapidly move into the sample. As time evolves an increasing number of carriers will suffer scattering events that will immobilize them for times on the order of t_t. The carrier packet grows asymmetrically, showing a leading edge that penetrates deeply into the bulk, while the maximum point of charge density moves slowly out of the generation region. For such cases the spread and the mean position have the same dependence; hence $\beta/l = $ const. Thus, the shape of the transient curve is independent of the transit time and a universal current shape results, as shown in Fig. 2.59. Therefore, the mean drift velocity of the propagating

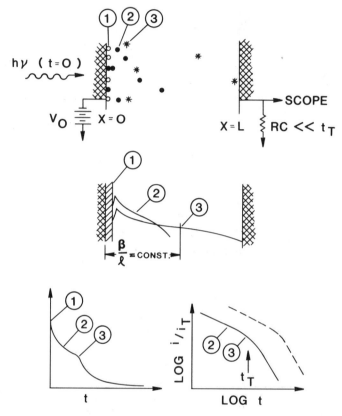

Fig. 2.59. Schematic representation of carrier propagation under ideal non-Gaussian conditions. Top: Position of representative carriers in the sample bulk at $t = 0$ (○), $t < t_t$ (●) and $t \sim t_t$ (∗). Middle: Charge distribution in sample bulk at $t = 0$, $t < t_t$ and $t \sim t_t$. Bottom: Current pulse in external circuit induced by charge displacement in linear units (left) and logarithmic units (right). Dashed line represents transient current for lower applied bias field, i.e., longer transit time (Pfister and Scher, 1978).

carriers must decrease with time and μ_d must show a dependence on the thickness of sample. Scher and Montroll (1975) assumed a distribution function that describes the time distribution as

$$\Psi(t) \alpha t^{-(1+\alpha_D)}, \qquad (2.73)$$

which was obtained from a consideration of hopping transport between localized states. They concluded that carrier transit pulses will show appreciable dispersion and that two regimes corresponding to slopes of $(1 - \alpha_D)$ and $(1 + \alpha_D)$ will occur on a log i vs. log t display for $t < t_t$ and $t > t_t$ respectively, as shown in Fig. 2.59; the sum of the slopes = 2. Since

ELECTRICAL TRANSPORT

α is independent of field and sample thickness, a universal pulse shape is realized. The physical meaning of α_D is that it is representative of disorder with $0 < \alpha_D < 1$; the smaller the value of α_D, the more dispersive the transport.

The earliest mobility measurement on GD a-Si:H were reported by LeComber and Spear (1970) as shown in Fig. 2.53. This study was later extended in a more comprehensive fashion by LeComber *et al.* (1972). The results for a series of samples deposited at different T_s is shown in Fig. (2.60). The data were interpreted on the basis of non-dispersive transport and the drift mobility on the basis of trap controlled transport assuming a linear density of states above and below the conduction band edge. Using Eq. (2.70), they found (from the high temperature region) that $\mu_{\text{ext}} \approx 1\text{-}10 \, \text{cm}^2 \, \text{s}^{-1} \, \text{V}^{-1}$, which is consistent with most of the estimates previously discussed. Since the activation energy for all the samples deposited at different T_s was found to be a constant at ~ 0.2 eV, they concluded that the localized states were in thermal equilibrium within 0.3 eV of the conduction band edge.

The lower temperature portion of $\ln \mu_D$ vs. $10^3/T$ data shown in Fig. 2.60 (also discussed in Section 2.5.5) is interpreted as hopping between localized states, since the probability of thermal release becomes smaller at lower temperatures. Further evidence that there is a transition from trap controlled

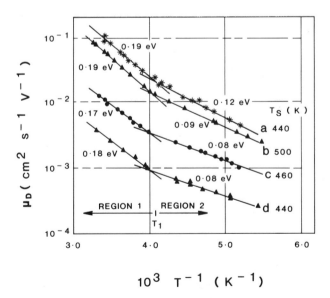

Fig. 2.60. Temperature dependence of the electron drift mobility for four a-Si specimens prepared at the values of T_s indicated (LeComber *et al.*, 1972).

mobility to hopping is provided by ln σ vs. $10^3/T$ data from the same samples (see Section 2.5.5). A further conclusion is that since the drift mobility was virtually independent of T_s, then the shallow states do not seem to be altered in terms of their density or their capture cross sectional areas. This could occur because the shallow states (band tails) are primarily associated with the lack of long range order. This result therefore provides indirect evidence for the existence of two types of states: inherent localized states due to the lack of long range order extending about 0.2 eV into the gap from the band edges; structural defect states situated in the central region of the mobility gap. The above work was confirmed by Moore (1977) and was further extended to include the measurements of hole mobility as a function of temperature, shown in Fig. 2.61.

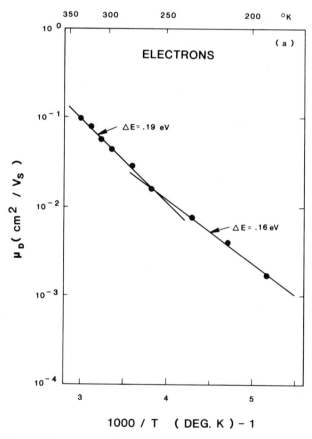

Fig. 2.61. Drift mobility as a function of temperature, plotted as μ_D vs. $1/T$. (a) Electron drift mobility in n-type material. (b) Hole drift mobility in p-type material (Moore, 1977).

ELECTRICAL TRANSPORT

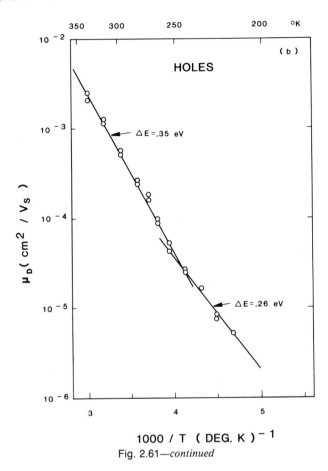

Fig. 2.61—*continued*

The above interpretation of a change in transport mechanism has been challenged by a number of workers. For example Fuhs *et al.* (1978) report that in their lightly P-doped GD a-Si:H films the electron transport is dispersive, since the injected current decreased as a function of time. Using Scher and Montroll (1975) theory, they find slopes for $t < t_\tau$, of 1.8 and for $t < t_\tau$ of 0.3, with a sum of the slopes that is slightly larger than the expected value of 2. t_τ is defined as a characteristic time and is related to the arrival of the leading edge of the carrier packet; if this were associated with a transit time, the resulting mobility would depend on thickness and on the applied field. The statistical character and the parameters underlying microscopic transport are contained in the dispersion parameter α_D and time t_τ. Scher and Montroll (1975) have indicated that the dispersive behavior can arise not only from hopping transport due to the random spatial distribution of hopping sites, but also from multiple trapping in a distribution of localized states. This can be distinguished by considering α_D to be a function of temperature. For transport dominated by hopping, α_D should be temperature independent. Fuhs et al. (1978) found this to be the case. However, this conclusion is open to doubt since Pollak (1977) has pointed out that hopping with a spatial distribution of sites will not lead to dispersive behavior and that an appropriate distribution of traps leads to dispersive transport, even with a temperature dependent α_D.

Tiedje et al. (1980) report that is GD a-Si:H the electron mobility is non-dispersive with mobility values that are significantly higher ($\sim 0.8 \text{ cm}^2 \text{ s}^{-1} \text{ V}^{-1}$) than shown in Fig. 2.60. The non-dispersive nature was confirmed since the carrier transit time was found to be inversely proportional to the applied field. This was not found to be the case for the sputtered a-Si:H material, which exhibited a lower value $\mu_D \approx 0.05 \text{ cm}^2 \text{ s}^{-1} \text{ V}^{-1}$. Figure 2.62 shows the temperature dependence of μ_D for GD a-Si:H, with the insets showing the shape of the current pulses (Tiedje et al., 1981). For the high temperature case the current decay has the abrupt fall off at the transit time, characteristic of conventional diffusion broadened transport, whereas at lower temperatures it exhibits a characteristic typical of dispersive transport. The electron transport undergoes a transition from dispersive behavior at low T to non-dispersive behavior at high T.

The temperature dependence of the hole drift mobility measured in a boron doped film is shown in Fig. 2.63 and, unlike for electrons, the holes show dispersion over the entire temperature range, with the dispersion increasing as the T is lowered.

In Fig. 2.64, α_D is plotted as a function of T for electrons and holes; a temperature dependence is observed, in contrast to the work of Fuh's et al. (1978). These results (Tiedje et al., 1981) were interpreted by the use of a multiple trapping model in which the mobility edge is considered to be sharp compared to the width of the localized state distribution. In this interpretation there was a tacit assumption that the localized states below the mobility edge fall off exponentially with energy and that the associated capture cross-sectional areas were identical. The release rates for carriers from these states are thermally activated and described by $\tau^{-1} = \nu \exp(-E/kT)$, where E is the depth of the localized state below the mobility edge and ν is an attempt to escape rate. These assumptions lead to a distribution of release times

$$\Psi(t) = \alpha_D \nu (\nu t)^{-(1+\alpha_D)} \quad \text{for } t > \nu^{-1}, \qquad (2.74)$$

where $\alpha_D = T/T_c$ and T_c is the characteristic temperature defining the conduction band tail. Equation (2.74), reproduces the hopping time distribution function proposed by Scher and Montroll (1975) in Eq. (2.73). The

ELECTRICAL TRANSPORT

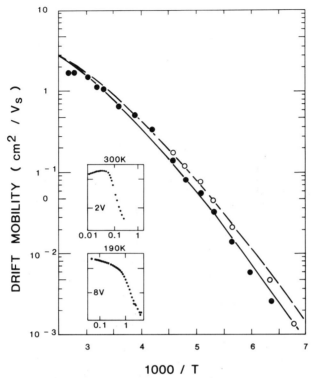

Fig. 2.62. Temperature dependence of the electron drift mobility for an electric field of 10^4 V/cm (closed circles) and 2×10^4 V/cm (open circles). The solid (broken) line is a fit to the low-(high-) field data with $T_c = 312$ K, $\mu_0 = 13$ cm^2/V s and $\nu = 4.6 \times 10^{11}$ s^{-1}. The insets are representative photocurrent decays at 300 and 190 K plotted as log I vs. log t; the time scale is microseconds (Tiedje et al., 1981).

transit time defined as the break point on the log I vs. log t curve is given by

$$t_t \cong \nu^{-1}\left(\frac{\nu}{1-\alpha_D}\right)^{1/\alpha_D}\left(\frac{d^2}{\mu_0 V}\right)^{1/\alpha_D} \quad (2.75)$$

for $T < T_c$.

The energy of the photoinjected electrons is determined by a competition between thermalization, which tends to push the electrons down in energy into the deepest localized states, and the DOS factor, which favors high energies. For $T < T_c$, thermalization wins and the photoinjected carriers (electrons) sink progressively deeper with increasing time; the transport becomes dispersive. For $T > T_c$, the electrons remain concentrated near the mobility edge and the charge transit exhibits non-dispersive behavior with $\mu_D = \mu_0(1 - T_c/T)$. Hence, a transition is to be expected from dispersive to

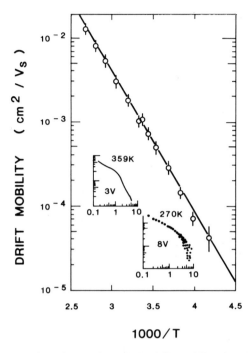

Fig. 2.63. Temperature dependence of the hole drift mobility for an electric field of 5.8×10^4 V/cm. The solid line is a fit to the data based on Eq. (2.75) with $T_c = 500$ K, $\mu_0 = 0.67$ cm^2/V s and $\nu = 1.6 \times 10^{12}$ s^{-1}. The insets are representative photocurrent decays at 359 and 270 K plotted as log I vs. log t; the time scale is microseconds (Tiedje et al., 1981).

non-dispersive transport with increasing T. The characteristic temperature of the conduction band tails can be estimated from the temperature at which the electron transport becomes non-dispersive or from the dispersion parameter, $\alpha_D(T)$. Based on fits to the experimental data, it was concluded that $T_c = 312$ and 500 K for the conduction and valence band tails, respectively.

In the intermediate temperature range close to T_c, where the transport changes from dispersive to non-dispersive, μ_D is given by

$$\mu_D \cong \frac{\mu_0 \alpha_D (1 - \alpha_D)}{(\nu t)^{1-\alpha_D} - \alpha_D^2}. \tag{2.76}$$

Using Eqs. (2.75) and (2.76), the values of the band tail widths determined above, and the experimental values of d and V, the extended state mobility can be determined by fitting the temperature dependence of μ_0. The best values for high quality GD a-Si:H films seem to be $\mu_0 \approx 13$ cm^2 s^{-1} V^{-1} for electrons and 0.67 cm^2 s^{-1} V^{-1} for holes.

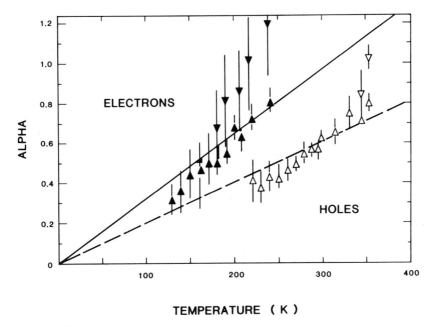

Fig. 2.64. Temperature dependence of the dispersion parameters α_D determined from the electron photocurrent decay before (closed triangles) and after (inverted closed triangles) the transit time, and for the holes before (open triangles) and after (inverted open triangles) the transit time. The lines are least-squares fits (of the form $\alpha_D = T/T_c$) to the data points, weighted by their relative accuracies. For electrons at high temperatures, $\alpha_D > 1$ corresponds to a current decay that is faster than t^{-2}, after t_t (Tiedje et al., 1981).

Although the above view of transport in a-Si alloys is commonly accepted, it may be possible to improve μ_{ext} if a suggestion by Silver et al. (1986) is valid. They suggest that the possible existence of charged T_3^+ and T_3^- defect centers produce long-range potential fluctuation that modulate the band edges and thereby reduce μ_{ext}. If these defects can be neutralized (e.g., during the growth process or by double injection), then it may be possible to raise μ_{ext} substantially above $10 \text{ cm}^2/\text{V-sec}$.

2.6 Photoconductivity

In the presence of photoexcitation the conductivity, σ_L, can be written as

$$\sigma_L = (ne\mu_n + n_0 e\mu_n) + (pe\mu_p + p_0 e\mu_p), \qquad (2.77)$$

where n and p are the electron and hole carrier concentrations under optical excitation respectively, n_0 and p_0 the thermal equilibrium concentrations in the dark, and μ_n and μ_p the conductivity mobilities of electrons and holes. The photoconductivity, σ_{ph}, is defined as

$$\sigma_{ph} = \sigma_L - \sigma_D. \tag{2.78}$$

In order to study the characteristics of a semiconducting material it is convenient to express the number of free carriers, e.g., electrons, as

$$n = G\tau_n, \tag{2.79}$$

where G is the volume generation rate and τ_n is the lifetime of the free carriers.

Photoconductors in general exhibit $\sigma_{ph} \propto G^\gamma$ where γ can be unity, greater than unity, less than unity, or equal to 0.5 (Rose, 1960). Amorphous silicon has exhibited all of these characteristics. The photoconductivity in GD a-Si:H has been studied by many (Anderson and Spear, 1977; Zanzuchi et al., 1977; Vanier et al., 1981; and in particular by Wronski and Daniel, 1981) who have made attempts to unravel the conditions under which the parameter γ can vary. Following Wronski et al. (1981), we consider a general type of distribution of defect states, as shown in Fig. 2.65. A distinction is drawn between trapping centers (N_{tn}) and recombination centers (N_1 and N_2). N_{r1} (N_{r2}) centers act as recombination centers for holes and P_{r1} (P_{r2}) centers act as recombination centers for electrons. The capture cross sectional area for electrons is C_{n1} (C_{n2}) and for holes it is C_{p1} (C_{p2}).

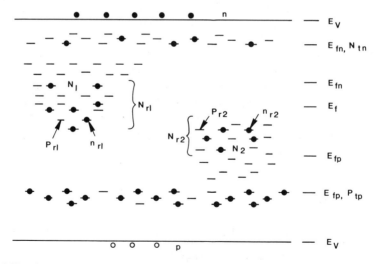

Fig. 2.65. Schematic energy-band diagram and distribution of states. n_r and P_r are the occupied and empty recombination centers respectively: $n_r + P_r = N_r$ (Wronski and Daniel, 1981).

PHOTOCONDUCTIVITY

As discussed in Section 2.5.6, the drift mobility measurements have shown that GD a-Si:H produced under different deposition conditions (e.g., T_s) have mobility values that are similar (see Fig. 2.60). In contrast, large changes in photoconductivity occur, for example, when the deposition temperature, T_s is altered. Therefore, the differences in σ_{ph} are associated with deeper lying states, which control the lifetime of electrons (τ_n) and holes (τ_p). For electrons, τ_n can be written as

$$\tau_n = \frac{1}{v[C_{n1}P_{r1} + C_{n2}P_{r2}]}. \tag{2.80}$$

Because of the continuous distribution of states in amorphous semiconductors, the occupancy of states, N_{r1} and N_{r2} will depend on the level of excitation. This leads to a non-linear dependence of σ_{ph} on the generation rate, G, and is represented by

$$\sigma_{ph} \propto G^\gamma. \tag{2.81}$$

The exact value of γ will depend on details of the distribution of states. The density of trapped electrons within the semiconductor is given by,

$$n_t(E) = N_t(E) \exp\left[\frac{(E_t - E_{fn})}{kT}\right]. \tag{2.82}$$

where we use the quasi-Fermi level, E_{fn}, rather than E_f, since we now have a non-equilibrium situation.

It should be recognized that the shallow trapped electrons do not directly influence the recombination mechanism. However, the transient response is affected by these states since the termination of illumination involves not only the return of the free electrons to thermal equilibrium, but the return of all the trapped carriers. If the recombination rate (n/τ_n) is less than the emission rate, the time taken to decay to half its value, τ_0, is given by,

$$\tau_0 = \left(1 + \frac{n_t}{n}\right)\tau_n. \tag{2.83}$$

We now discuss the various regimes of the γ-factor that have been exhibited by GD a-Si:H films.

Photoconductivities with $\gamma = 1.0$ have been associated with monomolecular recombination and observed in a wide variety of undoped a-Si:H films (Zanzuchi et al., 1977; Loveland et al., 1973). $\gamma = 0.9$ has been obtained over many orders of illumination intensity (f) (or equivalently generation rate, G), indicating that τ_n remains essentially constant over a wide range of E_{fn}. This is shown in Fig. 2.66 for a GD film produced at $T_s = 195°C$. Here $\eta\mu\tau \cong 10^{-6}$ cm^2 V^{-1} at $\lambda = 600$ nm (Zanzuchi et al., 1977).

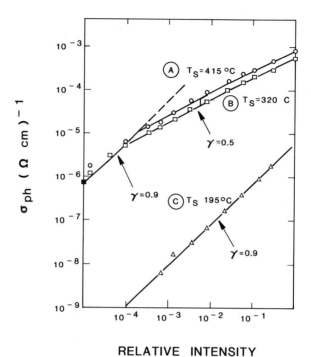

Fig. 2.66. Photoconductivity of rf discharge-produced a-Si films as a function of illumination for various substrate deposition temperatures (Zancuchi et al., 1977).

It has been reported that large γ values are observed in films which suffer photoinduced changes (Wronski and Daniel, 1981) in conductivity. As an example, Fig. 2.67 shows $\gamma = 0.9$, or $\tau_n \propto G^{-0.1}$, for a GD film made at 320°C which had undergone a 4 hr light exposure. Although τ_n changes slowly, τ_0 [from Eq. (2.83)], can change by orders of magnitude, as can be seen in Fig. 2.67. This indicates that τ_0 is determined by traps in the vicinity of E_{fn}; thus the trapping density $N_t = kTN_1(E_{fn})$. Therefore, Eq. (2.83) can be reexpressed as,

$$\tau_0 \approx \frac{n_t}{n}\tau_n = \frac{kTN_1(E_{fn})}{G}. \qquad (2.84)$$

Since τ_n remains virtually constant, then the total density of recombination sites (P_r) remains unaltered. Hence the N_1 states over the range of intensity investigated remains the same, which lead to $\tau_0 \propto G^{-0.1}$.

From Fig. 2.67, for illumination intensities in the range $10^{16} < G < 10^{18}$ cm^{-3} s^{-1}, n can be estimated to be 4×10^{10} and 3×10^{12} cm^{-3}, with E_{fn} located 0.55 eV and 0.45 eV respectively from E_c. This leads to an estimate

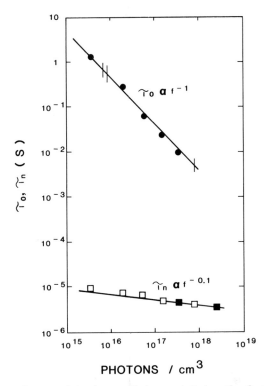

Fig. 2.67. Response time τ_0 and the electron lifetime τ_n plotted vs. the photogeneration rate. The results are for an rf discharge-produced film deposited at $T_s = 320°C$ after the film had been annealed at 200°C and subsequently exposed to 200-mW/cm² illumination for four hours (Wronski and Daniel, 1981).

of the DOS in this energy range of about $N_1 \cong 10^{17}$ cm^{-3} eV^{-1}. This DOS is somewhat larger than that measured using the other techniques, discussed in Section 2.3, and may be due to the possibility that an unoptimized or inferior material was investigated.

Bimolecular recombination between either trapped or free carriers is reflected by a value of $\gamma = 0.5$ and has been observed in both rf and dc GD a-Si:H films. This occurs when the density of electrons trapped in shallow states begins to exceed the density of recombination centers. Hence, $n_t(E_{tn}) \gg kTN_{t1}(E_{fn})$, and $P_r = N_t(E_{tn})$. As a result, τ_n and τ_0 are determined by the shallow traps, and the free electron density is given by

$$n = \left[\frac{GN_c}{vC_{n2}N_{tn}}\right]^{0.5} \exp\left[\frac{E_c - E_{tn}}{2kT}\right], \quad (2.85)$$

with $\tau_n \propto G^{-0.5}$ and $\tau_0 \propto G^{-0.5}$.

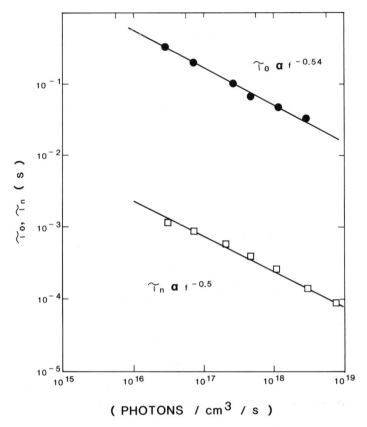

Fig. 2.68. Response time τ_0 and the electron lifetime τ_n vs. the photogeneration rate f (or G). The results here are for the film after an anneal at 200°C only (Wronski and Daniel, 1981).

The transition from recombination determined solely by deep lying states to this form depends on the densities and type of recombination centers as well as on E_{tn} and N_{tn}. The dependence of τ_n and τ_0 on G is shown in Fig. 2.68. Note that τ_n and τ_0 are now significantly larger in comparison with the results displayed in Fig. 2.67.

The change from a monomolecular recombination to bimolecular recombination mechanism has been reported by Anderson and Spear (1977). In Fig. 2.69, we show their results for σ_{ph} at room temperature as a function of $(E_c - E_f)$ for samples prepared with different amounts of phosphorus in the films. The figure shows that $\sigma_{ph} > \sigma_D$, except for the most heavily doped samples. The line P.G. = 1 is the point at which $\tau_n = t_t$ (the transit time) and for $(E_c - E_f)_0 \leq 0.53$ eV, P.G. > 1. (However, Rehm et al., (1977) find that for $(E_c - E_f)_0 \geq 1.2$ eV that P.G. > 1). The most notable point to consider is that for $(E_c - E_f)_0 < 0.6$ eV, $\gamma \cong 0.9$ and for doped samples γ changes to 0.5. We also note in Fig. 2.69 that the value of σ_{ph} saturates at $(E_c - E_f) \approx 0.4$ eV, which was attributed to a peak in the localized state density (see Fig. 2.27) causing a change in the recombination mechanism. However, it is certain that doping creates extra defect levels within the mobility gap; this has been recognized by Kagawa et al. (1983). In their experiment σ_{ph} was studied on samples constructed in the MOSFET (or field effect) configuration. The gate voltage is altered so as to produce an accumulation layer in the amorphous semiconductor, with the result that E_f is shifted towards E_c, which can be calculated as shown in Section 2.3.2.1. In this way the γ dependence as a function of $(E_c - E_f)$ can be measured without any change in the DOS spectrum, in contrast to when dopants are introduced. The σ_L vs. intensity was then measured, from which γ was determined. In Fig. 2.70a we show their results for γ vs. $E_c - E_f$ and note a continuous variation of γ from 0.9 to ≤ 0.5, rather than an abrupt variation, as suggested by the work of Anderson and Spear (1977). Further, Fig. 2.70b shows that

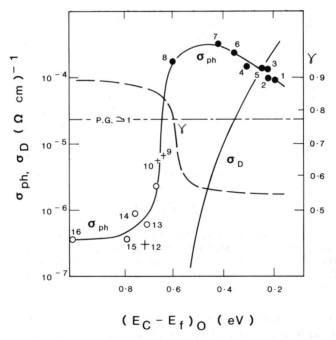

Fig. 2.69. Photoconductivity at 295 K plotted against $(E_c - E_f)_0$, the position of the dark Fermi level for phosphorus doped (●), undoped (+) and boron doped (○) specimens. The broken line represents the exponent γ (see right-hand ordinate) in the intensity dependence $\sigma_{ph} \propto G^\gamma$. σ_D is a typical dark conductivity curve. P.G. ~ 1 refers to unit photoconductive gain at a field of 3×10^3 V cm^{-1} (Anderson and Spear, 1977).

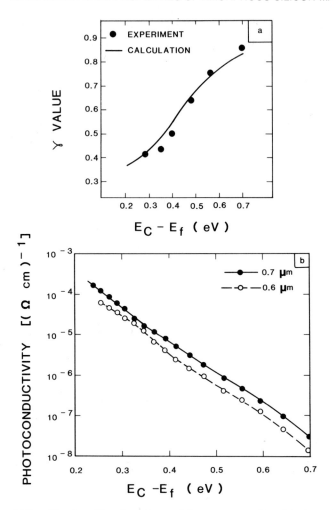

Fig. 2.70. (a) Fermi-level-position dependence of the exponent γ. Closed circles and solid line represent the experiment and calculated results, respectively. (b) Fermi-level-position vs. photoconductivity. Fermi-level-position is indicated by the energy below the conduction-band edge. Solid and dashed lines indicate analytic results for illumination at 0.7 and 0.6 μm wavelengths, respectively (Kagawa et al., 1983).

there is no saturation effect in σ_{ph} in comparison with the results of Fig. 2.69, which could be a further indication that doping creates extra defects.

There have been many reports that $\gamma \approx 0.7$ for large changes in illumination intensity (e.g., Madan et al., 1979). It does not seem possible that this can be explained on the basis of a mixture of monomolecular and bimolecular recombination. For an exponential distributed DOS Rose (1960) has suggested that the distribution of states, characterized by T_c, can be related to γ via the relation

$$\gamma = \frac{T_c}{T_c + T} \quad (2.86)$$

$\gamma \approx 0.7$ would result in a large T_c (~700 K) and is much too large to be consistent with the mobility measurements, which show that the characteristic energy for the conduction band tail is much smaller, $T_c \approx 300$ K (see Section 2.5.6). Further, Eq. (2.86) cannot explain the continuous change of γ with $(E_c - E_f)$, as shown in Fig. 2.70a. An attempt to arrive at a consistent explanation has been made by Kawaga et al. (1983), who recognized that there are two types of defects in a-Si:H, band tails and dangling bonds. Since the band tail density of states far exceeds the dangling bond density in the mid-gap region, the latter are neglected in the calculation. By considering the trap modulated Fermi Dirac functions under steady state, the generation rate (=recombination rate) is written as

$$G = R = \frac{N_R C_n n v_e C_p p v_h}{C_n n v_e + C_p v_h p} = \frac{N_R n p / \tau_n \tau_p}{(n/\tau_n) + (p/\tau_p)}, \quad (2.87)$$

where the C's are capture cross sections. Here the recombination rate is limited by the electron and hole capture rate, whichever is the smaller. If $C_p p v_h \gg C_n n v_e$ $(= p/\tau_p \gg n/\tau_n)$, then $R = G \propto n$. Hence $\gamma = 1$, corresponding to monomolecular recombination. For $\gamma = 0.7$ it follows that $n/\tau_n \gg p/\tau_p$. Therefore, if the recombination centers are donor like, $C_n/C_p \gg 10^3$, which is a typical ratio for the capture cross sections of charged and neutral centers (Lax, 1960). Using Eq. (2.84) C_p as a function of intensity (generation rate) can be found; using $C_p = 5.10^{-16}$ cm^2, Fig. 2.70a shows that σ_{ph} increases monotonically with a reduction in $E_c - E_f$. Further, because of the fit to their experimental data, it is thought that $C_n > C_p$; hence the recombination rate is limited by the capture rate of the recombination centers. If on the other hand, the recombination centers were acceptor like, since $n \gg p$, the good experimental fit could not be explained. The recombination is then clearly determined primarily by donor-like band tail states. The γ-variation between 0.35 and 0.85 can be explained by the model without invoking any transition in the recombination mechanism. Further, the increase in σ_{ph} with a decrease in $(E_c - E_f)$ can be explained simply by the fact that there are fewer donor states for electrons to recombine with.

McMahon and Madan (1985) have pointed out that for optimized materials with a low DOS, $(g(E_f) < 10^{16}$ cm^{-3} eV$^{-1})$ γ should be small for low intensities and increase to $\gamma = 0.9$ at higher intensities. Assuming a DOS of the form given by Eq. (3.54) [a variation of Eq. (2.22)], they calculate σ_{ph}

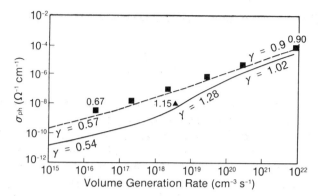

Fig. 2.71. Calculated photoconductivity for undoped (dashed curve) and lightly boron-doped (solid curve) layers using $g(E)$ spectra defined in the text; the corresponding experimental data with values of γ are shown (McMahon and Madan, 1985).

as a function of G as shown in Fig. 2.71, which seems to fit the experimental data quite well. They conclude that for a low DOS material the recombination rates for holes and electrons are expected to depend on intensity.

Superlinearity, corresponding to $\gamma > 1$, has also been observed. The electron lifetime can, in principle, be increased by orders of magnitude by the addition of recombination centers. This increase occurs at the expense in the lifetime of the other type of carrier. Figure 2.72 shows results for GD a-Si:H in which $\gamma > 1$ (Vanier et al., 1981). The explanation for this type of behavior is provided in Fig. 2.73. Here we show 3 sets of states in the energy gap: (1) hole traps in thermal contact with the valence band that have a small capture cross section (C_{n1}) for electrons; (2) recombination centers near the middle of the gap with a large capture cross section for electrons (C_{n2}) and electron traps; (3) states in thermal contact with the conduction band. At low levels of illumination the quasi Fermi level for free holes is situated between (1) and (2) so that most of the holes recombine with the more efficient electron recombination sites near (2). At higher light levels, E_{fn} and E_{fp} move apart sufficiently to encompass state (2). The result of incorporating states of type (1) is that there is a transfer of holes from (2) to (1). Since $C_{n1} \ll C_{n2}$, this leads to longer lifetimes and higher sensitivities. This superlinear dependence occurs only in the narrow region of temperature where holes are being transferred from state (2) to state (1).

2.7 Photoluminescence

Photoluminescence (PL) is the absorption of energy in matter and its re-emission as radiation. In amorphous semiconductors, radiation can be

PHOTOLUMINESCENCE

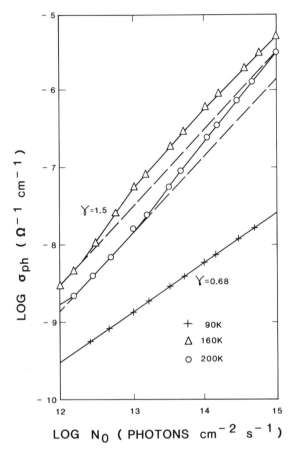

Fig. 2.72. σ_{ph} as a function of light flux at three different temperatures, showing a supralinear region (Vanier *et al.*, 1981).

emitted either by the relaxation of an excited defect or by an e-h pair bound together by their Coulomb attraction (an exciton). Extensive work using the PL technique has led to the conclusion that in chalcogenide glasses PL occurs via charged defects since the luminescent efficiency falls off rapidly with temperature. The electron (or hole) escapes from the neutral excited center and finds a non-radiative recombination channel. The emitted radiation shows a Stokes shift, which we shall discuss later.

In GD a-Si : H the photogenerated e-h pairs can separate and find a non-radiative recombination path for recombination. However, below 100 K the separation of carriers is inhibited and recombination occurs with the emission of radiation. Although PL has been observed in Si prepared by

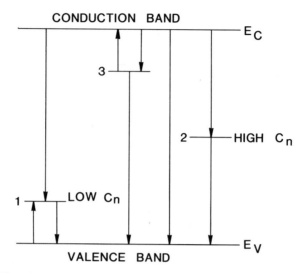

Fig. 2.73. Three state model used to explain supralinearity (Vanier et al., 1981).

evaporation (without hydrogenation) in ultra-high vacuum, most studies have been made on GD deposited material because of its technological importance. Figure 2.74 shows four well characterized photoluminescence peaks that have been observed in GD a-Si:H (see, e.g., Street, 1981); we shall discuss each in turn.

Generally, GD a-Si:H films that have been produced using optimum deposition conditions (see Section 2.1.2) show a single broad featureless PL peak at low temperature (curve A) that lies in the range 1.25–1.4 eV and has a width, typically ~0.3 eV (Engemann and Fisher, 1973). The slight variation of the peak energy position (1.25–1.4 eV) is thought to be due to changes in the deposition conditions, which in turn could effect the H content of the film and hence lead to materials with different band gaps.

The PL process involves an electronic transition; the energy released is given up as a photon together with phonons. Because of the strong electron–phonon interaction the atomic environment of the recombination center can be distorted when the electron occupancy changes (Stokes shift). Therefore, ideal PL spectra consist of zero phonon lines at the electronic energy of the transition, and a series of side bands at lower energies separated from the zero phonon line by integral units of the phonon energy. Distinct phonon side bands have not been observed in GD a-Si:H; the estimate of the phonon contribution is generally made by comparing luminescence and excitation spectra, an example of which is shown in Fig. 2.75 (Street, 1981). In this type of experiment the phonon coupling shifts the luminescence bands to energies

PHOTOLUMINESCENCE

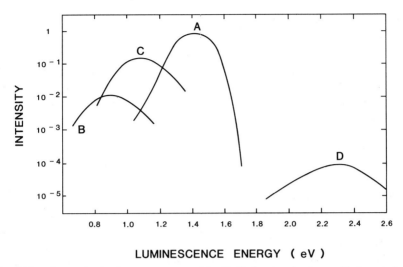

Fig. 2.74. Composite luminescence spectra of a-Si:H showing the four well characterized peaks (Street, 1981).

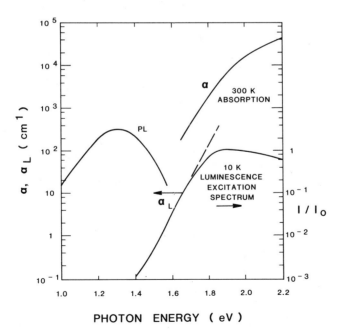

Fig. 2.75. The luminescence excitation-spectrum of a-Si:H showing the derived absorption coefficient, α, compared to the room temperature optical absorption (Street, 1978).

above and below the line of zero phonon energy. The object is then to calculate the absorption coefficient that corresponds to the luminescence, which is then compared to the absorption. An estimate of the size of the shift is obtained from Fig. 2.75, which shows that 0.4–0.5 eV separates the luminescence peak from the absorption, α, data. The distortion energy of the recombination is approximately half this value, or 0.2–0.25 eV. The observation that the excitation spectrum is apparently indistinguishable from the fundamental absorption edge implies that the recombination centers are part of the conduction and valence bands. Based on the above, Fig. 2.76 shows the luminescence process (Street and Biegelson, 1984). The absorption of above band gap radiation creates e–h pairs that thermalize quickly ($\sim 10^{-12}$ s) to the band tails. (At low temperatures, where luminescence is usually observed, diffusion is negligible after the thermalization process.) Since there is evidence of a Stokes shift, and these materials exhibit a smaller drift mobility for holes than electrons (Section 2.5.6), then it is speculated that the holes within the band tails become self trapped, with a binding energy ~ 0.3–0.4 eV. Radiative recombination then occurs by tunnelling between the band tails.

The luminescence intensity is related to the completing rates for radiative, P_R, and non-radiative, P_{NR}, recombination by the expression

$$Y_L = \frac{P_R}{P_R + P_{NR}}. \qquad (2.88)$$

The presence of non-radiative centers can provide competition for the luminescence process, which manifests itself as a decrease in intensity.

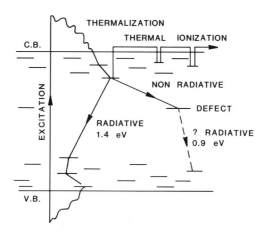

Fig. 2.76. Schematic diagram of the luminescence processes showing the radiative and non-radiative processes (Street and Biegelson, 1984).

PHOTOLUMINESCENCE

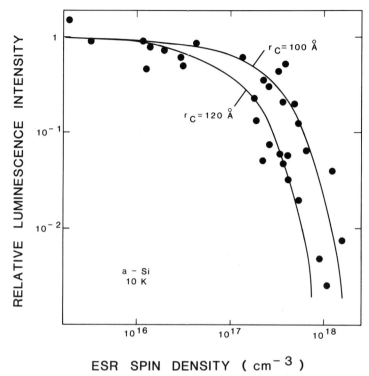

Fig. 2.77. Plot of luminescence intensity vs. spin density for a variety of samples prepared under different deposition conditions. The solid line is the fit to the non-radiative tunneling model (Street *et al.*, 1978b).

This is shown in Fig. 2.77, where the luminescence intensity is shown to vary with the ESR spin density (Street *et al.*, 1978b). The ESR spin density originates from dangling bonds, which are the sites for non-radiative recombination. Note that PL intensity is a maximum for the lowest defect density material; this is another indirect argument for correlating the 1.3 eV PL peak with transitions involving band tail states.

The PL process can be quenched either by increasing the temperature or by the application of an electric field (Engemann and Fisher, 1976; Nashashibi *et al.*, 1977); Tsang and Street, 1978; Paesler and Paul, 1980). For example, Fig. 2.78a shows that for $T > 50$ K, the luminescence is rapidly quenched and Fig. 2.78b shows that similar behavior is observed with the application of an electric field (Engemann and Fischer, 1976). The decrease in PL due to the field is attributed to the separation of the e-h pairs, and an increase in temperature allows the carriers to diffuse apart and move through the material until some non-radiative event occurs.

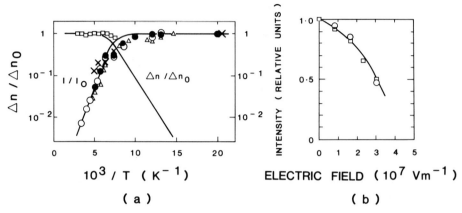

Fig. 2.78. (a) Temperature dependence of normalized photoluminescence intensity for glow discharge silicon deposited at 520 K (○), 440 K (△), 400 K (●) and 320 K (×). The photoconductivity (□) is independent of temperature in the temperature range where the luminescence intensity varies strongly. (b) Electric field quenching of luminescence (Engemann and Fischer, 1976).

Electroluminescence from forward biased p–i–n junctions fabricated from GD a-Si:H shows a peak at 1.27 eV (Pankove and Carlson, 1976), very close to the value found from the PL data. It is also reported that the PL diminishes in intensity and shifts to lower energies as hydrogen evolves from the films by annealing. This occurs because the effusion of H leaves behind extra defect states that create sites for non-radiative recombination processes to occur.

The luminescence, peak B in Fig. 2.74, in GD a-Si:H is attributed to extra defect states created by doping that shifts the luminescence band to a lower photon energy, as shown in Fig. 2.79a and b (Rehm et al., 1976). For either type of dopant the total luminescence intensity decreases with increasing concentration. Several models have been proposed to explain this. One model assumes the existence of internal fields due to charged dopants that could quench the PL signal for fields in excess of 10^5 V cm^{-1}. This is similar to the results of Fig. 2.78. However, specific evidence against this model is provided by the luminescence decay data. Austin et al. (1979) report that doping enhances the non-radiative component of the decay but the radiative component is unchanged, which argues against electric field quenching. However, doping can introduce additional defects that act as non-radiative centers (see Section 2.8); it is this aspect that can provide an alternative explanation for the quenching of the luminescence. It has been found by using a light induced ESR technique (LESR, Knights et al., 1977b) that there is a rapid increase of spin density with doping, which can be correlated to defects of the dangling bond type.

PHOTOLUMINESCENCE

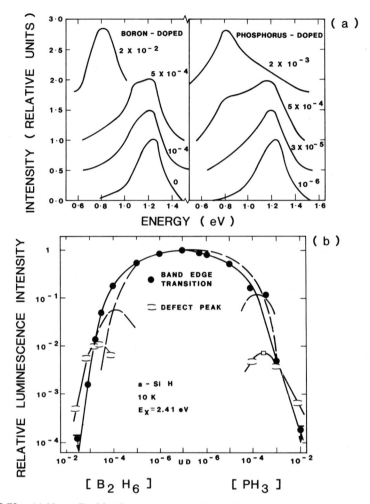

Fig. 2.79. (a) Normalized luminescence spectra for *p*- and *n*-type glow discharge-deposited silicon. Doping concentration as shown (from Rehm *et al.*, 1976). (b) Intensity of luminescence of doped samples (Street *et al.*, 1981a).

Further evidence against the internal electric field argument is provided by the results of Fig. 2.80, where we note that the PL intensity exhibits a maximum for compensated (nearly equal P and B doping) samples (Street *et al.*, 1981a). This suggests that the non-radiative recombination path is reduced in the presence of compensation. Therefore, according to the internal electric field model, compensation should reduce the luminescence intensity because the ionization of the dopants should be complete, leading

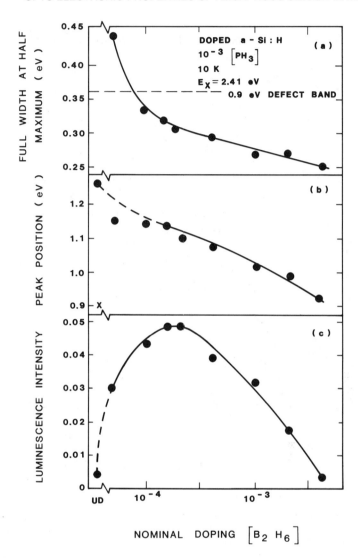

Fig. 2.80. Luminescence intensity, peak position and line width in a-Si:H samples doped with $10^{-3}[PH_3]$ and compensated with differing amounts of boron (Street et al., 1981a).

to large internal fields. However the LESR results show that the dangling bond spin density decreases with compensation. It is also concluded that with compensation the creation of additional dangling bonds may occur in the form of impurity complexes, as it does in III-V and II-VI compound materials.

As shown in Fig. 2.79a, a luminescence peak at 0.8 eV appears as a shoulder for GD a-Si : H films when they are doped either with B or P. This also appears to be true when the samples are partly doped with Li. However, the shoulder can be made to disappear when the samples are annealed at 400°C for a few minutes. Here it was noted that since E_f did not move with respect to the conduction band edge in this annealing range, then any change in the PL spectra is due to the creation of new network defects. It is speculated that diffused Li could form percipitates, and hence could deform the amorphous network around the percipitates, thereby creating strained or broken bonds which could be centers for non-radiative recombination.

The photoluminescence peak at 1.1 eV (peak C in Fig. 2.74) has been attributed to oxygen impurities and is observed when oxygen is deliberately added during deposition or by the post deposition oxidation of GD a-Si : H samples with a columnar structure (Street et al., 1978b; Street and Knights, 1980). This oxygen band usually appears as a weak shoulder on the low energy side of the peak A in Fig. 2.74. The exact transitions are somewhat speculative since oxygen is expected to reside in a twofold configuration and seems incapable of generating localized states in the gap. It is speculated that the oxygen luminescence originates from singly or threefold coordinated sites. The recombination mechanism would then involve a transition between a conduction band electron and a hole trapped at a negatively charged defect.

Finally, a peak D in Fig. 2.74 at ~2.3 eV has been found, where the PL intensity is about 10^{-4} times the value of the peak at A. It has been suggested that the luminescence band originates from surface contamination and is not an intrinsic property of a-Si : H.

2.8 Doping in Glow Discharge a-Si Based Alloys

Until n- and p-type doping had been demonstrated in GD a-Si : H (Spear and LeComber, 1976), it was generally believed that amorphous semiconductors could not be doped, since the local valency requirements could be satisfied by bond rearrangements. Although much of the credit must go to Spear and LeComber (1976), who showed that the room temperature dark conductivity could be changed by orders of magnitude with the addition of P or B to the GD a-Si : H, doping was first achieved by Chittick et al. (1969), who observed a change in σ_D by a factor of 40 with the addition of 200 ppm PH_3 to the gas phase for samples deposited at room temperature. Since the DOS for films deposited at an elevated temperature shows a much reduced value (see Section 2.3.2.1), it was this aspect that enabled Spear and LeComber (1976) to achieve large changes in σ_D with doping.

Generally, n- or p-type doping is achieved by the addition of PH_3 or B_2H_6 to the SiH_4 in the gas phase, respectively. In Fig. 2.81, σ_D and the conductivity activation energy, E_σ, are plotted as functions of the PH_3 and B_2H_6 concentrations for GD a-Si:H and GD a-Si:F:H type films (Spear and LeComber, 1976; Madan et al., 1979). The most heavily n-type doping results in $\sigma_D \sim 10^{-2}$ and ~ 10 $(\Omega \text{ cm})^{-1}$ for a-Si:H and a-Si:F:H type alloys respectively. These are used as n^+ layers and provide low resistance, linear (ohmic) contacts to the intrinsic layer (see Sections 3.1.3 and 3.1.4). Since the integrated DOS over the mobility gap for the two types of alloys is virtually the same, the reason for the improvement in σ_D for the fluorinated n^+ layer has to be sought elsewhere. It has been shown that the fluorinated n^+ layer is in fact microcrystalline, unlike the amorphous intrinsic layer (Tsu et al., 1981). Similar high values of conductivity in GD P-doped a-Si:H alloys have also been obtained when high powers were used to decompose the gas mixture of SiH_4, H_2 and PH_3 (Usui et al., 1979 and Hamasaki et al., 1980). The use of high power leads to an increase of the H flux at the growing surface, which results in the microcrystallization (μ.c.) of the film, as discussed in Section 2.2.2 (Veprek and Maracek, 1968). Microcrystalline (n^+) layers have a beneficial aspect in the improvement of solar cell performance, as we will discuss in Chapter 3.

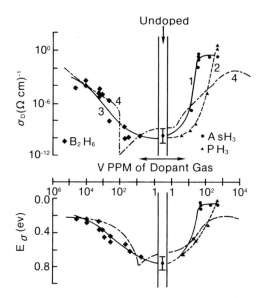

Fig. 2.81. The room temperature dark conductivity, $\sigma_D (\Omega \text{ cm})^{-1}$, and the conductivity activation energy E_σ (eV), plotted as a function of vppm of AsH_3 (curve 1), PH_3 (curve 2) and B_2H_6 (curve 3) in the premix gas ratio of $SiF_4/H_2 = 10/1$. The p- to n-transition indicated refers to an a-Si:F:H alloy (Madan et al., 1979). Curve (4) refers to the doping characteristics for the a-Si:H alloy (Spear and LeComber, 1976).

Although extrinsic conductivity (doping) has been demonstrated using various elements such as P, As, B and Ga, and employing the simultaneous addition of dopants during the growth, doping has also been achieved by the use of a post-deposition technique such as ion implantation (Müller et al., 1977). However, as discussed in Section 2.3, not all of the dopant atoms enter the tetrahedral configuration and the doping efficiency is consequently low (Spear, 1977). Further, fundamental changes in the electro-optical characteristics occur with doping, as discussed below.

2.8.1 The Effect of Phosphorous Doping

The influence of P doping on the transport properties of GD a-Si : H based alloys has been interpreted in various ways: (a) transport in the conduction band for T above 200 K (Overhof et al., 1980); (b) transport in a donor band (Dohler, 1979) changing to a transport in the conduction band for $T > 400$ K (Ast et al., 1980); (c) parallel path conduction through the donor and conduction bands (Jones et al., 1977). In a study of Schottky barrier (SB) heights by Viktrovitch et al. (1981), strong evidence was presented for the existence of transport through a donor band lying 0.3 eV below the conduction band edge for sufficiently large P doping.

Further, P doping results in a decrease in the photoluminescence signal (see Section 2.7), indicative of an increase in the DOS, as confirmed by the use of the primary photocurrent technique (Crandall, 1980). It has been shown that even 1 vppm PH_3 doping of a GD a-Si : H alloy can decrease the $\mu\tau$ product of holes by nearly two orders of magnitude from its initial value of $\sim 1.10^{-7}$ cm^2 V^{-1} (Kirby et al., 1983), as discussed in Section 3.1.4.3.

The simultaneous use of conductivity and thermopower measurements can elucidate the transport mechanisms in doped a-Si : H (see Section 2.5). Figures 2.82 and 2.83 show S and σ as a function of $10^3/T$ for samples containing different amounts of P (Jones et al., 1977). For low doping levels (20 vppm PH_3), the activation energy derived from the plots of both S and σ vs. $10^3/T$ are the same within experimental error. It is concluded that the transport takes place predominantly by electrons in extended states and hence the slope gives the activation energy, $(E_c - E_f)_0$ (see Section 2.5.5). This result supports similar conclusions from drift mobility and conductivity (LeComber and Spear, 1970; LeComber et al., 1972), photoconductivity (Spear et al., 1974) and Hall mobility (LeComber et al., 1977) measurements.

Further, as the P doping increases, as for sample numbers 7, 8 and 11 in Figs. 2.82 and 2.83, it evidently becomes increasingly difficult to ascribe an activation energy to the $S(1/T)$ data. These data (Jones et al., 1977) has been interpreted on the basis of two parallel conducting paths, one corresponding to extended state conduction above E_c and the second centered at a level E_M

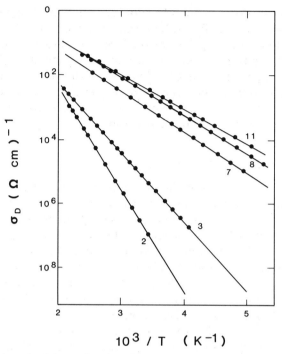

Fig. 2.82. Dark conductivity as a function of $1/T$ for samples prepared with different amounts of doping. Samples 2, 3, 7, 8, 11 prepared with PH_3/SiH_4 gas ratio of 20, 51, 300, 1000 and 3×10^4 vppm respectively (Jones *et al.*, 1977).

situated in the band tail, where the transport mechanism occurs by phonon assisted hopping. The total conductivity is given by

$$\sigma = \sigma_c + \sigma_M, \qquad (2.89)$$

where

$$\sigma_c = e\mu(E_c)g(E_c)kT \exp\left(-\frac{(E_c - E_f)_T}{kT}\right), \qquad (2.90a)$$

and

$$\sigma_M = e\mu(E_M)g(E_M)kT \exp\left(-\frac{(E_M - E_f)_T}{kT}\right). \qquad (2.90b)$$

The thermopower is calculated from the sum of the weighted contributions

$$S = S_c \frac{\sigma_c}{\sigma} + S_M \frac{\sigma_M}{\sigma}, \qquad (2.91)$$

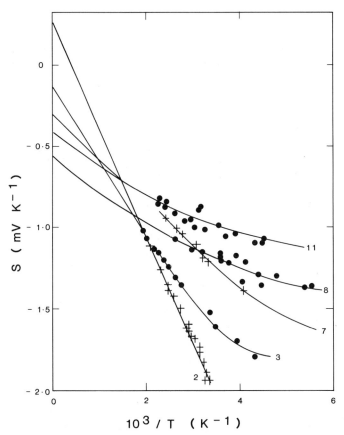

Fig. 2.83. Observed thermoelectric power S plotted as a function of reciprocal temperature for five a-Si specimens doped with various phosphorous concentrations as indicated in the figure caption of Fig. 2.82 (Jones et al., 1977).

where

$$S_c = -\frac{k}{e} \frac{(E_c - E_f)_0}{kT} - \frac{k}{e}\left[A_c - \frac{\delta_F}{k}\right] \quad (2.92a)$$

and

$$S_M = -\frac{k}{e} \frac{(E_M - E_f)_0}{kT} \frac{k}{e}\left[A_M - \frac{\delta_F}{k}\right]. \quad (2.92b)$$

Using this model, Jones et al. (1977) provide a good fit to the experimental data, as shown by the full lines in Figs. (2.82) and (2.83), and find, for lightly doped samples, that $(E_c - E_M) = (E_c - E_A) \cong 0.20$ eV and $W \cong 0.10$ eV. This suggests that for the lightly P-doped samples, there is an increasing contribution to the transport process from tail states as the temperature is decreased. Further, as the doping increases, an excellent fit is now provided with $(E_c - E_M) \cong 0.13$ eV, which suggests that the center of the hopping path has moved towards E_c, in agreement with the Hall effect measurements (LeComber et al., 1977). Hence, with an increasing donor concentration there is a transition from tail state hopping at E_A to donor hopping centered at $E_c - E_M = E_c - E_D = 0.13$ eV.

A similar type of study has also been performed by Beyer et al., (1977) as shown in Fig. 2.84. Note that the activation energy measured from the conductivity (E_σ) exceeds that measured from the thermoelectric effect (E_s). Further, they noted that with P doping the value of the preexponant, σ_0, varied over orders of magnitude and S_0 showed an anomalously large value of 0.73 mV K^{-1}. However, from Eq. 2.55, $S_0 = -k/eA_c$ for extended state conduction; if μ and $g(E)$ vary linearly with energy, then $A_c = 3$, which leads to a much smaller value of $S_0 \approx 0.25$ mV K^{-1}.

As mentioned above, an order of magnitude reduction of the prefactor σ_0 [Eq. (2.40)] with doping, shown in Fig. 2.85, has also been observed by Rehm et al. (1977), who explained that the activation energy of the conductivity

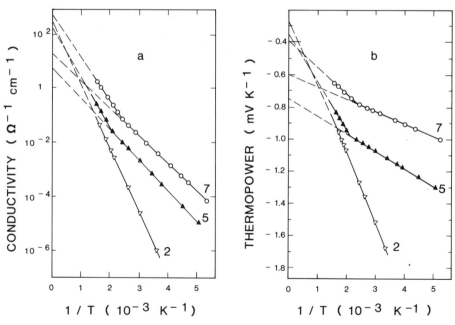

Fig. 2.84. (a) Logarithmic plot of the conductivity vs. inverse temperature $1/T$. (b) Dependence of the thermoelectric power S on $1/T$. The three samples 2, 5 and 7 were prepared from silane containing 1, 250 and 10^4 vppm of phosphine (Beyer et al., 1977).

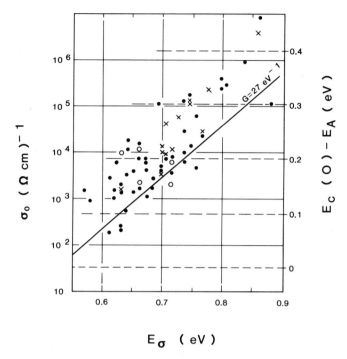

Fig. 2.85. Graph of ln σ_0 vs. E_σ for over 60 a-Si glow discharge specimens prepared in an inductively coupled (●) and capacitatively coupled (×) unit. Calculated $E_c(0) - E_A$ values are shown on the R.H. ordinate (Spear et al., 1980).

increases with T, since E_f shifts away from E_c (the so-called statistical shift). This occurs because of the rapidly rising DOS that becomes populated above E_f, and the opposing condition that the neutrality condition requires, that E_f move towards the midgap position. This effect would be more acute for increasing P concentration, since E_f approaches a rapidly increasing DOS, shifting E_f away from E_c as T increases. Another contribution could be the variation of the band gap with temperature, which is expected to dominate for undoped or lightly doped samples where E_f lies near the minimum in the DOS spectra.

Using the DOS taken from the field effect data (Madan, 1973; Madan et al., 1976) and shown in Figs. 2.27 and 2.28, Jones et al. (1977) calculated δ_F [in Eq. (2.92)] as a function of $(E_c - E_f)_0$ for the heavily doped samples, and showed that $\delta_F \cong 0.25$ meV K^{-1} for $(E_c - E_f)_0 = 0.2$ eV. Assuming $A_c \cong 3$–4, then $S_0 \approx 0.6$ mV K^{-1}, which is now closer to the experimental value of 0.73 mV K^{-1}, discussed above. However, a great uncertainty exists in the evaluation of A_c, which relies upon knowledge of $\mu(E)$ [or $\tau(E)$] and in the $g(E)$ spectra for $E > E_c$. Also, the field effect results, as discussed in Section 2.3.2.1, can overestimate $g(E)$, which adds further uncertainty to the evaluation of δ_F.

Beyer et al. (1977) have shown that a much larger statistical shift with T will occur if the DOS is assumed to be much smaller than that deduced from the field effect data of Figs. 2.27 and 2.28. This statistical shift could account for both the large values of S_0 and the large variation in σ_0 with the conductivity activation energy (E_σ), as shown in Fig. 2.85 (Spear et al., 1980). These results have prompted an intensive theoretical study by Dohler (1979) and Overhof et al. (1981, 1983). For example, Overhof et al. (1983), using Eqs. (2.40) and (2.55) defined a quantity Q given by

$$Q = \ln \sigma - \frac{e}{kS} = \ln \sigma_0 + A_c. \qquad (2.93)$$

Here Q represents a transport property that does not contain the dominant term $(E_c - E_f)_T = (E_c - E_f)_0 - \delta_F T$. By plotting Q vs. $10^3/T$ for differently doped samples, they found that the data could be fit to $Q = 1.8 - 1.25\Delta/kT$, as shown in Fig. 2.86, where $0.04 < \Delta < 0.12$ eV. The difference in the activation energies, E_σ and E_s, is interpreted on the basis of long-range statical fluctuations in potential, first proposed by Fritzche (1971), which may be due to the presence of a random distribution of charged centers. Some evidence for this exists, since the NMR data of Reimer et al. (1980) indicate an inhomogeneous distribution of hydrogen in a-Si:H film. Therefore, they assume that for lightly doped samples the fluctuations may be due to structural inhomogeneities; for heavily doped samples the charged donors could initiate additional potential fluctuations. Δ is thus regarded as the magnitude of the fluctuations within the film. By plotting Q vs. T, Dohler (1979) concludes that Q would be the same function of T for all samples if $g(E)$ and $\mu(E)$ were independent of doping. However, he notes that the slope increases with increasing doping and hence $g(E)$ and $\mu(E)$ cannot be quantities that are independent of doping. Further, he concurs with the above conclusions that at high T the predominant conduction path is above E_c and at low T an increasing contribution from the band tails results.

2.8.2 The Effect of Boron Doping

A systematic study of the changes in opto-electronic behavior in B-doped a-Si:H has been performed by Tsai (1979), where the regime from GD a-Si:H to a-B:H films was investigated. In Fig. 2.87, we show the results of α vs. $h\nu$ as function of X_g ($\equiv B_2H_6/SiH_4$). Evidently, the addition of B results in a reduction in E_g. However, it should be recognized that this

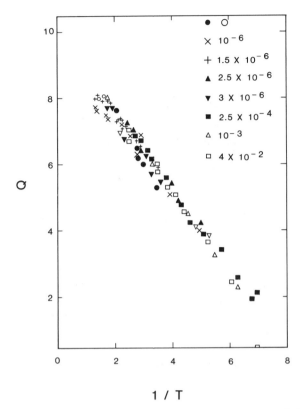

Fig. 2.86. Q vs. $1/T$ for phosphorus doped (indicated by PH_3/SiH_4 gas ratio in the diagram) a-Si:H samples (Overhof and Beyer, 1983).

conclusion requires that the quantity B in Eq. (2.32) remain constant throughout the entire regime. Thus, the use of this definition should be treated with caution, as discussed in Section 2.4. Nevertheless, the absorption curves do shift to lower photon energies with increasing incorporation of B.

The structure of the films was investigated by the use of infra-red spectroscopy and the H content determined from the effusion measurements discussed in Section 2.2. The study concluded that the hydrogen concentration of the cathode type films decreased from about 18 at.% for a-Si:H film ($B_2H_6 = 0$) to about 10% for a-Si:B:H films ($SiH_4 = 0$). Further, the spin density increased from $< 10^{16}$ cm^{-3} ($B_2H_6 = 0$) to 5.10^{18} cm^{-3} when B_2H_6 constituted 20% of the gas flow. Correspondingly, the ir spectra indicated strong absorption at 2560 cm^{-1} due to the B-H stretching modes, which increased with increasing B incorporation, whereas the band at 2000–2100 cm^{-1} (due to the Si-H stretching mode) decreased in intensity. An

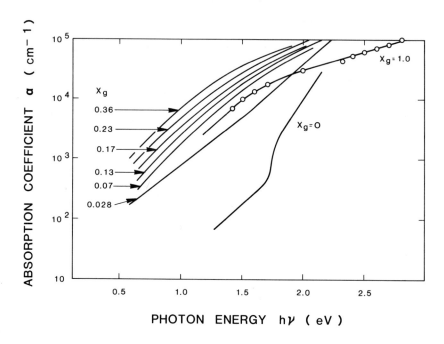

Fig. 2.87. Optical absorption of films prepared at 270°C with different Boron concentrations in the gas phase, X_g, as indicated in the figure (Tsai, 1979).

absorption band below 1100 cm^{-1} was observed and was attributed to the B–B vibration, whereas the absorption band near 1900–2000 cm^{-1} was interpreted as a B–H–B bridge bond. However, Shen and Cardona (1981) did not find evidence for B–B bonding, nor for the bridging bonds in similarly prepared samples containing 9 at.% B. Greenbaum et al. (1982) have concluded from their NMR studies that more than 99% of the B is threefold coordinated and the doping efficiency is less than 0.1%. The creation of these defects has an influence on the basic conduction mechanism, as illustrated in Fig. 2.88, where we show (Jan et al., 1980) ln σ vs. $1/T$ for GD films deposited with different B_2H_6/SiH_4 starting gas ratios. Undoped samples were found to be slightly n-type. Therefore, with the addition of small amounts of B, the n-type conductivity was still retained. Here, all the samples prepared with $B_2H_6/SiH_4 = 10^{-5}$ ppm exhibit a single activation energy (E_σ and E_s) which is consistent with conduction via extended states. For $B_2H_6/SiH_4 > 10^{-4}$, however, neither the conductivity nor the thermoelectric power show a single activation energy and the data were interpreted using a two channel model (see Section 2.8.1) with a defect level now situated 0.42 eV above the valence band edge.

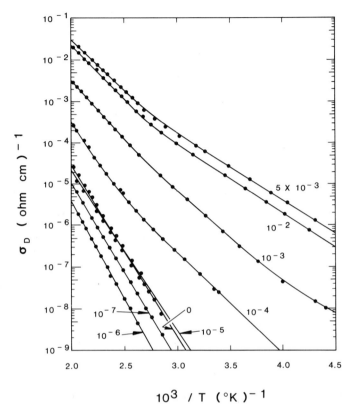

Fig. 2.88. Temperature dependence of the electrical conductivity for eight B doped samples of a-Si : H. The B_2H_6/SiH_4 ratio used in the preparation is given beside each curve (Jan et al., 1980)

2.8.3 The Effect of Compensation

It is generally found that the spin density, as measured by ESR in undoped materials (where E_f is near midgap) equals the density of dangling bonds, which can be $< 10^{15}$ cm^{-3} for device quality GD a-Si : H type alloys (see, e.g., Street and Biegelson, 1984). When the samples are doped p-type or n-type such that E_f moves towards the band edges, the dangling bond signal disappears. This is not an indication that the material has a low defect state density, but, on the contrary, as the photoluminescence data in Fig. 2.79a and b show, the PL intensity drops with either form of doping. This is an indication of an increased non-radiating recombination path due to the presence of extra defects. The decreasing ESR signal is postulated to be due to a decrease in the fraction of singly occupied defects. However, as shown in Fig. 2.80, the PL intensity can be made to increase with compensation; the corresponding ln σ vs. $10^3/T$ for these samples is shown in Fig. 2.89 (Street et al., 1981a). (An interesting observation was also made during the compensation experiments: the minority dopant species was found to be incorporated into the film more efficiently. It was concluded the deposition process has a tendency to equalize the two dopant concentrations). Full compensation occurs for 10^{-3} PH$_3$ and 4.10^{-4} B$_2$H$_6$ as shown in Figs. 2.80 and 2.89. For Boron concentrations exceeding this value it was found that the thermopower was positive and for concentrations less than this it was negative. Referring to the data of Fig. 2.80, it was concluded that the PL involves band edge transitions rather than being associated with the 0.9 eV defect peak (Section 2.7), because the T dependence of the peak was found to be larger in the compensated sample than for the 0.9 eV peak. The mechanism was then associated with defects. Further, it was noted that for

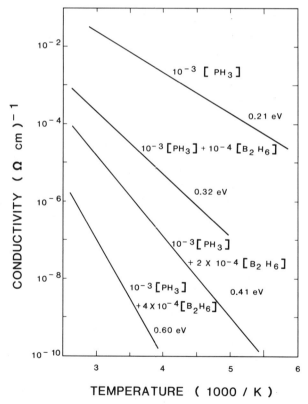

Fig. 2.89. DC conductivity vs. inverse temperature for various doped and compensated n-type samples. The activation energies are indicated (Street et al., 1981a)

the nominally compensated material with E_f at midgap position, the LESR signal disappears along with the ESR signal. Hence, the dangling bond density is greatly reduced; in this sense the compensated material could be considered to be identical to the undoped material. However, as can be seen from Fig. 2.80, since the peak in the luminescence signal is shifted in energy to lower values, then the compensation process does indeed introduce new localized states that act as hole traps. This result is based on the fact that the luminescence radiative decay in compensated material is similar to that for undoped samples. The conduction band tail then does not change. Therefore, the shift of the luminescence peak is due to the presence of additional states above the valence band. It is conjectured that these extra states give rise to an intrinsic absorption band, which could then account for the broadening and shift of the absorption edge in B doped samples, as shown in Fig. 2.87. Since there is also evidence from photoemission data that states above the valence band are produced by PH_3 doping, and that during the compensation process the B and P concentrations are equalized within the film, then it is possible that states above the valence band arise from B-P complexes that would act as hole traps. However, it appears that compensation is not fully effective in restoring the luminescence intensity; therefore, extra non-radiative recombination centers must be introduced.

2.9 Surface States on Amorphous Silicon

As discussed in Sections 2.2 and 2.3, the detailed structure and distribution of electronic states in the bulk of GD a-Si:H is only approximately known. This seriously complicates the understanding of surface states. In contrast, the surface of crystalline Si and their respectively reconstructions are relatively well understood. On unoxidized surfaces the surface state density is high ($\geq 10^{14}$ cm^{-2} eV^{-1}) and E_f in non-degenerate crystalline Si is controlled (pinned) by these surface states. Because the number and energy of the bulk defect states is known, the amount of band bending at the crystalline Si-vacuum interface, as measured by photoemission, correlates well with the bulk and surface state densities calculated or measured experimentally using various techniques. The photoemission measures the position of E_f with respect to E_v, or a core level, as a function of distance from the surface. This is accomplished by altering the excitation energy and thereby adjusting the mean escape depth of the emitted photoelectrons. Himpsel *et al.* (1983) reported that when the crystalline Si (111) surface is oxidized the number of surface states is reduced significantly. Intensive research is presently underway in an attempt to understand the nature of defects within ~5 Å of the Si/SiO$_2$ interface, where it has been shown that SiO$_x$ is formed. Here, the oxidation stages of silicon: Si^{3+}; Si^{2+}; Si^+, occur at a ratio of 0.4; 0.3; 0.3 (Hollinger *et al.*, 1984). In GD a-Si:H it might be expected that many surface states would be eliminated by hydrogen passivation, and to some extent, this seems to be the case. Surface state densities ranging from 4×10^9 cm^{-2} eV^{-1} (Hirose *et al.*, 1979) to as high as 1.5×10^{13} cm^{-2} eV^{-1} (Viktorovitch, 1981; Nielsen and Gredin, 1983) have been reported. A large density of surface states, although still much lower than the 1 state/atom found on crystalline Si surfaces, is very significant, as these states are likely to mask the bulk properties, even in good quality a-Si:H. As in the case of the bulk DOS, it is also expected that the surface DOS depends strongly on the preparation conditions.

Early photoemission experiments by von Roedern *et al.* (1979) and Williams *et al.* (1979) concluded that the surface state density of unoxidized as well as oxidized GD a-Si:H was "low" because the positions of E_f, determined spectroscopically near the surface (with respect to a core level or with respect to E_v) were in reasonable agreement with the changes in E_σ, when intrinsic and doped (P or B) materials were compared. Figure 2.90 shows the position of E_f determined spectroscopically with respect to E_v or the Si 2p core level (right-hand scale) for a surface prepared *in situ*. Von Roedern *et al.* (1979) also noted that these relative energy positions were changed very little when the *in situ* prepared surfaces were exposed to oxygen, and took this as an indication of a "low" surface state density at the unoxidized surface.

As discussed in Chapter 1, the major distinguishing feature of an amorphous semiconductor is the presence of a continuum of defect states. As discussed in Section 2.3.2.1 and shown in Figs. 2.27 and 2.28, the bulk DOS for GD a-Si:H yielded a minimum DOS at E_f to be $\sim 10^{17}$ cm^{-3} eV^{-1}, which was probably an overestimate since the field effect analysis employed neglected the surface state contribution. A detailed study by Lachter *et al.* (1982) shows that the field effect response can be dominated by the interface states at the a-Si:H/insulator interface. They were able to obtain large field effect responses when an insulator such as Si_3N_4 was grown *in situ* before the a-Si:H layer was deposited, which presumably minimized the interface state density. The field effect produced a DOS spectra that differed substantially from the DOS derived from zero bias Schottky diode admittance measurements, as shown by Viktorovitch (1981). The latter method enables a direct measurement of the surface state density; it was reported to be 10^{13} eV^{-1} cm^{-2}. It is important to note that sputtered a-Si:H does not show a field effect response unless the interface is carefully prepared (Weisfield *et al.*, 1981). It is therefore expected that surface or interface state densities on a-Si:H with densities between 10^{12} and 10^{13} eV^{-1} occur frequently. Even the diode admittance measurements seem to be sensitive to surface states (and thus yield high bulk DOS values) when the bulk DOS near E_f is below

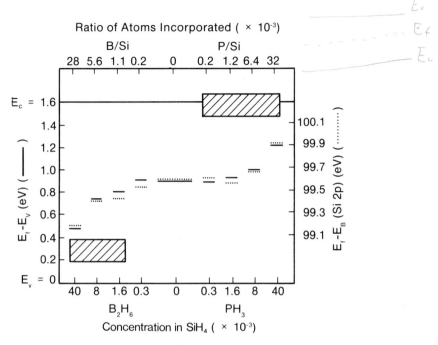

Fig. 2.90. Position of the Fermi level, E_f in the gap of a-Si:H, as determined spectroscopically for a range of P and B doped samples. The hatched regions indicate the energies at which transport takes place, as deduced by combining the conductivity activation energies with E_f (Von Roedern et al., 1979).

10^{16} cm^{-3} eV^{-1}. On the other hand, space charge limited current measurements yield bulk densities of localized states as low as $3 \times 10^{+14}$ eV^{-1} cm^3 (Weisfield, 1983, discussed in Section 2.3.2.3).

To deduce the transport mechanisms in a-Si:H alloys, great reliance is generally placed on the behavior of the dark conductivity, σ_D, as a function of T, as discussed in Section 2.5.1. Plots of $\ln(\sigma_D)$ vs. $10^3/T$ (shown in Figs. 2.49 and 2.51) generally lead to straight lines for good quality material. However, the presence of interface states can and does influence the conductivity measurements, since these are generally performed on thin samples in a coplanar configuration. Interface states could result in space charge layer widths that could be on the order of the thickness of the sample, and hence lead to values of σ_D and E_σ that cannot be ascribed to the bulk alone (Ast and Brodsky, 1980; Solomon and Brodsky, 1980).

The model of a variable conductance due to interface charges is supported by an investigation by Solomon et al. (1978), who studied the samples prepared in a MOSFET (or field effect) configuration (see Section 2.3.2.1).

Starting with an annealed sample, a strong electric field (5×10^4 V/cm) was applied across an insulating substrate held at $T = 150°C$. After a few hours the sample exhibited a larger apparent conductivity and a lower E_σ than in the "annealed" state. It was deduced that the electric field drifted charges toward the a-Si:H surface, which increased the band bending; an accumulation layer was formed. The application of a negative electric field for a few hours decreased the conductivity and increased E_σ, as expected. After a few hours it was noted that E_σ decreased again and, correspondingly, an increase in σ_D occurred. It was conjectured that there was a removal of positive charges out of the surface region, which resulted in a decrease of the band bending. After a long time the accumulation of negative charges was such that the bands were bent upwards (depletion layer), with the result that the measured E_σ now corresponded to that of holes.

Surface effects can also be modulated by adsorbates. Measurements performed by Tanielian (1982), shown in Figs. 2.91a and b, demonstrated conductance changes with time in n- and p-type layers when the samples were exposed to different gases after removal from the growth chamber (a native oxide initially formed at the surface). Upon exposure to H_2O vapor the conductance increased for the n-type material (Fig. 2.91a) and decreased for the p-type (Fig. 2.91b) samples. As seen in Fig. 2.91b, the effect of NH_3 is qualitatively similar to that obtained with H_2O treatment. The effect of H_2O vapor is to increase the conductance of an n-type film while it reduces the conductivity of a p-type film; this indicates that this adsorbate acts as an electron donor on the surface, which is similar to observations made on many crystalline semiconductors (Brattain and Bardeen, 1953; Jantsch, 1965). Consequently, an accumulation layer is formed via partial charge transfer from the H_2O molecule to the a-Si:H material. It was concluded that there are two different reaction rates between the adsorbed H_2O and a-Si:H film. There is a fast process, probably associated with the formation of a donor-like state close to the conduction band, that most likely represents the presence of weakly bound H_2O on the surface of a-Si:H. There is also a slow process, probably associated with the formation of an acceptor-like state close to the middle of the gap; it most likely arises from the more strongly bound compound formed by the reaction of H_2O with the SiO_x surface. In summary, Tanielian (1982) concluded that H_2O, NH_3, CO and $(CH_3)_2O$ act as electron donors, while O_2 and evaporated overlayers of Se act as electron acceptors. N_2, Co_2, C_2H_6 and SF_6 have little or no effect.

In section 2.2 we pointed out that GD a-Si:H films deposited at low T_s possess a columnar structure (Knights and Lujan, 1979) and hence are porous (Fritsche and Tasi, 1979a). Street and Knights (1981) have used these types of films to study the a-Si:H/oxide interface and concluded that the increase in ESR spin density with time is associated with dangling bonds at the surface

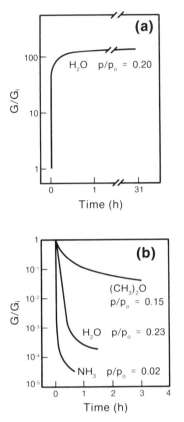

Fig. 2.91. Adsorbate induced changes in the conductance of (a) n-type and (b) p-type samples (Tanielen, 1982).

of columns that are prone to change because of the porosity. The spin density was measured over a period of time for samples that were exposed to ambient air. Samples with columnar structure exhibited an increase in the spin density, but homogeneous samples did not show any change. The absence of a time dependence in the spin density in samples (without columnar morphology) with post deposition oxidation shows that the effect is not due to changes in the bulk properties, whereas the increase in the spin density with time for the porous samples was associated with dangling bonds at the surface of the columns. They deduced that the maximum increase in the spin density corresponded to a bulk density of 5×10^{17} cm^{-3} for a sample with an internal surface area per unit volume of 1.6×10^6 cm^{-1}, leading to a surface state density of 3×10^{11} cm^{-2}. Dangling bonds with similar densities are sometimes observed at the interface of Si and SiO$_2$ (Caplan et al., 1979).

Recent work involving photoemission measurements on a-Si:H has aided in elucidating surface effect phenomena, but as yet these are not conclusive enough to yield the exact densities and distribution of surface states. Miller et al. (1981) investigated the states near the top of the valence band using ultraviolet photoemission spectroscopy (UPS) and did not find any states in excess of 4×10^{13} cm^{-2}. Since this study was performed on unhydrogenated a-Si, it might therefore be expected that a larger surface DOS existed because of the absence of H, which tends to passivate defect states. Since the lower limit of detection is on the order of 10^{13} cm^{-2}, a direct observation of a lower surface state density in a-Si:H may prove to be difficult. An alternative technique is to determine the surface Fermi levels and compare their respective position with the position of E_f as determined from electrical transport measurements. Nielsen and Gredin (1983) measured the surface Fermi level position in n^+, i and p^+ materials and found that the energy difference between E_f and the valence band maximum was -1.25 to -1.30 eV for n^+ and -0.4 eV for the p^+-layers. These results are in reasonable agreement with the earlier results of von Roedern et al. (1979) and in excellent agreement with the open circuit voltage reported for amorphous p-i-n type a-Si:H solar cells, discussed in Chapter 3. This demonstrates that surface states do not substantially effect the position of the bulk E_f at the vacuum surface of p^+ and n^+ material, or at the n^+/i or the p^+/i interfaces. They concluded that the surface state densities of $(2-6)\,10^{12}$ eV^{-1} cm^{-2} were situated above E_f and 8.6×10^{11} eV^{-1} cm^{-2} below E_f for the i-layer vacuum interface.

Figure 2.92 shows the results of XPS (X-ray Photoemission Spectroscopy) measurements on the Fermi level position (Wagner et al., 1983). It should be stressed that the position of E_v was not measured experimentally but derived indirectly by using the conduction band edge ($E_c = 0$) as a reference energy, and by equating $E_v = E_c - E_g$, where E_g is the optical band gap. The bulk positions of E_f was obtained from the dc conductivity data discussed in Section 2.5.1. It was assumed that the $\ln(\sigma_D)$ vs. $10^3/T$ curves with activation energies E_σ (or ΔE), for n-type materials, corresponded to $E_c - E_f$, and for p-type materials the activation energy was $E_f - E_v$, thus assuming extended state transport for both carriers. The surface Fermi level, E_{sf}, was determined spectroscopically and $|E_{sf} - E_c|$ or $|E_{sf} - E_v|$ of the oxidized surface was deduced to be ≈ 0.2 eV for the heavily n- and p-doped samples. In this procedure a margin of uncertainty has to be allowed in the accuracy of the determination of $E_f - E_{sf}$ since statistical shifts of the Fermi level due to temperature (discussed in Section 2.8.1) were ignored (Dohler, 1979). This is important because E_σ for intrinsic material was derived from measurements above room temperature, whereas for the heavily doped samples the data were taken below room temperature. Here the difference of $E_f(100°C) - E_f(-50°C)$ is about 0.1 eV (Gruntz et al., 1981; Ley, 1984).

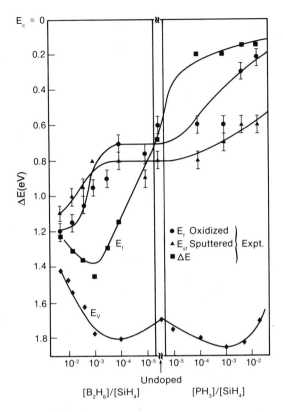

Fig. 2.92. Variation of surface Fermi level, E_{sf}, bulk Fermi level E_f, valence and conduction band edges for E_v and E_c as a function of volume fraction of B_2H_6 and PH_3 in SiH_4 (Wagner et al., 1983).

In addition, the assumption of extended state transport in heavily doped a-Si:H is still under debate (see Sections 2.8.1 and 2.8.2). Nielsen et al. (1983) and von Roedern (1979) have shown that heavy P doping introduces states near the top of the valance band, and it is likely that B doping does the same (discussed in Section 2.8.2). Therefore, the value of $E_f - E_v \approx 0.4$ eV in p$^+$ material observed spectroscopically appears to be a real limit of how far E_f can be shifted towards E_v with doping. Therefore, an uncertainty of ≈ 0.2 eV exists in the value of $|E_f - E_{fs}|$ and hence the values derived from the surface DOS by Wagner et al. (1983), (2.2×10^{12} eV^{-1} cm^{-2} for n-type material, 5×10^{12} eV^{-1} cm^{-2} for p-type material and 1.2×10^{13} cm^{-2}eV^{-1} for sputter cleaned surfaces) must be considered uncertain to within a factor of two. These values are well within the limits derived from the electrical measurements described earlier.

2.10 Hydrogenated Amorphous Silicon Using Techniques Other Than Glow Discharge Decomposition

2.10.1 Chemical Vapor Deposition (CVD)

Amorphous Silicon type films have been produced by many variations of the CVD technique, such as: (a) conventional CVD, with and without post hydrogenation; (b) HOMCVD, which is a hot gas/cold substrate technique; (c) laser induced CVD; (d) photo induced CVD (see, e.g., Pankove, 1984).

Conventional CVD has been attempted with the use of SiH_4 and higher silane gases. Practical deposition rates in a conventional pyrolitic system require that the deposition temperature exceed 500°C (Taniguchi et al., 1978). Although the band gap is a little lower, the spin densities of CVD films made in SiH_4 gas for $T_s > 500°C$ is in excess of 10^{18} cm^{-3} (Hasegawa et al., 1981). Due to the high temperatures ($T_s > 500°C$) required to achieve appreciable deposition rates, the amount of bonded H is small (Booth et al., 1980). This necessitates posthydrogenation in a H_2 plasma in order to develop optoelectronic properties that closely resemble the properties of GD a-SiH films (Sol et al., 1980; Suzuki et al., 1980). However, a high deposition rate at lower T_s can be achieved by the use of higher silanes (Gau et al., 1981). Systematic studies involving parametric variations of T_s, the relative proportions of silane and various polysilanes, the gas phase residence time, and axial positions, concluded that the films produced in this way generally lead to inferior properties (Ellis et al., 1984), with the DOS as measured by the SCLC technique about 4×10^{16} cm^{-3} eV^{-1} (Hagedus et al., 1984), which is considerably higher than that obtained for GD a-Si:H films (see Section 2.3). Solar cells constructed in a pin type configuration using this deposition approach have only produced maximum conversion efficiencies of $\eta \approx 4.0\%$, with the $\mu\tau$ products for holes measured to be about 2×10^{-9} cm^2/V.

In the HOMOCVD approach a hot gas/cold substrate is used (Scott et al., 1981). Here there are two important variables: T_s and the gas temperature, T_g. For $T_g < 650°C$ the deposition rate is found to be activated with an energy of 53.9 K cal/mol. Keeping T_g constant at 625°C, the a-Si:H growth rate is essentially constant for substrate temperatures up to 300°C. It was found that for $T_s = 225°C$ the concentration of H is about 5 at.% and the infra-red spectrum shows a pronounced peak at 2000 cm^{-1}. For samples deposited at a lower T_s (= 110°C) the concentration of H increases to ~21% and the infra-red spectrum now shows a peak at 2080 cm^{-1}. [These results are similar to GD a-Si:H films (see Section 2.2) where the effect of increasing T_s was shown to alter the structure of the film from a dihydride to a monohydride configuration.] It was hypothesized that SiH_2 is the a-Si:H precursor for growth, whose formation is the rate determining step. To account for the large reduction in the H/Si ratio within the film, SiH_2 is envisaged to insert itself into suface Si–H bonds. This serves to sweep out hydrogen within the advancing film–gas interface by creating surface silyl (SiH_3) groups, as shown in Fig. 2.93. Each silyl Si–H bond can then undergo SiH_2 insertion, as shown by reaction (a), and/or hydrogen can be eliminated, as shown by reaction (b). The measured spin densities have indicated a value of 2×10^{15} cm^{-3} for $T_s \sim 255°C$ (Scott *et al.*, 1982), with little or no Staebler-Wronski effect (see Section 3.1.1). The films have shown conductivity changes under AM-1 illumination of about 3 orders of magnitude, and have been readily doped (Meyerson *et al.*, 1983).

Another variant of the CVD technique is the photo-CVD. The deposition is caused by a H_g-photosensitized reaction of the SiH_4 gas:

$$H_{go}(^1S_0) + h\nu(253.7 \text{ nm}) \rightarrow H_g^*(^3P_1), \tag{2.94}$$

$$H_g^*(^3P_1) + SiH_4 \rightarrow Si + 2H_2 + H_{go}(^1S_0), \tag{2.95}$$

where H_{go} and H_g^* are the ground and excited electronic states (Saitch *et al.*, 1983). It is expected that this approach should lead to minimal damage of the films, since the high-energy radiation or charged particles (which may be encountered in the GD process) are absent in the plasma. For a substrate temperature of 250°C, $\sigma_D \sim 10^{-7} (\Omega \text{ cm})^{-1}$ and $\sigma_L(\text{AM-1}) \sim 10^{-4} (\Omega \text{ cm})^{-1}$. *p*-type films with and without the use of C have been demonstrated (Inoue *et al.*, 1984), as is the case for *n*-type doping (Inoue *et al.*, 1983).

2.10.2 Reactive Sputtering Techniques

Sputtering is a thin film deposition technique in which material is ejected from a target by ion bombardment from a plasma. The structure and physical properties of the deposited film depend not only on the mechanism of thin film nucleation and growth, but also on the nature of the plasma and its complex reactions with the growing layer. Various types of sputtering techniques have been employed, such as rf, dc and planar magnetron. As expected, the film properties primarily depend upon the: partial pressure and flow rate of the sputtering gas; cathode-anode distance; substrate biasing; target material; deposition temperature (see, e.g., Thompson, 1984).

Amorphous silicon films are generally deposited by the rf sputtering of a polycrystalline Si target in an Ar atmosphere, typically at a presure of 5×10^{-3} torr. Hydrogen incorporation into the films is achieved by adding H_2 gas to the sputtering gas. (The hydrogen partial pressure normally used is in the range of 10^{-4} to 10^{-3} torr.) In this way deposition rates of 1–5 Å s^{-1} can be achieved. Matsuda *et al.* (1982) have shown that the SiH_4 emission

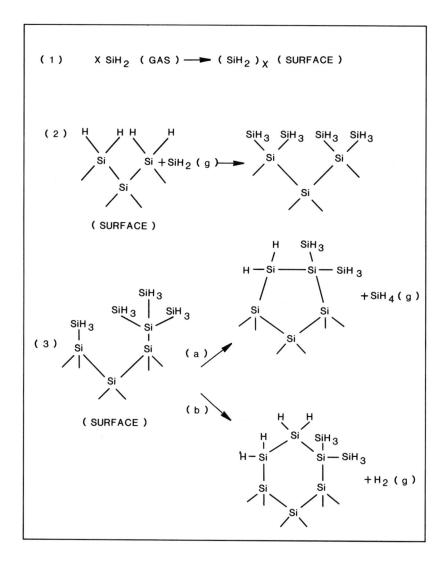

Fig. 2.93. Postulated sequence of surface reaction to account for the large H–Si ratio change upon deposition from SiH_2 (1) Polymerization; (2) SiH_2 insertion into surface Si–H bonds causing a hydride shift into the inserting species; (3) disproportionation reactions. Pathway (b) should increase in importance relative to (a) as substrate temperature increases (Scott *et al.*, 1981).

line from the OES technique is clearly observable in the plasma, as are the H_2, H and Ar lines. They speculate that the hydrogen gas introduced into the sputtering chamber is excited and decomposed into H free radicals that react with the sputtered Si atoms without the plasma close to the Si target, producing SiH_2 molecules. After suffering several collisions with other species, the SiH_2 molecules reach the surface of the substrate and are solidified into a-Si:H through surface reactions.

As discussed in Section 2.5.5, the sputtering deposition technique has proven to be a useful tool in the investigation of changes in the atomic arrangement and electronic structure of the material. This is best illustrated in Fig. 2.51, where $\ln \sigma$ vs. $10^3/T$ is shown for samples whose H concentration varied from 0 at.% H to 20 at.% H (Anderson and Paul, 1981). With increasing inclusion of H, the conductivity at room temperature showed a large decrease and an activated behavior (curve f); this is consistent with a decrease in the mid-gap DOS to about $10^{16} \text{cm}^{-3} \text{eV}^{-1}$ (Weisfield et al., 1981). Concomitant with this is an increase of the optical band gap (Moustakas, 1979) produced by alloying. However, as shown in Fig. 2.94, for H partial pressures in excess of about 1×10^{-3} torr there is no discernable change in the optical absorption. This is consistent with the photoemission studies (von Roedern et al., 1977) which show that the changes in the a-Si:H films induced by H seem to saturate for H concentrations in excess of 10 at.%. Specifically, it was found that for films prepared at $T_s = 350°C$ the hydrogen induces two peaks in the valence band, occurring at 5.3 and 7.5 eV below the top of the band. This structure is identified as arising from silicon monohydride; the spectra show a recession of the top of the valance band by 0.4 eV for 10% H in the sputtering gap. Beyond this value, no further change occurs.

The photoluminescence for heavily hydrogenated sputtered films exhibits a peak at about 1.2 eV (Brodsky et al., 1977); this peak can be made to shift to higher energies with increasing H concentration. The shift of the luminescence spectrum as a function of the partial pressure during the deposition process has been attributed to the removal of dangling bonds by H or an enlargement of the optical band gap (Moustakas and Paul, 1977). The low DOS in these types of films is reflected in a high photoconductivity (Moustakas et al., 1977). Successful n-type and p-type doping has been achieved with the introduction of PH_3 and B_2H_6 respectively in the reactive gas, as shown in Fig. 2.95 (Anderson, 1978). Schottky barrier diodes with near ideal behavior ($n = 1.0$, $J_0 = 7 \times 10^{-11}$ Å cm^{-2}), and p-i-n junctions have also been fabricated with the following dark characteristics: $n = 1.57$; $J_0 = 8 \times 10^{-12}$ A cm^{-2} (Moustakas and Friedman, 1982). Although the films show opto-electronic behavior that closely resemble those of GD a-Si:H, nevertheless some significant differences seem to exist. For example, the

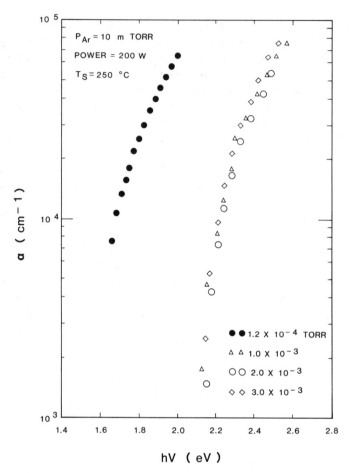

Fig. 2.94. Absorption coefficient vs. photon energy for a number of a-Si films deposited at different partial pressures of hydrogen (Moustakas, 1979).

minority carrier diffusion lengths seem to be smaller, i.e., 3000 Å (Moustakas et al., 1981) and the electron carriers exhibit dispersive transport (Tiedje et al., 1980; Tiedje et al., 1981) with a lower mobility of 0.05 cm^2 s^{-1} V^{-1}. It has been pointed out by von Roedern and Madan (1985) that the sputtered material may be more disordered than the GD variety because of the violence involved in the sputtering process. The increase in the disorder in the sputtered material manifests itself as a shallower Urbach edge, since E_0 (see, Section 2.4) for sputtered material has been measured to be 70–80 meV, whereas for GD a-Si:H, E_0 is found to be smaller, in the 50–60 meV range. An indication of this is also given by the work of Longeaud et al. (1984),

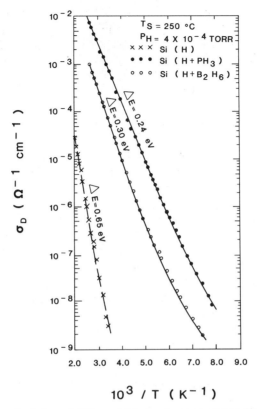

Fig. 2.95. Dark conductivity vs. $1/T$ for an intrinsic n-type and p-type a-Si samples, deposited by sputtering as described in the text (Moustakas, 1979).

who find that although the $\mu\tau$ for electrons increases with H inclusion, μ remains dispersive with low values of 10^{-3} cm^2 s^{-1} V^{-1}. They conclude that H decreases the DOS in the mid-gap region but has little effect on the conduction band tail. Because of the dispersive nature of the electron mobility it is then possible that the band tail is indeed wide and, correspondingly, the valance band tail may also be wider. Because of this, the diffussion length is small; hence, relatively poor performance, especially in the fill factor, results for solar cells.

2.10.3 Miscellaneous Techniques

In the ion plating technique, a metal or semiconductor is evaporated in and through a glow discharge in a hydrogen plasma, The method is to be distinguished from ion implantation by the use of ion energies typical of the glow discharge process (500–2000 eV). The energy level of the evaporated material (typically 0.1 eV) is increased radically through interaction with the glow discharge plasma. By evaporating Si through a H plasma, films containing up to 25% H have been reported (Cocks *et al.*, 1980). Band gaps similar to those for GD a-Si:H were obtained and the infra-red spectra showed peaks at 2000 and 2090 cm^{-1} (Cocks *et al.*, 1980; Ceaser *et al.*, 1981), indicative of a defective film (see Section 2.2).

There have been many attempts to evaporate a-Si, generally with some form of H present. Successful hydrogenation has been achieved using a theta pinch plasma source (Tong *et al.*, 1981) where it was found that the films were photoconductive [$\sim 10^{-5}$ (Ω cm)$^{-1}$] and heat resistant. However the dark conductivity was high, $\sim 10^{-7}$ (Ω cm)$^{-1}$. a-Si:H films have also been fabricated by the e-beam evaporation of Si in the presence of atomic hydrogen (Dellafera *et al.*, 1981) as well as in the presence of molecular hydrogen (Dellafera *et al.*, 1982). High rates of deposition (≈ 0.5 μm/min) have been achieved using an activated evaporation technique (Anderson and Biswas, 1985) with the demonstration of *n*- and *p*-type doping. Variations on the sputtering approach (e.g., Singh *et al.*, 1984) or the glow discharge approach (e.g., Okoneiwski *et al.*, 1985) have been attempted. However, none of these techniques have as yet produced a-Si:H films whose properties approach those obtained by the GD technique.

2.11 Some Properties of Halogenated Amorphous Silicon Alloys

2.11.1 Fluorinated Amorphous Silicon

As discussed in Section 2.5.5, the low DOS obtained in GD a-Si:H is primarily due to the inclusion of H. Since F is more electronegative than H, then low DOS materials, in principle, could be fabricated with the use of halogens. The first attempt to produce fluorinated amorphous Silicon-based materials utilized the rf glow discharge of SiF$_4$ and H$_2$ gas mixtures (Madan *et al.*, 1979; Ovshinsky and Madan, 1979; Madan, 1984); this led to materials with a low DOS, high photoconductivity and low dark conductivity (see Fig. 2.54) that did not exhibit the Staebler–Wronski effect (see Section 3.1.6). Further, doping with phosphorous led to σ_D in excess of 10 (Ω cm)$^{-1}$, as shown in Fig. 2.81 (see Section 2.8).

The field effect data (see Section 2.3.2.1) in Fig. 2.96 shows $g(E)$ spectra for GD a-Si:F:H films deposited at $T_s = 250°$C for different gas ratios $r =$ SiF$_4$/H$_2$. For the corresponding films, Fig. 2.54 shows ln σ vs. $10^3/T$ data; Fig. 2.97 shows the F concentration within the films as determined by

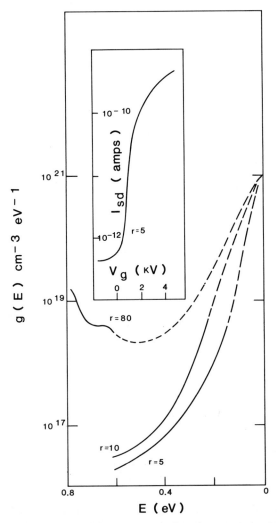

Fig. 2.96. The localized density of states $g(E)$ [$cm^{-3}\,eV^{-1}$] plotted for different gas ratios $r(=SiF_4/H_2)$. The inset shows a typical $i_{sd}(V_g)$ curve and the arrow indicates the assumed flat-band position (Madan et al., 1979).

the SIMS technique (see Section 2.2) and plotted as a function of r; note that as r is decreased, the F concentration increases.

The change in conduction mechanism (Fig. 2.54) with r becomes understandable in view of the field effect data shown in Fig. 2.96. As r decreases from 80 to 10, the predominant transport mechanism changes from variable range hopping type conduction to a well defined activated process (see

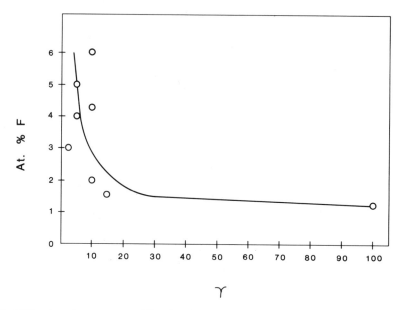

Fig. 2.97. Atomic percentage of fluorine determined by Auger analysis, plotted as function of r ($= SiF_4/H_2$) (Madan et al., 1979).

Sections 2.5.1 and 2.5.3). From the field effect response it was noted that these samples exhibited n-type conduction; for $r = 10$ the DOS at E_f was several orders of magnitude lower than that for the case of films deposited with $r = 80$. Since the DOS is high for large values of r, then the dominant conduction process is expected to be via localized states in the vicinity of E_f. As r is decreased, as shown in Fig. 2.96, the DOS decreased sufficiently for the conduction to change to an activated process. The change in the DOS with r is attributed to the inclusion of F as well as H. As shown in Fig. 2.97, the F concentration increases with decreasing r, which is correlated with a decrease in the DOS, as seen in Fig. 2.96. To attribute the reduction of DOS entirely to F is not appropriate, since these films also possess H concentrations of up to 4%, which is much less than the H concentration found in GD a-Si:H films (see Section 2.1.2). Nevertheless, F can act as a bond terminator; we discuss this below.

For low DOS films it was found that σ_L (or σ_{ph}) under AM-1 excitation exceeded 10^{-5} (Ω cm)$^{-1}$, whereas, as can be seen from Fig. 2.54, $\sigma_D \leq 10^{-10}$ (Ω cm)$^{-1}$ for the corresponding films. The films clearly show a highly photoconductive response. As shown in Fig. 2.81, one of the more remarkable features was the doping efficiency, especially with P or As. It was found that the conductivity of GD a-Si:F:H (P) films was three orders of magnitude better than that obtained using the GD of SiH_4 and PH_3 gas mixtures, and was due to the presence of microcrystallization (μc). [It should be emphasized that undoped a-Si : F : H alloys were found to be amorphous (Tsu *et al.*, 1981).] Similar high conductivities in P doped a-Si : H alloys have since been reproduced using dilute SiH_4/H_2 gas mixtures with high decomposition powers; $\sigma > 1$ $(\Omega\,\text{cm})^{-1}$ can also be produced from B doped a-Si : H using highly diluted SiH_4/H_2 gas mixtures to yield $\mu c - p^+$ layers (Nakamura *et al.*, 1984).

Attempts to unravel the role of F from that of H have been made via the use of reactive sputtering techniques. In Section 2.10.2 it was shown that by sputtering Si in a reactive H environment, good quality a-Si : H could be produced. These results showed that upon hydrogenation the dark conductivity at room temperature could be reduced by orders of magnitude and that $\ln \sigma$ vs. $10^3/T$ could be made to exhibit an activated process (see Fig. 2.51). It was suggested that hydrogenation caused a decrease in the DOS in the gap region as well as at the edge of the valence and conduction bands. In a study of sputtered a-Si : F fabricated from Ar–Si : F_4 gas mixtures, Gruntz *et al.* (1981) found experimental evidence that the top of the valence band recedes with the incorporation of F in the same way that H affects the a-Si : H alloy and confirmed the theoretical results of Ching *et al.* (1979).

In Fig. 2.98 a plot of $\ln \sigma$ vs. $10^3/T$ is given for sputtered a-Si : F and a-Si : D (deuterated) films; the affect of annealing is also displayed (Matsumura *et al.*, 1980). The dark conductivity curve for a-Si : F yield an activated semiconducting behavior, indicative of a low DOS in the gap. Upon annealing, σ_D does not change for the a-Si : F film, which is not the case for a-Si : D films. For the latter, when the annealing temperature exceeded 350°C, the effusion of Deuterium leads to a high DOS, as indicated by the hopping type conductivity shown in Fig. 2.98. This is in agreement with the work of Fritzsche *et al.* (1979) on hydrogen effusion, as discussed in Section 2.2. (From Rutherford backscattering experiments, the F concentration was determined to be about 20%). Further, they found that the sputtered a-Si : F films were only slightly less photoconductive $[10^{-8} - 10^{-7}\,(\Omega\,\text{cm})^{-1}]$ than the sputtered a-Si : D films under an illumination intensity of 50 mW/cm². For the case of sputtered a-Si : F the addition of 4500 ppm PF_5 to the reactive gas changes σ_D from 10^{-10} to 10^{-4} $(\Omega\,\text{cm})^{-1}$; this result is remarkably similar to what has been achieved for the sputtered Si : H alloy (see Section 2.10.2), with the added advantage that the fluorinated films were found to be heat resistant to much higher temperatures. Although *p*- and *n*-type doping of sputtered a-Si : F has been achieved by the addition of B and P atoms, the photoconductivity of a-Si : F is considerably lower than that of GD a-Si : H or GD a-Si : F : H films. The reason for this may be related to the μc of the sputtered a-Si ; F films. Using TEM and Rutherford back scattering

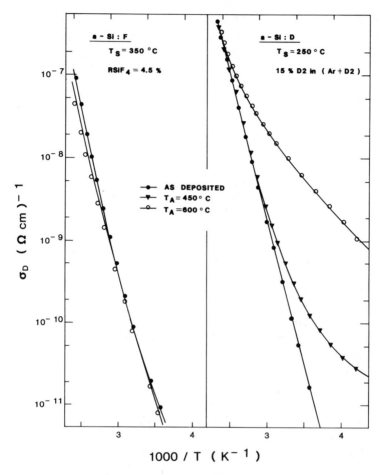

Fig. 2.98. σ_D vs. $10^3/T$ plots for a-Si:F and a-Si:D films (Matsumura et al., 1980).

techniques it was found that: (1) a-Si:F consisted of small grains of a-Si networks; (2) there was a considerable inclusion of SiF_4 as well as SiF_2 molecules; (3) SiF_4 molecules were localized at the grain boundaries (Matsumura et al., 1981). It was suggested that the enlargement of the grain size is the key factor controlling the properties of the a-Si:F films.

The GD a-Si:F:H alloys show a rich infra-red spectra. Fang et al. (1980) have identified the following F related ir peaks from a study of sputtered a-Si:F:H films fabricated from Ar, H_2 and SiF_4 gas mixtures: 1010 (SiF_4, stretching): 930 cm^{-1} (Si–F_2, Si–F_3, stretching): 828 cm^{-1} (Si–F, stretching): 510 cm^{-1} (Si TO mode induced by F): 380 cm^{-1} (SiF_4, band bending), 300 cm^{-1} (Si–F, Si–F_2, wagging).

SOME PROPERTIES OF HALOGENATED AMORPHOUS SILICON ALLOYS

The ESR spin density for a-Si:H and a-Si:F:H films are shown in Fig. 2.99 (Shimizu et al., 1982). They conjecture that the spin density is determined by the fact that H (and not F) terminates the dangling bonds. From their NMR data they conclude that it is the dispersed SiH units that relax the strained structure, leading to a reduction in the dangling bond density.

Using Raman scattering, Tsu et al. (1981) have investigated sputtered Si, GD a-Si:H and GD a-Si:F:H alloys, and observe that the Raman line width is 110, 80 and 70 cm^{-1} respectively. They concluded that the decrease in the line width for GD a-Si:F:H, in comparison with the GD a-Si:H alloy, is

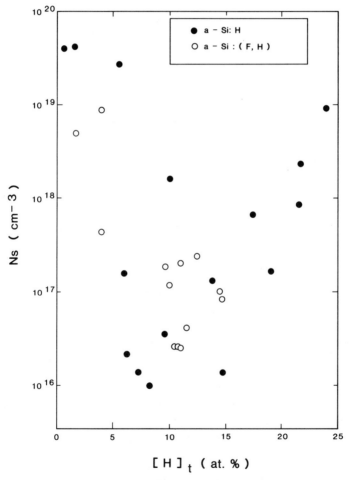

Fig. 2.99. ESR spin density, N_s, vs. the total H content, $[H]_t$, for various a-Si:H (closed circles) and a-Si:(F, H) (open circles) films (Shimizu et al., 1982).

due to a decrease in the internal strain of the film, and attributed this change to the role of F. Further, by using Rayleigh scattering techniques and comparing the fluctuation of the dielectric constant of GD a-Si : H with GD a-Si : F : H, Tsu et al. (1981) note a decrease in this constant for the fluorinated case. They attributed this to the large electronegativity of F and the ionic nature of the Si=F bond. They concluded that F occupies active sites.

Chemical Vapor Deposition (CVD) of SiF_2 gas, resulting in a-Si alloys, has been studied by Janai et al. (1981). They used a transport reaction such as

$$2SiF_2 \underset{1200°C}{\overset{600°C}{\rightleftarrows}} Si(solid) + SiF_4, \qquad (2.96)$$

Films fabricated in this way exhibited a low F concentration of 1.5% with a spin density of 3×10^{19} cm^{-3}. They found that the ESR signal remains constant with increasing deposition temperature, whereas the F content in the film decreases. They concluded that the F in CVD a-Si : F films does not remove dangling bonds. It should be recognized that a-Si : H films produced from the CVD of SiH_4 gas also has similar problems (see Section 2.10.1), since as-deposited films reveal H concentrations of only about 0.2% and the quality of the film is only improved by post-hydrogenation in a plasma. In contrast, films of a-Si : F produced from thermal CVD in SiF_2 gas and atomic H (Matsumura et al., 1985) yielded spin densities of 6.10^{15} cm^{-1} and exhibited high photosensitivity.

The exact role of F in a-Si : F or a-Si : F : H type alloys is as yet unclear. Delal et al. (1981) suggests that a-Si : F : H films grown from GD in SiF_4 and H_2 gas mixtures are subjected to strong reactive ion etching, with the net growth being determined by a balance between deposition and etching rates. They suggest that reactive ion etching is the major distinguishing feature between the growth of GD a-Si : F : H alloys, in comparison with GD a-Si : H alloys. Their model implies that: (a) the growth rate depends upon electrical bias; (b) under certain growth conditions, the films may be subjected to strong ion bombardment, thus promoting μc; (c) if F is to provide reactive ions, then it will still be necessary to satisfy dangling bonds; (d) it is possible that the weakly bonded Si atoms are etched and hence produces fewer dangling bonds for H to satisfy; (e) it is possible that F could also etch the weakest Si–Si bond and hence promote an increase in the band gap. It is suggested that high conductivity n-type doped films (n^+ layers) can only be produced when the power level on the rf target is very high ($\sim$ several W/cm^2). The Auger data indicated that there was no evidence of F incorporation within the film; the F concentration was less than 1%. This is in contrast to the work of Madan et al. (1979) who found that high conductivity n-type doped films could be obtained with low power levels (typically ~ 0.1 W/cm^2) with the F content in the films in the range 2 to 4% (see Fig. 2.97). It is also suggested by some authors that the F is included in the films via the inclusion of SiF$_4$ (Fang *et al.*, 1980) and hence, under these conditions, it is not surprising that the F remains inactive in tying up dangling bonds. It is, therefore, possible that the negligible F content observed in the work of Delal *et al.* (1981) could be via the inclusion of SiF$_4$ rather than by active incorporation of F.

An attempt to explain some of the above divergent results, at least for the rf glow discharge in gas mixtures of SiF$_4$/H$_2$ and SiF$_4$/SiH$_4$/Ar, has been made by Potts *et al.* (1984). They emphasized the importance of a Paschen-like curve for characterizing the glow discharge reactor for the particular gas being used (see Section 2.1). The curves for SiF$_4$ and H$_2$ exhibited a minima at 0.2 and 0.7 torr respectively. Since the curve for conditions using SiF$_4$/H$_2$ = 8/1 was found to be quite similar to SiF$_4$, it was then quite likely that for $p < 0.4$ torr, the only charged fragments present in the plasma are those that result from the dissociation and ionization of SiF$_4$ gas alone. For $p > 0.4$ torr they reason that the plasma chemistry should change significantly as the reactive H species become available. The influence of these changes becomes clear when they plot infrared transmission spectra for films grown at different pressures, and note that the intensity of the related modes (830, 940 and 1010 cm^{-1}) decreases as the pressure increases. (The H$_2$ concentration of all the films investigated, in this case, was measured to be in the range 7-10%). However, no results regarding the electronic properties of the films were reported). Their conclusion is that the amount of F concentration, and presumably its activity in the film, is controlled by the plasma conditions. This can perhaps explain some of the divergent results which have been reported.

In an unrelated work involved with the interfacial properties of SiO$_2$/Si for the fabrication of crystalline MOSFET devices (see Section 2.3.2.1 and 3.2), Ho *et al.* (1981) have reduced the interfacial state density to less than 10^{11} states/eV-cm^2 near midgap by annealing in a H$_2$ ambient at 450°C. However, there still remained an interfacial state density of 10^{12} states/eV-cm^2 near the conduction and valence band edges. They reasoned that by replacing hydrogen with a halogen such as Cl or F, and introducing this into the oxide to form stable Si-Cl or Si-F bonds at the interface, then additional interfacial states could be removed. [There is also support for this notion from the theoretical work of Sukurai *et al.* (1980).] Using Cl, the interfacial state densities were reduced to lower than 10^{10} states/eV-cm^2 near midgap. Further, the magnitude of the mobility of the MOSFET treated with Cl was 225 cm^2 s^{-1} V^{-1}, while that of the MOSFET prepared with a normal thermally grown oxide was 192 cm^2 s^{-1} V^{-1}.

In conclusion, the precise role of F has yet to be determined, as is in fact the case for H. Nevertheless, excellent quality GD a-Si:F:H films have been demonstrated and present investigations using other gases and mixtures such as SiH_2F_2, SiF_4/SiH_4, Si_2F_6, etc. (Koinuma et al., 1985; Nakayama et al., 1985) may uncover the role of F. Further, as discussed in Section 2.12, encouraging results seem to be emerging when F is employed in the synthesis of new types of alloys.

2.11.2 Chlorinated Amorphous Silicon Films

Aside from the use of H or F as a dangling bond terminator, attempts have also been made using Cl. Several groups have studied films of the type a-Si : H : Cl using the rf GD of $SiCl_4$ and H_2 gas mixtures (see, e.g., Plattner et al., 1979; Iqbal et al., 1982; Wu et al., 1983; Augelli et al., 1982). The infra-red results have indicated that the Cl has been successfully adopted into the network (Plattner et al., 1979) and that the DOS measured from the field effect technique (see Section 2.3.2.1) is on the order of $10^{16}\,cm^{-3}\,eV^{-1}$ (Augelli et al., 1985). It has been reported that as more Cl is included, the subband gap absorption (see Section 2.4) is reduced (Mostefavi et al., 1985), indicative of a lower DOS. However, it was found that when $SiCl_4$ is added to the SiH_4 and H_2 gas mixtures, the sub-band gap absorption increases, which could be interpreted as an indication of an increase in the DOS. It has been speculated that because of the larger size of Cl in comparison with H and F, the satiation of the dangling bonds could be impeded.

The photoluminescence results (Fortunato et al., 1981) shown in Fig. 2.100 indicate that for films prepared with $r[SiCl_4/(SiCl_4 + H_2)] \geq 0.1$, only one peak is observed, similar in shape and intensity to those found for GD a-Si : H type films (see Section 2.7). As shown in the inset, this peak (a) corresponds to a transition between the conduction and valence band tails. As is to be expected, when r is altered the PL peak and intensity shift to lower energies, which is indicative of a higher DOS than in GD a-Si : H or a-Si : H : F.

The $\ln \sigma$ vs. $10^3/T$ plots for these films exhibit three regions of interest (Augelli et al., 1982). Between 90 K and 250 K the conduction behavior is interpreted on the basis of hopping, with an activation energy in the range of 0.10 to 0.16 eV. Between 250 K and 400 K, E_σ is characterized by an energy range between 0.23 to 0.47 eV. For $T > 500\,K$, E_σ is greater than 1 eV. These data were interpreted (Bruno et al., 1983) on the basis of a two phase model (crystalline islands interspersed in an amorphous phase), since it has been noted from the Raman data (Iqbal et al., 1982) that the films exhibit a peak at $519\,cm^{-1}$, indicative of μc. It has also been found that the grain size is strongly influenced by the deposition conditions. For example, in films prepared under conditions of high power and high deposition pressures the grain size was approximately 40–70 Å, which changed to 80–200 Å when the films were deposited in the low power and low deposition pressure regime.

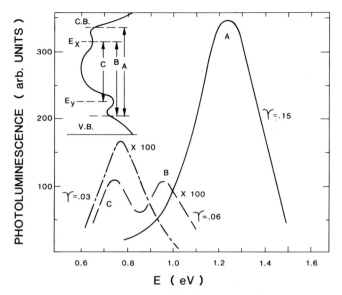

Fig. 2.100. Photoluminescence vs. photon energy for three samples grown at $T_s = 320°C$, rf power = 75 W and different values of r (Fortunato *et al.*, 1981).

This is perhaps understandable, since it has been argued (Veprek *et al.*, 1981) that the μc is promoted whenever the deposition conditions are on the borderline of the deposition and etchant regimes.

2.12 Narrow and Wide Band Gap Alloys

The band gap of an amorphous semiconductor can generally be altered at will. For example, alloying a-Si with Ge or Sn leads to a reduction in the band gap, whereas the addition of C or N can increase the band gap. Recent research activity has concentrated primarily on a-SiC:H and a-SiGe:H alloys, although other materials such as a-SiN:H and a-SiSn:H have gained some attention as well. To date, all such hydrogenated binary alloys exhibit electronically inferior properties compared to GD a-Si:H alloys. In the following, we review some of the experimental work on alloys.

2.12.1 Amorphous Silicon-Germanium Alloys

Using various deposition techniques such as reactive sputtering (Weisz *et al.*, 1984) and glow discharge (Chevalier *et al.*, 1977), it has been found that the addition of Ge to Si generally leads to a deterioration of the photoelectronic properties (Dong *et al.*, 1981; McKenzie *et al.*, 1985). Various suggestions as to why this occurs have been put forward, such as: (a) preferential bonding of H to Si rather than to Ge (Paul *et al.*, 1981); (b) the existence of weakly bonded hydrogen in Ge rich films, leading to heterogeneties.

It is almost a universal result that when Ge is added to Si there is a decrease in the optical band gap (Chevalier *et al.*, 1977); the ESR spin density increases (Paul *et al.*, 1981) and there is an increase in the sub-bandgap absorption (McKenzie *et al.*, 1985) as shown in Fig. 2.101. Note that the slope (E_0) of the exponential edge decreases with increasing Ge content; this is interpreted as an increase in the width of the valence band tail (see Section 2.4).

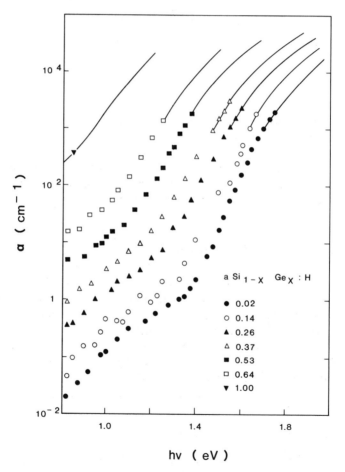

Fig. 2.101. Absorption coefficient, α, as a function of energy for different alloy compositions of a-Si$_{1-x}$Ge$_x$:H (McKenzie *et al.*, 1985).

McKenzie et al. (1985) have attempted to optimize the properties of GD a-$Si_{0.5}Ge_{0.5}$: H type alloys using SiH_4 and GeH_4 gas mixtures. They found that the opto-electronic properties are optimized at $T_s = 300°C$, which yields $\sigma_{ph} \cong 10^{-8}$ $(\Omega\,cm)^{-1}$ for an incident photon flux of $N_0 = 10^{15}\,cm^{-2}\,s^{-1}$ at a wavelength of $\lambda = 600$ nm. As T_s is increased from 220°C to 380°C the concentration of H decreases from 15% to 4% and the infra-red absorption peak centered at $2000\,cm^{-1}$ increases in importance relative to the peak at $2090\,cm^{-1}$. It is concluded that raising T_s leads to a sharpening of the band tails and hence to a decrease in the DOS at the edges of the valence and conduction band tails. This is interpreted as a reduction in the topological disorder. On the other hand, the role of H is: (a) to compensate the dangling bonds, resulting in a lower mid-gap DOS; (b) to reduce the overall average coordination number and hence facilitate a smaller topological disorder. These contrasting effects can produce an optimum a-Si : Ge type alloy.

Rudder et al. (1984) report that there is an improvement in the photoconductivity for a-$Si_{0.47}Ge_{0.53}$: H alloys ($E_g = 1.4$ eV) using a dual magnetron sputtering approach. They report that as the Ge concentration is increased there is a significant increase in the Ge–H stretching vibrational mode ($1980\,cm^{-1}$) and conclude that the separated plasmas at the Si and Ge targets, together with a greater hydrogen incorporation, results in a more effective bonding of H to both Si and Ge. Note that for this type of material $\sigma_L \cong 10^{-6}$ $(\Omega\,cm)^{-1}$ ($N_0 = 10^{16}\,cm^2\,s^{-1}\,V^{-1}$ for $\lambda = 6329$ Å), $\sigma_D \cong 10^{-8}$ $(\Omega\,cm)^{-1}$ and that σ_L is much lower than that obtained in GD a-Si : H which is a wider band gap material with a lower absorption at $\lambda = 6328$ Å. The magnetron sputtering approach has also been utilized to investigate the whole range of a-$Si_{1-x}Ge_x$ alloys (Saito et al., 1984). The band gap decreases with increasing Ge concentration, the deposition rate increases and, for $x > 0.6$, the hydrogen concentration in the films is observed to decrease. In the range $0 < x < 0.1$ σ_D decreases and for $x < 0.1$, it increases monotonically. For $x < 0.6$ ln σ_D vs. $10^3/T$ data generally show an activated behavior, but for $x > 0.6$ hopping conduction dominates. This is attributed to the decrease in H concentration, which implies that there is a large number of dangling bonds. The minimum observed in the σ_D vs. x plot is attributed to a reorganization of the tetrahedrally bonded structure and a relaxation of the long-range order. (It is interesting to note that similar minimum has also been found (see Section 2.12.2) for the a-Si : Sn case, where a n- to p-type transition was given as the reason.) For $x > 0.3$ the photoconductivity becomes insignificant and seems to be related to the dominance of the $2090\,cm^{-1}$ ir peak associated with the SiH_2 vibrational centers, which in turn are a source of recombination centers.

The fluorinated approach, which has worked successfully in the a-Si case (see Section 2.11.1), has also been attempted for narrow band gap alloys using gas mixtures of SiF_4, GeF_4 and H_2 (Oda et al., 1984). With a final composition of $Si_{0.7}Ge_{0.3}$, $E_g = 1.4$ eV, $\sigma_L = 9.10^{-5}$ $(\Omega\,cm)^{-1}$ under AM-1 illumination and a low ESR spin density (N_s) of 7.10^{15} cm^{-3}, are obtained. Although the F concentration of the films was less than 1%, the improved properties of the films are attributed to the different growth kinetics involved and may be due to the chemical etching action by fluorine. Despite a low N_s, the authors report that the PDS technique (see Section 2.4) indicates a significant sub-bandgap absorption, indicative of a large number of defect states.

As stated before, the addition of Ge leads to an increase in the DOS, as shown in Fig. 2.102, where $g(E_f)$ is plotted against E_{04} for sputtered a-Si:Ge (Weisfield, 1983). (Here E_{04} is defined as the energy at which the absorption coefficient $\alpha = 10^4$ cm^{-1}.) It is interesting to note that the alloying may introduce a defect reducing feature, since a-Si:Ge alloys seems to exhibit a lower $g(E_f)$ value than GD a-Si:H films of the same band gap.

The dominant defect state for these types of alloys is considered to be the dangling bond, either the neutral T^0 (or D^0) or the doubly occupied T^- (or D^-) level. In Fig. 2.103 we reproduce the band diagram as envisaged by McKenzie et al. (1985). Here E_v is increased and E_c is decreased at the same rate with increasing x (addition of Ge). Further, the correlation energy $E(T^-) - E(T^0)$ (see Section 2.3) was assumed to remain unchanged at 0.35 eV for Si and 0.1 eV for Ge. It is speculated that as x is increased and as E_f approaches the valence band, more dangling bond states become positively charged, with the consequence that they act as efficient recombination centers for electrons, which explains the decrease in the photoconductivity. Further increase of dangling bond density with x will simply accentuate this effect.

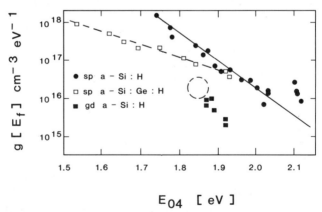

Fig. 2.102. Density of states at the Fermi level, $g(E_f)$, plotted as a function of the "band gap" E_{04} for sputtered (●) and GD a-Si:H materials (■); also included are the results of sputtered a-SiGe:H material (□). The circle designates the expected $g(E_f)$ values for CVD a-Si:H material (Weisfield, 1983).

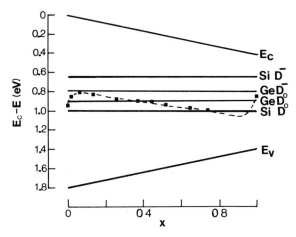

Fig. 2.103. The variation of E_v, E_c and the $E(D^0)$ and $E(D^-)$ for Ge and Si dangling bonds as x is varied from 0 to 1. The dashed line and points indicate the variation of E_f with x deduced from the conductivity activation energy $(E_c - E_f)_0$ (McKenzie et al., 1985).

2.12.2 Amorphous Silicon-Tin Alloys

Attempts to decrease the band gap have been made by the addition of Sn to a-Si based alloys using sputtering (Verie et al., 1981) and GD (Mahan et al., 1984; Nakamura et al., 1982) deposition techniques. A detailed study has been employed for films produced by the GD technique using $SiH_4 + H_2$ mixed with $SnCl_4$ or $Sn(CH_3)_4$ gas sources (Mahan et al., 1984). Figure 2.104a and b show σ_D and E_σ (or ΔE) as a function of E_g, which in turn is related to the Sn concentration. The activated behavior of σ_D changes to hopping for alloys with $E_g \leq 1.6$ eV, since the $\ln(\sigma_D)$ vs. $1/T$ plot exhibits curvature at low T. Figure 2.104c illustrates the $\eta\mu\tau$ product derived from σ_L, measured at 600 nm. The photoresponse drops drastically with Sn incorporation and remains low as more Sn is added. As discussed in Section 2.5.5, intrinsic a-Si:H exhibits n-type extended state conduction with $E_\sigma \cong 0.8$ eV, indicating that E_f is located above the midgap position. With the initial addition of Sn, localized states are created above the valence band, which cause an initial increase in E_σ. With further Sn incorporation the conduction mechanism is changed to hole hopping within these states. (This interpretation was confirmed by measuring positive Seebeck coefficients (up to 2000 μV/K) for a-SiSn:H films with low band gaps.) Structure in $g(E)$ (using the SCLC technique), with the addition of small amounts of Sn, was observed and attributed to an increase in the dangling bond defect density, which were believed to be located near or slightly above the midgap position. However, a significant increase in states above the valence band edge was also noted and

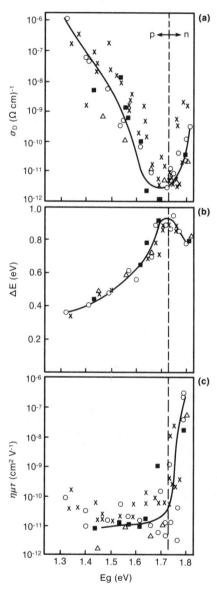

Fig. 2.104. Dependence of (a) dark conductivity, σ_D, (b) activation energy, E_σ and (c) photoresponse ($\eta\mu\tau$ measured at 600 nm with a photon flux of $3.10^{15}\,\text{cm}^{-2}\,\text{s}^{-1}$) as a function of band gap, E_g, for a-SiSn:H alloys. The transition from p- to n-type behavior is noted (Mahan et al., 1984).

attributed to acceptor state formation. It was this rapid increase in acceptor levels which accounted for n- to p-type changes in the conduction mechanism with the addition of Sn. [Evidence for dangling bonds in a-SiSn:H made by a sputtering technique has also been given by Morimoto et al. (1984).] From this study it was found that the DOS spectra can be changed somewhat by manipulating deposition conditions. Nevertheless, the DOS in these types of alloys remained significantly larger than in the GD a-Si:H type alloys; this presently restricts the use of this alloy for device purposes.

In summary, for narrow band-gap semiconductors such as a-Si:Ge:H and a-Si:Sn:H, the microscopic defect levels that are responsible for the observed degradation in the electronic properties have not as yet been identified. Generally, the band tails exhibit significant variations, and in most cases are shallower than for a-Si:H material (von Roedern et al., 1984, 1985). It appears that only optimized a-SiGe:H materials have yielded bandtails that are almost as sharp as in the a-Si:H case ($E_0 = 52$ to 56 meV). Although the tail distribution in a-SiGe:H alloy can be relatively sharp, nevertheless these materials possess significantly higher DOS at E_f ($>10^{17}$ cm^{-3} eV^{-1} for a band gap of 1.5 eV), which poses a considerable limitation on device performance, especially in the solar cell applications area. This will be discussed in Chapter 3.

2.12.3 Amorphous Silicon-Carbon Alloys

Extensive work has been reported (Sussman et al., 1981; Tawada et al., 1982) on hydrogenated amorphous silicon carbon (a-SiC:H) alloys, a material that exhibits a wider band gap with increasing carbon incorporation. But, interest has been primarily limited to the use of the material as a p-type window layer (Madan, 1981; Tawada et al., 1982) in a-Si:H $p-i-n$ type solar cells.

The addition of C leads to an increase of the band gap (Tawada et al., 1982), an increase in the ESR spin density (Morimoto et al., 1982) and, correspondingly, an increase in the density of states in the mid-gap region, as shown in Fig. 2.105 (Schmidt et al., 1985). Further, from the sub-band gap absorption data, the slope (E_0) of the Urbach tail decreases and a shoulder is observed at 0.8 eV which increases with increasing C concentration. It has been noted that the C incorporation should result in an increase of disorder, since the bond lengths differ for C–C and Si–Si bonds; the bond lengths are 1.54 Å and 2.24 Å respectively (Schmidt et al., 1985).

The structure of the films, and hence the opto-electronic properties also seems to be dictated by the choice of the source gases. Tawada et al. (1982) have compared the properties of a-SiC:H layers prepared by the GD technique using $SiH_4 + CH_4$ and $SiH_4 + C_2H_4$ gas mixtures. Films using both gas mixtures can lead to a band gap variation in the range 1.8 to 2.8 eV

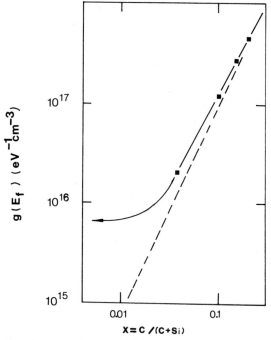

Fig. 2.105. Density of states at the Fermi level, $g(E_f)$, as a function of C/(C + Si) content (Schmidt et al., 1985).

that depends upon the ratio of the starting gas mixtures. However, the former gas mixture resulted in films which show about two orders of magnitude better photoconductivity and improved doping characteristics in comparison with the films prepared by the use of the latter gas mixture. Results for E_g, σ_{ph} and σ_D as a function of the former gas mixture for undoped and B doped films are shown in Fig. 2.106. The reason for the difference is attributed to structural aspects as deduced from their infra-red results, which indicate that the use of ethylene (C_2H_4) tends to promote C_2H_5 groups as well as SiH_2 formation within the films. In contrast, using methane, the SiH_2 group is reduced by a factor of three and carbon is bonded primarily in CH_3 type configurations. Because of the reduction in SiH_2, which is the source of recombination centers (see Section 2.2) the improved photoresponse is explained. It is then concluded that methane-based films exhibit more of an ideal amorphous SiC type alloy in contrast to ethylene-based films, which show an organo-silane type structure.

ESR studies of films prepared using SiH_4 and CH_4 gas mixtures have shown that as the C content increased from 0 to 20%, the spin density, N_s, increased from 10^{16} to greater than 10^{18} cm^{-3}, with a corresponding decrease in the

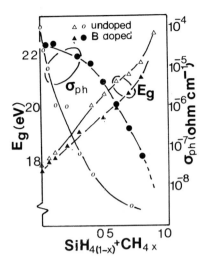

Fig. 2.106. E_g and σ_{ph} as a function CH_4 to SiH_4 gas ratios (Tawada et al., 1982).

photoconductivity (Morimoto et al., 1982a). In this work, it was also noted that N_s could be lowered as the power in the GD was decreased. It was concluded that the increase in N_s with increasing C content is primarily due to an increase in SiH_2 and CH_n bonding.

Some understanding has been achieved recently for GD a-Si:C:H films. It has been found that under certain deposition conditions there is evidence for graphite type bonding in carbon deficient a-SiC:H films (Mahan et al., 1985). Further, the graphite-type bonding can be minimized by significant H attachment to C via CH_n ($n = 2, 3$) bonding; this results in a-SiC:H films with low gap state densities and sharp Urbach tails. Figure 2.107 illustrates the $\eta\mu\tau$ product, derived from the photoconductivity measured at 600 nm, plotted vs. the optical gap, E_g. Also shown in the figure are the $\eta\mu\tau$ data taken from Bullot et al. (1984) and from Morimoto et al. (1982a). It should be noted that the considerable scatter in the experimental data of Mahan et al. (1985) is due to variations in the deposition parameters employed. The two most important parameters affecting material quality are the discharge power and the deposition chamber pressure. As the RF power is increased from 4 W to 20 W there is a significant increase in both the C and bonded H contents of the a-SiC:H films. This is reflected in an increase in E_g, as shown in the inset of Fig. 2.107. Further, $g(E_f)$ and $g(E)$, as measured respectively by the SCLC (see Section 2.3.2.3) and PDS techniques (see Section 2.4), are both decreased and the Urbach tail is sharpened to values approaching that of GD a-Si:H alloys. However, there is no basic change in the infrared spectrum with increasing RF power, i.e. the same infrared modes (indicating CH_n type

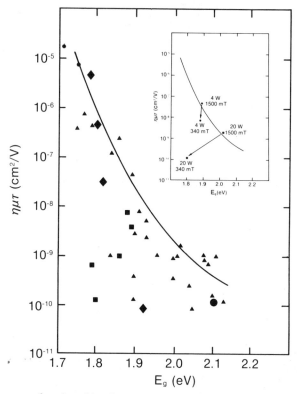

Fig. 2.107. $\eta\mu\tau$ as a function of band gap for a-SiC:H films. The symbol (■) denotes films deposited at low (340 mT) chamber pressures. Also shown are results of Bullot et al., 1984 (♦) and Morimoto et al., 1982a (●). The insert illustrates the change in $\eta\mu\tau$ as the deposition parameter ratio D (see text) is changed (Mahan et al., 1985).

bonding) appear for both 4 W and 20 W samples. The most significant change in the overall a-SiC:H film properties occurs when the chamber pressure is reduced from 1500 mT to 340 mT, especially for the high power (20 W) samples. In particular, the following changes are observed as the chamber pressure is lowered to 340 mT: the C content increases from 14 at.% to 20 at.%, while the amount of H bonded to C decreases sharply; $g(E_f)$ as measured by SCLC increases by a factor of 3; the Urbach tail undergoes significant broadening; E_g decreases sharply; $\eta\mu\tau$ with respect to the best $\eta\mu\tau$ curve is reduced by almost 10^4, as shown in the inset of Fig. 2.107; the shape of the infrared response alters significantly in that the 600–900 cm^{-1} peak observed in all the other samples deposited at high (>1500 mT) pressure disappears and is replaced with a broad featureless peak extending over 300 cm^{-1}.

NARROW AND WIDE BAND GAP ALLOYS 159

The inferior properties of the low pressure films, particularly those deposited at high discharge power, were attributed to the presence of a graphitic type component in a-SiC:H. In this context it should be noted that hydrogenated amorphous carbon (a-C:H) contains a considerable graphitic component that can be increased when H is eliminated from the amorphous network. This occurs upon an increase of the deposition parameter ratio D [= power density/pressure (Bubenzer et al., 1982)]. At larger values of D more energetic ion bombardment of the growing film is presumed to occur, which would break C–H bonds and allow C to bond in the graphitic phase, the lowest energy state of the C atom in the solid. Along with this increase in graphitic coordination in a-C:H there is a reduction in E_g, the infrared CH_n mode at ~2900 cm^{-1} is reduced, an infrared frequency corresponding to the C–C stretch mode of graphite appears at 1600 cm^{-1}, and the midgap $g(E)$ increases (presumably due to the breaking of C–H bonds and the leaving of C dangling bonds at the edges of graphitic regions, or perhaps the formation of graphitic sp^2 states located near midgap). Similar trends were observed for a-SiC:H samples deposited at 20 W, and to a lesser extent for those deposited at 4 W, as the chamber pressure was lowered or the ratio D was increased. The one exception is the 1600 cm^{-1} infrared mode; it is not observed in the low pressure a-SiC:H samples, but does appear for 20 W low pressure films made using 100% CH_4, indicating that the deposition conditions are appropriate for extensive graphitization to occur in pure a-C:H or perhaps C rich a-SiC:H films. Hence, certain deposition conditions can promote graphitization. The study concludes that the relatively good photoconductivity, lower $g(E_f)$ and sharp Urbach edges in material produced using high power and high pressure, is correlated with CH_n type bonding (2900 cm^{-1}). Nevertheless, the photoconductivity is still low compared to device quality GD a-Si:H films. This may be due not only to the SiH_2 polymer bonding, but also to CH_n type bonding; both could result in a sizeable number of recombination centers that could contribute to the loss in photoconductivity with C incorporation.

2.12.4 Amorphous Silicon–Nitrogen Alloys

Relatively little work has been reported on a-SiN:H type alloys prepared using the GD technique in SiH_4 and NH_3/N_2 gas mixtures. With the addition of N, band gaps of up to 3.0 eV have been reported (Alvarez et al., 1984; Karcher et al., 1984). However, in agreement with the above discussion, the addition of N also tends to degrade the opto-electronic behavior of these alloys.

Bauer et al., (1984) have shown that there is a marked difference in the transport properties of a-SiC:H and a-SiN:H type alloys. This is shown in

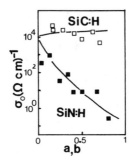

Fig. 2.108. σ_0 (Ω cm^{-1}) plotted for a-SiC : H and a-SiN : H vs. neutral gas composition (Bauer et al., 1984).

Fig. 2.108, where σ_0 (the preexponant in Eq. 2.40) is plotted as a function of neutral gas composition for both alloys. From their photoconductivity measurements and the fact that the photoluminescence peak for a-SiC : H shifts to much higher energies in comparison with a-SiN : H, they conclude that the rapid drop in σ_0 is primarily due to a decrease in the mobility for the a-SiN : H case. This suggests that there is a higher degree of disorder present in a-SiN : H, which seems to be supported by the observation of broader infra-red peaks.

3

Opto-Electronic Applications of Amorphous Silicon Based Materials

3.1 Solar Cells	163
3.1.1 Ideal Conversion Efficiency	164
3.1.2 Estimation of the Conversion Efficiency in Amorphous Silicon Based Alloys	168
3.1.3 Schottky and MIS Solar Cells	188
3.1.4 *p–i–n* Type Junctions	219
3.1.5 Tandem Cells	256
3.1.6 Stability	259
3.1.7 Large Area Solar Cells	268
3.1.8 Economic Considerations	274
3.2 Thin Film Transistors for Display Applications	276
3.2.1 Basic Requirements	278
3.2.2 Theory of Operation	280
3.2.3 Performance of a-Si Alloy TFTs	291
3.2.4 Dynamic Performance	297
3.2.5 Dual Gate TFTs	299
3.2.6 Vertically Stacked TFTs	300
3.2.7 *p*-Channel TFTs	303
3.2.8 Problems Encountered in Large Area Displays	305
3.3 Miscellaneous Applications	309
3.3.1 Threshold and Memory Switching	309
3.3.2 Linear Image Sensors	310
3.3.3 Optical Recording	312
3.3.4 Charge Coupled Devices (CCD)	313
3.3.5 Other Applications	316

Amorphous silicon based alloys have emerged from almost total obscurity within the last decade to having established a secure beachhead in the technological marketplace. Although the technology is still in a fairly early stage of development, the applications and products emerging from this

Table 3.1. Applications of Amorphous Silicon (from LeComber, 1986)

(a) Products Commercially Available

Device	Product
Photovoltaic cell	Calculators, watches, battery chargers, etc.
Photoreceptor	Electrophotography, LED printer
Photoconductor	Color sensors, light sensors, etc.
Image sensor	Contact-type image sensor, electronic write boards
Solar control layer	Heating reflecting float glass
Anti-reflecting/antistatic layer	Television screens
Thin-film field-effect transistor (FETs)	Displays, televisions

(b) Other Proposed Applications

Image pick-up tubes	Passivation layers
Position sensors	Charge-coupled devices
FETs for logic circuits	Strain gauges
FETs for ambient sensors	Photolithographic masks
Fast detectors and modulators	
Diodes	
Bipolar transistors	
Optical Waveguides	
Optical recording	
LEDs	

versatile semiconductor are growing by the day, as shown in Tables 3.1a and 3.1b (LeComber, 1986).

As discussed in Chapter 2, a-Si alloys possess a low density of localized states, which enables them to be doped n- and p-type. This feature, together with their high photoconductivity, direct band gap, and the ability to be deposited over large areas, affords us many possibilities of application that would have been difficult (and uneconomical) for conventional crystalline semiconductors to address. One major drawback is their inherently low mobility, which limits the applications to areas where speed is not of paramount importance.

Of all the applications shown in Table (3.1), the most overwhelming area of interest is the solar cell, which is one of the most demanding of all electronic devices since very efficient extraction of both types of carriers is required. Further, the technology must demonstrate that these cells can be highly efficient (10–15% range), are stable under harsh conditions and that large panels can be produced inexpensively. From a fundamental physics point of view the solar cell under operation has a space charge region and

SOLAR CELLS   163

is charge depleted. In contrast, a field effect transistor, another major application area, is usually operated in the accumulation regime and the conductivity modulation of only one type of carrier is usually required.

The bulk of this chapter reviews the physics of the solar cell as well as the thin film transistor. We end the chapter with some other applications such as switching, recording and image sensing.

3.1 Solar Cells

A solar cell is a photovoltaic device designed to convert sunlight directly into electrical power. It is attractive from an energy point of view as it represents a potentially inexhaustible source of energy, unlike depletable resources such as coal, gas and oil. Solar cells were originally designed for the space program and were constructed from high quality single crystal Si materials from which high efficiency devices could be made. The importance of solar cells cannot be understated, since their potential for large scale power generation capable of meeting terrestrial energy requirements is immense. However, if they are to be competitive with other methods of power generation, their cost must be reduced substantially and the operating conversion efficiency increased. There have been several approaches to the solution of this problem; many semiconducting materials have been utilized such as: single crystal Si; polycrystalline Si; GaAs; ZnP; CdTe; $CuInSe_2$; CdS/Cu_2S (see, e.g., 18th IEEE Photovoltaic Conference Proceedings, Las Vegas, 1985.) Single crystal Si has been by far the most intensively studied and has attained a maximum conversion efficiency in excess of 20% under AM1 (Air Mass) conditions (Green, 1985). However, because of the inherently high cost involved in the fabrication of these cells further large scale terrestrial usage is not envisioned, since it appears that they may not be able to meet the stringent cost goals (less than $1 per peak watt) that are required to compete with other forms of energy production.

It is for this reason that two types of approaches are now under consideration: low cost flat plate collectors; concentrator systems. In the former approach the major factors affecting the performance criteria are the support structures, the cost of land, material costs and the availability of raw materials. The primary materials under consideration are amorphous silicon based alloys, $CuInSe_2$ and CdTe, all of which have exceeded a conversion efficiency of 10%. In the latter approach the primary cost factors to consider are the tracking mechanisms and concentrators involving lenses and cooling systems; the material under consideration for this approach is GaAs, where one of the goals is to reach 35% conversion efficiency. The inclusion of a tracking mechanism introduces a disadvantage, as moving parts are now introduced into the total photovoltaic system.

In this chapter we are concerned with the problems of using a-Si based alloys for power generation. Since its advent (Chittick *et al.*, 1969), rapid and impressive progress has been made in the basic scientific understanding of the material, the achievement of high photovoltaic conversion efficiencies and their commercialization in an array of products. Moreover, amorphous semiconductors offer us a good opportunity for attaining high conversion efficiencies (>20%) by using a multijunction stack cell approach. It is this potential ability which puts amorphous semiconductors at the forefront of present research and development activity.

In this chapter we discuss the four types of a-Si solar cells: the metal-semiconductor (Schottky Barrier—SB) cell; the metal–insulator–metal (MIS) structure; the p^+-n^+ cell; the p^+-i-n^+ type cell. We will first emphasize the problems of optimizing the conversion efficiency and the selection of the best material. Next, the various types of cells are discussed, with emphasis on preparation techniques, doping effects, current–voltage, I(V), characteristics and transport properties. We end the section with a discussion of large area and tandem cells, the important problem of device stability, and economic considerations.

3.1.1 Ideal Conversion Efficiency

The $I(V)$ curve under illumination passes through the fourth quadrant. Therefore, power can be extracted from the cell, as shown in Fig. 3.1. The actual operating point on a solar cell $I(V)$ curve is determined by the load resistance R_L, whose value should be such that the device is biased at the maximum power point (I_m, V_m) in Fig. 3.1. The energy conversion efficiency, η, of the solar cell is defined by

$$\eta = \frac{J_{sc} \cdot V_{oc} \cdot (FF)}{P_{in}}, \qquad (3.1)$$

where J_{sc} is the short circuit current density, V_{oc} is the open circuit voltage, P_{in} is the incident power and FF is the fill factor, which is defined by the relation

$$FF = \frac{(I_m V_m)}{(I_{sc} V_{oc})}. \qquad (3.2)$$

The conversion efficiency of a practical solar cell is limited by intrinsic and extrinsic losses. The latter include losses due to reflection, series resistance, incomplete extraction of photogenerated carriers and the non-radiative recombination of carriers. The intrinsic loss includes the inability of a cell with a specific E_g to properly match the broad solar spectrum. There have

SOLAR CELLS

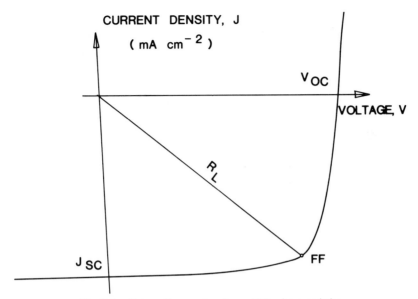

Fig. 3.1. Solar cell current-voltage, $I(V)$, characteristics.

been many predictions of the upper limit to the conversion efficiency. The approach taken by Schockley and Queisser (1961) considered the limitations imposed by thermodynamics and the fundamental radiative processes. They approximated the solar spectrum as a blackbody at 6000 K and used the principle of detailed balance to calculate the intrinsic radiative current. Henry (1980) used an extension of their approach and calculated the efficiency using a graphical technique. Using a standard AM1.5 terrestrial solar spectrum, the short circuit current was calculated by assuming that every photon of energy in excess of E_g creates an electron-hole pair that can be extracted. The number of photons, n_{ph}, available for the creation of carriers as a function of E_g is shown in Fig. 3.2. The losses are illustrated for a material with $E_g = 1.35$ eV, the optimum gap for power conversion in crystalline type materials. The shaded areas correpond to the three types of intrinsic losses: (i) the area labelled $hv < E_g$ is lost because photons are not absorbed by the material; (ii) the area labelled $hv > E_g$ is lost because carriers generated by photon absorption lose all the energy in excess of E_g as heat; (iii) the area labelled $W < E_g$ is lost because the radiative recombination limits the work performed per photon. The remaining unshaded area equals the power/unit area that the solar cell delivers. It is evident that the unshaded region (the power generating region) is zero for either very large or small values of E_g. A plot of the solar cell efficiency versus E_g is given in Fig. 3.3. (The slight oscillations present in the figure are due to the atmospheric

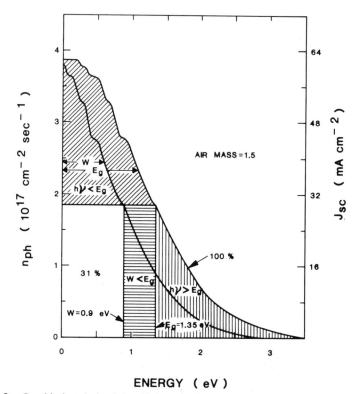

Fig. 3.2. Graphical analysis of the efficiency of an ideal solar cell. The outer curve is E (eV) vs. n_{ph}, the absorbed photon flux. The inner curve is W, the work per absorbed photon vs. n_{ph}. The area under the outer curve is the solar power per unit area. The shaded areas equal losses. The unshaded area is the power per unit area delivered to the load (Henry, 1980).

absorptions that are present in the solar spectrum.) Note that the conversion efficiency has a broad maximum and is not critically dependent upon E_g. Therefore, semiconductors with gaps between 1 and 2 eV, such as crystalline Si, amorphous Si, GaAs, and CuInSe$_2$, are potentially useful materials. They have been incorporated into different types of cell configurations that can be roughly subdivided as follows:

(1) homojunctions, that employ n- and p-type semiconductors of the same species;
(2) heterojunctions, that utilize dissimilar semiconductors;
(3) Schottky barriers (SB), that involve a junction between a semiconductor and a metal;
(4) metal–insulator–semiconductors (MIS) structures, in which an insulator is inserted between the metal and the semiconductor.

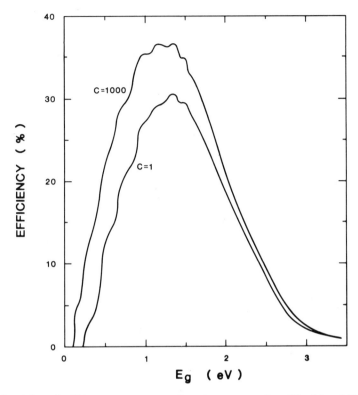

Fig. 3.3. Solar cell efficiency vs. energy gap for solar concentrations (C) of 1 and 1000 suns (Henry, 1980).

With the creation of a junction either in or on the semiconductor, a rectifying structure is produced. Under incident illumination, photocurrent flows through the diode provided that photo absorption occurs in or near the junction region. Electron-hole pair creation at depths greater than about a minority carrier diffusion length from the junction edge is likely to result in recombination (loss) prior to extraction. The photon absorption that occurs within the solar cell is governed by the absorption coefficient, α (cm)$^{-1}$, of the material, that depends upon: the physical structure of the material; the joint density of states, the nature of the band gap (direct or indirect); the wavelength, λ. The absorption coefficient of some prominent crystalline solar cell materials is shown in Fig. 3.4. For crystalline Si, which has an indirect band gap, the gradual change of α with λ means that the photon absorption occurs from the surface to depths of several hundred microns. On the other hand, for GaAs, a direct gap material, the steep absorption edge limits absorption to within 2.5 μm of the illuminated surface.

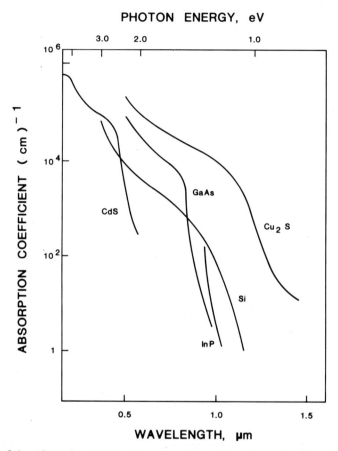

Fig. 3.4. Absorption characteristics of some crystalline solar cell semiconductors.

3.1.2 Estimation of the Conversion Efficiency in Amorphous Silicon Based Alloys

Several attempts have been made to derive the conversion efficiency of amorphous Si based alloy solar cells. Estimates have ranged between 6% and 22% (Carlson *et al.*, 1976; Debney, 1978; Crandall *et al.*, 1979; Delal, 1980; Kuwano *et al.*, 1982a; Swartz, 1982; Carlson, 1983; Madan, 1985a). The major difficulty in deriving these estimates lies in formulating the exact mechanism of carrier transport. The primary difference, as discussed in Chapter 1, between the a-Si : H based alloy and the crystalline material is that the former possesses a nonuniform DOS composed of inherent band tails and defect states due to dangling bonds or impurities. As discussed in Section 2.3,

SOLAR CELLS

the exact DOS is not known and importantly, little is known about the capture cross sectional areas of the defects. This presents a problem in determining the recombination mechanisms, which in turn determines the upper limit to the various photovoltaic parameters, such as V_{oc} and FF. Estimation of the short circuit current has also not been without controversy, since geminate recombination has been proposed as a limiting mechanism (Crandall et al., 1979) for charge collection.

In the following an attempt is made at deriving (or at least estimating) an upper limit to the three factors that determine the conversion efficiency. We first consider one of the main differences that exists between a-Si based alloys and crystalline Si type solar cells. In crystalline Si the indirect band gap necessitates relatively thick devices ($\sim 100\,\mu$m) for sufficient absorption to take place. Here the space charge region is on the order of 1 μm thick. Because of the large mobility of electrons ($\mu_e \sim 1300\,\text{cm}^2\,\text{s}^{-1}\,\text{V}^{-1}$) and holes ($\mu_h \sim 500\,\text{cm}^2\,\text{s}^{-1}\,\text{V}^{-1}$), the photogenerated carriers are able to diffuse to the junction where they are separated and can thereby generate a photocurrent. Since a-Si:H behaves optically like a direct band gap semiconductor (see Section 2.4) with low electron and hole mobilities (see Section 2.5.6), it then has a low diffusion length (see Section 3.1.4.7b) and, therefore, a large portion of the observable current is due to the carrier generation that occurs within the space charge region (SCR).

3.1.2.1 Geminate Recombination

As discussed above, since a large portion of the effective carrier generation is limited to the absorption that occurs within the space charge region, then it is important to consider the concept of geminate recombination, which is usually a limiting factor to charge collection is some amorphous semiconductors such as Se (Pai and Enck, 1977).

Geminate recombination is important when the photoexcited electrons and holes do not escape their mutual Coulomb attraction. For simplicity, we assume that the more mobile electrons possess all the kinetic energy of the pair after excitation. There are two important parameters: r_0, the average distance that the photoexcited electron moves away from the hole before it thermalizes; r_c, the effective range of the Coulombic attraction between the electron-hole pair. Clearly, r_0 depends upon the frequency of the exciting light. Since the photoexcited electron can escape from the Coulomb field of the hole if it has energy within kT of the conduction band edge, then r_c is given by

$$r_c = \frac{e^2}{4\pi\varepsilon kT}. \tag{3.3}$$

For a-Si:H, $\varepsilon \sim 12\,\varepsilon_0$ so that $r_c \sim 50$ Å at room temperature (Adler *et al.*, 1980; Silver *et al.*, 1980). When an electric field, F, is present, Onsager (1938) has shown that the quantum yield Y, (defined as the number of charges separated divided by the total number created) can be approximated by

$$Y = Y_0 e^{-r_c/r_0}\left[1 + \tfrac{1}{2}\cdot\left(\frac{e}{kT}\right)r_c F + \cdots\right]. \qquad (3.4)$$

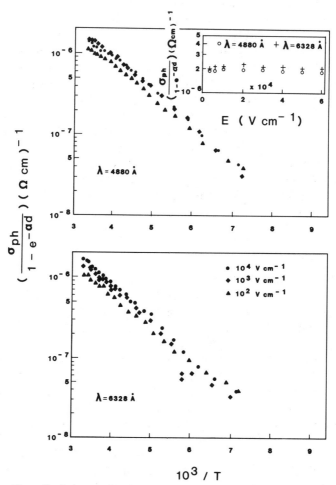

Fig. 3.5. Normalized photoconductivity as a function of reciprocal temperature plotted for applied fields in the range 10^2 to 10^4 V cm^{-1} for incident illumination of (a) $\lambda = 4880$ Å and (b) $\lambda = 6328$ Å; α is the absorption coefficient and d the thickness of the film. The inset shows the normalized photoconductivity at room temperature for fields up to 6×10^4 V cm^{-1} (Madan *et al.*, 1980).

When $r_0 < r_c$, (a) the photoconductive yield is reduced considerably and (b) the presence of an electric field does not significantly increase the yield until F is on the order of kT/er_c. Using Eq. (3.3), the critical field, $F_g \cong 4\pi\varepsilon(kT)^2/e^3$. In order to evaluate the quantum efficiency (or yield) it is necessary to determine the distance, r_0. Knights and Davis (1974) proposed that $r_0 \sim (\hbar\omega - E_g - e/4\pi\varepsilon r_0)^{1/2}$, where ω is the frequency. Except for situations where $\hbar\omega \cong E_g$, the term $e^2/4\pi\varepsilon r_0$ can be neglected. This, together with Eq. (3.4), suggests that for $r_0 < r_c$ the photoconductive yield should be a rapidly varying function of the frequency of the light. Furthermore, the activation energy for the photogeneration, $e^2/4\pi\varepsilon r_0$, should depend upon the frequency. This has been observed in a-Se by Pai and Enck (1977).

To summarize, there are three criteria that determine when geminate recombination is important: (a) a rapidly increasing photoconductive yield saturating at $Y = 1$ as the frequency of light increases; (b) a decreasing activation energy for the photoconduction as the frequency of light increases; and (c) an increase in photoconductive yield at applied fields in excess of 10^3 V/cm at room temperature or 10^4 V/cm at 100 K (Adler et al., 1980; Silver et al., 1980).

Figure 3.5 shows the normalized photoconductivity for different wavelengths as a function of the electric field and temperature for an a-Si:F:H alloy (Madan et al., 1980). Note that none of the above criteria required for geminate recombination are applicable for this type of alloy.

Yamaguchi et al. (1980) have considered the impact of geminate recombination in relation to the performance of a-Si:H p^+-i-n^+ type devices. By assuming an exponential DOS with $g_0 = 10^{16}$ cm^{-3} eV^{-1} [see Eq. (3.9)], they calculated the electric field as a function of distance, x(Å), for a 5000 Å thick intrinsic layer, as shown in Fig. 3.6a. Taking $r_c = 46$ Å and the variation of field with distance, $F(x)$, from Fig. 3.6a, the quantum yield $Y(x)$, is calculated for 25 Å $< r_0 <$ 150 Å, as shown in Fig. 3.6b. (Note that the increase in $Y(x)$ at the p^+/i and the n^+/i interfaces is due to an increase in the field at these junctions.) Since $Y(x, \hbar\omega)$ is virtually constant over the entire width of the device, then we can conclude that the maximum conversion efficiency should be attained for device thickness of about 5000 Å. By fitting their theory to the collection efficiency curves with a device thickness of 5000 Å and a zero field diffusion length of 500 Å, it is found that the zero field carrier generation probability is between 0.53–0.75 for incident photon energies in the range of 1.8–3.0 eV. However, in p^+-i-n^+ junctions where the internal electric fields are in excess of 10^4 V/cm, the free carrier generation probability is enhanced, at least to 0.6 to 0.8 on the average, and reaches 0.8–0.96 in the high electric field regions in the vicinity of the p^+/i and i/n^+ interfaces. [It should be emphasized that Yamaguchi et al. (1980) interpreted

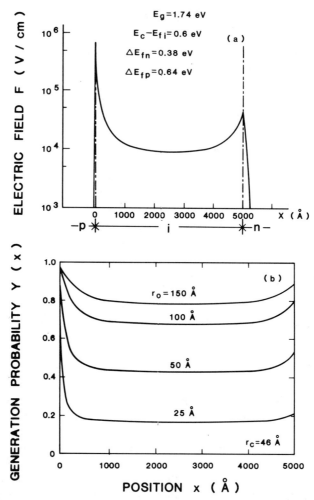

Fig. 3.6. (a) Internal/electric field distribution $F(x)$ in the i-layer of an a-Si:H p–i–n junction. (b) free carrier generation probability $P_G(x, \hbar\omega)$ as a function of thermalization distance r_0 (Yamaguchi et al., 1980).

the data on devices with conversion efficiencies of only 5%. However, since efficiencies have now exceeded 12%, then it is to be expected that the fit of their theory to the collection efficiency curves of more efficient devices would lead to even larger quantum yields and optimum thickness in excess of 5000 Å (see Section 3.1.4.6).]

Further confirmation of the lack of geminate recombination is provided by the work of Ackley et al. (1979), who used picosecond spectroscopy

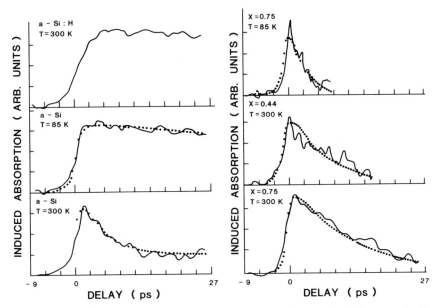

Fig. 3.7. (a) Induced absorption $\Delta\alpha$ vs. delay for a-Si:H at 300 K and unhydrogenated a-Si at $T = 85$ K and 300 K; (b) $\Delta\alpha$ vs. delay for a-$AS_2S_{3-x}Se_x$ at $T = 85$ K and 300 K. Full lines are experimental data, dotted lines are fits (Ackley et al., 1979).

technique on hydrogenated and unhydrogenated a-Si, and chalcogenide films of composition $As_2S_{3-x}Se_x$ ($0.25 < x < 0.75$). The time dependence of the induced absorption in a-Si, a-Si:H and $AsS_{3-x}Se_x$ at various temperatures is shown in Fig. 3.7. All the samples showed instantaneous (faster than 0.5 ps) absorption that persisted in a-Si:H beyond 200 ps. Further, there was no change in the spectra when the sample was cooled to 85 K. (The absorption coefficient of about 30 cm^{-1} corresponds to a cross section of $3 \cdot 10^{-17}$ cm^2.) Non-hydrogenated samples exhibited an initial rapid temperature decay of the induced absorption, but after the initial relaxation no further decay was observed up to a maximum delay time of 100 ps. This indicates that the carriers probably thermalize in a time less than 0.5 ps (see Section 2.7). This is particularly evident in a-Si:H where the instantaneous induced absorption does not show a fast decay component. The thermalized carriers, which are localized, absorb light by transitions to high energy states in the same band where the density of states is large, and are accessible because the k-conservation rule does not apply. The absorption decays when the carriers recombine or fall into deep traps where their absorption cross section is smaller. The difference in behavior between the three types of materials, particularly with respect to the temperature dependence of the decay times, was explained on the basis of Onsager's (1938) theory. The data for the chalcogenide samples can be understood if during the thermalization process it is assumed that $r_0 < r_c$ (≈ 80 Å at 300 K, ≈ 300 Å at 85 K) and the carriers behave as trapped excitons, which leads to an estimate for r_0 of ~3-5 Å. Conversely, for a-Si and a-Si:H materials, it is proposed that $r_0 > r_c$ (~50 Å at 300 K). However, the absorption decay for unhydrogenated Si at lower temperatures is explained in terms of defect levels with a decay time given by $\tau = (bN_t)^{-1}$ where $b = 4\pi rD$ and N_t is the trap density ($= 10^{20}$ cm^{-3} for unhydrogenated a-Si). Here, r is the average distance between traps and D is the diffusion constant. Using these parameters a reasonable fit to their data was obtained, as shown in Fig. 3.7. Using the same reasoning for the hydrogenated a-Si case (in which case the trap density, N_t is lowered by ~10^3) the relaxation time is then expected to be about 100 times larger.

Using a delayed collection field technique Mort et al. (1981) have studied thick samples ($> 15 \mu$m) fabricated with the following deposition parameters: (a) substrate temperature—230°C; rf power—18 W; 5% SiH$_4$ dilution in He; (b) 230°C; 2 W; 100% SiH$_4$. It was found that samples of type (b) yielded a zero field quantum efficiency of 0.55 with $r_0 \sim 80$ Å. However, for samples of type (a) the corresponding numbers were 0.40 and $r_0 \sim 50$ Å. This provides further confirmation of the previous point made in Section 2.1, device quality materials are usually prepared by using conditions approximating that of (b). The above quantum efficiency of $Y \sim 0.55$ is expected to increase in the presence of an electric field, as discussed earlier.

Indirect evidence that geminate recombination is virtually non-existent in an a-Si p–i–n solar cell is provided by the fact that the actual short circuit current density obtained is very close to the theoretical maximum, assuming $Y \approx 1$ (see Section 3.1.4.7a).

3.1.2.2 Short Circuit Current Density

The above discussion suggests that geminate recombination is unimportant, especially in a solar cell device in which large electric fields ($> 10^4$ V cm^{-1}) are generally present. Assuming that $Y(x) \sim 1$, knowledge of α and the incident power as a function of the wavelength λ, then the short circuit current density, J_{sc}, can be calculated using

$$J_{sc} = \int en_{ph}(\lambda)(1 - \exp -\alpha(\lambda)d) \, d\lambda, \tag{3.5}$$

where the number of photons, $n_{ph}(\lambda) = 10^{15}P(\lambda)/1.6(1.24/\lambda)$; $P(\lambda)$ is the power of the incident illumination and d is the thickness of the absorber. To derive an estimate of the upper limit to J_{sc}, optical or resistive losses are

SOLAR CELLS

not considered; also it is assumed that the free carrier contributions only arise from incident photon energies exceeding E_g. For ease of calculation the AM-0 solar spectra is approximated as a 5800 K blackbody. Thus, the power (mW cm^{-2} μm^{-1}) can be written as

$$P(\lambda) = \frac{0.861}{\lambda^5} \cdot \frac{1}{(\exp 2.51/\lambda - 1)}, \qquad (3.6)$$

and the total integrated power is 135.7 mW cm^{-2}. The absorption coefficient reproduced from Eq. (2.32) is

$$\alpha(\lambda) = \frac{B^2(h\nu - E_g)^2}{h\nu}, \qquad (3.7)$$

with $B = 700\,\text{eV}^{-1/2}\,\text{cm}^{-1/2}$. In Fig. 3.8 (Madan, 1985) we plot the short

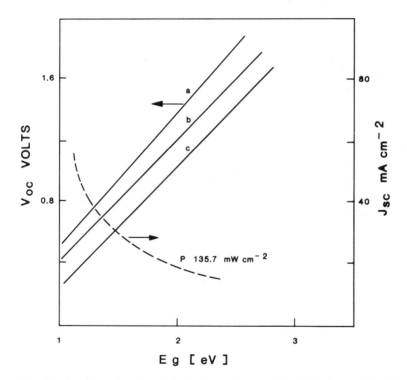

Fig. 3.8. V_{oc} plotted as a function of the band gap for materials with different minima in the DOS spectrum: (a) 10^5 cm^{-3}, (b) 10^{14} cm^{-3}, (c) = 10^{17} cm^{-3} with $R = 10$ (see Section 3.1.2.3). The plot also shows J_{sc} as a function of band gap. The total incident power is 135.7 mW cm^{-2} (Madan, 1985).

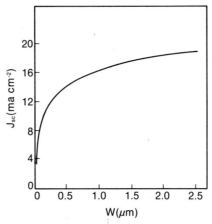

Fig. 3.9. J_{sc} plotted as a function of the collection width (Czubatyj et al., 1980).

circuit current density as a function of E_g. Note that for $E_g = 1.7$ eV, $J_{sc} = 29$ mA cm^{-2}, which reduces to 21.3 mA cm^{-2} when normalized to an incident power of 100 mW cm^{-2} (corresponding closely to AM-1 conditions). Since the AM-1 spectral response is shifted more towards the blue end of the spectrum when compared to the AM-0 condition, then this will lead to a slightly larger value of J_{sc}. However, J_{sc} will be smaller than this value in practical devices due to extrinsic losses such as reflection and absorption by the contacts, and the unavoidable series resistance effects. In addition, there will be a loss due to incomplete absorption as well as due to a small minority carrier diffusion length, which we will be treating later. Considering for the moment only the collection width, W_c, which is a function of the width of the space charge region as well as the minority carrier diffusion length, J_{sc} can be written as (Czubatyj et al., 1980),

$$J_{sc} = e \int n_{ph}(\lambda)\{1 - \exp(-\alpha(\lambda)W_c)\}\, d\lambda. \tag{3.8}$$

J_{sc} as a function of the collection width is shown in Fig. 3.9 for one-light-pass absorption. It should be noted that for a 7000 Å thick cell, $J_{sc} \sim 15$ mA cm^{-2} is possible. More complete internal absorption can be achieved with the use of reflectors and the texturing of substrates, leading to larger current densities; we will discuss this in more detail in Section 3.1.4.6.

3.1.2.3 Open Circuit Voltage

In crystalline semiconductors the open circuit voltage, V_{oc}, is generally limited by the direct band-to-band recombination of photogenerated carriers.

However, a nonuniform DOS within the mobility gap, which is present in amorphous semiconductors, leads to recombination via these defect states and provides an upper limit to V_{oc}. The treatment we employ now is based on the work of Simmons and Taylor (1971), and has been applied to a-Si:H alloys by Tiedje (1982), who found that V_{oc} is limited to approximately 1.0 V. For simplicity, it is assumed that a-Si:H can be represented by a DOS given by [see Eq. (2.22)]

$$g(E) = g_0 \cosh\left(\frac{E - E_f}{E_{ch}}\right), \quad (3.9)$$

where E_{ch} defines the slope of the DOS spectra, and that the DOS is comprised of donor and acceptor type states only. The acceptor type states are defined as negatively charged when occupied and neutral when empty; conversely, the donor states are positively charged when empty and neutral when occupied.

Uniform illumination results in a constant generation rate per unit volume, G, of electron-hole pairs and the occupancy statistics for the traps may be derived using either of two points of view (Simmons and Taylor, 1971). The rate equations for the conduction and valence bands (Shockley-Read approach, 1952), can be employed or, alternatively, the rate equation for a particular trap level can be developed.

Under non-equilibrium steady state conditions the rate at which electrons enter the conduction band is equal to the rate at which electrons exit. This can be represented for electrons by

$$\frac{dn}{dt} = G - \int_{E_v}^{E_c} \bar{n} g(E)[1 - f(E)] \, dE + \int_{E_v}^{E_c} e_n g(E) f(E) \, dE = 0. \quad (3.10)$$

Similarly, for holes the rate equation can be written as

$$\frac{dp}{dt} = G - \int_{E_v}^{E_c} \bar{p} g(E) f(E) \, dE + \int_{E_v}^{E_c} e_p g(E)[1 - f(E)] \, dE = 0, \quad (3.11)$$

where $\bar{n} = K_n n$ and $\bar{p} = K_p p$; e_n and e_p represent the emission probability from traps for electrons and holes respectively; K_n and K_p represent the capture rate constants for electrons and holes. Using Eqs. (3.10) and (3.11) the probability of occupation, $f(E)$, of a trap at energy level E is given by

$$f(E) = \frac{\bar{n} + e_p}{\bar{n} + \bar{p} + e_n + e_p}. \quad (3.12)$$

The second point of view considers the rate equation for a particular trap. In steady state the occupancy of any trap is constant. Thus, the four processes that fill and empty the traps are in balance. Hence

$$\bar{n}g(E)[1 - f(E)] - e_n g(E)f(E) - \bar{p}g(E)f(E) + e_p g(E)[1 - f(E)] \, dE = 0, \quad (3.13)$$

which can be rewritten and leads to Eq. (3.12).

Although the two approaches may appear to be no more than just an alternative means of deriving the same statistics, it should be noted that Eqs. (3.10) and (3.11) apply to the traps taken collectively, whereas $f(E)$ as derived by Eq. (3.13) is relevant for the traps taken individually, and shows that the statistics are independent of the means of stimulation. The latter could be a more powerful approach, especially in the case of a-Si:H alloys, since different types of defects could co-exist at the same energy (inherent dangling bonds, impurities, etc.); hence, a different assignment can be made for each species.

By differentiating between a trap and a recombination level such that for $E > E_{tf}^n$ the traps are empty and for $E < E_{tf}^n$ the traps are occupied, the modulated Fermi–Dirac function for trapped electrons can be determined by

$$f(E) = \frac{Rn}{Rn + p} \cdot \frac{1}{1 + \exp(E - E_{tf}^n/kT)}, \quad (3.14)$$

where R is the ratio, K_n/K_p. The free carrier concentration, n, in the conduction band can be defined in terms of the electron quasi-Fermi level, E_{fn}, define by

$$n = g(E_c)kT \exp\left[-\left(\frac{E_c - E_{fn}}{kT}\right)\right]. \quad (3.15)$$

Hence, E_{tf}^n and E_{fn} from Eqs. (3.14) and (3.15) are related by

$$E_{tf}^n = E_{fn} + kT \ln\left(1 + \frac{p}{n}\right), \quad (3.16)$$

which shows that $E_{tf}^n > E_{fn}$.

Similarly, the Fermi level for trapped holes can be derived and represented by

$$1 - f(E) = \frac{p}{Rn + p} \cdot \frac{1}{1 + \exp(E_{tf}^p - E/kT)}, \quad (3.17)$$

such that for $E > E_{tf}^p$ the traps are filled and act as recombination centers and for $E < E_{tf}^p$ the traps are empty. Further, $E_{tf}^p < E_{fp}$, where E_{fp} is the quasi-Fermi level for free holes.

The number of trapped electrons, n_t, and trapped holes, p_t, can be defined for the density of states assumed in Eq. (3.9) by

$$n_t = \int g(E)f(E)\,dE, \tag{3.18}$$

where $f(E)$ is given by Eq. (3.14), and

$$p_t = \int g(E)[1 - f(E)]\,dE, \tag{3.19}$$

where $[1 - f(E)]$ is given by Eq. (3.17).

Since charge neutrality must prevail, i.e., $n_t = p_t$, then n, p, E_{tf}^n and E_{tf}^p can be found and the position of E_{fn} and E_{fp} can be derived. Therefore, using $V_{oc} = E_{fn} - E_{fp}$, the value of the open circuit voltage can be found.

To evaluate V_{oc} the rate constants K_n and K_p need to be determined. In high mobility solids, where the mean free path is large compared to the radius of the charged recombination centers, the recombination coefficient is given by

$$K = Cv, \tag{3.20}$$

where v is the thermal velocity and the capture cross section, C, is determined by the rate of energy loss to phonons. However, if the mean free path of the electrons is small, which is the case for an a-Si:H type alloy, then the electrons must diffuse towards the recombination center. Here K is not directly a function of v, but depends instead on the diffusion coefficient, D, given by,

$$D = \frac{\mu k T}{e}. \tag{3.21}$$

This assumes that the rate of energy loss is fast compared to the rate of diffusion past the recombination center. The diffusion limited recombination kinetics were first analyzed by Langevin (1903), who noted that the current flow into a charged recombination center for unit carrier density is proportional to K. Thus, for spherical symmetry

$$K = 4\pi r^2 \mu F, \tag{3.22}$$

where F is the field produced by the center and is given by

$$F = \frac{e}{4\pi \varepsilon r^2}. \tag{3.23}$$

Thus, the rate constant is related to the mobility by

$$K = \left(\frac{e}{\varepsilon}\right)\mu. \tag{3.24}$$

However, if the recombination center is neutral, then

$$K = 4\pi Dr, \quad (3.25)$$

which leads to [using Eq. (3.21)]

$$k = 4\pi \left(\frac{\mu kT}{e}\right) r. \quad (3.26)$$

Tiedje (1982) calculates $K_p = 2.10^{-9}$ cm^3/s using the diffusion coefficient for a free hole, where $D_h = 0.018$ cm^2/s and r is the capture radius of a neutral valence band tail state, assumed to be 2 Å. It is further assumed that the capture probability for electrons by the hole occupied version of the same state will be larger, both because the free electron diffusion coefficient ($D_e = 0.33$ cm^2/s) and the capture radius of the charged state will be about an order of magnitude larger: $K_n = 4.10^{-7}$ cm^3/s.

In Fig. 3.8 we show V_{oc} as a function of E_g for various values assumed for the minimum in the DOS, g_0. Here R was assumed to be 10. The upper limit of V_{oc} was found to be 1.05 V for a material with $E_g \cong 1.7$ eV.

3.1.2.4 Extended Analysis

A more complete analysis for the derivation of the full $I(V)$ characteristic of a solar cell involves the solution of the continuity and Poisson equations. An analysis that incorporates the effect of a nonuniform DOS has been attempted by Sichanugrist *et al.* (1984) and Hack *et al.* (1983). In these studies the DOS was assumed to be asymmetric and given by

$$g(E) = \frac{g_0}{2}\left[\exp\left(\frac{E}{E_{ch1}}\right) + \exp\left(-\frac{E}{E_{ch2}}\right)\right], \quad (3.27)$$

which is a modified version of Eq. (3.9). Here E_{ch1} and E_{ch2} are the characteristic energies of the acceptor and donor like states defined earlier. Using this, the net charge, ρ, can be computed and hence the width of the field region (the depletion width), W, can be defined by (Sichanugrist *et al.*, 1984),

$$W = \frac{1}{q}\int_{\phi_n}^{\phi_p} \frac{d\phi}{[(2/\varepsilon_s\varepsilon_0)\int_0^\phi \rho(E)\,dE + c^2]^{1/2}}, \quad (3.28)$$

where ϕ_p and ϕ_n are the boundary potentials that were obtained from the experimentally determined conductivity activation energies of the *p*- and *n*-type samples, and *c* is a constant. Using $\phi_p = 0.55$ eV and $\phi_n = -0.45$ eV, Fig. 3.10 shows the calculated energy profiles for $g_0 = 10^{16}$ cm^{-3} eV^{-1}, $E_{ch1} = 0.059$ eV and $E_{ch2} = 0.095$ eV. (It should be recognized that more realistic values are: $\phi_p = 0.3$ eV, $\phi_n = -0.2$ eV, $E_{ch1} \approx 0.025$ eV and

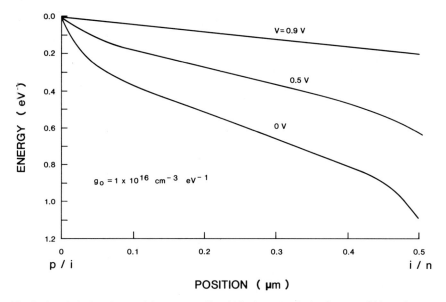

Fig. 3.10. Calculated potential energy profile within the p–i–n device for several bias voltages (Sichanugrist et al., 1984).

$E_{ch2} = 0.55$ eV (see Sections 2.5 and 2.8). This would alter the energy profiles shown in Fig. 3.10.)

In order to find the photocurrent the following continuity equations for electrons and holes have to be solved simultaneously in the undoped layer:

$$D_p \frac{d^2 \Delta p(x)}{dx^2} - \mu_p F \frac{d \Delta p(x)}{dx} - \mu_p \Delta p(x) \frac{dF(x)}{dx} - R_{rec}(x) + G(x) = 0 \quad (3.29a)$$

$$D_n \frac{d^2 \Delta n(x)}{dx^2} + \mu_n F \frac{d \Delta n(x)}{dx} + \mu_n \Delta n(x) \frac{dF(x)}{dx} - R_{rec}(x) + G(x) = 0 \quad (3.29b)$$

It is assumed that $D_p(D_n)$ and $\mu_p(\mu_n)$ are independent of the electric field. Here, $\Delta p(x)$ and $\Delta n(x)$ are the excess carrier densities for holes and electrons and $F(x)$ is the electric field. Further, $R_{rec}(x)$ is the recombination term and is a function of $\Delta p(x)$, $\Delta n(x)$ and is given by the Shockley–Read type recombination mechanism:

$$R_{rec}(x) = \frac{n(x)p(x) - n_i^2}{\tau_p[n(x) + n_i] + \tau_n[p(x) + p_i]}, \quad (3.30)$$

where $n(x) = \Delta n(x) + n_0(x)$, $\Delta p(x) = \Delta p(x) + p_0(x)$, $n_i^2 = n_0(x)p_0(x)$. τ_p

and τ_n are the hole electron and lifetimes and n_0 and p_0 are thermal equilibrium concentrations of electrons and holes, respectively. These quantities are assumed to be constants of the material and independent of $n(x)$, $p(x)$, and x. Since under strong illumination p_0, n_0, and n_i are much smaller than the excess carrier concentration $\Delta p(x)$ and $\Delta n(x)$, then Eq. (3.30) reduces to,

$$R_{\text{rec}}(x) = \frac{\Delta p(x) \Delta n(x)}{\tau_p \Delta n(x) + \tau_n \Delta p(x)}. \quad (3.31)$$

If the cell is illuminated through the p-layer, then $G(x)$ (carrier generation) can be expressed as,

$$G(x) = \int [1 - R_f(\lambda)] N_0(\lambda) e^{(-\alpha_p(\lambda) d_p)} \alpha_i(\lambda) e^{-(\alpha_i(\lambda)x)} d\lambda, \quad (3.32)$$

where R_f is the reflection coefficient at the front surface and d_p is the thickness of the p-layer. $\alpha_p(\lambda)$ and $\alpha_i(\lambda)$ are the absorption coefficients of the p- and i-layers respectively.

The boundary conditions are determined by the hole-electron recombination velocity such that at the p^+/i ($x = 0$) interface it is given by S_{no}, and at the n^+/i ($x = d$) interface by S_{po}. A further assumption is that $\Delta p(x)|_{x=0} = 0$ and $\Delta n(x)|_{x=d} = 0$. Hence,

$$D_n \frac{d \Delta n(x)}{dx} \bigg|_{x=0} = \Delta n(0)[S_{no} - \mu_n F(0)], \quad (3.33a)$$

$$-D_p \frac{d \Delta p(x)}{\Delta x} \bigg|_{x=d} = \Delta p(d)[S_{po} - \mu_p F(d)], \quad (3.33b)$$

and $\Delta p(0) = \Delta n(d) = 0$. If the cell is illuminated through the p-layer, as is generally the case with high efficiency cells, the photocurrent can then be calculated using

$$I_{\text{ph}} = -eD_p \frac{d \Delta p}{dx} \bigg|_{x=0} + eS_{no} \Delta n(0). \quad (3.34)$$

Finally, the I-V characteristics of illuminated pin cells are calculated by assuming that the principle of superposition holds such that,

$$I_d(V) = I_0 \left[\exp \frac{eV}{QkT} - 1 \right] \quad (3.35)$$

where I_0 is the reverse saturation current and Q is the diode quality (ideality) factor. By taking I_0 and Q as constants (although they should be related to the $\mu\tau$ products), Figs. 3.11a and b show V_{oc}, J_{sc}, FF and η as a function of the $\mu\tau$ products when the cells are illuminated through the p^+ and the n^+

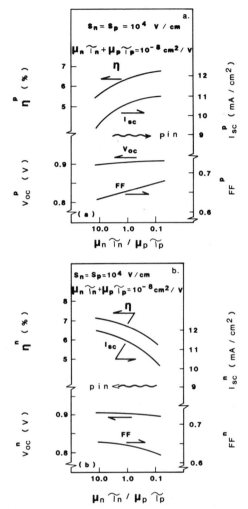

Fig. 3.11. Open circuit voltages $V_{oc(p)}$, $V_{oc(n)}$; the short circuit current densities $I_{sc(p)}$, $I_{sc(n)}$; the fill factors $FF_{(p)}$, $FF_{(n)}$ and the conversion efficiencies η_p, η_n when the cell is illuminated through its p and n layers, respectively, as a function of $\mu_n \tau_n / \mu_p \tau_p$ (Sichanugrist et al., 1984).

layers, respectively. Note that if $\mu_n \tau_n = \mu_p \tau_p$, the FF are identical for the devices when the illumination is either through the p^+ or the n^+ layer; if $\mu_n \tau_n > \mu_p \tau_p$ the parameters improve when illuminated through the n^+ layer and if $\mu_n \tau_n < \mu_p \tau_p$ the parameters improve when illuminated through the p^+ layer.

Figure 3.12 shows the photovoltaic parameters with $\mu_n \tau_n$ and $\mu_p \tau_p$ as the variables and Fig. 3.13 shows the computed photovoltaic parameters with S_n

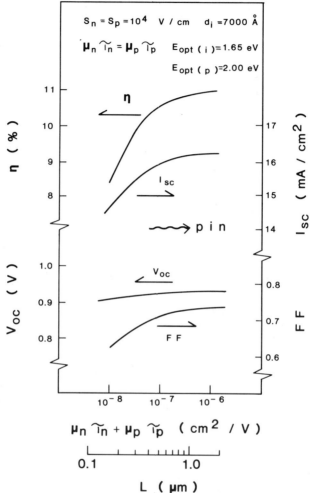

Fig. 3.12. Photovoltaic parameters obtained when the cell is illuminated through its p layer as a function of $\mu_n \tau_n + \mu_p \tau_p$ (Sichanugrist et al., 1984).

as the variable. There is a considerable drop in the FF with increasing S_n. It is then concluded that not only the diffusion length of the minority carriers, but also the interface recombination velocity, affects the performance of the solar cell.

The above discussion is necessarily an estimate of the photovoltaic parameters, since the full details of the recombination mechanisms involved in a-Si:H type alloys have not been incorporated. The more complete analysis presented by Sichanugrist et al. (1984) used Shockley-Read Hall

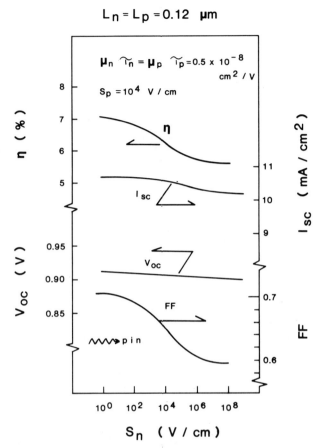

Fig. 3.13. Photovoltaic parameters obtained when the cell is illuminated through its *p* layer as a function of S_n (Sichanugrist *et al.*, 1984).

statistics rather than the trap modulated Fermi–Dirac functions for each of the different species of traps, which are bound to exist since impurities and inherent defect levels are generally present in the material. Further, the capture cross section has been shown to vary with energy (Okushi *et al.*, 1982), which will have a bearing on the recombination kinetics and hence on the final photovoltaic parameters. The effect of electron affinity changes that could be present in the p^+, i, and n^+ layers has not been considered as yet and will also alter the outcome of the analysis. Nevertheless, if we consider the upper limits of $V_{oc} \sim 1\,\mathrm{V}$, $J_{sc} \sim 22\,\mathrm{mA\,cm^{-2}}$ and $FF \sim 0.80$, then $\eta_{max} \cong 18\%$. However, if the DOS, especially the band tail widths, can be altered, then higher efficiencies are theoretically possible.

3.1.2.5 Tandem Cells

In the foregoing discussion the optimum conversion efficiency for a single a-Si:H type solar cell was found to fall in the range 12–20%. The uncertainty of nearly a factor of 2 is primarily due to the lack of consensus on the parameters governing electronic transport in a device configuration. Nevertheless, if the average of the forecasted efficiencies is taken to be 15%, this is sufficiently high for the meaningful development of solar cells for terrestrial power applications (Armstrong-Russel et al., 1983).

One approach that would significantly improve the conversion efficiency is the use of a multijunction or tandem cell techniques. Here the cells are connected in series and stacked on top of each other in the configuration $p^+i_1n^+p^+i_2n^+, \ldots$, with the widest band gap material (i_1) facing the incident illumination. In this type of arrangement the low energy photons are directed to cells fabricated from materials with relatively narrow band gaps and the high energy photons are directed to wide band gap cells, where their energy is not dissipated by creating electron-hole pairs with energies much in excess of the band gap. The thicknesses and the values of the band gaps are arranged such that the currents generated from the constituent cells are the same.

Since by far the most useful amorphous material for solar cell applications is presently the a-Si based alloy with a band gap of 1.7 eV, then we are restricted to form multijunction stack cells based on this semiconductor. Calculations leading to the design of multijunction stack arrays have been performed by Kuwano et al. (1981), who predict a conversion efficiency of 24% with a stack consisting of 1.45 eV, 1.7 eV, and 2.0 eV band gap materials (see Section 2.12).

In what follows we expand on our earlier discussion (Sections 3.1.2.2 and 3.1.2.3) and consider, in particular, the voltage addition that is to be expected as a function of the material quality of the semiconductors involved. Using the incident solar power relation given by Eq. (3.6) and the absorption coefficient given by Eq. (3.7), J_{sc} generated for an ideal 2 cell and 3 cell

Table 3.2. Calculated Short Circuit Density (Normalized to 100.0 mW/cm^2) and Open Circuit Voltage for a 2 Cell and a 3 Cell Multijunction Stack Arrangement (Madan, 1985)

Number of cells	Band Gap (eV)			J_{sc} (mA cm^{-2})	$g_{min} = 10^5$ cm^{-3} $V_{oc}(V)$	$g_{min} = 10^{14}$ cm^{-3} $V_{oc}(V)$
	Eg_1	Eg_2	Eg_3			
2	2.245	1.7	—	10.8	2.64	2.21
2	—	1.7	1.0	21.3	1.56	1.25
3	2.245	1.7	1.35	10.8	3.15	2.74

SOLAR CELLS

multijunction stack is given in Table 3.2). Note that for a 2 cell configuration there are two possible structures, resulting in either a high voltage or a high current device, with semiconductors having band gaps of either 2.24 eV or 1.35 eV to be coupled with the 1.7 eV band gap material. The voltage addition to be expected from the cells in the multijunction stack is also shown, and is calculated by the method given previously. (Note that the generation rate used was appropriate to that part of the solar spectrum incident on the cell.) These calculations are based on an assumed minimum DOS of 10^5 cm^{-3}, as indicated in Table 3.2. This also shows the effect of raising the minimum in the DOS to 10^{14} cm^{-3} in all the semiconductors involved in the stack. Note that there is a considerable drop in the open circuit voltage addition due to increased recombination (Madan, 1985). However, any change in the minimum in the DOS as well as in the capture rate constants of the traps will also affect the voltage addition in the stack.

In practice, an increase in the minimum DOS will be reflected in a decrease in J_{sc}, as well as a decrease in the *FF*. This is primarily due to the shrinkage of the space charge region (or depletion width) as well as an expected decrease in the minority carrier diffusion length. Further, the calculations indicate there does not appear to be much of an advantage, in terms of the conversion efficiency, to be gained in going from a 2 cell to a 3 cell array since the voltage increase in the 3 cell configuration is virtually cancelled out by the current reduction. The slight improvement in the conversion efficiency may well be counteracted by the added complexity and cost involved in the fabrication of a 3 cell stack.

As discussed in Section 2.12, the band gap of amorphous semiconductors can be altered by alloying. However, this introduces compositional disorder in addition to the positional disorder inherent in amorphous semiconductors. The effect of this point is now considered. The slope of the band tail, E_{ch}, is associated with the extent of the disorder and is related to E_0, the Urbach parameter defined in Section 2.4. We now rewrite the DOS distribution defined by Eq. (3.9) as

$$g(E) = g_0 \cosh\left(\frac{E - E_f}{E_{ch}}\right) + g_{min}, \tag{3.36}$$

where g_{min} represents a flat distribution of states in the mid-gap region corresponding to deep defects (Madan, 1985b; von Roerdern and Madan, 1985). In Fig. 3.14 V_{oc} as a function of E_g is shown for various values of the disorder parameter, E_{ch} (or E_0). Note that V_{oc} falls off rapidly because of a larger E_{ch} due to increased recombination. (In this context, it is interesting to note that the trade-off between voltage loss and increased J_{sc} has been reported for the case of unalloyed sputtered a-Si : H solar cells, where Morel and Moustakas (1981) reported that when E_g was altered from 1.60 to

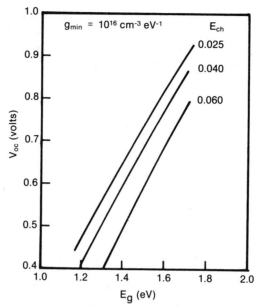

Fig. 3.14. V_{oc} plotted as a function of the band gap for materials with different minima in the DOS spectrum: (a) 10^5 cm^{-3}, (b) 10^{14} cm^{-3}, (c) 10^{17} cm^{-3}. $R = 10$ and total incident power is 135.7 mW cm^{-2} (Madan, 1983).

1.95 eV, the larger band gap material had a better overall performance because of its larger V_{oc}.) These considerations also reveal that the quest for low g_{min} values is perhaps insufficient and that attention should also be paid to the extent and width of the band tails, if the alloying techniques are to succeed.

From the above discussion it would seem that for a 2 cell stack we must develop new device quality amorphous semiconductors with an E_g of either 2.25 eV or 1.0 eV, and for a 3 cell stack an E_g of either 2.25 eV or 1.35 eV. However, the above calculations were based on the total absorption of the appropriate part of the solar spectrum incident on the constituent cell. If this condition is relaxed then the cell configurations can be optimized using different thicknesses as a variable, resulting in slightly different band gap requirements in the multijunction stack arrays.

3.1.3 Schottky and MIS Solar Cells

A-Si Schottky barrier (SB) and Metal–Insulator–Semiconductor (MIS) cells were the first investigated. Reports of conversion efficiencies of 2.5% (and a later paper by Carlson and Wronski (1976) reporting a conversion efficiency of 5.5% using a Pt/a-Si:H SB structure) sparked considerable interest in the use of these structures for solar energy conversion. Although the major interest now is in p^+-i-n^+ type device, the SB type cells are useful from a diagnostic point of view, especially for determining the electronic quality of the semiconductor employed.

3.1.3.1 Schottky Barrier Cells

A Schottky barrier, or more precisely, a metal–semiconductor junction, is generally formed by depositing a high work function metal onto an n-type semiconductor or, conversely, a low work function metal onto a p-type semiconductor. This creates a space charge region in the semiconductor, resulting in a rectifying contact. The barrier arises because of the existence of a contact potential between the metal and the semiconductor (Rhoderick, 1978; Shaw, 1981). According to Mott (1938), the barrier height, ϕ_b, should be equal to the difference between the work function of the metal, ϕ_m, and the electron effinity, X_s, of the semiconductor. However, contacts to covalently bonded crystalline semiconductors do not generally produce a simple relation between ϕ_b and ϕ_m. One reason for this is the existence of interface states that can arise because of interdiffusion, metal-induced gap states, chemical reactions, foreign atoms and an insulating layer on the surface. All can give rise to levels that influence the properties of the semiconductor in a region near the interface. These can result in a large density of localized states at or near the interface, whose energy levels are distributed within the gap. These may be present in densities as great as 10^{15} cm^{-2}, or about one state for each surface atom in the crystal. The electrical properties of the semiconductor are influenced to a considerable depth into the material by electrons occupying the interface states, since in a crystal any excess charge in these states must be compensated by changes in the free carrier concentration in order that charge neutrality be maintained. For example, negatively charged surface states on an n-type crystal repel free electrons from the surface, leaving positively charged ionized donors to neutralize the effect of the surface charges. This causes a depletion layer, with a resultant built-in potential at the free surface. A similar situation arises for positively charged states on a p-type semiconductor, with the concommitent downward bending of the bands and the formation of an accumulation layer. If the band bending is sufficient an n-type surface (inversion layer) can be produced on a p-type semiconductor, and vice versa.

Generally, two types of interface states are readily classified. The localized states located at the semiconductor/insulator interface are called fast states since their occupancy can change rapidly with changes in the bulk carrier concentration. For example, if the carrier concentration is altered by photon absorption, the density of charge in the fast surface states can readjust in times on the order of a microsecond. (These levels are responsible for the recombination of excess carriers within a device.) However, if foreign atoms or the like are located within the oxide (insulating) layer or at the oxide/air interface, the charge in these states may require many seconds to readjust under non-equilibrium conditions. Generally, it has been found that the fast interface state density is on the order of 10^{12} cm^{-2} on crystalline Si and Ge, while the slow interface state density is between 10^{11} to 10^{13} cm^{-2} and depends upon the thickness of the oxide layer (see, e.g., Many et al., 1965).

Surface states influence the rectifying behavior of the metal–semiconductor contact. For example, contact to crystalline Si is usually made after cutting, polishing and chemically etching, followed by the evaporation of a metal in a relatively dirty vacuum of 10^{-6} torr. The procedure of etching leads to the formation of a thin insulating layer, usually SiO$_2$. Deposition of metals onto Si covered with this type of native oxide leads to a barrier height that is measured to be about $\frac{2}{3}$ of the band gap of Si. Bardeen (1947) pointed out that the surface states located in the gap *pin* the Fermi level. Consequently, the barrier height becomes independent of the work function of the metal used. Present models for this pinning effect are decidedly more complex and specific (see, e.g., Shaw, 1981).

Intimate contacts can be formed by cleaving the semiconductor in ultrahigh vacuum, followed immediately by evaporating the metal onto the cleaved surface. This type of junction, although ideal (but not generally abrupt) because of the absence of an insulating layer between the metal and

Table 3.3. Barrier Heights of Intimate Contacts to Crystalline Si (Thanakalius et al., 1976).

		ϕ_b (eV)	
Metal	ϕ_m (eV)	from J/V characteristics and photoelectric measurements	from C/V characteristics
Al	4.17	0.61	0.70
Fe	4.58	0.63	0.98
Co	4.97	0.61	0.81
Ni	5.10	0.59	0.74
Cu	4.55	0.62	0.75
Ag	4.41	0.68	0.79
Pt	5.30	0.71	0.82
Au	5.10	0.73	0.82
Pb	4.25	0.61	0.72

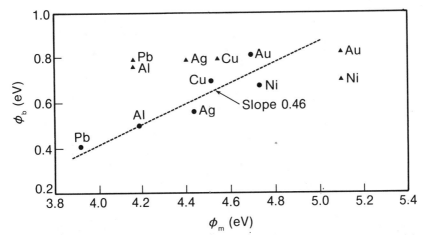

Fig. 3.15. Barrier height on etched (●) and cleaved (△) Si surfaces (Turner and Rhoderick, 1968).

the semiconductor, is not completely understood. Intimate contacts to crystalline Si have been studied by, among others, Thanakalius et al. (1976). In this work crystalline Si samples were cleaved along the (111) plane in a vacuum of 5.10^{-11} torr, followed by the evaporation of various metal films. The work function of the metal was measured without breaking the vacuum. The results are given in Table 3.3., where we note that the barrier height is not a unique function of the work function of the metal.

Metal-insulator-semiconductor (MIS) type structures are easier than SB structures to model because the insulator can isolate the metal from the semiconductor to such an extent that it is a reasonable approximation to regard the interface states as a property of the particular semiconductor-insulator combination. Contacts formed in this way lead to results different from those shown in Table 3.3 (intimate contacts). Figure 3.15 shows that a correlation between ϕ_b and ϕ_m exists when etched crystalline Si surfaces are allowed to age; SiO_2 layers are allowed to form prior to the deposition of various metals (Turner et al., 1968).

In summary, it appears that for crystalline Si, ϕ_b is proportional to ϕ_m in the presence of an insulating layer but in its absence ϕ_b is not simply related to ϕ_m. The same conclusions seem to hold for GaAs (Smith, 1969).

A proposed explanation for these observations is given in terms of the electronegativity of the metal, X_m. It is found experimentally that ϕ_b only increases with X_m when the material is ionically bonded, such as SiO_2. When metals are contacted to covalently bonded materials, such as Si or Ge, ϕ_b is almost independent of X_m. The dependence of ϕ_b on X_m is measured by the interface index $S = d\phi_b/dX_m$ and a plot of S as a function of ΔX, the

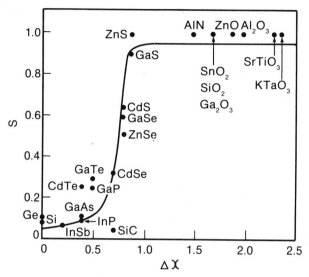

Fig. 3.16. The interface index, S, vs. ionicity, ΔX (Kurtin et al., 1969).

semiconductor ionicity, is shown in Fig. 3.16 (Kurtin et al., 1969). Note that the degree of ionicity seems to determine the functional dependence of ϕ_b on X_m. However, there are enough exceptions to the behavior shown in Fig. 3.16 as to make the relationship between S and ΔX quite suspect.

Brillson (1978) has shown that the results of Fig. 3.16 can be equally well fitted to ΔH instead of ΔX, where ΔH is the heat of formation. Covalent semiconductors have a lower heat of formation than ionic semiconductors and are thus more likely to react at the metal–semiconductor interface. When contacted to Si, most metals, including all the transition metals, form silicides, especially after heat treatment. A vast majority of these silicides exhibit metallic conductivity, in which case the silicide–silicon junction behaves very much like a metal-semiconductor contact, exhibiting rectification. The silicide–silicon interface is formed some distance below the original surface of the Si and it is, therefore, essentially free from contamination. Contacts formed in this way generally show stable electrical characteristics that are very close to ideal (e.g., Rhoderick, 1978). It has been shown that the presence of an oxide at the surface can severely inhibit silicide formation. Crider et al. (1980) have shown that the formation of platinum silicide of the type PtSi can be completely halted, even in an O_2 partial pressure of 10^{-7} torr.

As is evident from the foregoing discussion, metal–semiconductor contacts are difficult to understand; the behavior of the junction is complicated by interface states, the presence of a native insulator, surface preparation conditions and chemical reactivity. Contacts to amorphous semiconductors are perhaps even more problematical because of the present inadequate understanding of the properties of the bulk material. In the usual preparation of a-Si/metal type cells, there is undoubtedly a native oxide formation that occurs. Hence, in the remainder of this section we treat SB and MIS cells as having virtually similar structures.

3.1.3.2 Characterization of Schottky Barriers

Rectifying behavior in a a-Si based devices, in the simplest form, is achieved by first depositing the back contact. Here a thin n^+ layer of a-Si:H alloy is deposited onto a low work function metal such as Mo, Cr or stainless steel. Although ϕ_m exceeds X_s for a-Si:H alloys, nevertheless an n^+ layer forms a low resistance, linear (ohmic) contact with the metal (see, e.g., Shaw, 1981). This occurs because with the addition of 1% PH_3 in the gas phase it is found that not all the P atoms enter a tetrahedral configuration (see Section 2.8). The nontetrahedrally coordinated P atoms create extra states within the gap region with the consequence that the n^+ layer becomes highly defective. Therefore, although the n^+ layer contact to a metal such as Cr will in principle produce a barrier region, this region is so thin that field emission tunnelling from the semiconductor to the metal occurs and a low resistance contact results. A similar situation occurs for a p^+ layer deposited onto a metal or a TCO (Transparent Conducting Oxide).

After the deposition of the n^+ layer, an undoped layer of a-Si:H, which is usually slightly n-type, is deposited and the simplest type of rectifying Schottky contact is provided by depositing metal layers such as Al, Mo, Cr, Ni, Au, Pd, Pt, etc. The metal layer is usually deposited after the a-Si:H layer has been exposed to ambient conditions. Here a native oxide will always form between the semiconductor and the metal. To our knowledge, no study of intimate metallic contacts to a-Si:H has ever been made.

The quality of a SB type cell is determined by the forward and reverse bias characteristics. The barrier height can be determined by studying the reverse saturation current density, J_0 (see, e.g., Rhoderick, 1978). The motion of an electron in going from a semiconductor to a metal is governed by the processes of diffusion and drift. When it reaches the metal it must have a large component of momentum normal to the interface, and this defines a narrow allowed cone of momentum whose axis is normal to the boundary. This latter process is governed by the number of Bloch states in the metal that communicate with the semiconductor. These two processes, drift and diffusion carrier transport through the depletion region, and the transfer across the interface, are effectively in series. The current is primarily determined by whichever presents the greater resistance to current flow. In the diffusion theory (DT), which should be applicable for low mobility solids, the bottleneck is presented by the barrier region, whereas in the thermionic emission theory (TET) the transport of carriers from the semiconductor into the metal limits the current. The latter theory is usually applicable for high mobility solids.

In the limit when ideal TET is applicable, the current-voltage characteristics can be described by (see, e.g., Rhoderick, 1978)

$$J = A^* T^2 \exp\left(-\frac{e\phi_b}{kT}\right)\left(\exp\frac{eV}{kT} - 1\right), \quad (3.37a)$$

where A^* is the Richardson constant.

When DT is applicable, the characteristics are given by

$$J = e\mu g(E_c) F_{max} \exp\left(-\frac{e\phi_b}{kT}\right)\left(\exp\frac{eV}{kT} - 1\right), \quad (3.37b)$$

where F_{max} is the electric field at the semiconductor–metal interface. (A more complete description incorporating both viewpoints has been developed by Crowell and Sze (1966).) Equations (3.37a) and (3.37b) hold as long as ϕ_b is not a function of the bias voltage. However, ϕ_b may depend on the bias voltage via: (1) image force lowering of the barrier; (2) interface states; (3) the presence of an insulating layer. Then, assuming a linear relationship between ϕ_b and V

$$\phi_b = \phi_{b0} - (\Delta\phi_{bi})_0 + \beta V, \quad (3.38)$$

where the subscript zero denotes zero bias conditions and $\Delta\phi_{bi}$ is due to image charge lowering of the barrier. It can be shown that (see, e.g., Rhoderick, 1978)

$$J = J_0\left[\exp\left(\frac{eV}{QkT}\right)\right][1 - e^{-eV/kT}]. \quad (3.39)$$

If $\partial\phi_b/\partial V =$ constant, then Q will be a constant and is referred to as the ideality factor. By plotting $\ln\{J/1 - \exp[-eV/kT]\}$ vs. V, the slope will be e/QkT, for all V. Then

$$\frac{1}{Q} = \frac{kT}{e}\frac{d}{dV}\ln\left\{\frac{J}{(1 - \exp[-eV/kT])}\right\}; \quad (3.40)$$

for $V \geq 3kT/e$

$$\frac{1}{Q} \simeq \frac{kT}{e}\frac{\partial \ln J}{\partial V}. \quad (3.41)$$

For an ideal diode, $Q = 1$; in practice $Q \neq 1$ is virtually always observed

SOLAR CELLS

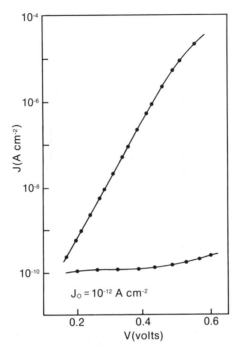

Fig. 3.17. Forward and reverse dark J-V characteristics for a MIS a-Si:F:H type device (Madan *et al.*, 1980).

and the amount of deviation from unity generally represents losses within the device. An example of the forward and reverse I-V characteristics of a SB (or rather a MIS) type cell is shown in Fig. 3.17 (Madan *et al.*, 1980a). Note that the rectification ratio at $V \sim 0.5$ V is 10^5, indicating good barrier formation. The curvature that is apparent in the J-V characteristic for forward bias voltages exceeding 0.4 V is due to the resistive nature of a-Si:H material. (This would be an inhibiting factor for solar cell performance were it not for the highly photoconducting nature of the material.)

Figure 3.18a shows the results of a microprobe analysis of a-Si:F:H n^+/i type devices (Madan *et al.*, 1984). Note that for both types of devices there is an abrupt transition of the P concentration from the n^+ to the i layer, with slight inadvertent P contamination within the intrinsic layer. Further, the O and C contamination is $<0.5\%$ for both types of devices. It should be noted that more recent results reveal O and C contamination at lower levels of 2×10^{18} and 1×10^{17} cm^{-3} respectively, due to the better vacuum systems now being employed. This leads to better opto-electronic properties such as minority carrier diffusion lenths as long as 1 μm (Nakano *et al.*, 1986).

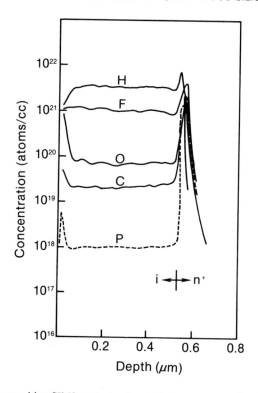

Fig. 3.18. (a) Composition SIMS analysis of an MIS-type structure (Madan et al., 1980).

As noted in Section 2.1, the preparation conditions have a profound efffect on the bulk properties of the material and the differences are mirrored in the DOS; this is also reflected in the diode characteristics, as observed by Deneuville et al. (1979) and shown in Table 3.4. The Q value changes with deposition conditions and reaches a minimum of 1.14 for $T_s = 300°C$ and a Silane gas deposition pressure of 0.06 torr. They reason that the decrease of Q with increasing deposition temperature is due to a decrease in the number of defect states, which then minimizes the recombination current. They also attempted to distinguish between the DT and TET theories by plotting $\ln J_0$ vs. $10^3/T^2$ and $\ln J_0$ vs. $10^3/T$ respectively, and obtained straight lines for both cases, deducing $\phi_b = 1.05$ eV and 0.95 eV respectively. It should be recognized that in order to distinguish between the two theories the data have to be taken at much lower temperatures, but since the current levels exhibited by the device are low, experimental verification of the dominant transport mechanism becomes difficult.

SOLAR CELLS

Table 3.4. Static and Transport Diode Parameters as a Function of Substrate Temperature, T_s and Silane Gas Pressure, P for Low Gas Flow Rates (Denneuville and Brodsky, 1979).

T_s (°C)	200		250		300	
P (torr)	0.06	0.3	0.06	0.3	0.06	0.3
V_{oc} (V)	0.44	0.50	0.52	0.50	0.39	0.48
Q	1.35	4.2	1.28	4.2	1.14	3.7
J (A/cm^2)	2.0×10^{-10}	1.2×10^{-8}	6.0×10^{-11}	1.2×10^{-8}	5.2×10^{-10}	2.7×10^{-8}

3.1.3.3 Barrier Height Determination

One of the easiest ways to determine ϕ_b is to plot the extrapolated reverse saturation current $\ln J_0$ against $10^3/T$. Generally, $\phi_b \simeq 1.0$ eV for Pd or Pt contacts. ϕ_b is lowered when low work functions metals are used, such as Al or Cr (e.g., Carlson, 1977; Madan et al., 1980c). Alternatively, by employing DT [Eq. (3.37)] and using an extrapolated value of J_0, ϕ_b can be calculated if the other parameters are known. Wronski et al. (1976) using $g(E_c)kT = 1.10^{20}$ cm^{-3}, $\mu = 5$ cm^2 s^{-1} V^{-1} and $F_{max} = 4.10^4$ V cm^{-1}, calculated that $\phi_b = 0.97$ eV for Pd/a-Si:H structures. Madan et al. (1980b) used a similar expression except that F_{max} was calculated based on a realistic DOS spectrum given by Eq. (3.9). They obtained $\phi_b = 1.0$ eV, which was in agreement with ϕ_b obtained from the plot of $\ln J_0$ vs. $10^3/T$ (Madan et al., 1980a) for their Pd/a-Si:F:H SB type cell.

Another way to experimentally determine ϕ_b is to consider the characteristics of the diode in the presence of light. Since V_{oc} can be approximated by,

$$V_{oc} \cong \frac{Q'kT}{e} \ln\left[\frac{J_{sc}}{J_0} + 1\right], \qquad (3.42)$$

where Q' is the diode quality factor under illumination. (Q' is generally different from its value in the dark because of a potential drop in the interfacial layer, which becomes negligibly small due to interface states that may take on a positive charge). Using $Q' = 1.0$, and plotting V_{oc} as a function of temperature, the intercept at $T = 0$ K gives ϕ_b, which for Pd/a-Si:F:H SB junctions is 1.12 eV (Madan et al., 1980a), in close agreement with the previous result.

A more direct measure of ϕ_b is provided by internal photoemission. In order for photoemission to be observable, the semiconductor must have a low value of optical absorption below E_g, otherwise, as shown in Fig. 3.19, the carriers generated in the semiconductor (process A) will overwhelm the photoemission current (process B). Using this technique, Wronski et al., (1980)

198 OPTO-ELECTRONIC APPLICATIONS OF AMORPHOUS SILICON MATERIALS

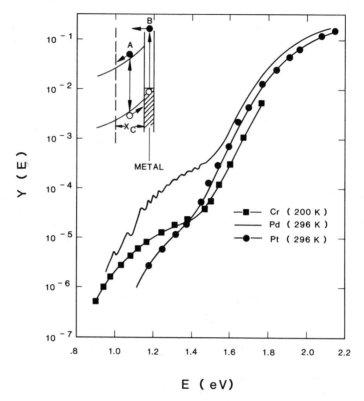

Fig. 3.19. Photoelectric yield $Y(E)$ vs. photon energy E for semitransparent metal contacts on SiH_x films. The temperatures at which the measurements were made are indicated in the figure. The interference fringes measured for the Cr and Pt were of the same magnitude as for the Pd—for sake of clarity they were averaged out. The schematic Schottky barrier in the inset shows the internal photoemission process B and the semiconductor absorption process A. The region of the semiconductor over which photogenerated carriers are collected is indicated by X_c (Wronski et al., 1980).

determined barrier heights for Cr, Pd and Pt junctions on a-Si:H SB devices. In Fig. 3.20, we show their results replotted using $[Y(E)]^{1/2} = A(E - \phi_b)$, where $Y(E)$ is the collection or quantum efficiency. (A is a constant determined by the number of photons absorbed in the metal film and by the escape probability of an excited electron in the metal.) Values of ϕ_b are determined from the intercepts of the straight lines with the energy axis. The values found were 0.7, 0.98 and 1.12 eV for Cr, Pd and Pt respectively. These are consistent with the barrier heights inferred from reverse saturation currents. Recent work (Mariucci et al., 1987) has also indicated the resonant tunneling through localized states in the conduction band tail will also affect ϕ_b.

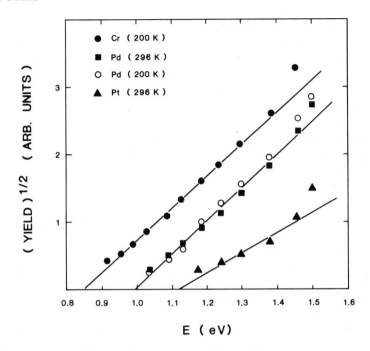

Fig. 3.20. Plot of $[Y(E)]^{1/2}$ vs. E. The intercepts of the straight lines with the E axis indicate the potential barrier heights for the three metals, Cr, Pd and Pt. The temperatures at which the measurements were made are indicated in the figure (Wronski *et al.*, 1980).

3.1.3.4 Annealing Effects on Schottky Barrier Contacts

The annealing of rectifying contacts causes a change in the electrical characteristics. In order to understand this effect we first recap the effects of annealing on crystalline Si/metal junctions. Chino (1973) first reported that Al/crystalline Si junctions heated above 450°C showed a significant change when the Si was *n*-type, and these changes could be ascribed to an increase in ϕ_b. Basterfield *et al.* (1975) had attributed this to the fact that Si can be taken up into a solid solution by Al by an amount determined by the solubility limit at the temperature of interest. Upon cooling, the dissolved Si recrystallized onto the *n*-type Si for form an Al doped *p*-type layer, since Al acts as an acceptor. Krishna *et al.* (1981) heat treated a-Si:H/Al diodes and reported that the effective barrier height changed from 0.75 eV in the unannealed state to 0.98 eV after a heat treatment of one hour at 200–250°C, with the diode quality factor changing from 1.5 to 1.2. The diode properties degraded after a 300°C anneal. This was attributed to hydrogen effusion from the a-Si:H material, leading to an increase in the DOS (see Section 2.2).

The increase in ϕ_b with annealing (for temperature below 250°C) was attributed to the formation of a p^+ region below the metal contact. However, conflicting results have been reported by Ishihara et al. (1982). Annealing effects on Al/a-Si:H and Ni–Cr/a-Si:H interfaces were studied and the heat treatment was carried out in the temperature range 100–350°C for 30 minutes in a vacuum of ~2 × 10^{-5} torr. For the NiCr contact case there was no visible change at the interface between NiCr and the a-SI:H material, even if the specimens were heat treated to 350°C. However, for the case of Al electrodes they reported that the surface became pitted at an anneal temperature of 170°C and the conductivity increased to beyond 1 $(\Omega\,cm)^{-1}$ because of crystallization.

Annealing of metal/a-Si:H junctions may be particularly important if the metal is capable of forming a silicide. As discussed earlier (see Section 3.1.3.1), most metals, including all the transition metals, can form silicides with Si after an appropriate heat treatment. Silicide formation between a-Si:H and Pd or Pt has been reported by Tsai et al. (1981), and was studied by interference enhanced Raman spectroscopy, which yields information regarding the interfacial structure. Figure 3.21 shows the Raman spectrum of Pd, Pt and Au for metal/a-Si:H/n^+ device structures after an anneal at 200°C for 15 minutes. The spectral features for the as-deposited sample, curve (d), are dominated by a broad peak at 480 cm^{-1} due to vibrations of the a-Si:H network. However, for the case of Pd contacts several sharp lines appear between 90 and 210 cm^{-1}, which are attributed to a crystalline silicide phase having a composition near Pd_2Si. The diode quality factor, Q, changed from 1.23 for the as-deposited film to 1.05 after annealing, and the stability of the diode was improved due to the silicide growth. Since all the above devices had been exposed to air prior to metallization, then Auger electron spectroscopy was employed to locate the position of the oxide. Figure 3.22 shows a compositional depth profile for a diode annealed at 300°C, which consisted of a 60 Å thick Pd film evaporated onto an unetched a-Si:H film. It appears that the oxygen present at the interface becomes relocated above the silicide layer after the heat treatment.

The case of Pt contact, as shown in Fig. 3.21 [curve (b)], also exhibits sharp lines at 80 and 140 cm^{-1}, indicating the presence of crystalline platinum silicide compounds. (These features apparently become evident only after annealing at 200°C.) Another aspect of the Pt contact behavior on a-Si:H material is that, unlike the Pd case, the two sharp features show the simultaneous presence of Pt_2Si and $PtSi$.

In contrast, for the case of a Au metal layer contact, sharp spectral features are absent, except for a line near 520 cm^{-1}, indicative of the crystalline Si structure. Therefore, Au does not interact with a-Si:H to produce stoichiometric compounds, as in the case of Pd and Pt contacts. Rather, Au causes

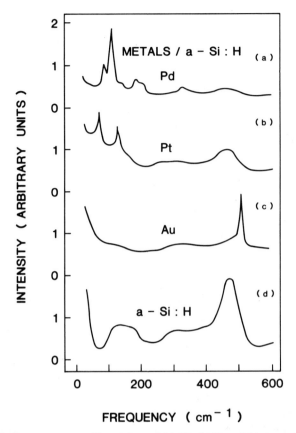

Fig. 3.21. The Raman spectra of 2 and 6 nm of (a) Pd, (b) Pt and (c) Au on 10 nm a-Si:H after 15 minute anneal at 200°C, compared to that of a-Si:H (d). All spectra were obtained using the IERS configuration (Tsai et al., 1981).

the amorphous silicon to crystallize at a low temperature of 200°C. Auger depth profiles and SEM photographs have shown that the Si diffuses into and through the Au, appearing near the top Au surface even at room temperature. The diffused Si remains amorphous at room temperature and crystallizes upon annealing at 200°C. The Au–Si phase agglomerates to form regions surrounding the crystalline Si islands.

Pt contacts to a-Si:H have also been studied by Goldstein et al. (1980) using AES (and LEED) techniques. a-Si:H samples near 1 μm thick were deposited onto polished Mo substrates and, prior to each measurement in a vacuum of 5×10^{-10} torr, the samples were sputter cleaned and heated to 225°C to produce as atomically clean, stable and reproducible set of starting conditions as possible. Their results suggest that there is complete absence

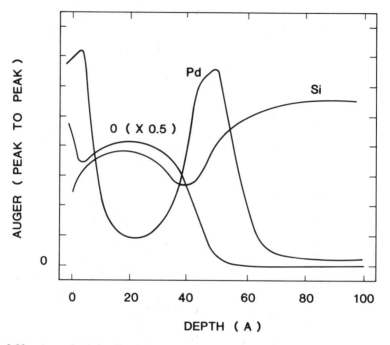

Fig. 3.22. Auger depth profile of a Pd/a-Si:H diode annealed to 300°C for 15 minutes (Tsai et al., 1981).

of a silicide line in the Auger spectrum. It was concluded that Si does not diffuse through and react with Pt at room temperature, but instead Pt forms in complete monolayers. Their results on Pt/oxide/a-Si:H layers involving a 20 Å thick oxide layer suggests that the behavior is somewhat different from what Tsai et al. (1981) reported, i.e., at coverages of less than a monolayer the Pt atoms initially are uniformly absorbed, but with further deposition the Pt agglomorates to form islands.

3.1.3.5 Schottky Barriers on p-Type a-Si:H

In the above discussion metal semiconductor contacts on slightly n-type a-Si:H alloys were considered. Greeb at al. (1982) have shown that a space charge region in a-Si:H can be formed with the use of low work function metals, such as Y and Yb, on slightly B doped p-type a-Si:H layers. The rectification ratio for Yb/a-Si:H exceeded 10^4 at $V \sim 0.5$ V with $\phi_b \sim 0.97$ eV and $Q = 1.32$. They found that ϕ_b increased with decreasing work function and seemed to saturate for $\phi_m \leq 3.5$ eV. From their studies of the dependence of ϕ_b on the Boron content in the a-Si:H active layer it was found that ϕ_b decreases with increasing B concentration. They attributed the saturation effect to the entrance of E_f into an energy range in which the DOS is high (band tail states); this is in harmony with observations made on n-type a-Si : F : H alloys (Madan et al., 1980b) and discussed more fully in Section 3.1.3.10.

3.1.3.6 Capacitance-Voltage Measurements on Schottky Barriers

The width of the depletion region, W, is generally measured in a SB configuration using the Capacitance-Voltage (C-V) technique. In crystalline semiconductor devices the barrier capacitance technique has been a useful tool for the investigation of the impurity distribution. Such measurements when performed as a function of frequency can also give information on deep lying centers in the gap (Roberts and Crowell, 1970).

In a C-V measurement the differential capacitance, C per unit area, of the depletion region is determined by the change in charge, q, for an incremental change in the surface energy, E_s, produced by the applied potential. Thus

$$C = \frac{dq}{dE_s}. \tag{3.43}$$

Because of the non-uniform DOS, q will be a function of E_s, which in turn is a complicated function involving $g(E)$. W and the potential variation in the space charge region can be evaluated using Eq. (3.28).

In this type of measurement an ac signal is superimposed across a device that is dc biased (forward, zero or reverse bias). The deep states present in the material affect the frequency response of this type of measurement in a complicated fashion. The charges located in the deep states will then only follow the changes in the ac voltage if the measuring frequency, $\nu > \tau^{-1}$, where τ is the thermal release time from the deepest trap to the conduction band edge, and can be described by

$$\tau = (N_c v_{\text{th}} C)^{-1} \exp\left(\frac{E_c - E}{kT}\right), \tag{3.44}$$

where N_c is the effective density of states at the conduction band edge, v_{th} is the thermal velocity, and C is the capture cross section of states situated at a level E. For example, $\tau \sim 500$ s for $E_c - E \simeq 1$ eV (Spear et al., 1978).

In Fig. 3.23 we show the frequency dependency of the dark capacitance taken at four different temperatures. Note that the capacitance saturates at high frequencies and at low temperatures to a value equal to the geometric capacitance (Tiedje et al., 1980). In Fig. 3.34 we show the C-V in reverse bias. Note that the capactiance increases with bias (Spear et al., 1978). This

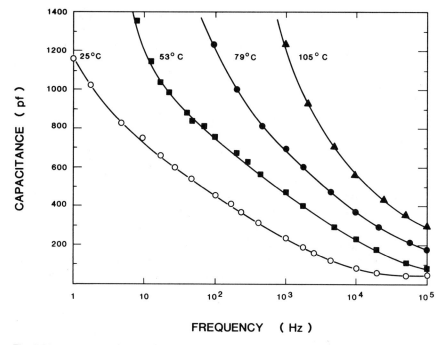

Fig. 3.23. Frequency dependence of capacitance, $C(\omega)$, at four different temperatures (Tiedje et al., 1980).

unusual behavior is caused by changes that occur in the charge distribution with reverse bias, since this will cause a sharp rise in the barrier profile. This behavior was confirmed theoretically by Shur et al. (1980), who assumed a DOS distribution given by Eq. (3.9) with $g_0 = 1.5 \times 10^{17}\,\text{cm}^{-3}\,\text{eV}^{-1}$, shallow band tails and $E_{ch} = 70\text{--}100$ meV. They predicted a depletion width less than 2000 Å. It was concluded that for such material under reverse bias, because of the rapidly rising DOS, the charge will become increasingly located at the metal/a-Si:H interface with the consequence that the capacitance will increase. Calculations by Singh and Cohen (1980) also confirmed these findings.

However, Tiedje et al. (1980) do not find an increase of C with reverse bias (0.5 V), which could be due to the utilization of better material (Hirose et al., 1979), since their data were interpreted on the basis of smaller DOS, with $g_{min} = 3 \times 10^{15}\,\text{cm}^{-3}\,\text{eV}^{-1}$ with $E_{ch} = 70$ meV. Further, from their studies they concluded that the potential profile is as shown in Fig. 3.25, with the barrier steeply sloped at the metal/a-Si:H interface ($x = 0$) and decaying with a long tail into the film.

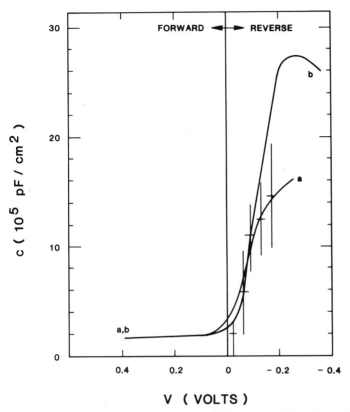

Fig. 3.24. (a) C–V calculated from the density of states of Fig. 2.28 (Singh and Cohen, 1980); (b) C–V calculated from Spear *et al.*, 1978 using the density of states of Fig. 2.28.

3.1.3.7 MIS Type Devices

The light output characteristics of SB devices can be altered substantially by the insertion of an insulator that forms an MIS device. The conversion efficiency of an MIS device can be 30–40% greater than that of a comparable SB device; this occurs primarily because of an increase in the open circuit voltage. One of the first attempts to do this using a-Si:H type materials was made by Wilson *et al.* (1978) using the structure Ni/TiO$_x$/a-Si:H/n^+/metal.

MIS structures rely upon ultra-thin interfacial layers such that the wave function of the charge carriers can penetrate the insulating layer, appear on the metallic side, and provide for current flow within the device. The interfacial layers should be such that

$$E_{gi} \geq X_s - X_i + E_{gs}, \tag{3.45}$$

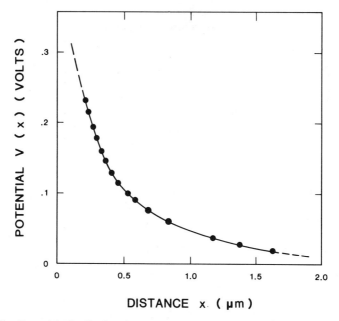

Fig. 3.25. Potential distribution in the depletion region derived from the frequency dependence of capacitance (Tiedje *et al.*, 1980).

where X_s and X_i are the electron affinities of the semiconductor and insulator respectively, and E_{gi} and E_{gs} are the energy band gaps of the corresponding materials. The inequality in Eq. (3.45) is required to guarantee that the device will operate as a minority carrier cell. This condition implies that an interfacial layer with a moderate band gap, but large electron affinity, could serve as an optimum insulating material.

Calculations of the forward $I(V)$ characteristics for a MIS Al-SiO$_2$-Si diode with 2 Ω-cm *p*-type base crystalline semiconductor have been performed for the ideal case of zero defect density at the oxide/semiconductor interface or within the oxide itself (Green *et al.*, 1974). Under forward bias a departure from the ideal diode behavior is calculated, indicating a changeover from semiconductor limited operation to tunnelling limited operation. Since the effect of the tunnel-dominated region would be equivalent to a series resistance, then when the oxide layer thickness, $\delta > 20$ Å, suppression of the photocurrent is to be expected. This should hold for MIS-type devices employing a-Si type alloys. Figures 3.26a and b show J_{sc} and V_{oc} as a function of the insulator thickness. Note that J_{sc} decreases for $\delta > 30$ Å and the device, according to the above interpretation, is in the tunnel controlled regime (Madan, 1985).

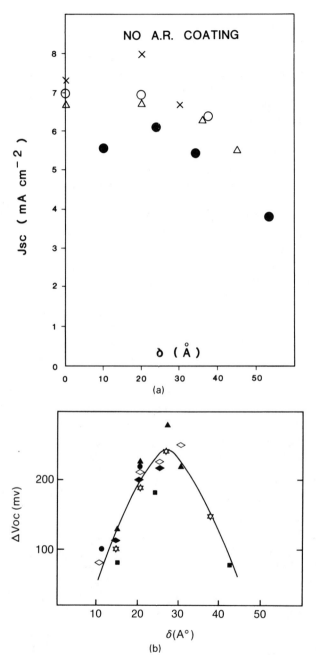

Fig. 3.26. (a) J_{sc} and (b) V_{oc} plotted against the oxide layer thickness in a MIS structure (Madan et al., 1981).

The maximum reported conversion efficiency for a MIS type device is 6.6% over an active area of 0.73 cm^2 (Madan et al., 1981). This was constructed by first depositing a thin, highly conductive n^+ layer onto a conducting substrate such as Mo in order to provide an ohmic back contact. Next, a 5000 Å thick active layer of a-Si:F:H was deposited and was followed by a thermal evaporation of a thin oxide (20–30 Å) layer; contact was made to the device using a 70 Å thick layer of a high work function metal, such as a Au:Pd alloy or Pt. Finally, a 350 Å thick layer of ZnS served as an antireflection coating. Oxides such as TiO_2, Nb_2O_5, SiO_2 and Ta_2O_5 were also attempted, but the most satisfactory and consistent results were obtained using Nb_2O_5. The reason for the use of an insulator becomes clear when we consider the change in V_{oc} with the oxide layer thickness, δ, shown in Fig. 3.26b. Note that for $\delta \sim 30$ Å, V_{oc} is enhanced by about 250 mV, without any deterioration in J_{sc} or in the fill factor. (Note that the conversion efficiency dropped to about 4% without the insulator.) The increase in V_{oc} becomes understandable in view of Eq. (3.42) in which J_0, the reverse bias saturation current, is suppressed when the insulator layer is inserted, since the electrons from the metal must tunnel first through the insulator into the a-Si:H material. The reduction in J_0 translates into a higher V_{oc}; note that the improvement in V_{oc} will depend upon the quality of a-Si:H/insulator interface as well as in the properties of the insulator itself.

Gutowicz-Krushin et al., (1981) have suggested that the introduction of the insulator could also enhance J_{sc}, primarily at the blue end of the spectrum, by reducing the thermal diffusion of electrons against the electric field. This was confirmed on a-Si:H (Abeles et al., 1981) and a-Si:F:H type devices (Madan et al., 1981; Madan, 1984), since there is an observed increase in the quantum efficiency at the blue end of the spectrum, as shown in Fig. 3.27.

Deliberate oxidation, as well as junction formation on "atomically clean" surfaces has been reported by Wronski et al. (1981) using conventional glow discharge a-Si:H films. An "intimate" contact was formed a few minutes after the a-Si:H film deposition by etching the surface in buffered HF, followed by rinsing in deionized water just prior to metallization. The oxidation was performed at 300°C prior to the evaporation of a semi-transparent electrode. The surface of the films were characterized using Auger Electron Spectroscopy for elemental surface characterization. By using photoelectron spectroscopy along with the Si Auger line shape, the nature of the oxide was investigated. It was found that the freshly etched surface exhibited an Auger line $(dN(E)/dE)$ peak at 90 eV, which is a characteristic of non-oxidized Si and is considered to be qualitatively similar to that of clean Si. The ratio of the intensities of the oxygen (510 eV transition) to that of Si (90 eV transition), was found to be between 0.03 to 0.10, which could be reduced to 0.005 by sputter cleaning. Upon oxidizing the

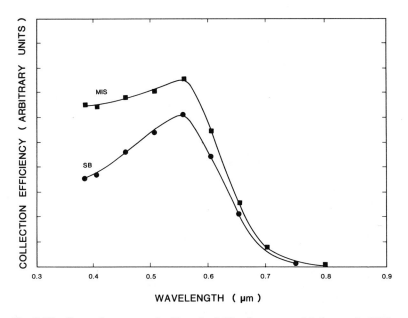

Fig. 3.27. Spectral response of a SB and a MIS cell structure (Madan *et al.*, 1981).

surface the amplitude of the oxygen signal increased while that of the Si line at 90 eV decreased, as shown in Fig. 3.28. (In this figure, curve A corresponds to a sputter cleaned sample, curve B to the oxidized state and curve C to the devices). Additional lines below 80 eV become evident after oxidation. It was concluded that the ratio of the amplitudes of the two peaks, $SiO_2(75)/Si(90)$, represented a measure of the state of oxidation of the surface. In Fig. 3.29, V_{oc} is plotted as a function of the $SiO_2(75)/Si(90)$ ratio for several Pd/a-Si : H junctions; note that there is an increase in V_{oc} with the degree of oxidation. The maximum enhancement in V_{oc} coincides with $\delta = 30$ Å, the oxide layer thickness, and is consistent with the results shown in Fig. 3.26b. However, the maximum is V_{oc} was only 0.7 V for the SiO_2 oxide, in contrast to $V_{oc} = 0.88$ V for the Nb_2O_5 case (Madan *et al.*, 1981). This low value of V_{oc} was attributed to a misalignment of the SiO_2 energy band with respect to the a-Si : H energy band (Wronski *et al.*, 1981).

3.1.3.8 Forward Bias *I–V* Characteristic in the Dark

Figure 3.30 shows the temperature dependence of the diode quality factor, Q, for a device with and without the oxide (Madan *et al.*, 1982). Note that at high temperatures Q saturates to a value that depends upon the oxide layer

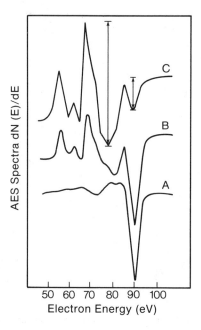

Fig. 3.28. Auger electron spectra for three different conditions of the a-SiH$_x$ surface: sputter cleaned (A), oxidized (B), and heavily oxidized (C). The arrows indicate the amplitudes of the SiO$_2$(75) and the Si(90) peaks (Wronski *et al.*, 1981).

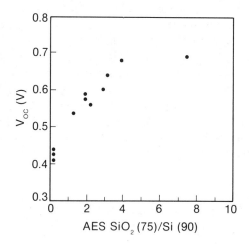

Fig. 3.29. V_{oc} as a function of the ratio of the SiO$_2$(75) peak to the Si(90) peak (Wronski *et al.*, 1981).

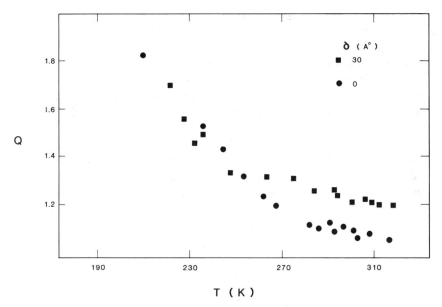

Fig. 3.30. The diode quality factor Q determined from forward bias conditions as a function of temperature for a typical a-Si : F : H device with and without the insertion of an oxide layer of thickness δ (Madan et al., 1980).

thickness, i.e., $Q = 1.05$ and 1.20 for $\delta = 0$ A (corresponding to the native oxide on an a-Si : F : H alloy) and 30 Å, respectively. At lower temperatures the value of Q for both devices coalesces to a value exceeding 1.5.

Since in a SB device the current is primarily composed of injection and recombination, with the former dominating at room temperature, then as the temperature is lowered we expect that the recombination current should dominate. Hence, Q increases as T decreases, and its value will reflect the extent of recombination that occurs within the space charge region.

An increase in Q with oxide layer thickness has been observed for crystalline Si MIS devices. Card et al. (1971) have obtained the following expression to describe the behavior of Q with δ:

$$Q = 1 + \frac{\delta\varepsilon_s}{W(\varepsilon_i + \delta e^2 D_s)}, \quad (3.46)$$

where ε_s and ε_i are the permittivities of the semiconductor and the insulator and D_s is the surface state density. For a-Si based MIS devices the above expression does not seem to be sufficient, since if reasonable values for the various parameters are assumed, such as $\varepsilon_s/\varepsilon_i = 4$, $\delta = 30$ Å, $D_s = 5 \times 10^{12}$ cm^{-2} eV^{-1} and $W = 4000$ Å, then $Q = 1.02$. The inability of Eq. (3.46) to explain the experimental result of $Q = 1.2$ for $\delta = 30$ Å is possibly due to the fact that despite the relatively low DOS, a-Si based alloys still posses a significant number of trapping levels within the mobility gap. Hence, in the derivation of Eq. (3.46) the effect of these trapping levels on the electric field at the a-Si:H/insulator interface, F_{max}, must be considered.

The effective barrier height, ϕ_b, depends upon δ, D_s and the forward bias voltage via the term F_{max}. ϕ_b can be written as (Madan et al., 1980b)

$$\phi_b = \phi_{0b} - \theta F_{max}, \quad (3.47)$$

where $\theta = \delta \varepsilon_s/(\varepsilon_i + \delta e^2 D_s)$ and ϕ_{0b} is the barrier height under flat band conditions. F_{max} is a complicated function that depends upon the doping density, N_D, δ, D_s and the DOS within the mobility gap.

Using diffusion theory, calculations have shown that F_{max} derived for an assumed density of states spectrum is given by

$$F_{max} = [\phi_b - (E_c - E_f) - V]f(\phi_b)^{1/2}, \quad (3.48)$$

where V is the applied bias, and

$$f(\phi_b) = \frac{2}{\varepsilon_0 \varepsilon} \left\{ eN_D + \frac{eg_0 E_{ch}^2}{\phi_b - (E_c - E_f)} \right\}$$

$$\times \left\{ \cosh\left(\frac{\phi_b - E_i}{E_{ch}}\right) - \cosh\left(\frac{E_c - E_f - E_i}{E_{ch}}\right) \right\} \quad (3.49)$$

where $g_0 = 1.5 \times 10^{16}$ cm^{-3} eV^{-1} [defined by Eq. (3.9)], E_{ch} is a characteristic energy described earlier, and E_i is the position of the intrinsic Fermi level. Q can then be calculated, leading to $Q = 1.05$ and 1.14 for $\delta = 5$ and 40 Å respectively. The decrease in Q with increasing surface state density is primarily due to the screening of the electric field by the interface states. Comparing these calculations with the experimental values of $Q = 1.05$ for $\delta = 0$ Å and $Q = 1.2$ for $\delta = 30$ Å, as shown in Fig. 3.30, it has been concluded that the density of surface states in these alloys is relatively small ($\leq 10^{12}$ cm^{-2} eV^{-1}). This conclusion is also consistent with the explanation used for the variation of the open circuit voltage with the work function of the metal (see Section 3.1.3.10), to be discussed later, as well as data on interface states (see Section 2.9).

3.1.3.9 Reverse Bias I-V Characteristics in the Dark

Figure 3.31 shows the current-voltage, $I_r(V_r)$, characteristics in reverse bias for typical SB and MIS type devices (Madan et al., 1980b). Note that there is a lack of saturation in current with reverse bias voltage, in contrast to what is implied by Eq. (3.37b). This can be attributed to one or more of the

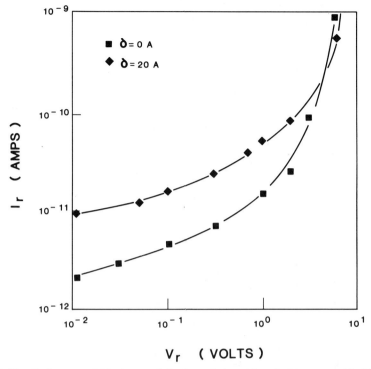

Fig. 3.31. Dark reverse $I_r(V_r)$ characteristics for a device with and without an oxide (Madan et al., 1980).

following causes: (a) image force lowering; (b) generation current; (c) barrier height lowering due to the presence of an insulating layer. The first two give the functional dependencies $I_r \alpha \exp[-(e/kT)(A - V)^{1/4}]$ (where A is a constant) and $I_r \alpha V_r^{1/2}$ respectively. However, these dependencies cannot account for the observed $I_r(V_r)$ characteristics. The last mechanism (c), can, since under reverse bias the field at the insulator/semiconductor interface increases and causes ϕ_b to decrease. The $I_r(V_r)$ characteristics have been simulated by using the analysis outlined in the previous section (Madan et al., 1980b; Madan, 1984). Assuming that a potential V_i is dropped across the oxide in an MIS type structure, it can readily be shown by using Eq. (3.49) that ϕ_b decreases with increasing reverse bias voltage and increasing oxide layer thickness. Figure 3.32 shows a simulation of the $I_r(V_r)$ characteristics for varying oxide layer thicknesses. Note that there are qualitative similarities between this and the experimental results displayed in Fig. 3.31, which suggests that the reduction of ϕ_b with V_r can account for the lack of saturation in the reverse characteristics.

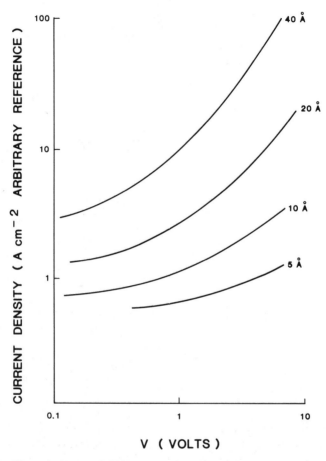

Fig. 3.32. Theoretical reverse $I_r(V_r)$ characteristics for increasing oxide layer thickness (Madan et al., 1980).

The above analysis also provides a means for investigating the quality of the photoactive material within the device. For example, the photoluminescence (PL) intensity of the active material generally decreases as a function of P doping (see Section 2.7). The reduction of the intensity has been attributed to the inclusion of non-tetrahedrally coordinated P, which introduces non-radiative recombination centers (see Section 2.8). In Fig. 3.33 the $I_r(V_r)$ characteristics have been derived when g_0, the minimum density of localized states [Eq. (3.9)], is increased. Figure 3.34 shows the experimentally obtained $I_r(V_r)$ curves when E_f of the active layer is altered by the addition of P (Madan, 1984). Once again we obtain qualitative similarity with the theoretically predicted curves shown in Fig. 3.33).

3.1.3.10 Variation of the Open Circuit Voltage with the Work Function of the Metal

The low DOS possessed by a-Si:H material is also reflected in an increase of V_{oc} with ϕ_m. This is shown in Fig. 3.35 for a variety of metal contacts on devices fabricated with and without the deliberate insertion of an oxide layer of thickness $\delta = 30$ Å (Madan et al., 1980b). Note that V_{oc} varies linearly

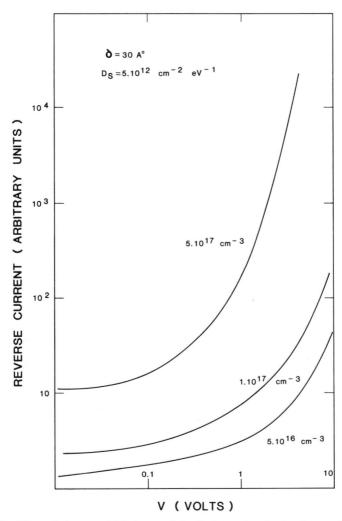

Fig. 3.33. Theoretical reverse $I_r(V_r)$ characteristics for increasing density of states (Madan et al., 1980).

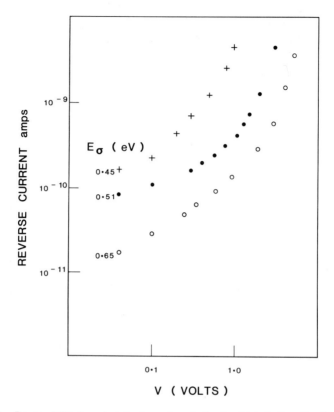

Fig. 3.34. Reverse $I_r(V_r)$ characteristics for various devices in which the Fermi level of the active layer was varied (Madan et al., 1980).

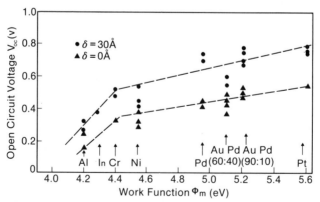

Fig. 3.35. V_{oc} plotted against the work function of metal for SB ($\delta = 0$ Å) and MIS ($\delta = 30$ Å) type devices (Madan, 1983).

SOLAR CELLS 217

for $\phi_m < 4.4\,\text{eV}$ and tends to show a sublinearity for $\phi_m > 4.4\,\text{eV}$. For $\phi_m < 4.4\,\text{eV}$ the slope of V_{oc} against ϕ_m can be described by

$$\Psi = \frac{\varepsilon_i}{\varepsilon_i + \delta e^2 D_s}. \tag{3.50}$$

For a low density of surface states, $D_s < 10^{12}\,\text{cm}^{-2}\,\text{eV}^{-1}$, Ψ should be 1.0, as observed in Fig. 3.35. As the work function of the metal increases ($\phi_m > 4.4\,\text{eV}$) the Fermi level of the semiconductor within the space charge layer enters the rapidly rising distribution of tail states near the valence band edge, with the result that the potential profile exhibits a sharp spike. This would result in the tunneling of majority carriers through the thin barrier and lead to saturation of the effective barrier height, translating into a weak dependence of ϕ_b on high work function metals such as Pd and Pt.

3.1.3.11 Schottky Barriers Using Other Deposition Techniques

Reactivity Sputtered a-Si:H. The diodic properties of reactively sputtered a-Si:H in a device configuration have been investigated by Thompson *et al.* (1978) as a function of the deposition parameters of the *i*-layer. As discussed in Section 2.10.2, the electronic properties of a-Si based alloys produced in this fashion are primarily functions of the H pressure, p_H, and substrate temperature, T_s. Nearly ideal forward bias current behavior was obtained with SB devices prepared in the presence of 4×10^{-4} torr of H_2 at $T_s = 200°C$, and led to a rectification ratio of about 10^4 with $Q = 1.25$. However, when p_H exceeds 4×10^{-4} torr, it was found that the diode characteristics deteriorated.

Morel *et al.* (1981) considered the diodic properties of reactively sputtered a-Si:H films prepared by the rf sputtering of a polycrystalline target at a power density of $2\,\text{W/cm}^2$, $T_s = 275°C$, an Ar pressure of 5 m torr and a partial pressure of hydrogen varying from 0.3 to 1.5 m torr. The hydrogen profile and content of the deposited films was measured by a nuclear reaction technique (see Section 2.2) and found to vary between 8 and 25%. The SB structures were fabricated on stainless steel substrates that were first coated with a thin n^+ layer. High work function metals such as Pt or Pd were used to complete the device structure. They found that the optical band gap increased systematically with hydrogenation, i.e., with 8 at.% H, $E_g = 1.68\,\text{eV}$ whereas with 20 at.% H, $E_g = 1.92\,\text{eV}$. Figure 3.36 shows the relation between V_{oc}, ϕ_b and E_g. The diodes approached ideality with an increasing degree of hydrogenation (increasing E_g). It is speculated that the data are consistent with a downward movement of the valence band, since with increasing hydrogenation there could be a sharpening of the band tail.

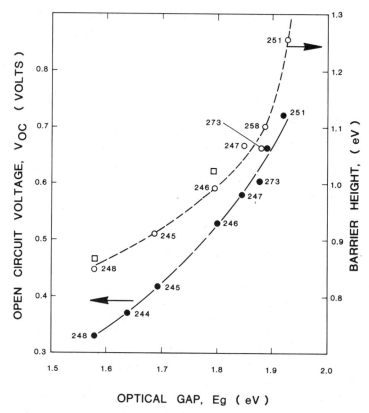

Fig. 3.36. Open circuit voltage under illumination vs. band gap for a-Si:H/Pt Schottky barriers (●). Barrier height vs. band gap from: (○)—J_0 values assuming a pre-exponential $e\mu g(E_c)F_{max}$ of 3×10^6 A/cm^2 (□)—J_0 vs. temperature (Morel et al., 1981).

Photochemical Vapor Deposition. As discussed in Section 2.10.1, a-Si:H films have been prepared by Hg-photosensitized decomposition of a-Si:H (Saitch et al., 1983). Au Schottky barriers were fabricated with $J_{sc} = 4$ mA cm^{-2}, $V_{oc} = 0.5$ V, $FF = 0.52$ and $\eta = 1\%$. The current density is expected to be more than 10 mA cm^{-2} after correcting for the transmissivity of the Au film and application of an anti-reflection coating. Highly efficient *p-i-n* cells (see Section 3.1.4) have been fabricated in a multichamber system. The structure used was glass/SnO$_2$/*p-i-n*/Ag; the *p*-SiC was prepared by the direct photo-CVD whilst the *n*- and *i*-layers were prepared by mercury sensitized photo-CVD. The thicknesses were 90 Å, 3500 Å and 300 Å respectively. The cell characteristics were $V_{oc} = 0.882$, $J_{sc} = 18.05$ mÅ cm^{-2}, $FF = 0.702$ and $\eta = 11.2\%$ for a small area device of 0.09 cm^2 (Konogai, 1986).

3.1.4 p-i-n Type Junctions

p–i–n junctions are now normally used to obtain high performance a-Si alloy solar cells. As discussed in Section 2.8, it is found that a-Si:H becomes highly defective when it is doped to form either *n*- or *p*-type layers. Therefore, the use of *p–n* junctions, as in the crystalline Si case, is precluded, since the space charge region in a-Si:H *p–n* junctions is extremely narrow. Instead, a structure of the type p^+-i-n^+ is normally employed, with the field separation region (space charge) occurring primarily in the *i*-layer. (The first letter, p^+ in this case, signifies the order of deposition onto the substrate.)

3.1.4.1 p-i-n Type Cells

In *p–i–n* device construction a thin p^+-layer (100–300 Å) is first deposited onto a conducting substrate, followed by the deposition of an a-Si:H intrinsic layer (5000 Å–7000 Å in thickness), and then by the deposition of a thin n^+-layer (<200 Å). The structure is completed with a metal reflector such as A_g or Al. p^+ layers are generally fabricated from SiH_4 and B_2H_6 gas mixtures, but as discussed in Section 2.8.2, E_g is narrowed with the addition of B, and hence these layers are unsuitable in a configuration when light enters the device through the p^+-type region. Here, the short wavelength light of the incident AM-1 solar spectra is absorbed and leads to a loss in the short circuit current. To circumvent this problem a wide band gap p^+-type material has been used successfully to improve the blue response of the cells (Tawada *et al.*, 1981; Madan, 1981). The intrinsic layer is deposited by a variety of techniques, as discussed in Section 2.1. n^+ layers are generally deposited with SiH_4 and PH_3 gas mixtures and have conductivities near 10^{-2} $(\Omega \text{ cm})^{-1}$ with $E_\sigma \approx 0.2$ eV (Section 2.8). However, as has been pointed out by Swartz (1982), microcrystalline (μc) (see Section 2.8) or highly conducting n^+ layers may be more useful, especially in a n^+-i-p^+ structure, since the back diffusion of holes can be minimized. By the same token, μc p^+-layers for a *p–i–n* type structure have been shown to be of importance and will be discussed in Section 3.1.5.

p–i–n type device structures have been deposited onto a variety of substrates, such as metallized glass, metallized plastics (polyemide), glass coated with transparent conducting oxide [Indium Tin Oxide (ITO) or Tin Oxide (SnO_2)], stainless steel and metallized coated ceramics. The choice of the substrate has an influence on the performance of the device. For example, it has been reported that the use of glass/ITO substrates can sometimes lead to a degradation of the cell, primarily in its FF and J_{sc}. This is attributed to an increase in the series resistance in the device structure (Kitagawa *et al.*, 1983). This does not happen with transparent conducting oxides such as SnO_2.

Further, it has been reported that a barrier can form between the SnO_2 layer and the p^+ layer, which could lead to reduced performance at the blue end of the solar spectrum [Kuboi, 1981; Sinencio and Williams, 1983)]. However, as discussed in Section 3.1.4.6, high efficiency devices are generally made on SnO_2 coated glass; hence, the interface problem between the p^+ and TCO may occur under very unusual processing conditions.

3.1.4.2 Dark I-V Characteristics

A better grasp of p-i-n device operation results from a detailed understanding of their dark I-V characteristics, since the performance is inextricably linked to the junction formation. As shown in Fig. 3.37, Okamoto et al. (1980) showed that for p^+-i-n^+ junction (curve A) the rectification ratio at 0.8 V exceeded 10^5. But, in contrast, the p-n junction (curve B) showed a much reduced rectification ratio due to the highly defective nature of the constituent layers, as discussed above.

From a study of the J-V characteristics of p-n and p-i-n junctions (Gibson et al., 1980) it was found that with increased P doping of the i-layer, both J_0 and the diode quality factor, Q increased. As shown in Fig. 3.38, Okamoto et al. (1983) have modelled the dark J-V characteristics of a p^+-i-n^+ junction as a function of the $\mu\tau$ product, and have concluded that the value of Q at low bias increases as the $\mu\tau$ product decreases. It should be noted that a decrease in the $\mu\tau$ product for holes results from the presence of a large number of defect states; this leads to increased recombination, resulting in

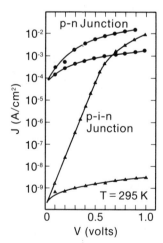

Fig. 3.37. Dark J-V characteristics of an ITO/p-i-n hetroface cell and a p-n junction (Okamoto et al., 1980).

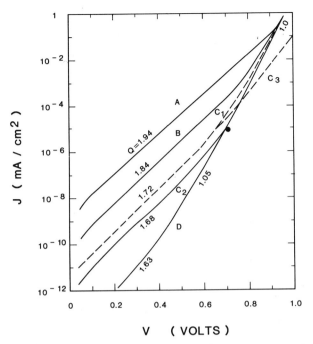

Fig. 3.38. Dark J-V characteristics for various values of $\mu_n\tau_n + \mu_p\tau_p$; curve A, 10^{-11} cm^2 V^{-1}; curve B, 10^{-9} cm^2 V^{-1}; curve C$_2$, 10^{-7} cm^2 V^{-1}; curve D, 10^{-5} cm^2 V^{-1} ($\mu_n\tau_n/\mu_p\tau_p = 1$) and $\mu_n\tau_n/\mu_p\tau_p$ (curve C$_1$, 0.01; curve C$_2$, 1; curve C$_3$, 100) ($\mu_n\tau_p + \mu_p\tau_p = 10^{-7}$ cm^2 V^{-1}). Here, $C_n = 10^{-5}$ Ω^{-1} cm^{-1}, $C_p = 10^{-7}$ Ω^{-1} cm^{-1}, $S_n = S_p = 10^4$ V cm^{-1}, $V_b = 0.9$ V and thickness, $l = 6000$ Å (Okamoto et al., 1983).

a larger value of Q. Their theory predicts that Q is expected to exhibit a transition to a value of 1 (injection) for bias voltages above 0.6 V. In the $J_{sc} - V_{oc}$ study of Konagai et al. (1982) on the photovoltaic properties of n^+-i-p^+ diodes, Q' was found to be 1.72, with a transition to 1.06 above 0.6 volt, thus confirming the above prediction of changeover in transport from recombination to injection.

McMahon et al. (1984) have performed a systematic study of p^+-i-n^+ junctions and found that the dominant transport mechanisms could be distinguished. Representative examples displaying these mechanisms are shown in Fig. 3.39. Generally, the different contributions to the dark log J vs. V current in a p^+-i-n^+ type device can be written as the sum of: injection; (diffusion) recombination; tunnel; shunt components. When recombination is evaluated only in the quasi-neutral regions, the injection (diffusion) component is given by (e.g., Grove, 1967)

$$J_{inj} \alpha e^{-E_g/kT}(e^{eV/kT} - 1), \qquad (3.51)$$

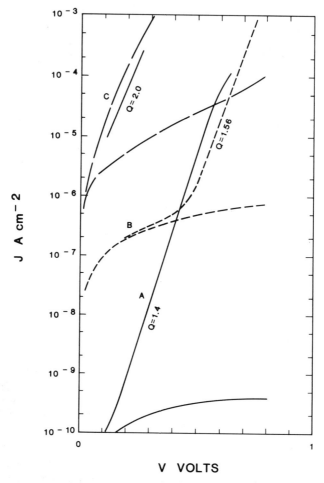

Fig. 3.39. Experimental dark J vs. V characteristics for three diodes exhibiting different transport behavior (McMahon et al., 1984).

and when evaluated over the space charge region, the recombination component is given by,

$$J_{\text{rec}} \alpha e^{-E_g/2kT}(e^{eV/QkT} - 1). \tag{3.52}$$

The tunnelling component is given by (see, e.g., Sze, 1969)

$$J_{\text{tunn}} \alpha e^{C\beta T_e \alpha eV}, \tag{3.53}$$

where C depends on the electric field, electron mass and tunnelling probability and exhibits a very weak dependence on temperature. β is the temperature coefficient term, typically $(2-4) \times 10^{-4}$ eV K^{-1}, and is defined by $E_g(T) = E_g(0) - \beta T$. Finally, the shunt component J_{shunt} will have a small T dependence and exhibit a linear reverse current.

In Fig. 3.39, curve A, the dark log J vs. V of an a-Si:H p^+-i-n^+ junction shows a rectification ratio exceeding 10^6 at $V = 0.8$ V, with $Q = 1.4$. A SIMS compositional profile showed that the Boron concentration in going from the p^+- to the i-layer (B_{p^+}/B_i) was in excess of 10^4 within about 200 Å of the i-layer growth, thus indicating good process control. They reported that a plot of log J_r, the current at a reverse bias at 0.8 V, vs. $10^3/T$ yielded an activation energy of $\Delta E = 0.83$ eV, corresponding to one-half the mobility gap. Since $\Delta E \approx E_g/2$ (see Eq. (3.52), it was concluded that recombination in the space charge layer plays a predominant role in the transport mechanism. The results of Fig. 3.39 (curve A) compare well with the predictions of Okamoto et al. (1983) shown in Fig. 3.38 only at low bias voltages. However, as noted in Fig. 3.39 (curve A) a transition corresponding to the $Q = 1$ region (injection) was not discernable, even when ln J_{sc} vs. V_{oc} [Eq. (3.42)] was plotted, thus implying that the recombination current plays a dominant role in a-Si:H p-i-n type junctions over the entire voltage region investigated. This was confirmed by performing a theoretical analysis to simulate ln J_{sc} vs. V_{oc} using a modified version of the DOS given by Eq. (3.9):

$$g(E) = g(E_c) \exp\left(\frac{E - E_g}{E_{\text{ch}}^c}\right) + g(E_v) \exp\left(\frac{-E}{E_{\text{ch}}^v}\right) + g_0. \quad (3.54)$$

V_{oc} was found via the analysis given in Section 3.1.2.2. V_{oc} and J_{sc} can be related by Eq. (3.42) by recognizing that for good devices, $J_{sc} \propto G^m$, where G is the generation rate, and hence $V_{oc} \propto Q' \ln G^m$. Theoretical plots of ln G vs. V_{oc} in Fig. 3.40 show that a straight line should be observable. Note also that as the DOS increases (larger E_{ch} and g_0) Q increases, representative of increased recombination.

Curve C in Fig. 3.39 shows the affect of doping the i-layer heavily with boron ($B_{p^+}/B_i \sim 10$). The rectification ratio is reduced greatly from the values found for diodes such as A; there is also an increase in the effective J_0 value. Further, because of the curvature in the forward characteristics, the value of Q is ill-defined, but, nevertheless, can be construed to be large (>2). The reason for this is probably due to the creation of recombination centers by Boron (Section 2.8). A plot of $\ln(J_0)$ vs. T led to a straight line that could be representative of a tunnelling mechanism, which is adequately described by Eq. (3.53). The excess current due to tunnelling leads to a reduction in the built-in potential and, consequently, to a reduction in V_{oc}. The increase in the density of states leads to a reduction in both the space charge region and the $\mu\tau$ products. A poor response to AM-1 intensity results.

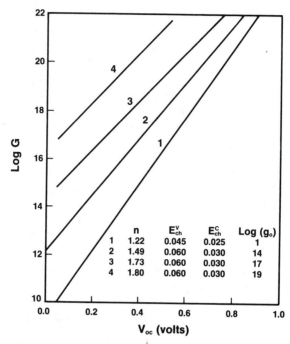

Fig. 3.40. Calculated V_{oc} as a function of generation rate, G, for different localized state distributions (McMahon *et al.*, 1984).

Curve B of Fig. 3.39 shows a diode that at large forward bias exhibits $Q = 1.56$ but for small bias voltages shows a large excess current. It was found from SIMS results that the Boron profile was abrupt ($B_{p^+}/B_i > 3000$), as for a diode of type A, whereas the EBIC (Electron-Beam-Induced-Current) technique, in contrast to the previous cases, showed the presence of various types of macrostructural defects (with some representative examples shown in Fig 3.41 (Yacobi *et al.*, 1984). The left-hand side of the figure corresponds to the secondary electron emission (SEI) modes and the right-hand side corresponds to the EBIC modes. In Fig. 3.41a a blistering effect is noted via EBIC, as indicated by a total lack of current collection (the dark region). Figure 3.41b shows the polishing marks still remaining on the stainless steel substrate. The SEI mode in Fig. 3.41c shows that the junction is ostensibly clean, but in the EBIC mode dark regions were noted, which may be due to incomplete junction formation for the pin holes that propagate part of the way through the film. The reverse I–V characteristic for diodes with macrostructural defects generally followed a $I \propto V$ (ohmic) response with a very small temperature dependence, thus confirming the shunting behavior.

SOLAR CELLS 225

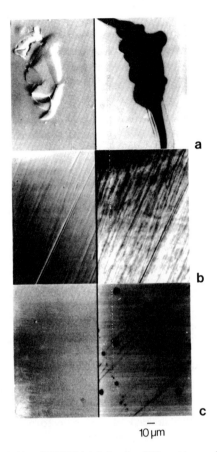

Fig. 3.41. SEI mode (left) and EBIC (right) showing different types of macrostructural defects (McMahon et al., 1984; Yacobi et al., 1984).

The reverse breakdown voltages of p^+-i-n^+ junctions have been studied by Gibson et al. (1980), and are shown in Fig. 3.42. They observe that there is an irreversible breakdown of the diodes that is sample dependent; this could be due to the macroscopic defects discussed above. In addition, they report a reversible "soft" breakdown at current values on the order of 10–100 mA cm^{-2}. This is a common feature and is attributed to the presence of a non-uniform DOS in the gap. It is expected that with increasing bias the DOS distributions of adjacent materials will begin to overlap and produce phonon assisted hopping between the two sides of the junction. A more gradual reverse breakdown characteristic will be expected before the onset of Zener or avalanche breakdown. Typical average irreversible breakdown voltages are on the order of 10 V at current densities between 10 and 100 mA cm^{-2}.

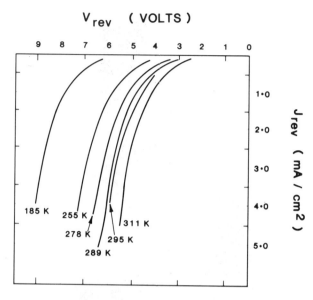

Fig. 3.42. Reverse *I-V* characteristics of a smoothly graded *p-n* junction at the stated temperatures (Gibson *et al.*, 1980).

3.1.4.3 The Influence of Doping

As discussed in the previous section, doping contamination of the active *i*-layer seriously affects transport in p^+-*i*-n^+ diodes. We now consider the impact of doping (at lower levels) on device performance in more detail. It was discussed in Section 2.8 that *p*-type conductivity can readily be achieved with the addition of B_2H_6 to the SiH_4 in the gas phase, but at the expense of a reduced absorption coefficient that leads to a lower effective band gap. It has been pointed out that while there were no discernable changes in $\alpha(h\nu)$ for n^+ layers (Madan and Ovshinsky, 1980), and since the doping efficiency for *P* and *B* was low, then extra defect states were created within the mobility gap of a-Si:H (*B* or *P*). This alters the transport properties, with some evidence, for the *P* case, that parallel path conduction at E_c and E_D (the donor level) takes place (see Section 2.8). Further, it has been found from a wealth of data that the doping of a-Si:H layer reduces the minority carrier diffusion length (Wronski *et al.*, 1982), and from luminescence (Section 2.7), ESR (Street *et al.*, 1978), and photoconductivity (Wronski *et al.*, 1982) measurements it is deduced that there is an increase in the DOS in the gap. These studies have concluded that charged dangling bonds could be produced as a direct consequence of doping, with a density that increases as the square root of the *B* or *P* concentration.

SOLAR CELLS

The influence of doping on the $\mu\tau$ products for electrons and holes is shown in Fig. 3.43 via time of flight and photoconductivity measurements (Street et al., 1983). The data fall along three lines corresponding to the i, n and p-type layers. Note that only a few ppm of doping reduces either $(\mu\tau)_n$ or $(\mu\tau)_p$ by 2 to 3 orders of magnitude. A model attempting to explain this is shown in Fig. 3.44, which illustrates that charged defects can trap carriers of one sign, unlike neutral defects, which may trap both electrons and holes. Since charged centers are expected to be very effective traps, the sensitivity to low doping levels is also explained. The reduction of $(\mu\tau)_n$ and $(\mu\tau)_p$ with B and P doping has also been reported by Kirby et al. (1983), who concluded that the doping influences the minority carrier lifetime but not the majority carrier lifetime. In contrast, Okamato et al. (1983) show from their analysis of p^+-i-n^+-type junctions that doping the photoactive layer with either B or P leads to a considerable reduction in the lifetimes of both types of carriers, as shown in Fig. 3.45. Here it should be noted that for B doping $\mu_n\tau_n > \mu_p\tau_p$; for P doping the opposite seems to result. We shall return to this point in a later part of this section.

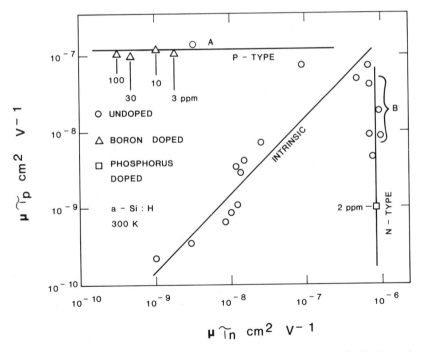

Fig. 3.43. Plot of $\mu\tau$ for electrons vs. holes for a series of undoped and doped a-Si:H samples. The gas phase doping levels are indicated (Street et al., 1983).

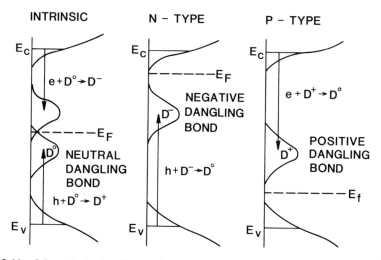

Fig. 3.44. Schematic density of states diagram showing the dangling bond states in doped and undoped samples, and the possible deep trapping transitions (Street et al., 1983).

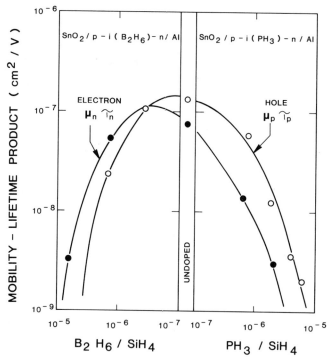

Fig. 3.45. Mobility-lifetime ($\mu\tau$) products of electrons ($\mu_n \tau_n$) and holes ($\mu_p \tau_p$) at 293 K as a function of gaseous composition during plasma deposition of the i layer (Okamoto et al., 1983).

SOLAR CELLS

The effect of P contamination of the active layer has been investigated by Wronski et al. (1982), who have noted from collection efficiency measurements performed on SB device configurations, that as the doping in the active layer increases from 0 to 1000 vppm of PH_3 in the gas phase, the minority carrier diffusion length decreases from 0.40 μm to 0.01 μm. They report that τ_p and τ_n show entirely different dependencies with doping, τ_p decreases from $\sim 10^{-7}$ s to 10^{-10} s with the addition of 1000 vppm PH_3, whereas τ_n decreases from 1.10^{-5} s to 4.10^{-8} s, indicating that the P doping introduces centers that appreciably enhance the capture probability for holes.

The effect of P contamination in the active layer is also reflected in changes in the dark I-V characteristics (Madan et al., 1982). As shown in Fig. 3.46, the diode quality factor, Q, at room temperature was found to be in the range 1.05–1.1 for a SB-type device when avoidable oxides were removed; Q increases to about 1.2 for an MIS-type structure. For the undoped samples Q increases from 1.0 to 1.6 as T is lowered. It should be recognized that in a SB the current is composed of an injection, recombination and tunneling (Mariucci et al., 1987), with the former dominating at room temperature since the built-in potential is small. When the temperature is lowered the injection current decreases as $\exp(-V_{bi}/kT)$, where V_{bi} is the built-in potential. At lower temperatures the recombination current eventually dominates and the diode quality factor increases well above unity. At room temperature, Q is also seen to increase with increasing P content within the i-layer of the p^+-i-n^+-type junction due to the states created by P. Here, when the temperature is lowered Q becomes very large (>2) and the transport cannot be associated simply with recombination. Rather, it can be interpreted by incorporating a second channel (see Section 2.8) for current flow via an impurity band located 0.3 eV below E_c.

In a two-channel conduction process (Jones et al., 1977) it is envisaged that as the P concentration increases a donor band forms through which hopping conduction, σ_{hop}, can compete with extended state conduction, σ_{ext}. Hence

$$I = I_d + I_r + I_{hop}, \tag{3.55}$$

where I_d, I_r and I_{hop} refer to the drift/diffusion, recombination, and hopping components to the total current, I.

$I_d + I_r$ can be combined and written as

$$I_d + I_r = I_{01} \exp\left(-\frac{e\phi_{eb}}{kT}\right)\left(\exp\frac{eV}{QkT} - 1\right) \tag{3.56}$$

It was concluded that for highly doped samples the hopping mode is important at low temperatures and at low values of forward bias, since few carriers are available to move over the barrier. If a parallel path exists at low

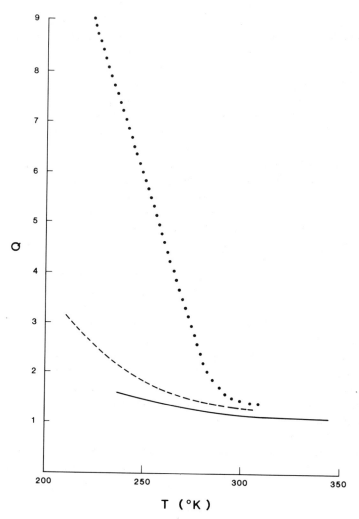

Fig. 3.46. Q as a function of temperature for undoped (———), 1000 vppm PH_3 (---) and 2000 vppm PH_3 (···) introduced into the premix in the fabrication of the active layer for a-Si:H alloys. All data taken for forward bias voltage of $=0.2$ V (Madan et al., 1982).

forward voltages, such as via tunnelling into the impurity band and hopping through it, transport will occur. However, as the forward bias increases, it reduces the field in the space charge region and drives the bands towards a flat band condition. Here more carriers are available to move over the barrier and fewer can tunnel into the impurity band. A transition is expected from the hopping to the drift/diffusion recombination regime. Little current is

SOLAR CELLS

expected until a specific voltage is reached, at which point the current increases rapidly to a value corresponding to the hopping current. With further increase in bias this component saturates and eventually the drift/diffusion-recombination component dominates. The hopping contribution can be written as

$$I_{\text{hop}} = G_{\text{hs}} V_s \exp\left(-\frac{eW_{\text{hop}}}{kT}\right), \quad (3.57)$$

where G_{hs} is the saturation conductance for the hopping process and W_{hop} is the hopping energy. Figure 3.47 shows an excellent fit to the experimental I-V data using the hopping component when the active layer is heavily doped with P. By plotting I, the current at fixed low forward bias voltage (0.2 volt), vs. $1/T$, the hopping energy was determined to be 0.09 eV, which agrees with the estimate derived from the two channel model of Jones et al. (1977).

From the above it is clear that doping of the photoactive layer will influence the photovoltaic characteristics. The increase in the P doping of the active layer leads to an increase in the DOS and a deterioration in critical parameters such as the minority carrier diffusion length (Wronski et al., 1982). This will increase the excess currents in the dark, which, from the

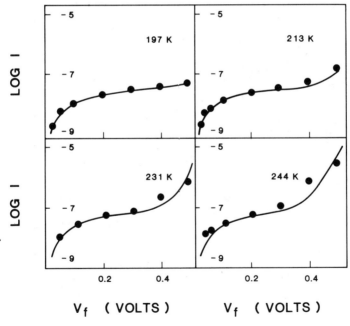

Fig. 3.47. Experimental (···) and calculated (———) results for a SB structure made with 2000 vppm PH$_3$ in the premix (Madan et al., 1982).

diode Eq. (43.2), results in a decrease in V_{oc}. Further, the depletion width decreases as the DOS increases [see Eq. (3.28)] which in turn lowers J_{sc} and hence the overall efficiency of the device.

The effect of heavy B doping is even more serious, since residual contamination deleteriously affects the material properties, with the photoconductivity being reduced dramatically (Anderson and Spear, 1977). This has the effect of introducing a series resistance, as well as reducing the width of the depletion region. The FF and J_{sc} are reduced: the efficiency of the device is thus degraded.

With a large increase in the DOS in n- and p-type a-Si alloys, high quality pn junction formation becomes difficult, as noted by Spear et al. (1976). However, there is evidence that small amounts of B included in the active layer can enhance the performance of the device (Sichanugrist et al., 1983; Konagai et al., 1984; Hack and Shur, 1984). Moustakas et al. (1982) have considered the effect of P as well as autocompensation on device performance. As shown in Fig. 3.48, the data reveal that reduced P contamination

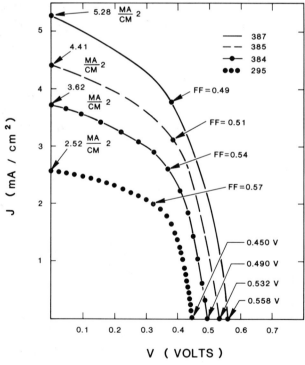

Fig. 3.48. Current-voltage characteristics of a number of a-Si:H$_x$/Pt Schottky barrier structures (Moustakas et al., 1982).

of the i-layer leads to an increase in J_{sc} and V_{oc} but only a slight decrease in the FF. The decrease in J_{sc} and V_{oc} with P contamination is related to the introduction of defects, while the increase in FF is related to an improved photoconductivity in the bulk of the device. The additional improvement in J_{sc} for sample 387 in Fig. 3.48 could be consistent with the findings of Street et al., (1981); for compensated samples (with E_f near mid-gap) the number of dangling bond centers is reduced. Because of the effects of residual contamination on device performance, techniques such as separated reaction chambers, as shown in Fig. 3.49, have been employed (Kuwano et al., 1981). Figure 3.50 shows the B and P profile for devices fabricated in a single chamber as opposed to a separated reaction chamber, where residual contamination, especially due to B, can be reduced by over an order of magnitude. As shown in Fig. 3.51, it is evident that the control of B results in improved device performance (Kuwano et al., 1982a). More recent results using glass/SnO_2/p-i-n/Ag show efficiencies in excess of 10%, as shown in Table 3.5.

The inclusion of small quantities of dopant could, in principle, improve the device (e.g., Sichanugrist et al., 1983). For example, Fig. 3.52a shows that the performance can be significantly enhanced if the photoactive layer is doped with small quantities of B (<0.35 ppm B_2H_6 introduced in the gas phase). The reason for this improvement was given by Nakamura et al. (1982), who showed that when $E_f = E_g/2$ (for the intrinsic layer) a maximum in the depletion width resulted. For $B > 0.4$ vppm in the gas phase, the deterioration in the performance of the device was attributed to the fact that the i-layer may have changed to p-type, which leads to a change in the minority carrier from holes to electrons. In addition, there would be a

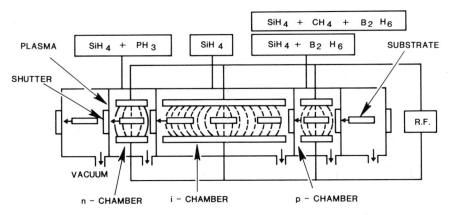

Fig. 3.49. Consecutive, separated reaction chamber apparatus for fabrication of a-Si:H films (Kuwano et al., 1981).

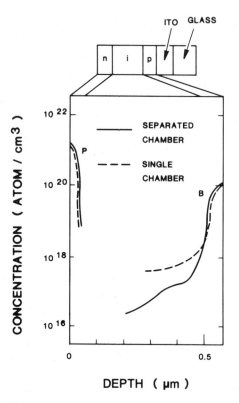

Fig. 3.50. Depth profile using ion microprobe analysis of impurity concentration in p–i–n diodes. Solid lines and broken lines represent a-Si diodes prepared by the consecutive, separated reaction chamber method and by the single reaction chamber method, respectively (Kuwano et al., 1981).

decrease in the depletion width due to the creation of extra defects. In order to significantly improve the cell efficiency (mainly via the FF) it is postulated that the spatial extent of the low field region in the i-layer must be reduced. This was achieved by B profiling the device such that $B_i = 0.15$ ppm was kept constant during the i-layer growth, whereas at the p^+/i interface B_p/B_i was varied. The characteristics of the cells thus obtained are shown in Fig. 3.52b. Structures produced in this way have led to $\eta = 9.45\%$ with $V_{oc} = 0.86$ V, $J_{sc} = 16.50$ mA cm^{-2} and $FF = 0.664$. The above results, particularly Fig. 3.52a, show that the device performance can be optimized with small inclusions of B in the active layer. (However, this type of result may be system dependent and depends on the original quality of a-Si:H layers, since if $E_f = E_g/2$ to begin with, then any addition of B could, in fact, degrade the device.)

SOLAR CELLS

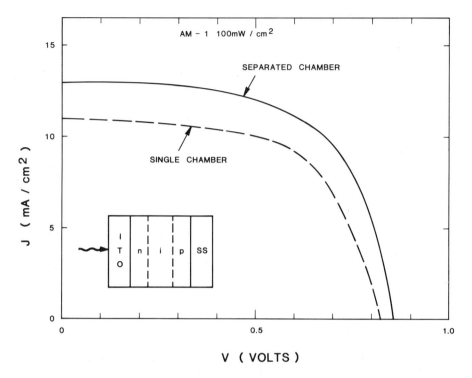

Fig. 3.51. Illuminated $I(V)$ characteristics of ITO/n-i-p/stainless steel a-Si:H solar cells deposited by the consecutive, separated reaction chamber method (solid line), and single reaction chamber method (broken line) (Kuwano et al., 1982).

Table 3.5. Small Area, Single Cell Conversion Efficiencies

Area (cm^2)	V_{oc} (volts)	J_{sc} (mA/cm^2)	FF	η (%)	Reference
1.0	0.915	18.1	0.705	11.7	Kuwano (1986)
1.0	0.872	17.3	0.694	10.5	Nakano et al. (1986)
1.48	0.85	17.2	0.69	10.1	Madan (1986)
0.25	0.887	16.64	0.70	10.9	Arya et al. (1986)
1.0	0.886	17.6	0.74	11.5	Uchida et al. (1986)
0.09	0.882	18.05	0.702	11.2	Konogai (1986)
4.15	0.865	16.10	0.73	10.2	Ullal et al. (1984)
1.09	0.840	17.80	0.68	10.1	Catalano et al. (1982)

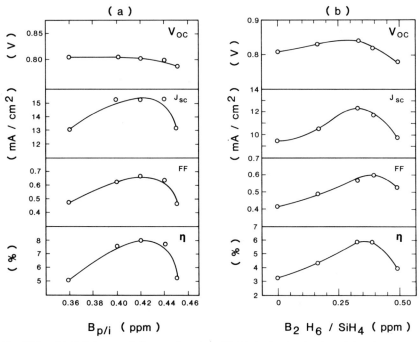

Fig. 3.52. (a) Cell characteristics as a function of B_2H_6 doping level at the p/i interface ($B_{p/i}$). (b) Cell characterization as a function of uniform doping level of B_2H_6 under AM1 (100 mW/cm^2) simulated insolation (Sichanugrist et al., 1983).

To recap, we have previously mentioned that B doping of the i-layer (Okamoto et al., 1983) could lead to a situation $\mu_n \tau_n > \mu_p \tau_p$, whereas P doping leading to $\mu_p \tau_p > \mu_n \tau_n$. Therefore, to maximize the device performance when illuminating through the p^+ layer, it is required that holes act as the minority carrier over the major portion of the device such that

$$\frac{\Delta n(x)}{\tau_n} \gg \frac{\Delta p(x)}{\tau_p}, \qquad (3.58)$$

and hence $\mu_p \tau_p > \mu_n \tau_n$. According to Fig. 3.45, this would be achieved when the active layer is doped slightly with P. However, when light enters through the n^+ layer, in order to minimize recombination, $\Delta p(x)/\tau_p \gg \Delta n(x)/\tau_n$, such that electrons act as the minority carriers; hence, $\mu_n \tau_n > \mu_p \tau_p$. Therefore, in this case the active layer should be doped with a slight amount of B. This is supported by some experimental evidence (for an ITO/p-i-n/ITO structure) that better performance is obtained in the case of n-side (p-side) illumination when the i-layer includes a small amount of $B(p)$ (Konogai et al., 1983).

3.1.4.4 p-i-n and n-i-p Cells

Amorphous Si solar cells are generally fabricated in the configuration-substrate/p^+-i-n^+/metal, rather than substrate/n^+-i-p^+/metal, since it has been found that the voltage output of the latter cell is 150–200 mV lower in comparison with the former. There have been many attempts to explain this phenomenon: the Dember potential (Han et al., 1981); self-field effect (Sakata et al., 1983); residual doping (Konagai et al., 1982); hydrogen effusion effects associated with the p^+ layer deposition process (Muller et al., 1982; Nakamura et al., 1984); tunnelling effects (McMahon and Madan, 1985). We will discuss them now.

Sakata et al. (1983) theoretically treat the differences that occur in photogenerated carrier densities in an intrinsic layer in the two types of configurations. Under steady state illumination, the I-V characteristics of an a-Si:H solar cell, together with the distribution of photogenerated free electrons and holes, were calculated by solving the continuity equation self-consistently. They assume that the electric field, $F(x)$, in a-Si:H intrinsic layers consists of $F_b(x)$, the built-in field and the self-field, $F_s(x)$, defined by

$$F_s(x) = \frac{e}{\varepsilon_s} \int (p(x) - n(x)) \, dx. \tag{3.59}$$

Using this, they find that the sign of the built-in field is positive for the substrate/n^+-i-p^+/metal configuration and negative when the configuration is changed to substrate/p^+-i-n^+/metal. The sign of F_s is positive in both cases, except for the n^+/i interface for the n^+-i-p^+ cell. The self-field then enhances the collection of carriers in the n^+-i-p^+ case and retards it for the p^+-i-n^+ case. The difference in F_s in the carrier collection results is considered to be the difference in the value of V_{oc} between the two types of cells. Also, by varying the mobility of electrons and holes in their calculation, the difference in V_{oc} (~0.20 V) between the two types of cells could then be explained. However, this interpretation is not substantiated by experiment and has been tested for devices using the configurations: (a) TCO|p^+-i-n^+|TCO and (b) TCO|n^+-i-p^+|TCO, and illuminating the junctions from either direction. The results displayed in Fig. 3.53 (Konagai et al., 1982) show that for each configuration there is virtually no difference in V_{oc} when the junctions are illuminated either through the n^+ or the p^+ layers. The small difference that did exist can be attributed to the slightly different generation rates. However a difference in V_{oc} of up to 180 mV exists between the two configurations. In another study, McMahon and Madan (1985) also investigated devices fabricated in configurations (a) and (b) defined above. They found that similar voltages resulted if the illumination level was adjusted so that identical current was generated by shining light either through the

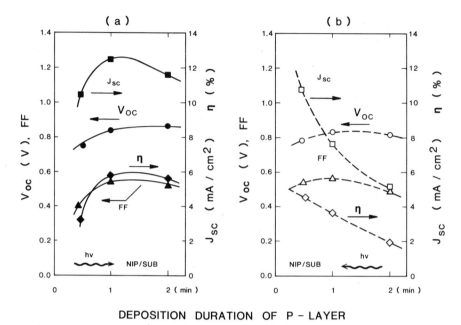

Fig. 3.53. Variation of photovoltaic parameters with duration of p-layer deposition for the n–i–p/substrate configuration. (a) Illumination through the n-layer; (b) illumination through the p-layer (Konogai et al., 1982).

p^+-layer or the n^+-layer. Nevertheless, a difference of about 80 mV existed between the two, with configuration (a) showing a higher V_{oc} than configuration (b).

Han and Anderson (1984) attempted to explain the difference by incorporating the Dember potential, but as pointed out by Sakata et al. (1983), the formalism used was only valid for crystalline materials.

The residual doping effect due to B alone is also unable to explain the difference, since, as can be seen from Fig. 3.54a and b, optimum device performance was obtained for either configuration for B concentrations of 1–3×10^{17} atoms/cm^3 in the i-layer (McMahon et al., 1985). Nevertheless, the n^+–i–p^+ configuration shows a poorer performance, especially in terms of V_{oc}, than the p^+–i–n^+ junction.

Another reason cited for the V_{oc} difference is that the growth mechanism of the p^+-layer on a substrate could be somewhat different from the case where it is deposited onto an i-layer (Muller et al., 1982). More specifically, it was conjectured that there is hydrogen evolution from the surface p^+-layer during the final stages of the deposition process. This would then result in different types of junction formation (e.g., a slight back barrier for the

n^+-i-p^+-type junction). Evidence for asymmetry in the junction formation was provided by the fabrication of a p^+-i-p^+-type junction and noting that the structure led to $V_{oc} = 300$ mV upon illumination. However, careful examination of the hydrogen evolution results of heavily doped B a-Si:H layers reveals that at a deposition temperature of 250°C the effusion of hydrogen is insignificant (Beyer et al., 1981). On the other hand, McMahon and Madan (1985) found that $V_{oc} < 15$ mV under AM-1 illumination for the p^+-i-p^+ structures that they constructed.

McMahon and Madan (1985) found from their study of n^+-i-p^+-type devices that the diode quality factor, Q, for the n^+-i-p^+ device was low (1.2) for small concentrations of B. They suggested that the transport was dominated by electrons; support for this was provided by the junction recovery technique. (In this type of experiment a diode is placed in a steady state forward bias condition so that charges are injected into the active intrinsic layer; the device is then switched into a reverse bias condition. It is found that for a p^+-i-n^+-type cell the recovered charge consists predominantly of holes, probably because the valence band tails are much shallower (Konnenkamp et al., 1985). Hence, this provides a useful technique for investigating shallow states near the valence band.) When B was added intentionally to the i-layer of the n^+-i-p^+-type device the value of Q increased (and approached that obtained in a p^+-i-n^+-type device) and the recovered charge, as measured by the junction recovery technique, also increased. This was explained on the basis of the shift of E_F towards the valence band, with the result that the barrier to hole current flow from the p^+-layer to the i-layer was reduced. However, it was also noted that for a B concentration of 4×10^{17} atoms/cm^3, two regions in the $\log J_{sc} - V_{oc}$ characteristic existed, from which the diode quality factor could be extracted: (a) a low voltage region in which $Q = 1.6$, which is representative of space charge recombination: (b) at higher voltages, $Q = 1$, which is indicative of electron injection. By extrapolating the latter region to $V_{oc} = 0$ it was evident that the built-in potential thus derived could be somewhat higher in comparison with curves obtained for an i-layer doped with $B = 4 \times 10^{16}$ atoms/cm^3. It was then concluded that the slight addition of B has the effect of (a) providing a shift of E_F towards the valence band, thus decreasing the barrier at the p^+/i interface for holes, and (b) increasing the barrier to electron current flow.

The latter point could be consistent in view of the work by Jackson et al. (1982), who have reported that the sub-band gap absorption (corresponding to defect levels) increased with P doping, but decreased with the simultaneous addition of P and B. Further, Street et al. (1981) report that: the minority dopant species tends to be incorporated in the film much more efficiently; the deposition process has a tendency to equalize the two dopant (B and P)

concentrations in the films, They concluded that there could be a cooperative reaction during the deposition and possibly a formation of a *B–P* pair. They also report that although the photoluminescence efficiency decreases with increasing *P* inclusion and can be made to recover with the simultaneous addition of *B*, compensation can reduce the number of non-radiative defects of the dangling bond type. However, a new hole trap is simultaneously introduced above the valence band.

McMahon and Madan (1985) speculate that the reduction in the DOS due to this cooperative action can result in changes in the potential profile at the p^+/i interface. With an initially large DOS, a large field at the p^+/i interface would result, with the effect that the carriers will tunnel through the resulting thin barrier at the p^+/i interface. With a slight introduction of Boron into the *i*-layer, it is suggested that there is a decrease in the DOS in the mid gap

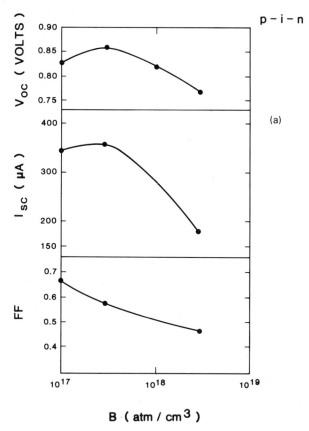

Fig. 3.54. Photovoltaic parameters as a function of *B* doping in the *i*-layer of configuration (a) *p–i–n*/substrate and (b) *n–i–p*/substrate (McMahon and Madan, 1985).

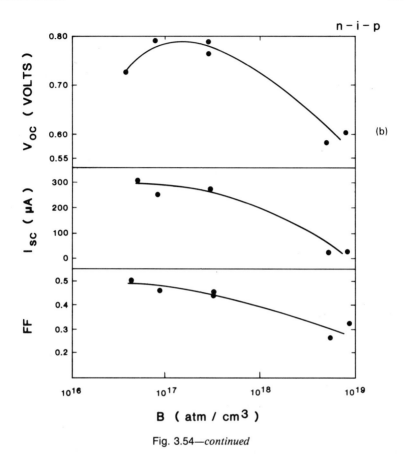

Fig. 3.54—continued

region. This has the effect of reducing the field at the p^+/i interface, with the consequence that tunnelling occurs at a higher energy and thus leads to an increased barrier for electron transport. This could ultimately translate into a higher V_{oc}, as noted in Fig. 3.54b. Confirmation of this model has been provided by Jeffrey et al. (1986).

3.1.4.5 Impurity Effects

From a compositional analysis of a-Si based materials it is generally found that there are relatively large concentrations of oxygen, carbon and nitrogen included with the films, as shown in Fig. 3.18. Delahoy et al. (1981) have studied the effect of air leaks, oxygen, nitrogen and monochlorosilane on solar cell device performance; they find that the addition of $N_2 + O_2$ gas mixtures to the silane discharge during the growth of the i-layer leads to a substantial reduction in J_{sc}. They attributed this to changes that occur within the bulk, and postulate that deep states are introduced, roughly 0.8–1.0 eV below E_c, which act as hole traps and reduce the $\mu\tau$ product for holes. With 2000 vppm in the gas phase $\mu\tau$ is reduced to 5.5×10^{-10} (cm^2 V^{-1}) from its initial value of 3×10^{-8} (cm^2 V^{-1}). This reduction in the $\mu\tau$ product corresponds to a decrease in the diffusion length of holes from 2800 Å to 380 Å.

One of the major sources of contamination is the gas feed stock itself, and it seems that monochlorosilane (SiH$_3$Cl) is a fairly common impurity in commercial tanks of silane. Delhoy *et al.* (1981) have also investigated the effect of this on solar cell performance and find that when the contaminant was removed from the gas tank the *FF* as well as J_{sc} improved substantially. The principal effect of SiH$_3$Cl in the plasma is to lower the Fermi level position in the film and hence limit the electron transport in the cell. As a consequence the photoconductivity is reduced, which could result in a degradation of the *FF*.

Generally, it is found that any impurity affects the device performance since the opto-electronic properties such as $g(E_f)$, E_0, E_{ch} are all degraded.

3.1.4.6 Optimization of Solar Cell Performance

In order to obtain as high an efficiency as possible, there are many interdependent parameters to consider. Most importantly, the *i*-layer must be a high quality material with a low DOS such that if it is incorporated into a SB configuration $Q < 1.1$, and if it is in a p^+-*i*-n^+ junction, $Q < 1.5$. To optimize its thickness, it must be recognized that the collection width is composed of a depletion width, W, and a zero field diffusion length, L_p^0 (see Section 3.1.4.7b), which under AM-1 leads to values on the order of 0.5 μm each. Hence, for a given quality of a-Si:H with specific values of W and L_p^0, the optimum thickness as far as J_{sc} is concerned should be $W + L_p^0 \approx 1\,\mu$m. However, the efficiency is also dictated by the *FF*, which depends upon the variation of the collection width with bias. Hence, the optimum conversion efficiency should therefore occur at a value smaller than 1 μm. In Fig. 3.55 we show the performance of solar cells with *i*-layer thickness as the variable; the optimum thickness is in the range of 0.7 μm, although it also can depend upon the details of the texture of the TCO employed, the optical properties of the p^+-layer (if illuminated through this layer), on the n^+-layer (if illuminated through this layer) and the reflective properties of the back metallic layer.

Use of Optical Windows. As discussed in Section 2.8, with increasing B doping the optical absorption of the p^+-layer increases. Hence, for cell configurations that utilize light entry through the p^+-type side, there is a loss

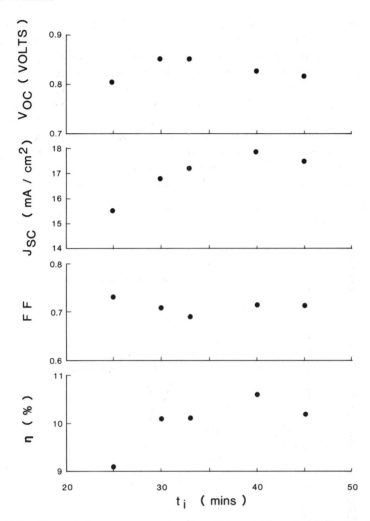

Fig. 3.55. Photovoltaic parameters as a function of i-layer thickness for devices (displayed as time of deposition; ~3 Å s^{-1}) constructed in TCO$|p^+$-i-$n^+|$Ag configuration (Hollingsworth et al., 1986).

of current in the device, especially at the blue end of the spectrum. Using gas mixtures of SiH$_4$, CH$_4$ and H$_2$, materials with band gaps ranging in energy from 1.76 eV to 2.2 eV have been fabricated. Using (SiH$_4$)$_{0.8}$ + (CH$_4$)$_{0.2}$, films of composition Si–C:H exhibit $E_g \cong 1.9$ eV and $\sigma_D \sim 10^{-10}$ (Ω cm)$^{-1}$. With the addition of PH$_3$ (1%), σ_D can be changed to 10^{-4} (Ω cm)$^{-1}$ and with 5% (B$_2$H$_6$), σ_D can be changed to 10^{-4} (Ω cm)$^{-1}$ without an apparent change in E_g (Tawada et al., 1981).

The collection efficiency (see Section 3.1.4.7a) of hetrojunction solar cells that have been fabricated in the configuration $SnO_2/p^+-i-n^+/Al$ (for p^+ layers deposited with different band gap materials) has been measured. As shown in Fig. 3.56, as the band gap of the p^+-layer is increased by the addition of C, the blue response increases, as is to be expected.

The optimum thickness of the p^+-layer can be obtained experimentally by paying close attention to the variations that occur in V_{oc} and J_{sc}. If the p^+-layer is too thin, V_{oc} decreases since the p^+-layer would not be able to sustain the depletion layer required for charge exchange between that layer and the i-layer; if the p^+-layer is too thick, J_{sc} decreases because of extra absorption, leading to a loss at the blue end of the spectrum.

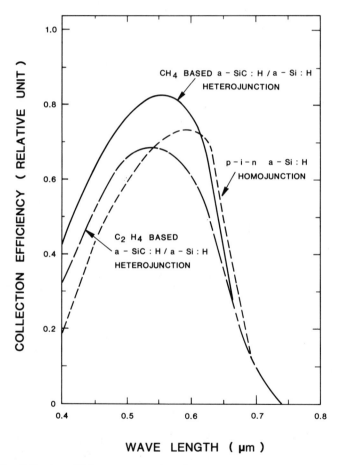

Fig. 3.56. Collection efficiency spectra of methane based and ethylene based a-SiC:H/ a-Si:H heterojunction cells, and a p–i–n a-Si:H homojunction cell (Tawada et al., 1981).

Another useful layer to employ is that of a μcp^+ (Section 3.1.5) instead of the wide band gap p-type version. Here the absorption coefficient is low (indirect band gap) and E_f is close to the valence band. Use of this material on stainless steel substrates has led to a V_{oc} in excess of 0.9 V (Nakamura et al., 1984).

Optical Trapping Effects. As shown in Fig. 3.9, the maximum photocurrent that can be generated when the light passes through the cell only once is of the order of 13 mA cm^{-2}. However, J_{sc} can be considerably enhanced by: the use of reflectors; texturing the surface to increase the total path length and thereby the total absorption. Table 3.6 shows the reflectance at an interface between a-Si:H and several metals (Ondris et al., 1980). The use of a Ag back contact is thus expected to lead to increased reflectance at the red end of the spectrum and thus increased absorption within the intrinsic layer.

A dramatic increase in J_{sc} can result from the texturing of the substrate. The key to this type of optical enhancement is to break the plane parallel symmetry of the thin film solar cell in order to permit trapping of the incoming light within the semiconductor. The breaking of symmetry is accomplished by texturing the semiconductor with a structure having characteristic dimensions nearly equal to the wavelength of the visible light within the material. This texture scatters incoming photons into optical modes that are trapped within the semiconductor, thereby increasing the path length for absorption. The maximum path length increase can be as much as $4n^2$ where n is the index of refraction (Yablonovitch and Cody, 1982).

Table 3.6. Reflectance at Interface between a-Si:H and Several Metals, Calculated from the Optical Constants (Ondris et al., 1980)

Metal	Reflectance at 600 nm	Reflectance at 700 nm
Ag	0.97	0.97
Al	0.76	0.69
Au	0.88	0.93
Cr	0.3	0.3
Cu	0.89	0.93
Fe	0.22	0.21
Mo	0.14	0.14
Ni	0.35	0.36

Using a lithographic technique, Deckman et al. (1983) produced a textured surface on a glass substrate, and the effect on the cell performance is shown in Fig. 3.57. Here the spectral response is compared with devices fabricated on flat Cr (unenhanced), flat Ag (weakly enhanced) and textured Ag (strongly enhanced). Note the marked improvement towards the red end of the spectrum for the textured Ag surface. One further beneficial aspect of this technology has been the reported improvement in the device yield, which has been attributed to the stress relief of a-Si:H films caused by the texturing.

A detailed study of the performance of TCO/p^+-i-n^+/metal has been reported by Walker et al. (1985). A comparison of ITO and SnO_2:F coated glass substrates, with and without a SiO_2 barrier, was made, as shown in Fig. 3.58. Region I corresponds to ITO substrates showing reduced solar cell performance; this was attributed to In diffusion into the p-i-n junction. Samples corresponding to region II do not incorporate SiO_2, with the consequence that the SIMS data revealed Na contamination within the structure emenating from the soda-lime glass substrate. High efficiency devices in region III correspond to textured TCO substrates with a SiO_2 barrier between the glass (soda-lime) and SnO_2:F.

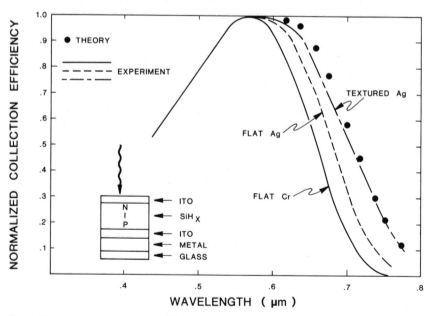

Fig. 3.57. Normalized collection efficiency for unenhanced (flat Cr), enhanced (textured Ag) and weakly enhanced (flat Ag) cells with an integral reflector structure. Normalized collection efficiencies are compared to remove differences in the antireflection properties of the cells (texturing can reduce from surface reflection). The theoretical model assumes $\eta = 22\%$ (Deckman et al., 1983).

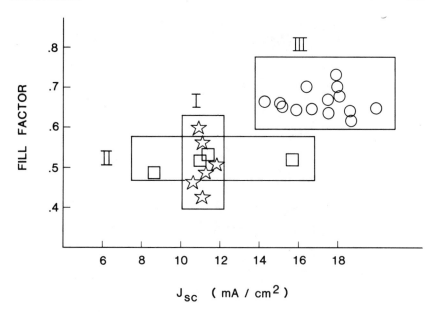

Fig. 3.58. *FF* vs. J_{sc} for a-Si:H *p-i-n* solar cells made on different types of TCOs layers (Walker *et al.*, 1986).

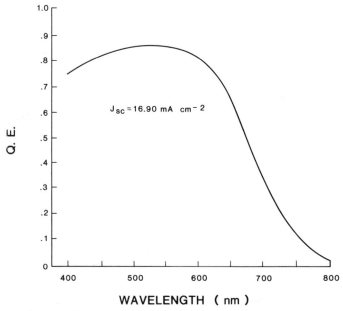

Fig. 3.59. Collection efficiency (quantum efficiency) as a function of wavelength for a *p-i-n* junction solar cell (Hollingsworth *et al.*, 1986).

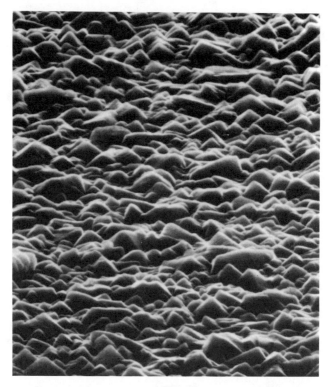

Fig. 3.60. SEM photograph of a textured SnO_2 : F deposited onto a SiO_2 coated glass (Walker et al., 1986).

In Fig. 3.59 we show the collection efficiency of a more optimized solar cell that utilizes the following structure: soda lime glass/SiO_2/SnO_2 : F/ p^+-i-n^+/Ag. Here a metal reflector (Ag), textured SnO_2 : F (Fig. 3.60), and a wide band gap p-type semiconductor are used. An excellent quantum efficiency over the visible spectrum is obtained, with $J_{sc} \approx 17$ mA cm^{-2}. In Table 3.5 data from many groups are summarized.

3.1.4.7 Some Useful Techniques for the Evaluation of Solar Cells

Solar cell performance is linked to many primary factors: the growth and process conditions of the p^+-, i- and n^+-layers; the interaction of the first layer with the substrate; the impurity and doping profiles of the layers involved. The optimization of the cell is accomplished by the type, quality and thickness of these layers. For diagnostic purposes the collection efficiency technique as well as diffusion length measurements has been found to be useful. We now discuss them in detail.

Carrier Collection Efficiency. The carrier collection efficiency (*CE*), $Y(\lambda)$, at different wavelengths is defined as the number of free carriers collected by the external circuit per incident photon:

$$Y(\lambda) = \frac{I_p(\lambda)}{eN(\lambda)}, \qquad (3.60)$$

where $I_p(\lambda)$ is the photocurrent density and N is the photon flux.

Spectral response measurements are generally made with a superimposed white light bias, of intensity corresponding closely to normal operating conditions (i.e., AM-1 illumination), with chopped (10–30 Hz) calibrated monochromatic light incident on the cell. Using a lock-in amplifier, the photocurrents generated by the cell as a function of the wavelength of the monochromatic light are measured under short circuit current conditions.

The electric field within the cell is important in determining the *CE*. Figure 3.61 shows the *CE* for a 0, -2 V and -4 V reverse bias (Wronski *et al.*, 1980). The *CE* increases markedly with reverse bias, especially over the red end of the spectrum where the absorption coefficient is low. This is due to an increase of the width of the space charge region. It has been noted that the collection efficiency in a *SB* device configuration can decrease at the blue end of the spectrum, as shown in Fig. 3.61 (Gutkowicz-Krusin *et al.*, 1981). This has been attributed to the diffusion of photogenerated majority carriers against the electric field at the metal-semiconductor junction. There they can be either emitted or recombine with minority carriers, causing a collection loss.

The *CE* as a function of wavelength has been theoretically derived [Gutkowicz-Krusin *et al.* (1981); Reichman (1981)] by solving the continuity equations to generate the steady state densities of photogenerated carriers. It was assumed that the total density of carriers at the boundaries is essentially unchanged under low levels of illumination (i.e., $\Delta n = \Delta p = 0$). Further, by choosing the potential within the space charge region as exponential, they calculated the spatial variation of Δn and Δp within the device, as shown in Fig. 3.62. (Note that for large values of αL, corresponding to strong absorption, the carrier densities are more sharply peaked than for small values of αL; L is the thickness of the sample.) From this, the current and hence the *CE* as a function of wavelength can be obtained; the fit to the experimental data for a Schottky barrier cell is shown in Fig. 3.63. (The measurements were made at low light intensities, $<10^{12}$ photons/cm^2 s^{-1}, without bias illumination.) Further, the dashed lines in Fig. 3.63 show the effect of the image force potential on the *CE*. It is clearly a minor effect, within the experimental uncertainty in the *CE* data. It is concluded that the fall off in Y at the red end of the spectrum is due to a decrease in the optical

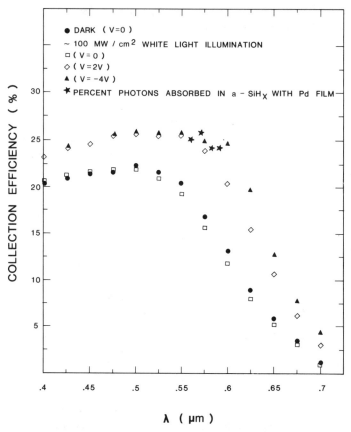

Fig. 3.61. The carrier collection efficiencies obtained on a Pd/a-SiH$_x$ cell structure for different values of reverse bias and 100 mW/cm^2 white light illumination. Also shown are the collection efficiency under short circuit conditions with no bias white light illumination and the results of optical measurements on the transmission of the Pd film (Wronski *et al.*, 1980).

absorption, α, at these wavelengths, whereas the fall off at the blue end of the spectrum is due to the thermal diffusion of electrons against the field.

An alternative explanation has been provided by Rothwarf (1982), who attributes the decrease in the collection efficiency at the blue end of the spectrum to the thermalization of the carriers against the field gradient. He postulates that the free electrons that are created by high energy photons will lose energy to phonons and have their velocity randomized by elastic collisions with impurities and other defects. The excess energy imparted to the electron is given by

$$\Delta E = \tfrac{1}{2}(hv - E_g) - qV(x). \tag{3.61}$$

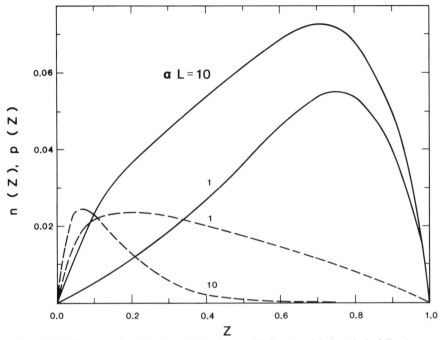

Fig. 3.62. Dimensionless densities of electrons (solid lines) and holes (dashed lines) as a function of position in the a-Si:H film, for two values of the absorption coefficient α (Gutkowicz-Krusin et al., 1981).

It is noted that there will be some distance, x_0, that the carriers can diffuse to before ΔE decreases below $(\frac{3}{2})kT$. Since the generation rate $G(x) = N(\lambda)\alpha(\lambda)\exp[-(\alpha x)]$, then the CE is given by

$$Y = \tfrac{1}{2}[1 + \exp(-\alpha x_0)]. \quad (3.62)$$

Therefore, for large values of α and x_0, corresponding to the blue end of the spectrum, CE and Y will decrease. The thermalization distance, x_0, in the absence of a field, will be ~90 A, which for an excess photon energy of 2 eV cm^{-1} for a-Si:H yields $Y \cong 0.70$ (for $\alpha = 10^6$ cm^{-1}). This implies that electrons created by such photons have reached the contact before thermalization. The implication is that the loss in efficiency at the blue end of the spectrum could be reduced by increasing the field at the semiconductor-metal interface or by the insertion of an insulator between the semiconductor and the metal such that it becomes blocking for electrons but not for holes. The effect of this on the CE is shown in Fig. 3.27, where the CE of a SB and a MIS type device are compared. Note the noticeable improvement at the blue end of the spectrum for the MIS case.

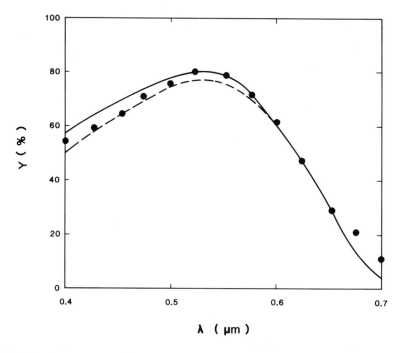

Fig. 3.63. Carrier collection efficiency as a function of wavelength. The solid line was computed using the exponential potential function. The dashed line shows the effect of the image charge potential. The experimental data are shown as circles (Gutkowicz-Krusin et al., 1981).

A useful parameter from a practical point of view is the carrier collection width, L_C, outside of which the recombination of all photogenerated carriers occurs. Figure 3.64 shows the results of Viktrovitch et al. (1981) for both sputtered and glow discharge a-Si:H samples. Note that L_C for the sputtered alloys is less than the depletion width, W, whereas $L_C \sim W \sim 5000$ Å for glow discharge samples. Further, they surprisingly conclude that L_C does not seem to be sensitive to variations in $\mu\tau$ for holes, which is consistent with their observation that this quantity ($<10^{-9}$ cm^2 V^{-1}) does not correlate well with their DOS measurements. This is not in accord with the notion that the hole is mobile until it recombines with a center near the middle of the gap. They speculate that the hole becomes immobile before recombination by hopping while relaxing through states above the valence band. This interpretation is not consistent with other data, but, nevertheless, one important aspect of the study is that $L_C \sim W \sim 5000$ Å for glow discharge films, which is consistent with the observation that the most efficient cells are on the order of 5000 Å thick (see Section 3.1.4.6).

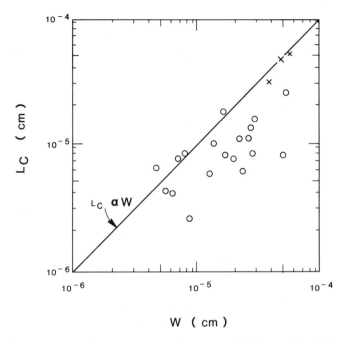

Fig. 3.64. Collection length, L_C, plotted as a function of depletion width, W, for sputter-produced (O) and plasma-decomposed (×) films (Viktorovitch et al., 1981).

As discussed in Section 3.5.6, the performance of a solar cell can be improved at the blue end of the spectrum by the use of an optical window (a-Si : C : B or $\mu c p^+$) and at the red end of the spectrum by the use of textured TCO as well as reflectors (Ag or Al). In Fig. 3.59 we show the CE as a function of wavelength for a device structure of the type glass/textured TCO/a-SiC/p^+-i-n^+/Ag. When this CE is convoluted with the known AM-1 spectra, $J_{sc} = 16.9 \text{ mA cm}^{-2}$.

Minority Carrier Diffusion Length Measurements. The measurement of the minority carrier (holes) diffusion length (L_p^0) is an important technique in optimizing solar cell performance. Several techniques have been employed, with values reported in the range 350 Å –2 μm. This wide discrepancy may be due in part to the preparation conditions and in part to the interpretation involved in the measurement technique employed. Staebler (1980), using a wedge type sample, reported $L_p^0 \sim 900$ Å and Snell et al. (1981a), using a junction recovery technique, reported $L_p \sim 2000$ Å. A technique that has received a considerable amount of attention is the surface photovoltage (SPV) method, and L_p^0 of up to 2 μm have been reported (Moore, 1982).

For single crystal and polycrystalline Si, methods such as SPV and measurements of J_{sc} (Wang et al., 1974) as a function of wavelength have been successfully employed to measure L_p^0. The method relies on the maintenance of a constant J_{sc} by adjusting the intensity of monochromatic radiation incident at various wavelengths. Under these conditions it can be shown that

$$F'[1 - R(\lambda)] = K(\alpha^{-1} + L_p^0). \qquad (3.63)$$

where $R(\lambda)$ is the reflectivity and K is a constant (Goodman, 1981). By plotting the light intensity F' against $1/\alpha$, the intercept at $F' = 0$ defines L_p^0. A variation of this technique is to employ the open circuit voltage condition. For a crystalline cell, V_{oc} is related to J_{sc} by Eq. (3.42); by plotting F' vs. $1/\alpha$ at constant V_{oc} the intercept at $F' = 0$ yields L_p^0.

For crystalline material the width of the space charge region, $W \ll L_p^0$ and Eq. (3.42) works quite well. However, in a-Si alloys W is on the order of L_p^0 (see Section 3.1.3.6), so use of the above equations to derive L_p^0 can produce erroneous results, as we discuss below.

Dresner et al. (1981) used the latter technique to measure L_p^0 on a-Si:H type alloys. In these experiments the sample structure was as follows: an n^+ layer 200 Å thick was deposited on a stainless steel substrate, followed by the deposition of a thick layer ($>2\,\mu$m) of undoped material. The sample was illuminated with monochromatic radiation and the surface photovoltage, ΔV, was measured with a vibrating Kelvin probe; L_p^0 was deduced from the intercept to be in the range 0.33–0.45 μm for GD a-Si:H prepared in the range 240°C $< T_s <$ 330°C. It was noted that although the saturated photovoltage was 75 mV, the measurements were performed at $V = 50$ mV, which would have led to a built-in potential of 25 mV. One further point to note is that these early measurements were performed without light bias incident on the cell; the technique thus overestimates the diffusion length, as it has been shown that the intercept (and hence L_p^0) is strongly influenced by light intensity (Hack and Shur, 1984) and by the width of the surface barrier (Hack et al., 1982). The effect of these parameters has been considered by Moore (1983), who made an attempt to develop a better analysis than that using Eq. (3.63), which is sensitive to both the internal electric field and electron and hole diffusion. The sample is assumed to be one dimensional, semi-infinite and is divided into a space charge region, SCR, $0 < x < W$, and a bulk region, $W \le x \le \infty$. The analysis utilizes the theory of Reichmann (1981), which neglects recombination within the SCR and assumes that the potential is set up by the space charge and hence is decoupled from the photo-generated carriers. (Poisson's equation is not solved simultaneously with the current flow equations.) The electron current in the SCR obeys the transport and continuity equations and is solved for using the boundary conditions that

SOLAR CELLS 255

at $x = W$, $n(W) = n_0$, and the potential, $\phi(W) = V_{bi} - V_{oc}$. At $x = 0$, $\phi(0) = 0$, $J_n(0) = ev_c(n_0 - n_0')$, where v_c is the interface recombination velocity, $n_0' = n_0 \exp[-(V_{bi}/V_T)]$ and $V_T = kT/e$. The hole current $J_p(0)$ at $x = 0$ consists of all the holes arriving from the bulk at $x = W$ and those that are generated within the depletion width, W. The short circuit current is then calculated from $J_{sc} = J_n(0) + J_p(0)$. However, the SPV experiment requires the use of open circuit voltage conditions. Thus, using Eq. (3.42), and assuming that J_{sc} is proportional to the light flux intensity, Moore (1983) obtains an expression that in the limit $\alpha^{-1} \to \infty$, the intercept at zero illumination intensity ($F' = 0$ on the α^{-1} axis) is given by

$$\frac{1}{\alpha} \approx L_p^0 \left[1 + \left(\frac{W}{L_p^0}\right)^2\right] \bigg/ 2\left(1 + \frac{W}{L_p^0}\right). \qquad (3.64)$$

If $W \ll L_p^0$, then $\alpha^{-1} = -L_p^0$, as for crystalline semiconductors [Eq. (3.63)]. Since the space charge width varies with light intensity, then the intercept can produce a wide range of values. In Fig. 3.65 intensive vs. α^{-1} is plotted for zero bias light. Note that the "diffusion length" can be as large as $L_p^0 \sim 1.55\,\mu\text{m}$, but with a bias of one sun superimposed, it reduces to a smaller value of $\sim 0.55\,\mu\text{m}$.

In the above analysis there are still some approximations, since by using Eqs. (3.63 and 3.64), the intensity is adjusted for a constant V_{oc} in order to determine L_p^0. Use of Eq. (3.64) implies the validity of the superposition principle, which may not be valid for all types of a-Si films (Rothwarf, 1978). Also, the analysis assumes that V_{oc} represents a flat band condition, which is not the case since the device would have to bias itself to some voltage (V_{oc}) necessary to develop a bucking current just able to oppose the light-generated current. Hence, performing SPV at V_{oc} leads, by necessity, to a small built-in potential. Therefore, a finite space charge region exists, rather than the assumed flat band condition. This has been considered by Hack et al. (1982) who, by using a generalized DOS spectrum, derived an effective field region which for $V_{bi} = 50\,\text{mV}$ can lead to a SCR width in excess of 1000 Å, even for a $g_{min} = 10^{16}\,\text{cm}^{-3}\text{eV}^{-1}$. This region would, of course, extend even further when g_{min} is reduced. Hence, under these conditions, a strong electric field can exist, and the intercept derived from the intensity vs. $1/\alpha$ at the V_{oc} condition measures a field assisted diffusion length. As discussed earlier, the intercept value of L_p^0 decreases with increasing light bias, and this seems to occur for materials that are considered to be "good". The influence of the SCR on the diffusion length was also considered by Hack et al. (1982), who showed that an observable diffusion length of $0.4\,\mu\text{m}$, after taking into account the various corrections, corresponds to a zero field diffusion length (L_p^0) of approximately $0.2\,\mu\text{m}$.

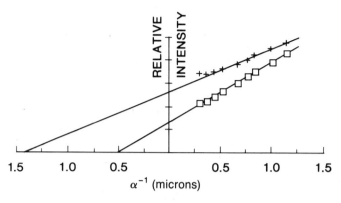

Fig. 3.65. Results of the surface photovoltage technique showing relative intensity against α^{-1} for cells not biased under light (+) and with biasing (□).

3.1.5 Tandem Cells

The reported conversion efficiency for single p^+-i-n^+ junctions using a-Si alloys has exceeded 11% (Nakano et al., 1984; Sakai et al., 1984), but so have the efficiencies of cells made from other non-traditional materials such as crystalline CuInSe$_2$ and CdTe. However, as discussed in Section 2.12, amorphous semiconductors can be made with various band gaps, and thereby offer a unique opportunity for attaining high conversion efficiency ($\geq 20\%$) via a multijunction or tandem cell approach.

We now review briefly solar cell performance using different hydrogenated binary alloys. Solar cells have been fabricated in a p^+-i-n^+ configuration using SiH$_4$ and Sn(CH$_3$)$_4$ for the deposition of an a-SiSn:H intrinsic i-layer. This led to a conversion efficiency of only 2.2%, with $J_{sc} = 8$ mA cm^{-2} and $V_{oc} = 0.60$ V (Kuwano et al., 1982b). The relatively low short circuit current density indicates a narrow space charge region and a low diffusion length for minority carriers. This is due primarily to an increase in the value of g_{min} (the minimum in the DOS) and wider band tails. Recent work on the material properties of a-SiSn:H type alloys, as discussed in Section 2.12, has confirmed that significant amounts of acceptor states are introduced into the material upon alloying. This leads to a drastic decrease in the $\mu\tau$ products of electrons and holes, as well as to a very significant broadening of the Urbach tail (see Section 2.12). Because of these negative factors the major emphasis has turned more towards the GD a-SiGe:H alloy. Here, defect densities on the order of $1-2 \times 10^{17}$ cm^{-3} eV^{-1} at E_f have been obtained for $E_g = 1.5$ eV, together with band tails that are sharp ($E_0 = 55$ meV). p^+-i(a-SiGe:H)-n^+ junctions have exhibited the following characteristics: $J_{sc} = 17.5$ mA cm^{-2}; $v_{oc} = 0.58$ V; $FF = 0.49$; $\eta = 5.0\%$ (Higaki et al.,

1984). The dark current-voltage characteristics have yielded a rectification ratio of 10^4 at $V_{oc} = 0.6$ V and a Q of 1.75, representative of recombination limited transport. The major limitation to device performance here also seems to be V_{oc}, since the highest voltage obtained is only about 0.6 V, probably a consequence of recombination due to a large $g(E_f)$ value, as discussed above. More recent results of a-Si:Ge:H ($E_g = 1.57$ eV) have led to cells with $\eta = 9.6\%$ (Catalano et al., 1987).

Relatively high efficiency (~9%) cells using a-Si:Ge:F:H have been reported by Yang et al. (1985); the band gap of the material was reported to be 1.5 eV. However no data with regard to the concentration of F or Ge within the film were given.

Attempts have also been made to use wide band gap materials in a device configuration. Using a structure $p^+(SiC)\text{-}i(SiN)\text{-}n^+$, cells of efficiency 5.02% with $V_{oc} = 0.80$ V and $J_{sc} = 10$ mA cm^{-2} have been reported (Kuwano et al., 1982). Also, relatively good quality material can be produced with the addition of Carbon and the resultant alloy ($E_g = 1.95$ eV) when inserted into a device led to $\eta = 6.3\%$ with V_{oc} of 0.92 V (Hollingsworth et al., 1987). One of the major losses, once again, is in V_{oc}, since this voltage is far below the voltage expected for a material with a band gap of 2.0 eV. For an efficient tandem cell, a requirement is that a shorting junction be established between the two cells. As discussed in Section 3.1.4.1, $p^+\text{-}n^+$ junctions fabricated from amorphous materials exhibit some rectification, which prevents the formation of a shorting junction in a tandem cell configuration. To overcome this problem $\mu c n^+$ along with a $\mu c p^+$ layer are used, which when combined together (Nakamura et al., 1984) produce a shorting junction. (It is of interest to note that V_{oc} in excess of 0.9 volts can be produced in a single junction, fabricated in the configuration $-\mu c p^+\text{-}i\text{-}n^+$ (Sasaki et al., 1986.)

In Table 3.7 we cite the performance characteristics of single and tandem junctions. Except for one reported case, the efficiency of tandem cells does not surpass the single cell efficiency (see Table 3.5) with currently available materials. This is probably due to the inherent limitations presented by the constituent materials. For increasingly complex alloys we expect an increase in positional as well as compositional disorder, with the concomitant development of a wider band tail (corresponding to an increase in E_{ch}). This will decrease V_{oc} because of increased recombination traffic, as shown in Fig. 3.14. These results necessarily imply that the quest for a low g_{min} is insufficient; sharp band tails also have to be achieved for a viable multijunction cell approach.

In the discussion of Section 2.12 it was found that the addition of any element except F to the a-Si:H based alloy leads to a degradation in the optoelectronic properties of the resulting film. Further, it appears that the Urbach tail is widened with the addition of other elements, which in turn implies an

Table 3.7. Experimental Data for 2 Cell and 3 Cell Multijunction Cells Using Different Amorphous Semiconductors

Band Gap			J_{sc}	V_{oc}		η	
Eg_1	Eg_2	Eg_3	(mA cm^{-2})	(volts)	FF	(%)	Reference
a-Si:H (~1.7 eV)	a-Si:H (~1.7 eV)	a-Si:Ge:H (~1.4 eV)	6.7	2.2	0.58	8.5	Nakamura et al. (1982)
—	a-Si:H (~1.7 eV)	a-Si:Ge:H (~1.4 eV)	9.7	1.4	0.57	7.7	Nakamura et al. (1982)
a-Si:N:H (~2.2 eV)	a-Si:H (~1.7 eV)	—	7.0	1.65	0.61	7.0	Kuwano et al. (1982)
a-Si:F:H (?)	a-Si:F:H (?)	a-Si:Ge:F:H (?)	7.0	2.55	0.71	12.7	Yang et al. (1986)
a-Si:H (1.7)	a-Si:H (1.7)	a-Si:Ge:H (?)	6.2	2.33	0.67	9.6	Sasaki et al. (1986)

SOLAR CELLS

increase in the width of the band tails. There are substantial indications that V_{oc} is limited by recombination within the tail states, which in turn is determined by the disorder and possibly by structural or electrical inhomogeneities due to variations of the H content. The mid-gap density of states affects device performance, mainly in J_{sc} and *FF*, via a decrease of the depletion width and the minority carrier diffusion length, and to some extent V_{oc}. The challenging frontier of materials research in amorphous semiconductors is the development of materials with sharp tail state distributions. This problem is far from trivial to solve, as has been pointed out by Von Roedern and Madan (1985). The steepest tails can be expected in materials with the least amount of disorder, whereas the deep localized states might be minimized when "some" additional disorder is present. Conceptually, optimized materials will be achieved more easily in unalloyed a-Si:H or a-Ge:H than in hydrogenated binaries. In addition, device engineering could circumvent some of the limitations encountered with unoptimized materials.

3.1.6 Stability

Stability is one of the major issues confronting the eventual commercialization of solar cells for large scale power applications. It is important to note that many types of cells, including crystalline Si, degrade under normal operating conditions (Werzer *et al.*, 1979). There has been an increasing amount of data (Tsuda *et al.*, 1984; Uchida *et al.*, 1983; Ullal *et al.*, 1984) showing that good stability can be attained in a-Si alloy devices, especially with the inclusion of ($B < 1$ ppm) in the *i*-layer (see Fig. 3.66a and b). However, a-Si alloys exhibit changes in their transport properties when exposed to light. Figure 3.67 shows data taken by Staebler *et al.* (1981) for two types of cells that exhibit stable and unstable behavior. For the unstable cells the degradation is seen primarily as a decrease in the *FF* and J_{sc}, but the device can be restored to its original performance level upon annealing at 175°C for 30 mins. The amount of degradation is a function of the operating conditions: it is greatest at V_{oc}, reduces at J_{sc}, and is virtually eliminated under far reverse bias conditions. Further, it appears that the changes are not solely induced optically, since Uchida *et al.* (1983) have shown that by applying a forward bias current of 10 mA cm^{-2} in the dark, a similar degradation in performance occurs. Therefore, it seems that the degradation is simply caused by the presence of excess carriers (whose origin may not be important). Degradation phenomena might thus be linked to a recombination mechanism.

The above observations have been associated with the so-called Staebler–Wronski (SW) effect (1980), who reported that application of light can significantly affect the electronic properties of the material, as shown by the

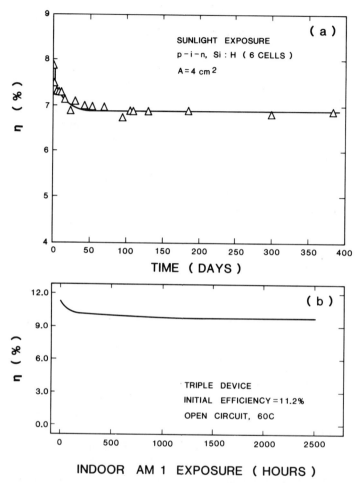

Fig. 3.66. (a) Variation of efficiency (%) vs. time (days) for six p–i–n a-Si:H, $A = 4\,\text{cm}^2$, solar cells exposed to normal sunlight (Ullal et al., 1984). (b) Conversion efficiency vs. light-soaking time for a dual-band-gap triple device (Yang et al., 1986).

changes in conductivity seen in Fig. 3.68. They were able to distinguish between two types of states: a fully annealed state attained by slowly cooling the material from approximately 500 K to room temperature in the dark (state A); a light soaked state, attained after the application of about 10^{21} photons/cm² to the material (state B). However, to attribute all the observed degradation phenomena in devices to the SW effect may not be entirely appropriate, since there will most likely also be other types of degradation phenomena operative. For example, Tawada et al., (1983) have shown that

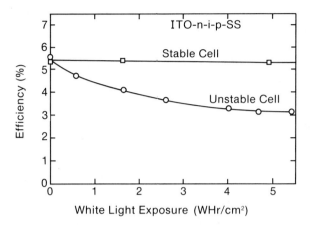

Fig. 3.67. Range of cell stabilities to white light exposure (Staebler et al., 1981).

the *CE* as a function of wavelength of a cell when exposed to Ar laser radiation is different from the *CE* after exposure to He-Ne laser radiation. In both cases the accumulated incident flux was 2×10^{21} photons/cm^2. Since the collection efficiency at the short wavelength, for the Ar case, is worse than for the He-Ne case, they concluded that the changes in the cell may be due to a combination of bulk as well as surface recombination effects. From their analysis they concluded that the light induced changes produce a decrease in the $\mu\tau$ products for holes as well as an increase in the surface recombination velocity.

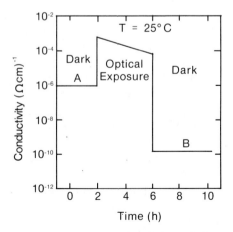

Fig. 3.68. Conductivity as a function of time before, during and after exposure to ~ 200 mW cm^{-2} of light in the wavelength range 6000–9000 Å (Staebler and Wronski, 1980).

There have been many investigations of photo-induced effects in a-Si:H films linked to material parameters. Changes have been observed in the carrier diffusion length (Dresner et al., 1981), unpaired spin density (Dersch et al., 1981), density of states in the gap (Lang et al., 1982) and infrared transmission (Carlson, 1982). The work of Tanielen (1982) showed that the large changes in conductivity could be due to changes in the position of the Fermi level. In general, it was found that light causes E_f to move toward the mid-gap position, the effect being most pronounced for moderate doping levels; vanishing for heavy doping.

The transition from state A to B (see Fig. 3.68) seems to be induced by any process that creates free carriers, including X-ray radiation (Pontuschka et al., 1982) and injection (double) from the electrodes (Staebler et al., 1981). Since degradation in a solar cell is accentuated at the open circuit voltage condition, then it seems that the A to B transition occurs upon recombination of excess free carriers in which the energy involved is less than the band gap. Adler (1983) points out that this transition is a relatively inefficient one and that the increase in the spin density takes place at a rate of 10^{-8} spins per absorbed photon. It is also possible that the A to B transition could be promoted by other recombination centers due to impurities such as oxygen, nitrogen and carbon, which are always present in a-Si:H type films (Nakamura et al., 1984).

The recovery from state B to state A seems to be complete upon annealing, suggesting that: (a) state B is metastable; (b) the photo-induced defects are distinct from the defects present in state A. Various explanations for the SW effect have been put forth. They can be roughly divided into surface effect phenomena, bulk effects, and a combination of the two.

It has been shown (Solomon et al., 1981) that charge accumulation regions exist near the free surface. Hence, in state A the conductivity is dominated by an accumulation channel at one of the surfaces. Upon illumination the barrier at the surface could be flattened, leading to a reduction in the dark conductivity. However, there appears to be a compelling evidence from a host of experimental results that the effect is related to the bulk material. Several models have been proposed to account for the A to B transition that involve either the creation of defects (Dersch et al., 1981) or the trapping of carriers (Crandall, 1981). In the former model the light generates new defects that act as additional traps and recombination centers. These two types of defects can explain the decrease in the minority carrier diffusion length, increase in the DOS, increase in spin density, and decrease in photoconductivity. Adler (1983) points out that it is difficult to understand from this model how excess carrier recombination with energies less than <1.6 eV can create two new defects and, further, why do only the photogenerated defects anneal away? He has proposed an alternative model based on evidence from

NMR (Remier et al., 1980) that a-Si:H films are inhomogeneous, consisting of regions with either high or low concentrations of hydrogen, and assumes that this may be responsible for both the $T_3^+ - T_3^-$ pairs and isolated T_3^0 centers. He assumes that the excess photogenerated electrons and holes can be trapped by the isolated T_3^0 centers (see Fig. 3.69) via the following process

$$T_3^0 + e^- \rightarrow T_3^-; \quad (3.65a)$$

$$T_3^0 + h^+ \rightarrow T_3^+. \quad (3.65b)$$

Since the T_3^0 centers are located in strained regions, then the T_3^+ and T_3^- centers are unlikely to induce significant atomic relaxations; hence the trapped carriers are not in equilibrium. The T_3^0 centers would act as conventional traps. However, spatially close $T_3^+ - T_3^-$ pairs represent the vast majority of charged gap states and can act as relatively shallow traps via the following reactions

$$T_3^+ + e^- \rightarrow T_3^0; \quad (3.66a)$$

$$T_3^- + h^+ \rightarrow T_3^0 \quad (3.66b)$$

In contrast, widely separated $T_3^+ - T_3^-$ pairs have energies closer to the mid-gap region and hence will not rerelease trapped carriers. The resultant T_3^0 center could then trap a carrier of the opposite sign and thereby act as a recombination center.

Adler (1983) considers that once the reactions given by Eqs. (3.66a and b) occur, there should be a change in the configuration from 120° and 95° bond angles, which are optimal for T_3^+ and T_3^- centers respectively, to 109.5° appropriate for the T_3^0 centers. This would, therefore, induce a shift in the effective density of states, increasing the density of T_3^0 centers, which are induced by the free carrier recombination process, as shown in Fig. 3.69. With this shift the trapped electrons fall below E_f and the trapped holes move above E_f, i.e., the carriers have recombined. The photoinduced state, B, however, is not in an equilibrium situation since the energy of the system can be reduced by the process

$$2T_3^0 \rightarrow T_3^+ + T_3^-, \quad (3.67)$$

which involves a distortion of bond angles from the two centers to go to 120° and 95° from their value of about 109°. This process must overcome a potential barrier that could be as large as 1.5 eV. The model is able to qualitatively account for the degradation mechanism and does not involve the creation of new defects but rather the repopulation of those defect centers initially present.

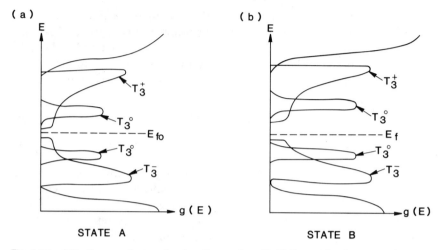

Fig. 3.69. Effective one-electron density of states for a-Si:H films under the assumption that dangling-bond defects are characterized by $U_{\text{eff}} < 0$ if complete atomic relaxations are possible, but with such relaxations retarded in certain strained regions, resulting in the presence of stable T_3^0 centers: (a) equilibrium state A; (b) light-soaked state B (Adler, 1984).

A model by Stutzmann *et al.* (1984) contends that it is the radiative tail to tail transitions that are the cause of the instability phenomena. The model used to explain the creation of metastable dangling bonds is shown in Fig. 3.70. Electrons and holes are created with a generation rate, G. The carriers thermalize into band tail states, where they contribute to the steady state concentrations, n and p, of localized excess electrons and holes, respectively. At room temperature, recombination of these carriers takes place, predominantly through non-radiative processes involving the dangling bond states with density N_s near mid gap as recombination centers. It is suggested that new dangling bonds are created following a direct non-radiative tail to tail recombination between an electron in a weak Si–Si antibonding state and a hole in a weak bonding state, with the result that the increase in the number of dangling bonds during illumination is given by

$$\frac{dN}{dt_i} \propto \frac{G^2}{N_s}, \qquad (3.68)$$

where t_i is the illumination time. Integration of this equation leads to an increase of N_s as a function of illumination time, t_i, and generation rate, G given by

$$N_s(t_i) \propto G^{2/3} t_i^{1/3}. \qquad (3.69)$$

SOLAR CELLS

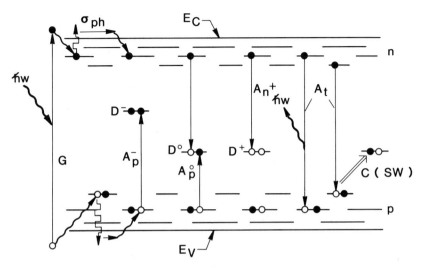

Fig. 3.70. Schematic model for the creation of metastable dangling bonds in a-Si:H (Stutzmann et al., 1985).

Hence, the creation of defect states should be rate limiting, since the creation rate decreases with the square of the existing defect density, N_s. Support for the validity of Eq. (3.69) is provided by the experimental results shown in Fig. 3.71. This could in part explain why highly defective samples generally do not show the SW effect.

The results of Fig. 3.71 are also in substantial agreement with the work of Lee et al. (1984), who have also measured the ESR spin density as a function of irradiation time. A simple atomic configuration consistent with these results is a weak Si-Si bond with a hydrogen atom at one of the back bonds. When a hole is trapped in this weak bond it becomes even weaker. A radiative event will then impart about 1 eV, which can then cause the weak bonds to break. The new configuration can become stabilized by a switching of the hydrogen atom from the back bond position into the weak bond before the local energy has dissipated.

Impurities such as O and C are expected to play a role as far as the degradation is concerned (Crandall et al., 1984; Carlson, 1982). For example, in the case of oxygen contamination the characteristic energy for annealing is 1.0 eV (Carlson et al., 1984), whereas for C impurities it is 0.4 eV (Crandall, 1981). Carlson et al. (1984) conclude that for C concentrations of about 10^{20} atoms/cm^3 a relatively small decrease in conversion efficiency initially occurs, but a more severe degradation in efficiency, after about 10^3 hours, results. The influence of oxygen impurities has also been investigated by Tsuda et al. (1984), who show that the reduction in photoconductivity,

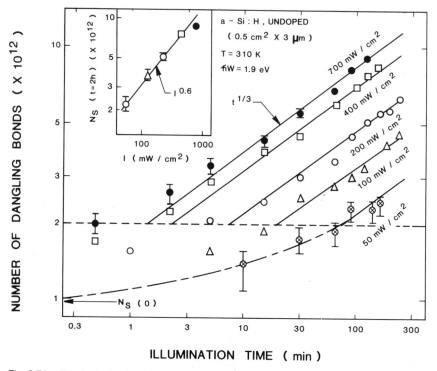

Fig. 3.71. Total spin density, $N_s(t)$, vs. illumination time for various light intensities. The inset shows the intensity dependence of N_s for a fixed illumination time (Stutzmann *et al.*, 1984).

σ_{ph}, with illumination can be decreased by minimizing the oxygen ($\sim 5 \times 10^{18}$ cm^{-3}) concentration within the film. This work also concludes that nitrogen can also accelerate the degradation effect. In contrast, Tsai *et al.* (1984) conclude that impurities do not influence the number of light induced defects below a critical concentration of 10^{20} cm^{-3} for oxygen and 10^{19} cm^{-3} for nitrogen.

The instability phenomena in these types of devices has been also linked to stress (Stutzmann *et al.*, 1985). Undoped a-Si:H generally possesses compressive stress (see Section 2.2), which can be measured using the substrate curvature technique (Campbell, 1970). They find that the number of metastable defects increased with the average stress, i.e., recombination induced breaking of weak bonds. However, this notion is contradicted by Tsuda *et al.* (1985), who did not find a correlation between the degradation of the device performance and the mechanical stress imposed upon it.

Despite these problems the degradation phenomena can be "engineered" out. The degradation appears to be self limiting, since the number of created dangling bonds saturates, which in turn means that the collection width decreases and eventually saturates to a smaller value. Therefore, if the device thickness is approximately equal to the smallest collection width attained, then the device should exhibit relative stability and the problem of degradation is largely circumvented. Figure 3.72 compares the degradation of a thick and a thin solar cell. Although the thicker cell shows a higher initial efficiency, the thinner cell stabilizes to a higher value after several hundred hours of test.

The overall efficiency for a thin device can be improved by several techniques. Using a single junction approach, reflectors and texturing effects, high conversion efficiency cells with excellent stability characteristics can easily be manufactured (see Section 3.1.4.6). The overall efficiency can also be improved somewhat by the use of vertically stacked junctions (pi_1npi_2n), with i_1 and i_2 active layers differing only in thickness. These approaches are effective in circumventing the stability problem to a large extent.

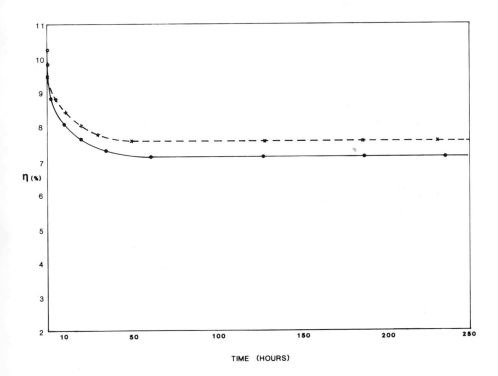

Fig. 3.72. Conversion efficiency as a function of illumination (AM 1.5 intensity) time at V_{oc} condition for p-i-n solar cells of different i-layer thicknesses; 7000 Å (- - -), 4000 Å (———).

3.1.7 Large Area Solar Cells

In order for the a-Si technology to become a practical source of electrical power, the cell performance achieved in small cells must be successfully scaled up to large areas. Also, practical solar cell structures must be utilized; Fig. 3.73a shows a series connected cell that can be manufactured in a monolithic fashion. The use of this design is one of the factors that can bring the cost of the panels down. The preferred way is to use a glass substrate on which a TCO is deposited. This is followed by the deposition of an a-Si:H single or tandem junction, topped off with a metal contact. The cells are scribed (or patterned) at appropriate places to effect a series connected array. This design has been used in many commercial panel applications; Fig. 3.73b shows a large area Arco Solar panel rated at about 6% efficiency. The structure of the monolithic series connected array can be designed (Fig. 3.73) such that it does not need metal grids because the current is low, equal to that produced in a one strip cell, whereas the voltage is high, equal to the sum of the voltages in all the strip cells. In this configuration the width, w, of the strip provides an upper limit (based on the resistance of the TCO) to the generated current. The maximum value of w can be calculated from the cell performance and material parameters from the following expression (Hanak, 1979):

$$w = \left[\frac{3fV_{oc}FF}{RJ_{sc}}\right]^{1/2}, \qquad (3.70)$$

where f represents the allowed fractional decrease in the power generated by the device, and R is the sheet resistivity of the front electrode. As an example, for $J_{sc} = 13$ mA cm^{-2}, $V_{oc} = 0.8$ volts, $FF = 0.65$, $R/l = 10$ and $f = 0.05$, w is equal to 0.08 cm. With gap widths between cells of 0.005 cm, the wasted area is then only 1%.

Detailed studies have been performed of the conversion efficiency as a function of area by Uchida (1982) and Uchida *et al.* (1981). They report that the conversion efficiency decreases as the cell area increases, as shown in Fig. 3.74, along with the various parameters that comprise the efficiency. The decrease in efficiency is due to reduced values of J_{sc} and FF because of: (a) series and shunt resistance that exist in the cell and (b) areal inhomogeneity of the a-Si:H layers, which could arise due to incorrect reactor design for large area depositions.

The effect of the series resistance is investigated by considering

$$P_m = P_c - \Delta P_s, \qquad (3.71)$$

where P_m is the output electrical power, P_c is the generated power of the cell and ΔP_s is the electrical power loss due to the series resistance in the contacts,

SOLAR CELLS

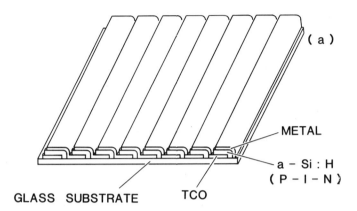

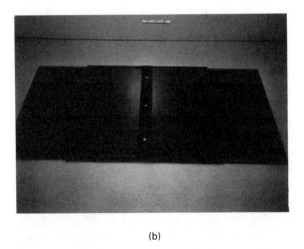

(b)

Fig. 3.73. (a) An efficient way to fabricate an amorphous silicon photovoltaic cell is to deposit a transparent conducting oxide on a glass substrate. An amorphous-silicon single or tandem junction is then deposited and finished off with a metal contact. The metal contact for each junction touches the transparent conducting oxide for the junction on its right, producing a series-connected array as an inherent feature of the manufacturing process. The cells are scribed and separated to produce arrays of the desired size. (b) Arco Solar's Genesis is a glass-substrate module with 6 percent conversion efficiency. The module's active area is 900 square centimeters (1 foot by 1 foot).

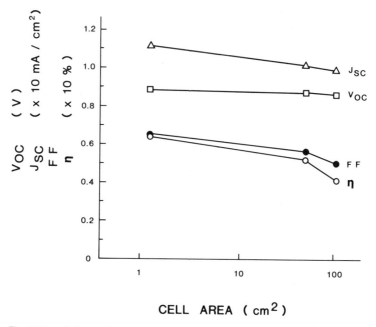

Fig. 3.74. Cell area dependence of output characteristics (Uchida et al., 1981).

including the TCO layer and the grid electrodes. Since large area single cells must employ a grid pattern for the top electrode, then in order to collect the current the cell is divided into units composed of two branch electrodes connected to the i-th point of the stem electrode and the TCO film, as shown in Fig. 3.75.

The effect of an areal inhomogeneity was investigated by making twenty-five 1.2 cm^2 cells on a 49 cm^2 substrate. J_{sc} varied between 11.8 mA/cm^2 and 9.8 mA cm^{-2}; the variation was attributed to changes in the thickness of the device. The junction was considered to be well formed since no variation in V_{oc} was observed.

From these studies Uchida et al., (1981) concluded that the decrease in FF and J_{sc} in enlarging the cell from 1.2 cm^2 to 49 cm^2 was due to various contributions: 50% of the loss arose from the series resistance; 30% was due to contact resistance between the TCO and the grid; 10% was due to the shunt resistance; about 10% was due to the areal inhomogeneity. Table 3.8 shows some data for large area interconnected cells. Note that $\eta \approx 7.0\%$ for areas in excess of 3000 cm^2 (e.g., Hamakawa, 1986) have been attained.

The fabrication of a complete monolithic photovoltaic panel requires a multi-step process and the integration of many well known technologies. The final cost of a panel will ultimately depend on cost reduction at every stage.

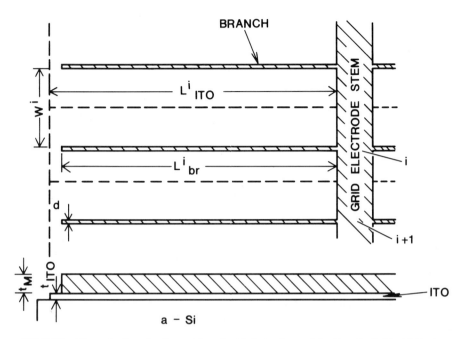

Fig. 3.75. The unit of conducting oxide and grid electrodes used for the calculation of power loss (Uchida et al., 1981).

Some of the key factors are: (1) the improvement in the quality of the TCO's sheet resistivity, transmission and texture; (2) novel patterning steps need to be implemented since the conventional approach using lasers, apart from causing some damage to the cell, is slow and can be one of the rate limiting steps in production; (3) further improvements in yield are required and can be traced to pinhole formation due to either process conditions in terms of gas phase nucleation or dust inclusion from the ambient (see Section 2.2).

Table 3.8. Large Area Solar Cell Conversion Efficiencies (taken from Hamakawa, 1986)

Area (cm^2)	V_{oc} (volts)	J_{sc} (mA cm^{-2})	I_{sc}	FF	η (%)	Institute
68.8	12.93		79.4 mA	0.664	9.9	TDK-SEL
100.0	12.10	17.0		0.662	9.7	Sanyo
419.3	23.0		183 mA	0.70	8.5	Solarex
1200.0	38.75		341 mA	0.684	7.53	Sanyo
3200.0	48.70		890 mA	0.50	6.70	Kanegafuchi

At present there are several different approaches being utilized in the production of a-Si alloy power panels. Some are:

Batch Process. The most conventional technique is the batch type process, which is inherently slow and suffers from the severe cross contamination of Boron and Phosphorous (see Section 3.1.4.3). This leads to a low throughput as well as low overall panel performance. Consequently, the cost of the panels is high.

Roll-to-Roll. In the roll-to-roll approach two types of substrates are under investigation: polyimide and stainless steel (Ovshinsky, 1984). This approach has some drawbacks. Firstly, a roll of material is fed through the system so that p-, i- and n-layers are deposited sequentially. This, by necessity, leads to cross contamination due to Boron (from the p-layer deposition section) and Phosphorus (from the n-layer deposition section), and thus to a degradation of the opto-electronic properties of the i-layer and consequently of the device (panel) performance (see Sections 2.8 and 3.1.4.3). Although various schemes exist to minimize the cross contamination, such as the use of gas curtains between the p- and i-layers, nevertheless, it should be recognized that even one part per million of Boron (or Phosphorous) severely affects the properties of the i-layer (see Section 3.1.4.3). Other problems with this approach are that the use of both substrates (stainless steel and polyimide) require the deposition of the TCO after the p-i-n deposition, which limits the processing temperature to 250°C. It is generally found that the TCO becomes more resistive at lower temperatures; the power delivered from the panel will thereby be decreased due to increase electrical losses. If this temperature is exceeded, H evolves from the a-Si:H layers and leads to a dramatic increase in the defect densities within the semiconductor (a-Si:H), and hence to a severe degradation in device performance (see Section 2.2). Using a polyimide substrate requires extensive prebake procedures in order to expel the H_2O component from within the substrate. Using an expensive stainless steel substrate prevents monolithic panel formation, since by definition, an insulating substrate is required. Using this substrate, the cells are cut into strips and subsequently joined to effect a series connected panel that slows down the production process and increases costs, and in effect becomes a batch type process. This approach ignores the advantages that a-Si alloy technology has to offer, that is, the ability to produce monolithic panels during the production process. One further disadvantage is that upon completion of the interconnects the panel is encapsulated with a polymer. However, virtually every polymer degrades under light, with the consequence that discoloration occurs and the transmission of light into the cell decreases, resulting in an extrinsic degradation phenomena.

Multichamber—Stationary. An approach that is commonly used is the multichamber approach shown in Fig. 3.49, in which the chambers are separated to prevent cross contamination between the *p*-, *i*- and *n*-layers (Kuwano *et al.*, 1982a). Each of these layers is deposited with the substrate stationary. Figure 3.76 shows the typical mass production process in the fabrication of a panel. After cleaning the glass substrate (step A), the substrate is moved in a batch mode to a stage involving SnO_2 deposition (typically an atmospheric pressure CVD process) and is then patterned using a laser cut process (step C). The substrates are then loaded into the a-Si deposition segment (steps D, E and F). The semiconductors are then patterned once again using a laser (step G), which is followed by the deposition of a metal contact (e.g., Ag or Al) and once again scribed using a laser (step I). The panels are then tested and encapsulated.

It should be emphasized that in the panel fabrication process the major rate limiting steps are the scribing and the deposition of the *i*-layer. The conventional scribing process involves 3 steps; during each step the rate of processing is limited by the speed of the laser cut, which is typically 60–240 cms per minute. Therefore, for a square foot panel involving, e.g., 30 cells, the processing time for each scribing step is between 4 and 15 minutes.

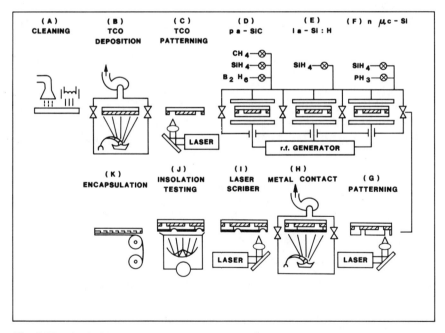

Fig. 3.76. Typical mass production process for monolithic panel production (Hamakawa, 1986).

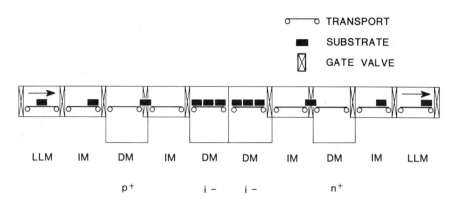

Fig. 3.77. Schematic of a GSI continuous multichamber p–i–n line (X = Gate valves; LLM, Load Lock Module; IM, Isolation Module; DM, Deposition Module). The figure shows substrates in various parts of the system; they are decelerated as they enter the rate limiting step (i-layer); the substrates are accelerated out of the i-layer for subsequent n-layer deposition (Madan, 1986).

One of the other limiting steps is the deposition of the i-layer, which is usually about 4000 Å–6000 Å thick, at a deposition rate of 1–5 Å s^{-1}.

Multichamber—Moving Substrate. To overcome the above limiting steps, we show in Fig. 3.77 a continuous multichamber in-line process that is part of a complete integrated facility with a glass in-panel out approach (Madan, 1986; von Roedern and Madan, 1986; Madan, 1987). The a-Si alloy segment is comprised of three building blocks: load lock module (LLM); Isolation Module (IM); Deposition Module (DM). All the chambers are provided with an individual transportation system so that the substrates can be moved at different speeds within the chambers in order to maximize throughput, while the IM's and the gate valves ensure zero cross contamination (between the p-, i-, n-layers), leading to a maximum performance from the panels. Another advantage of this type of approach is that an increase in the deposition rate of the i-layer can be readily accomplished by merely altering the speeds within each individual chamber.

3.1.8 Economic Considerations

The success of a-Si alloys as a power source depends not only on its stability but also upon economic considerations. A-Si alloys as power sources are being used in consumer applications such as watches, calculators and battery chargers. However, the major hope is its eventual use in terrestrial power applications.

SOLAR CELLS

Photovoltaic systems are currently being used, cost effectively, in an increasing number of remote stand-alone applications. These systems typically are between a few tens of watts and a few MW's. They provide power for a variety of items such as water pumps, navigation aids, communication equipment and residential electrification.

In many developing countries the electricity requirements are widely distributed and small in magnitude. Photovoltaics allow for the immediate delivery of electricity to remote regions without waiting many years to construct expensive grid extensions. A major advantage of photovoltaic cells is their modularity; as the load expands, so can the photovoltaic capacity. In the developing world, where there is no established electric grid, the alternative for electric power is the diesel generator. Photovoltaic cells are considered to be the "lowest cost" alternative in many applications. Estimates by the U.N. and the World Bank have indicated that a huge potential market exists for photovoltaics in developing countries. The U.N. has estimated that there are 10^6 villages without electricity. Their estimate of 30–50 KWP of electricity/village results in a potential market of 60,000–100,000 MWP.

The ultimate goal for photovoltaics is to compete economically with conventional methods of generating large quantities of electricity in a centralized power plant. Several studies (see, e.g. Armstrong-Russell *et al.*, 1983) have identified what is essential for this event to occur. The capital costs associated with an array can be divided into: (a) module cost, e.g., the solar cells and their interconnects; (b) encapsulation, mounting on a support structure; (c) land costs; (d) power conditioning costs. These costs are referred to as balance of systems costs, and careful consideration has to be given to minimize them. Total system costs are sensitive to module or cell assembly efficiency. Various studies have been performed as to the economic viability of flat plate collector systems. They have concluded that module efficiencies in the range of 10–15% have to be attained, which is well within the reach of a-Si alloys.

In this connection it is relevant to take heed of the experience of the Sanyo Corporation, who noted that even with an efficiency of 5%, the energy payback period, assuming only about 4 hours of insolation per day with a production output capacity of 10^9 W/yr, is less than one year, which is considerably less than the 20 year expected lifetime of the device. Their projected production cost is less than 15 cents per watt. The economics of a-Si alloy cell production must by necessity include throughput, which in turn implies high deposition rates. Several techniques for increasing the deposition rate have been attempted, particularly via the CVD of disilane (Konogai *et al.*, 1982; Itoh *et al.*, 1984; Matsuda *et al.*, 1983) and the glow discharge of $SiF_4 + SiH_4 + Ar$. The rate has been increased to beyond 10 Å s^{-1}, while the high quality electronic properties of the material have been retained. Using disilane gas, conversion efficiencies exceeding 7% have been obtained, with deposition rate > 30 Å S^{-1} (Itoh *et al.*, 1984).

Whether a-Si alloys, or for that matter other materials, will be ever utilized as large scale photovoltaic power sources remains to be seen. It depends on a host of factors that include political as well as business considerations. These factors are outside the scope of this book.

3.2 Thin Film Transistors for Display Applications

The operation of a field effect transistor (FET) depends upon the variation of the conductivity of a semiconducting layer produced by the application of an external electrical field. The device proposed first by Lillenfield (1930, 1935) did not produce significant changes in the conductivity with applied field. Shockley (1939) ascribed this to the presence of surface states at the insulator/semiconductor interface. After the discovery by Atalla *et al.* (1959) that the surface state density could be reduced at the insulator (SiO$_2$)-semiconductor (Si) interface, the first successful MOSFET (metal-oxide-semiconductor-field-effect-transistor) was reported by Hofstein *et al.* (1963). Subsequently, a very important technology developed because FET devices are used extensively in LSI (Large Scale Integration) and VLSI circuitry.

It has recently been appreciated that large area displays have significant market potential in the area of low power portable computers and small TVs, which are now in the process of being commercialized. Of immediate concern is the development of large area displays (< 12" × 12"). There is need for a technology that is able to deposit opto-electronic materials over large areas, since panel construction involving a number of submodules (e.g., crystalline Si wafers) would be technically and economically unattractive. Displays that use liquid crystals are normally illuminated from the rear and operate in the transmission mode. Since the liquid crystal is not a highly nonlinear element with a well defined threshold, then X and Y addressing can only be achieved on high information content (a minimum of about 250 × 250 picture elements) displays by incorporating a nonlinear switching element, such as TFT (thin film transistor) at each element of the display. Other schemes involving two terminal nonlinear devices such as varistors (Castleberry *et al.*, 1982) and metal-insulator-metal diodes (Baraff *et al.*, 1981) have been used effectively, but their use is limited to smaller arrays.

In Table 3.9 the minimum TFT requirements for a 500 line LCD (Liquid-Crystal-Display) with a resolution of 500 elements/inch (Luo, 1984) is shown. There have been many reports in the literature of TFTs employing

Table 3.9. 500 Line Matrix Display. (C_s Is a Storage Capacitance Often Used in CdSe Displays.) (Lou, 1984)

TFT Performance Requirements	
I_{on} (at $V_G > 0$)	5 µA
I_{off} (at $V_G = 0$)	
with C_s (~10 pF)	< 1 nA
without C_s	< 0.1 nA
Addressing time	$T = 33$ µs
Response time	$T_R = 3$ µs

various types of semiconductors such as CdS, CdSe, InAs, InSb, PbTe, PbSe and Te. Of these, CdSe has received the most attention, with a successful demonstration in 1973 of a 6" × 6", 20 lines/inch LCD display employing more than 14,000 TFTs (Brody et al., 1973). In spite of the early successes and the considerable progress since, many problems still remain. Among these are the difficulty in depositing device quality stoichemetric CdSe material over large areas, the inability to obtain a highly insulating native oxide and the difficulty of constructing stable injecting source and drain contacts. Another drawback is that this type of material utilizes a technology that incorporates few conventional steps such as those used in the processing of Si. Hence, attention has increasingly turned towards the development of Si films involving: (a) laser annealed (LA) crystalline Si (Nishimura et al., 1984); (2) small grained polycrystalline Si (Morozumi et al., 1984); (c) a-Si alloys. Table 3.10 (Thompson, 1984) compares the material parameters of these three technologies with its predecessor, CdSe. As can be seen from Table 3.10, although LA-Si offers good performance, its off-current is quite high. Further, this process is under question since the viability of uniform laser annealing over large area glass substrates has yet to be demonstrated. The processing temperature is limited to 600°C because of the use of an alkaline containing glass substrate; to circumvent this more expensive substrates are required, such as quartz. The same problems persist for polycrystalline Si. However, these problems are not encountered with a-Si:H, since the maximum temperature for processing is in the range of 250–300°C.

In the following sections the basic physics of a TFT using a-Si based alloys is considered. Application of this technology to large area displays is also discussed.

Table 3.10. Performance and Technology Issues of TFTs (Thompson, 1984)

	CdSe	a-Si:H	Poly-Si	LA-a-Si
Mobility cm^2 V^{-1} s^{-1}	~20	0.1–2.0	10–50	200–1000
I_{on}, A	5×10^{-6}	10^{-5}	5×10^{-6}	10^{-4}
Gate length, μm	$L = 10$	$L = 100$	$L = 10$	$L = 10$
I_{off}, A	2×10^{-10}	5×10^{-12}	5×10^{-12}	10^{-10}
Fabrication issues	Unconventional materials, i.e., CdSe, Al$_2$O$_3$ Photolithography?	Si$_3$N$_4$	Implant n^+ 600°C process	LA step High temperature process
	Yield?	Yield?	Yield?	Yield?

3.2.1 Basic Requirements

A normal display design incorporating a single transistor for each picture element (pixel) is shown in Fig. 3.78a (LeComber et al., 1981). $S_1, \ldots, S_n$ and $G_1, \ldots, G_n$ are the bus lines for the source and the gate electrodes, respectively. The drain contact is connected to each TCO square and the liquid crystal (LC) is sandwiched between the substrate (TFTs) and a TCO coated glass top plate that is normally grounded, as shown in Fig 3.78b. The LC is in series with the drain circuit and behaves electrically as a capacitor C_{LC} with a leakage resistance, R_{LC}. In one version of the TFT design, Fig. 3.78c (LeComber et al., 1981) shows a section through an individual device and Fig. 3.78d illustrates the design of an a-Si:H TFT in part of the matrix array. The first step in the fabrication consists of etching the TCO pattern that forms the back contact of the liquid crystal cell. A metal contact, such as Cr, is evaporated and etched to provide the gate electrodes, G, and the prepared substrate is then coated with a dielectric, such as Si$_3$N$_4$ and followed by the deposition of an a-Si:H layer. The unwanted a-Si:H is etched away, leaving small islands at each TFT site. Contact holes, denoted by A in Fig. 3,78d are etched through the nitride layer down to the TCO in order to provide for connection to the drain contact, D. Finally, a low work function metal, such as Al, is evaporated to form the source and drain contacts (although it is preferable to also deposit an n^+ layer between the a-Si:H and the metal contacts, whose effect will be discussed below). All these processes are performed using the standard photolithographic techniques used in the Si semiconductor industry.

THIN FILM TRANSISTORS FOR DISPLAY APPLICATIONS

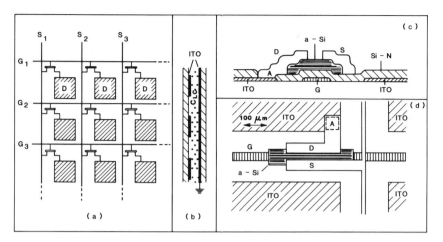

Fig. 3.78. Schematic layout of an addressable liquid crystal panel, (a) plan view, (b) side view, (c) shows the section through a transistor element and (d) shows a TFT in part of the matrix array (LeComber et al., 1981).

The addressing scheme involves writing one line at a time, corresponding to a TV line, in 63.5 µs. Therefore, in 63.5 µs gate G_1 in Fig. 3.78a is addressed while $G_2, \ldots, G_n$ are in the OFF mode. The information for G_1 is simultaneously applied to the source lines, i.e., zero potential for pixel elements in the OFF state and a potential, V_s, for pixels that are to be turned ON. After 63.5 µs the transistors that have just been addressed (G_1) are turned OFF and the charge is retained in the capacitor C_{LC}. In the next 63.5 µs video data are entered into the next row, G_2, and the information for that simultaneously applied to the source lines. When all the gate buses have been addressd the scan returns to the first bus line and the above process is repeated. However, it is advantageous that the polarity of the video signal in the next frame cycle is reversed (ac mode), since any dc component present tends to generally degrade the material.

In the above addressing scheme the requirements are that the resistance of the TFT in the ON mode (R_{ON}) should be low enough such that adequate charging of C_{LC} takes place in a time less than about 60 µs. Since C_{LC} is on the order of a few picofarads, then the requirement for the resistance in the ON state (R_{ON}) is that it should be less than 6 MΩ. Further, the OFF resistance (R_{OFF}) should be large enough so that the charge retained in the capacitor, C_{LC}, does not decay during the refresh time, which for 60 frames is, 16.7 ms. Hence, a prerequisite is that $R_{OFF}C_{LC} > 16.7$ ms and the minimum basic requirements are that $R_{OFF}/R_{ON} > 300$. However, to obtain an adequate gray scale, R_{OFF} should be much greater than 300 R_{ON} and R_{OFF}

should also be of sufficiently high impedance so that the LC element does not drift ON due to leakage effects (Brody et al., 1973; Luo, 1984).

In CdSe TFTs the above conditions are difficult to achieve and consequently a capacitor (Brody et al., 1983) is introduced, which results in the fulfilment of the above conditions.

Displays generally employ a twisted nematic mode of operation. The liquid crystal is sandwiched between two plates; alignment occurs at the surface and the two substates are then twisted 90° relative to each other. Because the liquid crystal is birefringent it has the ability to rotate the plane of the polarized light. When a small electric field is applied to the crystal perpendicular to the plates all the bulk molecules realign. This removes the twist from the cell and the vibration plane does not rotate, thus allowing for the transmission of light.

3.2.2 Theory of Operation

3.2.2.1 Generalized Theory

The general configuration of an inverted a-Si:H TFT is shown in Fig. 3.79a, where S and D represent the source and drain contacts, respectively, of width W separated by channel length L.

The electrostatic potential in the surface space charge layer at any point (x, y) is expressed as (Suzuki et al., 1982)

$$V(x, y) = V_0(y) + u(x, y). \quad (3.72)$$

$V_0(y)$ is the potential at the edge of the space charge layer where $\partial u/\partial x = 0$ is satisfied; $u(x, y)$ refers to the surface band bending. Here, $u(x, y) > 0$ for gate potential, $V_G > V_0(y)$ and $u(x, y) < 0$ for $V_G < V_0(y)$.

The conductance for an element of channel length dy and width W is composed of the flat band conductance $G^0_{y,y+dy}$ and the field induced conductance, $\Delta G_{y,y+dy}$ arising from the band bending $u(x, y)$.

The conductances can be expressed as

$$G^0_{y,y+dy} = \sigma_0 \frac{Wd}{dy}, \quad (3.73)$$

and the change in conductance by,

$$\Delta G_{y,y+dy} = \frac{\sigma_0 W}{dy} \int_0^{eu(y)} \left\{ \frac{(\exp eu/kT - 1)}{\partial u/\partial x} \right\} \frac{d(eu)}{e}. \quad (3.74)$$

$\sigma_0 \ (= e\mu_n n_b + e\mu_p p_b)$ is the flat band conductivity of the bulk a-Si:H layer and $u(y)$ is the surface potential. $n_b(p_b)$ are the concentrations of electrons and holes in the extended states and $\mu_n(\mu_p)$ are the corresponding mobilities.

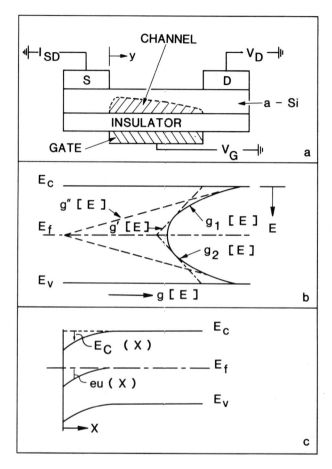

Fig. 3.79. (a) General configuration of an inverted TFT; (b) generalized DOS used in the analysis (see text); (c) band bending with the application of a positive gate voltage.

The differential resistance across the channel element $dV_0(y)$ for the source-drain current, I_{SD}, is equal to $I_{SD}\,dR$; by integrating over the channel from the source contact $[V_0(0) = 0]$ to the drain contact, $V_0(L) = V_D$, we obtain

$$I_{SD} = \frac{\sigma_0 W}{L} \left\{ dV_D + \int_0^{V_D} \left[\int_0^{eu(y)} \left(\frac{\exp eu/kT - 1}{\partial u/\partial x} \right) \right] \frac{d(eu)}{e} dV_0(y) \right\}. \tag{3.75}$$

If both the electric field, $\partial u(x, y)/\partial x$, and the surface potential $u(y)$ are obtained as a function of $V_0(y)$ for a given value of V_G and V_D, then I_{SD} can be calculated from Eq. (3.75).

The surface band bending $u(x, y)$ obeys the one dimensional Poisson equation:

$$\frac{\partial^2 u(x, y)}{\partial x^2} = -\frac{\rho(x, y)}{\varepsilon_s \varepsilon_0}, \qquad (3.76)$$

where ε_s is the dielectric constant of the a-Si:H layer. At $T = 0$ K the charge, $\rho(x, y)$ is related to the gap state density by

$$\rho(x, y) = (-e) \int_{E_f(y)}^{E_f(y)+eu(x,y)} g(E)\, dE + n_0 \left\{ \exp\left[\frac{eu(x, y)}{kT}\right] - 1 \right\}. \qquad (3.77)$$

If $\partial u/\partial x = 0$, then for $u(x, y) > 0$

$$\left.\frac{\partial u}{\partial x}\right|_x = -\left\{ \frac{2}{\varepsilon_s \varepsilon_0} \int_0^{eu(x,y)} d(eu) \left[\int_{E_f(y)}^{E_f(y)+eu(x,y)} g(E)\, dE \right.\right.$$

$$\left.\left. + n_0\left(\exp\frac{eu(x, y)}{kT} - 1\right) \right] \right\}^{1/2} \qquad (3.78)$$

The surface potential $u(y) + u_s$ is related to the voltage drop across the insulating film, V_i, and V_G by,

$$V_G = V_0(y) + V_i(y) + u(y). \qquad (3.79)$$

The surface space charge density, $Q(y)$, is given by,

$$Q(y) = -C_i V_i(y)$$

$$= -C_i[V_G - V_0(y) - u(y)] = \varepsilon_s \varepsilon_0 \left.\frac{\partial u}{\partial x}\right|_{x=0} \qquad (3.80)$$

$$= +\frac{\varepsilon_s \varepsilon_0}{C_i} \left\{ \frac{2}{\varepsilon_s \varepsilon_0} \int_0^{eu(y)} d(eu) \left[\int_{E_f(y)}^{E_f(y)+u(x,y)} g(E)\, dE \right.\right.$$

$$\left.\left. + u_0\left(\exp\frac{eu(x, y)}{kT} - 1\right) \right] \right\}^{1/2}. \qquad (3.81)$$

Hence, $u(y)$ can be uniquely determined as a function of $V_0(y)$ if $g(E)$ and V_G are given.

To understand the performance of a-Si alloy TFTs in more detail we use the $g(E)$ spectra for a-Si:H shown in Fig. 3.79b. Further, for sake of simplicity, approximations are used to derive the transfer characteristics of a TFT operating in the enhancement (accumulation modes) as well as the depletion (inversion) modes.

3.2.2.2 Approximate Solution

As discussed previously, an acceptable form of $g(E)$ has the donor and acceptor type states shown in Fig. 3.79b. For acceptor type states $g_1(E)$ is given by

$$g_1(E) = g'_a(E) + g''_a(E), \tag{3.82}$$

where

$$g'_a(E) = \frac{g'_{\min}}{2} \exp\left(\frac{E_f - E}{E'_{ch}}\right),$$

and

$$g''_a(E) = \frac{g''_{\min}}{2} \exp\left(\frac{E_f - E}{E''_{ch}}\right),$$

where $g'_{\min}$ and $g''_{\min}$ are the minimum in the DOS spectra and E'_{ch} and E''_{ch} define the slopes of the DOS spectra. In a similar fashion, the $g_2(E)$ states shown in Fig. 3.79b are donor type and are defined in a fashion similar to Eq. 3.82.

For the sake of convenience $T = 0$ K statistics are used. In the equilibrium case ($V_G = 0$) the negative charge within the semiconductor can be determined by the use of the Fermi integral:

$$Q^-_{\text{loc}}(0) = q \int_{E_f}^{E_v} g_1(E) \, dE, \tag{3.83}$$

and the positive charge by,

$$Q^+_{\text{loc}}(0) = q \int_0^{E_f} g_2(E) \, dE. \tag{3.84}$$

Since charge neutrality must prevail, it can be shown that for the symmetric DOS shown in Fig. 3.79b $E_f = E_v/2$.

Application of the positive V_G shown in Fig. 3.79a leads to the inducement of negative charge within the amorphous semiconductor. This results in band bending, as shown in Fig. 3.79c, with the result that the conductivity of the semiconductor increases, which in turn increases I_{SD} when a potential V_D exists between the two contacts. We first consider the case $V_D < V_G$, in which case $eu(x, y) \approx eu(x) = E_c(x)$. Also, our analysis is restricted to the case that $E_c(x)|_{x=0} = E_s < 0.6 \, \text{eV}$ for a symmetric DOS; in this case the induced charge resides primarily in the deep localized states. (Later we shall consider the case where E_s is sufficiently large such that charge becomes increasingly accommodated within the rapidly rising conduction band tail states.)

The net change in the negative charge is given by

$$\Delta Q_{loc}^-(V_G) = Q_{loc}^-(V_G) - Q_{loc}^-(V_G = 0) \qquad (3.85)$$

$$\Delta Q_{loc}^-(V_g) \approx e\left\{\frac{g'_{min}E'_{ch}}{2}\left[1 - \exp\frac{E_c(x)}{E'_{ch}}\right] + \frac{g''_{min}E''_{ch}}{2}\left[1 - \exp\frac{E_c(x)}{E''_{ch}}\right]\right\} \qquad (3.86)$$

High quality a-Si:H material is characterized by $E''_{ch} = 0.025$ eV, $E'_{ch} \approx 0.10$ eV (see Section 2.5.6), $g(E_c) \sim 1.10^{21}$ cm^{-3} eV^{-1} and $g'_{min} \sim 10^{15}$ cm^{-3}eV^{-1}. Therefore, the crossover from a DOS characterized by $g'_a(E)$ to $g''_a(E)$ occurs at approximately 0.2 eV below the conduction band edge. For a symmetric DOS, and a band gap of 1.7 eV, $E_f = 0.85$ eV. For a band bending of approximately 0.6 eV it is a good approximation to neglect the second term in Eq. (3.86). Then, for band bending such that $E_c(x) > 0.1$ eV and less than 0.6 eV, $\Delta Q_{loc}^- \gg \Delta Q_{loc}^+$, and Eq. (3.86) reduces to

$$\Delta Q_{loc}^-(V_G) \approx -e\frac{g'_{min}}{2}E'_{ch}\exp\frac{E_c(x)}{E'_{ch}}. \qquad (3.87)$$

The potential profile, or the shape of the barrier, can be calculated by solving Poisson's equation, Eq. (3.76). Since $eu(x) = E_c(x)$, then the field at x is given by,

$$F(z) = \frac{E_{ch}}{eZ'_0}\exp\frac{E_c(x)}{2E'_{ch}}, \qquad (3.88)$$

where

$$Z'_0 = \left[\frac{\varepsilon_s\varepsilon_0}{e^2 g'_{min}}\right]^{1/2} \qquad (3.89)$$

The extent of the space charge region, ω, is given by

$$\omega = \int_{E_s}^{0}\frac{dE_c}{qF(E_c)}, \qquad (3.90)$$

where at $x = 0$, $E_c(x) = E_s$ and at $x = \omega$, $E_c(x) = 0$. However, as the band bending increases, the induced negative charge within the amorphous semiconductor will be increasingly located at the semiconductor/dielectric interface. Hence the "effective" accumulation (or space charge) region shrinks with increasing gate voltage such that we can define the extent of this region by

$$D(E_s) \approx \frac{E_s}{F(x = 0)} \qquad (3.91)$$

The change in conductance, from Eqs. (3.73) and (3.74) is given by

$$G' = \Delta G_{y,y+dy} - G^0_{y,y+dy}$$

$$= \frac{W}{L} \frac{\mu_n n_b Z'_0}{E'_{ch} a} [\exp aE_s - 1], \qquad (3.92)$$

where $a = (1/kT - \frac{1}{2}E'_{ch})$.

The induced charge, Q^-_{ind}, in the amorphous semiconductor is given by

$$Q^-_{ind} = C_{eq} V_G, \qquad (3.93)$$

where $C_{eq} [= C_i C_s/(C_s + C_i)]$ is the equivalent capacitance of the insulator, C_i, and the amorphous semiconductor, C_s. $C_i = \varepsilon_i \varepsilon_0 / t_i$, where t_i and ε_i are the thickness and permitivity of the insulator, respectively, and $C_s = \varepsilon_s \varepsilon_0 / D(E_s)$. Typical parameters are, $t_i \approx 1\,\mu m$, $\varepsilon_i = 6.4$ (corresponding to Si$_3$N$_4$), $\varepsilon_s = 11.8$ (a-Si:H) and $D < 0.2\,\mu m$. Then $C_s \gg C_i$ and $C_{eq} \cong C_i$. Hence,

$$Q^-_{ind} = -C_{eq} V_g \approx -C_i V_G. \qquad (3.94)$$

Assuming an ideal insulator,

$$Q^-_{ind} = -C_i V_G = Q^-_{loc} = -\varepsilon_s \varepsilon_0 F(x)|_{x=0}. \qquad (3.94a)$$

$$C_i V_G = \varepsilon_s \varepsilon_0 \frac{E'_{ch}}{Z'_0} \exp \frac{E_s}{2E'_{ch}}, \qquad (3.94b)$$

where

$$E_s = 2E'_{ch} \ln\left(\frac{V_G C_i Z'_0}{\varepsilon_s \varepsilon_0 E'_{ch}}\right). \qquad (3.95)$$

By using Eqs. (3.92) and (3.95), $G'(V_G)$ can be calculated. It is apparent that the change in conductance of an a-Si:H TFT can be manipulated via inherent material properties (g'_{min}, E'_{ch}) as well as by geometric factors (W, L).

Turning now to the $I_{SD}(V_G)$ characteristics, including the variation of potential across the channel length, we have

$$I_{SD} = \frac{W\mu_n n_b Z'_0}{E'_{ch} aLe} \int_{V_1}^{V_2} [\exp aE_s(y) - 1] \, dV(y), \qquad (3.96)$$

where $E_s(y)$ is related to $V(y) = V_G - V_D y/L$ with V_G replaced by $V(y)$ in Eq. (3.95).

At $y = 0$, $V_1 = V_G$ and at $y = L$, $V_2 = V_G - V_D$. Using Eqs. (3.96) and a modified Eq. (3.95), the basic current-voltage characteristics, $I_{SD}(V_G)|_{V_D}$

and $I_{SD}(V_D)|_{V_G}$, corresponding to the accumulation mode (enhancement or n-channel FET mode) can be found. Note that as V_D is increased, $V(y)$ will decrease, becoming a minimum at the drain contact. Hence, the accumulation region will in principle exhibit the profile shown in Fig. 3.79a. With increasing V_D, the accumulation shrinks underneath the drain contact; at $V_G = V_D$, the accumulation disappears entirely. For $V_G > V_D$ the non-accumulated region will then shift towards the source contact. The drain voltage where the accumulation region shrinks to zero underneath the drain contact is referred to as the saturation voltage, or the threshold voltage, V_{th}. Increasing V_D beyond V_{th} does not result in any further increase of I_{SD}. I_{SD} can only increase with further increase of the gate voltage.

We now consider the case beyond the intermediate case, such that the band bending, E_s, is large enough so that the Fermi level resides within the rapidly rising conduction band tail states shown in Fig. 3.79b. The induced charge will now be distributed among the deep localized states (Q_{loc}^-), the conduction band tail states (Q_{trap}^-) and the mobile charge above the conduction band (Q_{free}^-). Figure 3.80 shows [for the symmetric $g(E)$ case, $E_f = 0.85$ eV] the distribution of charge as a function of band bending, E_s (eV). Note that for $E_s > 0.6$ eV the charge is increasingly situated in the shallow trapped states. We also show as an example, for $E_f = 0.67$ eV, where the transition now occurs, at $E_s \approx 0.5$ eV (Madan, 1983). [Shur and Hack (1984) then employed the Madan (1983) results for $T > 0$ K statistics and arrived at the results displayed in Fig. 3.81, which are identical to the results of Fig. 3.80.] With increasing positive gate bias there is a transition from charge trapped in deep states to shallow states, which has an affect on the I-V characteristics. This occurs because further application of a positive bias leads to accommodation of the charge defined by Q_{trap}^-, with the result that there is little change in the surface band bending, as seen in Fig. 3.80. Hence, there is an abrupt change in the I-V characteristics. Figure 3.82 shows the transconductance, defined by

$$g_m = \frac{\partial I_{SD}}{\partial V_G}\bigg|_{V_D}, \qquad (3.97)$$

taken from the transfer characteristics of Fig. 3.83 (Snell et al., 1981).

3.2.2.3 The Effect of Contact Resistance

In the above it was assumed that the source and drain contacts were ohmic. However, the actual contact resistance can substantially affect the performance of a-Si TFTs, as we shall now show. The source contact constitutes a

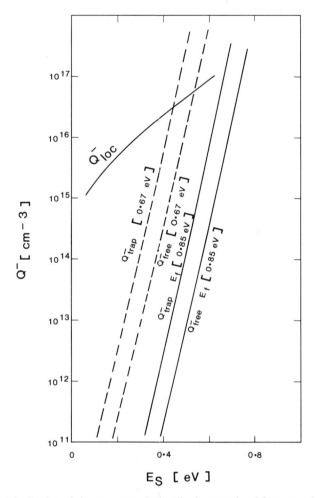

Fig. 3.80. Distribution of charge among the localized, trapped and free states (see text) as a function of band bending, E_s (Madan, 1983).

resistance, R_s, and alters the potential $V(y)$ as

$$V(y) = V_G - V_D \frac{y}{L} - I_{SD} R_s. \tag{3.98}$$

The drain contact resistance, R_D, (for $R_s = 0$) alters $V(y)$ according to

$$V(y) = V_G - V_D \frac{y}{L} - I_{SD} R_D \frac{y}{L}, \tag{3.99}$$

The effect of these two resistances is somewhat different; in the presence of drain resistance, V_D is lowered and thus the saturation voltage is larger. However, for $V_D > V_{th}$, the magnitude of V_D has no significant affect on the drain current. On the other hand, the impact of R_s is strongest when $I_{SD}R_s$ is on the order of V_D. Hence, to obtain optimal TFT performance it is desirable that the contact resistance be minimized; this is normally achieved by depositing an n^+ layer between the a-Si:H layer and the metal source and drain contacts.

3.2.2.4 Depletion Case

By applying a negative V_G, positive charge is induced in the semiconductor, resulting in the formation of a depletion layer. I_{SD} will therefore initially be reduced. However, with increasing negative bias an inversion layer will form, beyond which the device becomes accumulated with holes. Under these conditions it is referred to as a *p*-channel device.

I_{SD} is now given by Eq. (3.96) with μ_n and n_b replaced by μ_p and p_b, corresponding to the hole transport parameters. In the weak depletion case

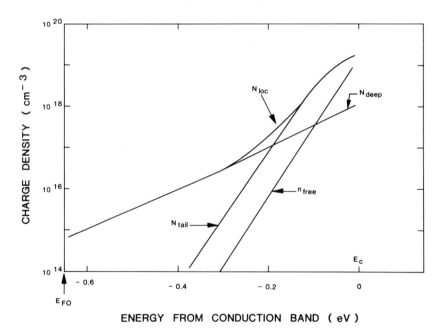

Fig. 3.81. Charge distribution in an a-Si:H FET as a function of Fermi level position. Q_{tail}-charge in localized tail states; Q_{deep}-charge in deep localized states; Q_{mobile}-charge induced into the band (Shur and Hack, 1984).

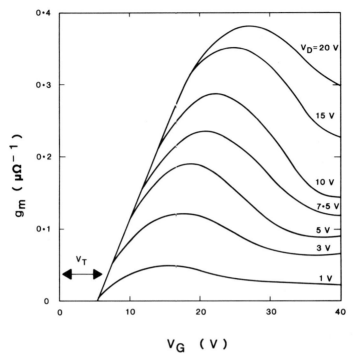

Fig. 3.82. Dependence of transconductance g_m on gate voltage V_G obtained from the transfer characteristics shown in Fig. 3.83. (Snell et al., 1981).

it can be shown that

$$I_{sd} \approx \frac{W}{L}[d - D(E_s)]\sigma_0$$
$$+ WZ_0'e\mu_n n_b \sum \frac{V_d}{l} \sum_l \left(\frac{1}{kT}\right)^l \frac{E_s^l}{l}$$
$$+ WZ_0'e\mu_p p_b \sum \frac{V_d}{l} \left(\frac{1}{kT}\right)^l \frac{E_s^l}{l} \quad (3.100)$$

3.2.2.5 Surface States

The performance of TFTs can be severely affected by the presence of surface states (see Section 2.9). Several different types of surface states have been identified for the most well characterized system-crystalline Si/SiO$_2$ (see, e.g., Grove, 1967). Surface states can be characterized as: (a) fast surface states, which are located at the oxide/Si interface; (b) charges residing within

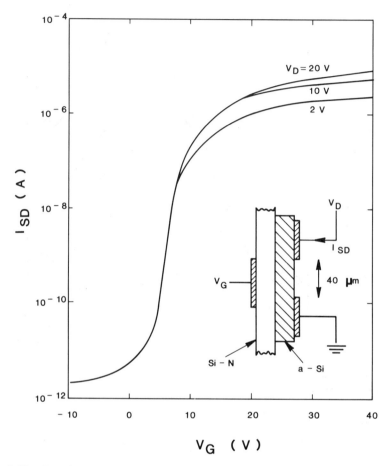

Fig. 3.83. Transfer characteristics of an a-Si:H TFT element. The drain current, I_{SD}, is plotted against the gate voltage, V_G, for three drain potentials, V_D (Snell et al., 1981).

the insulator due to ionic contamination or to traps that are ionized by radiation; (c) fixed surface state charge located at the interface between the oxide and Si.

All of these can affect the I-V characteristics, since the occupancy of the surface states will in general be altered by the gate bias. Also, instabilities in the performance of the device can arise because the flat-band voltage can change with time if the predominant effect of surface states is due to ionic species. Hence, one of the more important issues is the processing of the dielectric as well as the interface itself. We consider these aspects further below.

3.2.3 Performance of a-Si Alloy TFTs

The MOSFET structure was originally used to study the DOS (Spear and LeComber, 1972; Madan, 1973; Madan et al., 1976; Madan et al., 1977). As discussed in Section 2.3, a-Si:H has a remarkably low DOS; hence the conductivity can be altered by several orders of magnitude by changing the gate voltage. Recent research has concentrated on improving the accuracy of the measurements via better processing of the active layer, better understanding of the dielectric/a-Si:H interface, and geometrical considerations. TFT operation has been demonstrated in hydrogenated devices prepared from the rf glow discharge of SiH_4, arc discharge in SiH_4, rf discharge of SiH_4 and SiF_4 gas mixtures. Also, unhydrogenated a-Si:F devices were fabricated via sputtering in a SiF_4–Ar atmosphere (Matsumura et al., 1980). In attempts to improve performance, dielectrics such as SiO_2 and Si_3N_4 have been deposited from various starting gas mixtures.

A typical transfer characteristic of an a-Si TFT is shown in Fig. 3.83 (Snell et al., 1981). Here the dielectric used, Si_3N_4, was deposited from an rf glow discharge in SiH_4 and NH_3 gas mixtures. The a-Si:H film was approximately 0.4 μm thick and the source and drain contacts were Al (note that an n^+ layer between the a-Si and Al was not employed). The geometry of the TFT consisted of a channel length of 40 μm and width of 500 μm. Note that the drain current achieved was in excess of 1 μA for a drain voltage of only 5 V and that I_{off} was low, less than 10^{-11} A. Figure 3.84 shows the corresponding output characteristics $I_{SD}(V_D)|_{V_G}$; it is clear that potentials in the 10–15 V range can be used in the operation of the TFTs. (The advantage of this is the compatibility with conventional CMOS circuitry; the TFT can be driven directly by conventional low power logic circuits.) Figure 3.82 shows the variation of the transconductance, g_m, with gate voltage. For gate voltages in excess of the turn-on voltage (~ 5 V), the conductance of the device increases sharply. Further, g_m exhibits a maximum, in sharp contrast to crystalline TFTs, which generally show that g_m remains constant at large gate voltages. As previously discussed (see Section 3.2.2.1), this is due to the fundamental nature of the a-Si:H alloy. For large positive values of V_G the induced charge is located within the rapidly rising conduction band tail states. Hence, there is very little change of the surface band bending for large V_G, with the result that I_{SD} exhibits an abrupt change with V_G, and thus g_m exhibits a maximum.

For TFTs in the accumulation mode, I_{ON} is given approximately by [see, e.g., Grove (1967)]

$$I_{ON} = \left(\frac{\mu CW}{2L}\right)(V_G - V_{th})^2, \qquad (3.101)$$

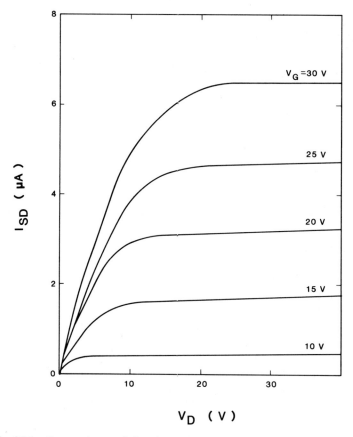

Fig. 3.84. Output characteristics of an a-Si:H TFT element (Snell et al., 1981).

so that

$$g_m = \frac{\mu CW}{L}(V_G - V_{th}), \qquad (3.102)$$

where C is the capacitance per unit area. The above argument is valid for the initial, rising part of the $g_m(V_G)$ curve (Fig. 3.82); the slope of the curve yields a mobility, the field effect mobility, μ_{fe}, found to be $0.4\,\text{cm}^2\,\text{s}^{-1}\,\text{V}^{-1}$, although much larger values of $1.9\,\text{cm}^2\,\text{s}^{-1}\,\text{V}^{-1}$ have been reported by depositing a-Si:H via the arc discharge of SiH_4 (Ishibashi et al., 1982).

Powell et al. (1981) have made a useful comparison of the transfer characteristics of the TFTs that have been reported in the literature. In order to eliminate differences in the device geometry and gate dielectrics, the sheet conductance, G, as a function of the total space charge density induced by

THIN FILM TRANSISTORS FOR DISPLAY APPLICATIONS 293

the gate field is plotted in Fig. 3.85 (Thompson, 1984). This plot is valid for low values of drain voltage, where there is a uniform field in the dielectric. But, unfortunately, most of the data reported from the literature employs large values of V_D. Nevertheless, there is general agreement that large ratios of I_{ON}/I_{OFF} can be obtained. A point worth noting is that the solid lines in the figure were obtained using Si_3N_4 as the dielectric; the performance using SiO_2 is shown by the dotted curves. The use of Si_3N_4 apparently produces a better response.

Some of the differences observed in Fig. 3.85 might be attributed to the different quality of the insulator and/or a-Si:H prepared in different laboratories. As discussed in Section 2.2, the properties of a-Si:H are controlled by many deposition parameters; this holds also for the dielectric layer. In an attempt to unravel whether the differences in oxide and nitride

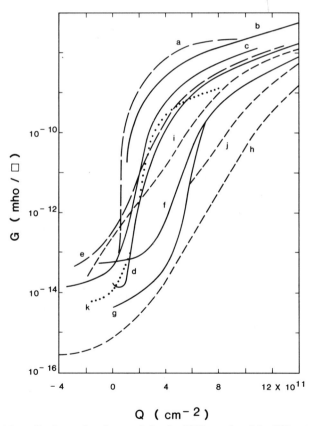

Fig. 3.85. Normalized transfer characteristics for TFTs produced in different laboratories compiled by Thompson (1984).

devices could be attributed to differences in the quality of a-Si:H or the dielectric, Thompson *et al.* (1982) used an inverted structure for the TFT, with a thermally oxidized doped single crystal Si wafer, to obtain a 1000 Å thick film of high quality SiO_2 material with a low surface state density. For the TFT with a nitride dielectric an additional 500 Å of Si_3N_4 was deposited by a CVD process on top of the SiO_2 layer. Source and drain contacts of Al were deposited next. Figure 3.86a and b show a comparison of the two devices. Noting that there was no significant difference, and that $\mu_{fe} \approx 0.2 \text{ cm}^2 \text{ s}^{-1} \text{ V}^{-1}$ irrespective of the choice of dielectric, they concluded that the interface between the a-Si:H layer and a good quality insulator is not a function of the insulator. They suggested that a-Si:H grows and nucleates on SiO_2 and Si_3N_4 in such a way that the electronic properties of the interfaces are identical.

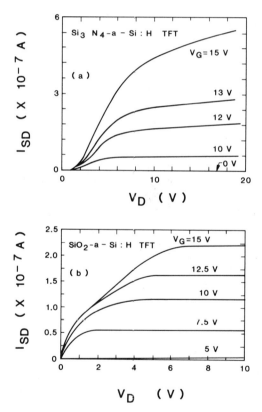

Fig. 3.86. (a) Transistor output characteristics for an a-Si:H–Si_3N_4 TFT. The channel length is 10 μm and the channel width is 50 μm. (b) Output characteristics for an a-Si:H–SiO_2 TFT. The channel length is 24 μm and the channel width is 50 μm (Thompson *et al.*, 1982).

Another factor that could partially explain the discrepancies found in Fig. 3.85 could be the effect of adsorbates on the exposed area of the a-Si:H between the S and D contacts in Fig. 3.79a. As discussed in Section 2.9, a-Si:H is extremely sensitive to the ambient conditions, which can produce either depletion or accumulation layers depending upon the type of adsorbant. In order to avoid these inherent instabilities, Powell *et al.* (1981) proposed the deposition of an additional dielectric SiO_2 (prepared from the plasma decomposition of tetraethoxysilane) between the two top contacts. This would avoid the possibility of surface band bending by passivating the a-Si:H layer. They also reported that in their structure the current showed no significant decay with time, up to a maximum measured delay of 1 minute. They concluded that electron trapping, which can cause source-drain current decay and hysterisis, was at a low level.

Processing can have a profound influence on TFT characteristics (Thompson, 1984). Using a single chamber to deposit the dielectric and subsequent a-Si:H layer can have deleterious consequences for the device performance. If there is insufficient purging between the dielectric deposition and a-Si layer deposition, a residual contaminant could be incorporated into the semiconducting layer. As discussed in Section 3.1.4.5, residual contaminants generally create extra defect states within the mobility gap, thus leading to a lower I_{ON} and a reduce μ_{fe}. Another effect that can occur for Nitrogen contamination is that this impurity acts as a *n*-type dopant, which would increase I_{OFF}. This can present a serious problem, since it has been pointed out (Thompson, 1984) that despite extensive pumping and purging between the dielectric and a-Si:H layer depositions, SIMS analysis revealed that there was still a large concentration of nitrogen in the a-Si:H close to the interface, as shown in Fig. 3.87.

McKenzie *et al.* (1983) reported that a ratio of $SiH_4:NH_3$ of 1:4 in the gas phase yielded pinhole-free Si_3N_4 films with large breakdown strengths ($>1\,MV\,cm^{-1}$). They also concluded that the most stable devices were obtained whenever there was a rapid changeover of the gases, accomplished within about 30 s.

Ishibashi *et al.* (1982) contend that silicon oxynitride is more suitable as the dielectric than SiO_2 or Si_3N_4. In their TFT structures the dielectric was prepared using a reaction gas mixture of SiH_4, N_2 and CO_2 deposited at $250°C$ on a molybednum gate contact. After depositing the gate insulator the substrate was annealed in a vacuum at $500°C$. This was followed by an a-Si:H layer deposited using the arc-discharge decomposition technique (Matsumura *et al.*, 1982). These devices had an extremely high μ_{fe} of $1.9\,cm^2\,s^{-1}\,V^{-1}$, with $I_{ON}/I_{OFF} \sim 10^7$; the drain current varied by six orders of magnitude for $V_G = 3\,V$.

The above devices generally utilized Al in intimate contact with the a-Si:H

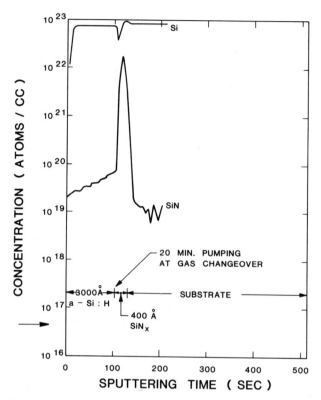

Fig. 3.87. SIMS profile of N(SiN) in a-Si$_3$N$_4$ and a-Si:H near the interface, the deposition system was pumped down for 20 min between the nitride and the following a-Si:H deposition (Thompson, 1984).

layer. However, as discussed in Section 3.1.3.10, even a low work function metal such as Al forms a rectifying contact. This can be clearly seen in Fig. 3.35, where it is shown that in a solar cell configuration Al forms a Schottky Barrier, resulting in the development of an open circuit voltage. A rectifying contact tends to suppress I_{ON}, as pointed out by Thompson et al. (1982) and confirmed by McKenzie et al. (1983), who noted that the use of a n^+ layer between the contact and the a-Si:H layer increased I_{ON} by a factor of 3 compared to devices that did not ultilize a n^+ layer. The output characteristics of the device are shown in Fig. 3.88, with I_{ON} about 20 μA for applied voltages in the range 10-15 V. Fractions of a milliamp could be obtained if the supply voltages were increased to 30-40 V. This increased performance was only partially due to the use of a n^+ layer. More importantly, the enhanced performance was liked to the change in the geometry, since much shorter channel lengths (4 μm) were employed. Although the performance

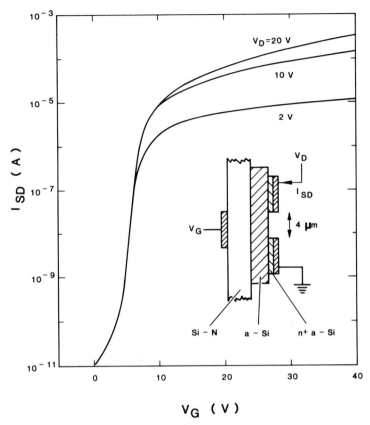

Fig. 3.88. Transfer characteristics of an a-Si:H TFT element. The drain current I_{SD} is plotted against the gate voltage V_G for three drain potentials V_D (McKenzie et al., 1983).

was vastly improved compared to the results of Fig. 3.83, this type of geometry could present a problem when large area displays are to be fabricated, as discussed below.

3.2.4 Dynamic Performance

The above results show that a-Si:H TFTs have sufficiently large ON currents and low OFF currents. However, there is one other criterion required to drive a display, the turn-ON time, given by

$$t_{ON} = \frac{L^2}{\mu V_D}. \tag{3.103}$$

To test that the devices function with the addressing scheme given before, Snell *et al.* (1981) and McKenzie *et al.* (1983) have simulated conditions approaching those in the LCD panel by connecting the drain contact to a 10 pF capacitor. This closely approximates the capacity, C_{LC}, of a 1 mm × 1 mm liquid crystal. In their experiments the voltage applied between source and ground is varied between +10 V and −10 V with a period of 3 ms. When 15 V gate pulses were applied they observed that C_{LC} charged or discharged accordingly, and that the high OFF resistance of the TFT ensured that C_{LC} remained charged for times on the order of hundreds of milliseconds. An important factor is the rise time of the drive potential, V_{LC}, (~3–6 V) that exists across the LCD panel. As can be seen from Fig. 3.89 (where V_{LC} is plotted against the width of the gate pulse) a V_{LC} of 3 V is achieved in 30 μs for a TFT with a 40 μm channel length. As predicted by Eq. (3.103), much faster times are obtained (~5 μs) when the channel lengths are reduced. It was suggested that these low voltage levels would be suitable

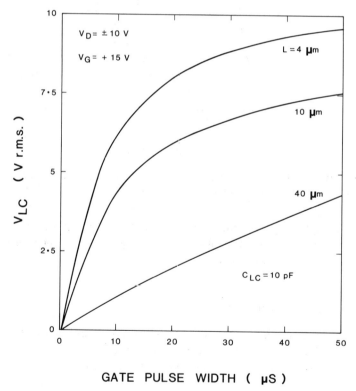

Fig. 3.89. Voltage level V_{LC} across the 10 pF capacitor plotted against the width of the gate pulse necessary to obtain the value of V_{LC} (McKenzie *et al.*, 1983).

for low threshold twisted-nematic liquid crystal displays and, for a frame time of 25 ms, an array consisting of a few thousand lines could in principle be addressed.

One of the major limitations of a-Si:H type devices is the low mobility (see Section 2.5.6), which by necessity limits the dynamic performance to the 10–20 MHz range. While an increase in mobility may prove to be difficult to achieve, an increased frequency response can be obtained by device engineering and geometric considerations. A reduction in the channel length can vastly improve the t_{ON} of the device; by reducing L from 10 μm to 1 μm, the frequency of operation can be increased by a factor of 100, provided that the transconductance increases and the gate capacitance decreases as the gate length is reduced. We now consider two types of novel structures: a dual gate and a vertical TFT type structure.

3.2.5 Dual Gate TFTs

Dual gate TFTs, shown in Fig. 3.90 (Tuan *et al.*, 1982), have been developed in order to increase the ON current. Here both dielectrics were 3000 Å thick and deposited at 350°C using the rf decomposition of SiH_4 and NH_3 gas mixtures. The undoped layer of a-Si:H was 1 μm thick, sandwiched between the gate, and the source and drain contacts were NiCr. By applying a gate bias to both the gates, two conducting channels are formed. Figure 3.91 shows the response of the device for both the individual channels and the two together. In Fig. 3.91c the arithmetic sum of the two individual characteristics is plotted as the solid lines; the drain current in the dual gate mode is always larger than the sum of the two drain currents obtained from

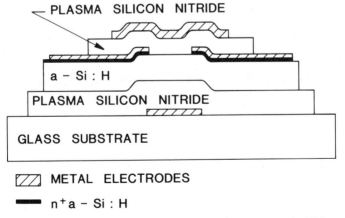

Fig. 3.90. Schematic of a dual gate a-Si:H TFT (Tuan *et al.*, 1982).

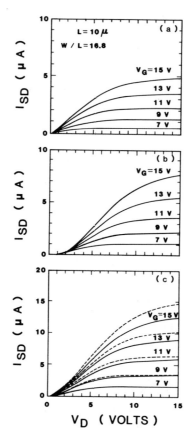

Fig. 3.91. (a) The output characteristics of the top channel with the bottom electrode grounded. (b) The output characteristics of the bottom channel with the top gate electrode grounded. (c) Solid lines—the sum of the two sets of characteristics shown in (a) and (b). Dotted lines—the measured output characteristics of the TFT operated in the dual-gate mode (Tuan et al., 1982).

single-gate operation. Note that the divergence increases for larger gate voltages; this is attributed to the overlap of the accumulation regions and by the potential redistribution along the channels, especially near the contact regions.

3.2.6 Vertical Stacked TFTs

Although devices with channel lengths of 4 μm with t_{ON} of 5 μs have been demonstrated, there may be a problem in utilizing them because of the limitation imposed by the photoetching process, which involves delineating the source and drain contacts. To overcome this engineering aspect Uchida *et al.* (1984) have proposed and demonstrated a vertical type TFT whose schematic structure is shown in Fig. 3.92. It is similar to the VMOS (Holmes *et al.*, 1974) used in single crystal Si. The gate electrode is at the left-hand side and the top and bottom contacts constitute the source and drain. The channel is formed at the sidewall of a-Si layer and the current flows perpendicularly to the substrate surface. The structure should inherently lead to smaller gate capacitances and larger I_{ON}. Figure 3.93 shows the characteristics for a 1 µm length channel prepared in the vertical TFT mode. The figure also compares the results with $L = 5$ µm and 20 µm prepared simultaneously, but in the conventional horizontal type configuration. Although I_{ON} is increased, I_{OFF} also increases significantly for the vertical type TFT, purely because of geometrical factors.

Yaniv *et al.* (1984) have also successfully demonstrated the operation of these devices with a channel length of 1 µm and width of 150 µm; the results are shown in Fig 3.94. They find that currents larger than 10 µA are achieved for $V_D = V_G = 6$ V, with a low $I_{OFF} \simeq 5 \times 10^{-11}$ Å. Further, the field effect mobility was found to be $1\,\text{cm}^2\,\text{s}^{-1}\,\text{V}^{-1}$. However, higher

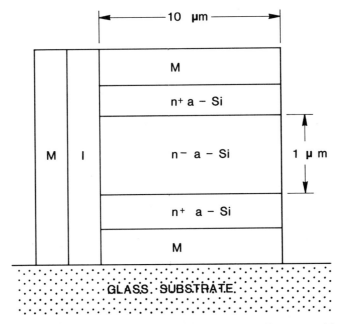

Fig. 3.92. Cross section of the basic TFT with a vertical channel. The upper and lower metal electrodes are source and drain electrodes. The left-hand-side metal electrode is the gate electrode (Uchida *et al.*, 1984).

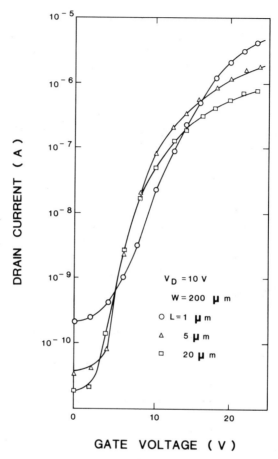

Fig. 3.93. Experimental results for the on-off transition characteristic with various channel lengths L on the same substrate. The TFT with $L = 5\,\mu$m and $20\,\mu$m are of the horizontal type and the TFT with $L = 1\,\mu$m is of the vertical type (Uchida et al., 1984).

switching speeds were not observed, probably because of the electrode overlap capacitances. In an attempt to increase the frequency, a different device structure was constructed, as shown in Fig. 3.95. Here the gate metal again provides for the vertical structure (Yaniv et al., 1985); the source and drain contacts were deposited in a single step by means of an angled or shadowed evaporation. The use of this structure allows for a decrease in the total capacitance to 0.20 pF, which leads to an improved frequency response of 2.5 MHz. Despite an increase in the frequency response, the device did not have good characteristics, but it was anticipated that values of $\mu_{\text{fe}} = 1\,\text{cm}^2\,\text{s}^{-1}\,\text{V}^{-1}$ and a threshold voltage of 2 V could be achieved.

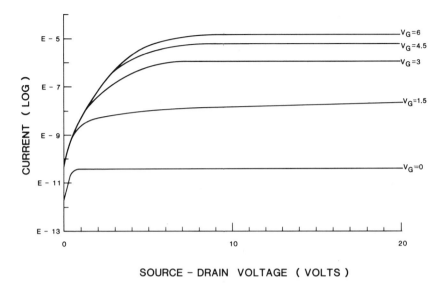

Fig. 3.94. Current voltage characteristics for the 1 μm channel length a-Si:H TFT (Yaniv et al., 1984).

3.2.7 p-Channel TFTs

The above discussion has centered on TFTs employing undoped a-Si:H alloys operating in the enhancement or n-channel FET modes. When a negative rather than a positive gate bias is applied, a positive charge is induced in the a-Si:H and a depletion layer forms. With increased bias an inversion layer develops (hole accumulation), with a concommitant increase in the conductivity. The TFT is now in the p-channel FET mode. Figure 3.96 shows an example of this, with n- and p-channel modes developing in the

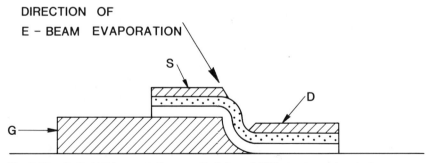

Fig. 3.95. Optimized low capacitance vertical channel TFT structure using shadow evaporation of the source and drain contacts (Yaniv et al., 1985).

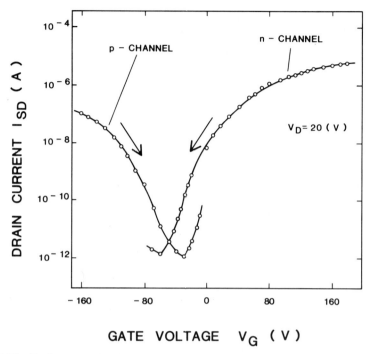

Fig. 3.96. Drain current, I_{SD} plotted against V_G. Arrow indicates direction of voltage change $L = 20\,\mu\text{m}$ (Matsumura et al., 1980).

same TFT structure. Note that $I_{ON}/I_{OFF} \sim 4 \times 10^6$ and 10^5 for the n- and p-channel modes, respectively. The field effect mobilities for the electrons and holes were found to be 0.1 and $2 \times 10^{-3}\,\text{cm}^2\,\text{s}^{-1}\,\text{V}^{-1}$, respectively, a result consistent with the discussion in Section 2.5.6.

The characteristics of the TFT can be altered by doping with P or B, as shown in Fig. 3.97a and b, (Matsumura et al., 1980a). For low levels of P (or PH_3 in the gas phase) the device operates in both the n- and p-channel modes. With increasing concentrations of PH_3 the device characteristics degrade because of the introduction of defect levels (see Section 2.8). This is also the case for the B_2H_6 doped samples shown in Fig. 3.97b.

Amorphous Si:F TFTs have also been fabricated by sputtering in a SiF_4 and Ar atmosphere (see Section 3.1.3.11). They exhibited characteristics similar to a-Si:H type devices, with I_{ON}/I_{OFF} of 10^5, improving to about 10^6 when annealed at a temperature of 500–600°C in a N_2 atmosphere. Even larger ratios of 10^7 have been obtained for devices prepared by the arc-discharge of $SiH_4 + SiF_4$ or $SiF_4 + H_2$ gas mixtures (Matsumura et al., 1980).

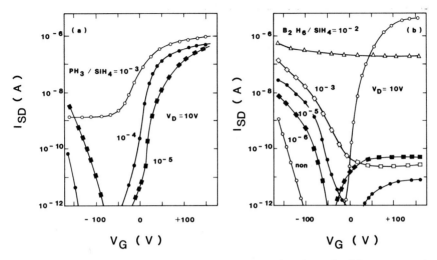

Fig. 3.97. I_{SD} vs. V_G for (a) phosphorous and (b) boron doped samples (Matsumura et al., 1980).

3.2.8 Problems Encountered in Large Area Displays

In a display involving an excess of 250,000 pixels, which would be the case for a 12″ × 12″ display, many problems are anticipated. Some are engineering oriented and others are attributable to the intrinsic process of a-Si:H and dielectric deposition. Some of the problems envisaged are (not prioritized): (a) the ability to deposit device quality a-Si:H layers over large areas; (b) as in (a) but for dielectrics; (c) pin holes in the dielectric or within the a-Si:H layer (d) photolithography capability over large areas; (e) variability in stress; (f) substrate flatness.

For any display application the yield must be 100%, since one inoperable TFT can lead to a line error, which is clearly visible. The yield problem is also encountered in the production of large panel a-Si alloy solar cells (see Section 3.1.7), where pin holes create shunt resistances, leading to an overall loss of performance. Pin holes can arise from a number of sources, such as dust particles from the ambient, dust within the system (which could arise from an incorrect choice of operating conditions that promote gas phase nucleation), improper cleaning of the substrate and highly localized stresses. Great strides are now being made in these areas in order to understand and control these faults by better processing methods and the use of environmental controls such as clean rooms, ultra high vacuum systems, filtering of particulates from the gas, the use of deposition conditions that do not promote gas phase nucleation and reduce inherent stress (see Section 2.2).

Nevertheless, there is one advantage that the TFT technology has over solar cell technology; it is the inclusion of redundant elements and laser scribing of faulty transistors, an approach used in VLSI technology (Yamano *et al.*, 1985).

The question of yield is also related to the number of processing steps, of which there are many. The number of masking steps can be reduced by the self alignment technique proposed by Kodama *et al.* (1982). Figure 3.98 shows the sequence, beginning with the glass substrate. First, a patterned gate electrode metal is formed using a standard photolithographic process. Then, as shown in Fig. 3.98a, the dielectric and a-Si:H alloy layers are deposited in succession, followed by a positive photoresist coating that is

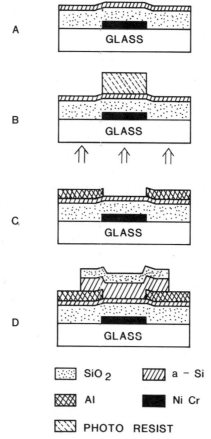

Fig. 3.98. Schematic cross-section showing the fabrication process for a-Si:H TFT (Kodama *et al.*, 1982).

then illuminated by a Hg lamp through the a-Si alloy layer from the glass substrate side, as shown in Fig. 3.98b. (In order to avoid intolerably long exposure times, it is necessary that the a-Si:H layer be thin (< 1000 Å) so that sufficient light reaches the photoresist.) After depositing the Al contact layer on the top surface, the source and drain contacts are delineated by a lift-off procedure as shown in Fig. 3.98c; self-alignment is achieved. Finally, as shown in Fig. 3.98d, the final a-Si:H alloy and dielectric layers are deposited and patterned. One further advantage of the self-alignment technique is that there is no gate-top contact overlap, thus reducing the capacitance. This in principle should lead to optimized operating frequencies.

The question of the large area deposition of a-Si alloys with good optoelectronic properties has been effectively resolved, since areas larger than those being envisaged for displays are routinely fabricated and commercialized in solar cell applications (see Section 3.1.7). By the same token, the problems associated with the deposition of a dielectric over these large areas are not anticipated to be severe.

As has been pointed out by Thompson (1984), much of the technology for producing large area arrays (10 × 8 inch) already exists. Furthermore, as the masks have to be very large and thus are directly written from a pattern generator without optical reduction, there is a limit to the placement accuracy of a feature on a mask, which means that larger tolerances are required for alignment. Hence, real feature sizes on masks will have to be at least 10 μm; this would have an effect on the frequency response of the TFT, as discussed above.

Major problems that are anticipated are the large area alignment of the substrate and the flatness of the mask. Distortions in the substrates may either be intrinsic or could be introduced during the processing at elevated temperatures from the stress inherent in the a-Si:H layer (see Section 2.2). Jones (1985) has calculated that for a typical cell spacing of 8 μm and warp values of 50 microns/inch, there is a high probability that some of the TFTs will experience compressive stress. In the worst case Jones (1985) calculates, assuming a Youngs modulus of 8×10^5 kg/cm^2 for the glass, that an individual transistor will experience an intrinsic stress of up to 2×10^6 dynes/cm^2. In order to test this, TFT structures were made and subjected to compressive stresses, up to 5×10^6 dynes/cm^2. The results are displayed in Fig. 3.99; the compressive stress resulted in changes in all the parameters: OFF conductance; V_{th}; ON-conductance. The increase in the OFF conductance was attributed to an increase in the number of dangling bonds.

Despite the above concerns, LeComber *et al.* (1981) have concluded that the reproducibility and uniformity within the array does not pose a problem, since in a small array (7 × 5) all of the TFT's exhibited threshold voltages that fall within ±0.4 V of the mean value.

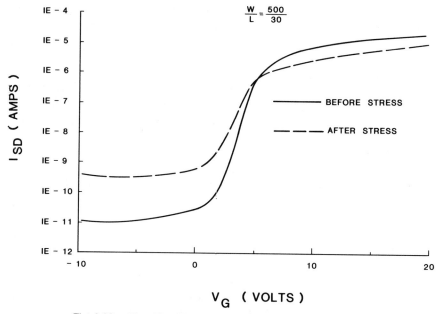

Fig. 3.99. The effect of stress on a-Si:H TFT (Jones, 1985).

In short term stability tests Ast (1983) reports that I_{SD} decreases with time at constant gate voltages because of the (a) trapping of electrons in gap states and (b) trapping of electrons in the gate dielectric. The duration of the first effect is limited to time scales of less than 0.5 s, while the long term decrease ($t > 0.5$ s) is due to the insulator. He finds that this is more pronounced for Si_3N_4 deposited at low temperatures (250°C) and speculates that it might be reduced by depositing the dielectric at a higher temperature. Tuan et al. (1982) find that initially the drain current increases and is then followed by a slow decrease. They attribute this to the injection of charge from the accumulation layer at the interface into the Si_3N_4; the charge could then reside at a Si dangling bond in the nitride layer (Robertson et al., 1984). Thompson (1984) points out the trapped charge within the nitride could be reduced by an application of a negative bias. Since the TFT LCD display operates in the ac mode, then fortunately the above instabilities would not pose a fundamental problem in their operation. This may explain why long term stability tests, performed over 2 years and involving an excess of 10^9 operations, have not shown instability phenomena (LeComber, 1985).

It is of interest to note that an a-Si:H addressed LCD with a monochromatic TV is already on the market. Sanyo (Yamano et al., 1984) have also fabricated a 5" color LCD TV receiver (250 lines by 666 horizontal pixels) incorporating a 3-color mosaic of color filters. Ugai *et al.* (1984) have also reported a 7.2 inch diagonal color display panel.

3.3 Miscellaneous Applications

3.3.1 Threshold and Memory Switching

Switching devices (threshold and memory) have also been demonstrated using a-Si (see Chapter 5 for a detailed discussion of these phenomena). Elemental evaporated a-Si constructed in a sandwich configuration (Ti-Si-Ti) exhibits a switching effect (Dey and Fong, 1979); the mechanism is attributed to a mixture of electronic as well as thermal processes (an electrothermal model) (Day, 1980). Threshold switching in GD a-Si:H, when sandwiched in a configuration of n^+-i-n^+, was reported by den Boer (1982). For an intrinsic layer of 5 μm, an unswitched device required 40 V–100 V to "form", after which switching occurred at a lower voltage; this is similar to what is observed in devices utilizing chalcogenide materials (see Chapter 5). As in the chalcogenides, there is a delay time (~ 1 ms) associated with the switching event that decreases with overvoltage ($V > V_{th}$, where V_{th} is the threshold voltage).

Investigation of the device after the forming process via SEM revealed that a permanent spot of several microns in diameter was visible. This filamentary region is likely to have a composition (perhaps crystalites) that is different from the rest of the film. It is suggested that the device is locally heated to temperatures beyond the T_s of the a-Si:H film. It is this region of high conductivity that may explain the high OFF to low ON resistance in the device.

Memory device action has also been reported using the structure metal/ p-n-i/metal by Owen *et al.* (1982). The i-layer was added to stabilize the device operation although the switching could be observed without the use of i-layer. As in the chalcogenide case, it was found that the first switching operation was unique (a forming process) with the following cycles exhibiting, reproducibly, a considerably lower threshold voltage. For a device area of 0.01 cm^2, the OFF resistance was on the order of 1 MΩ, and the ON resistance about 50 Ω with a threshold voltage of 4 V. Investigation of the dynamic response revealed that there was little, if any, delay time associated with the switching event and the change in resistance could be accomplished quickly; i.e., a 10 V, 100 ms pulse switches the device into a nonvolatile conducting state. Although the exact mechanism for switching is not understood, it is suspected that there might be the formation of a microcrystalline filament that could be ~1 μm in diameter. This was conjectured from the ON-state conductance, which implies a conductivity in the region of $1\text{-}10\,(\Omega\,\text{cm})^{-1}$, typical of a microcrystalline n^+ layer (see Section 2.11.1). However, at present no reliable evidence for the presence of a microcrystalline filament is available. Indeed, there is some evidence that a form of charge trapping is involved in the operation of the device (Silver *et al.*, 1986).

3.3.2 Linear Image Sensors

The high photoconductivity (see Section 2.6) exhibited by a-Si : H type layers has been exploited with the development of linear image sensors (Kaneko, 1984; Ozawa *et al.*, 1985). The advantage of this technology over the ordinary facsimile equipment (which uses a Si monolithic sensor whose width is the same size as the document) is that it removes the necessity for the use of optics and thus results in compact scanners.

A contact linear image sensor can make the equipment compact because of the avoidance of magnification lenses. The sensor unit consists of an illuminator, a compact optical guide, and a long photosensor array (210 mm long for the ISO-A4 reader) connected to the scanner. In the example shown in Fig. 3.100, a pair of LED arrays is used for the illumination and a rod lens is used for the optical guide. Light from the LED arrays illuminates the document and the reflected signal is guided by the rod bus array. The photosensor array scanned by the scanning circuit converts the reflected light signal to a sequential electric signal. Since the optical path from the document to the photosensor array is only 17 mm, this can result in a compact scanner.

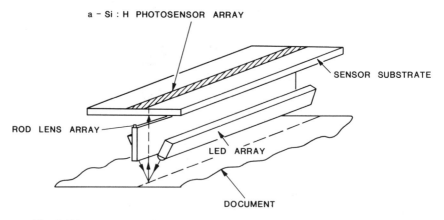

Fig. 3.100. Exploded view of contact linear image sensor (S. Kaneko *et al.*, 1982).

MISCELLANEOUS APPLICATIONS

These devices use a-Si:H as a photodiode sandwiched between ITO (Indium–Tin–Oxide) and a metal electrode. The structure normally employs blocking contacts at both electrodes to prevent carrier injection, as shown in Fig. 3.101 (Kaneko, 1984). In the operation mode the ITO is positively biased and the carriers are generated near the interface between the Si_3N_4 and a-Si:H layers. The electrons enter the Si_3N_4 and the holes drift towards the top (Al) contact. Here the Si_3N_4 acts as a hole blocking layer and the B-doped (p-type) a-Si:H layer prevents the injection of electrons.

Figure 3.102 shows the I-V characteristics under dark and light conditions and also compares the results obtained without the use of blocking layers. As is to be expected, the dark characteristics are significantly suppressed by the use of blocking layers.

The spectral response of the photosensors is nearly uniform over the visible region and the photoresponse time is less than 0.1 ms, which is satisfactory for a high speed facsimile technology. Arrays for A4 and B4 size scanners are now being mass produced.

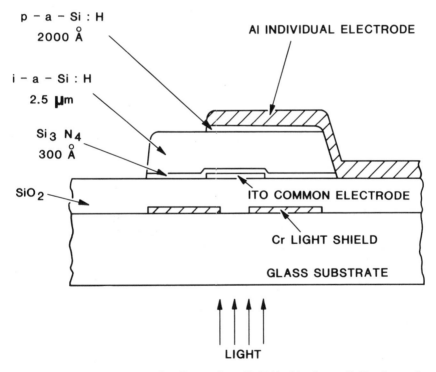

Fig. 3.101. Photodiode array using Si_3N_4 and p-a-Si:H blocking layers (S. Kaneko et al., 1982).

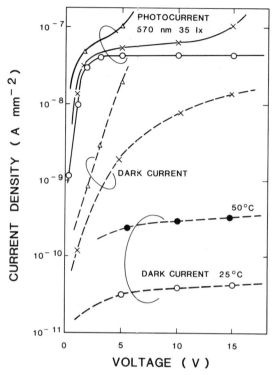

Fig. 3.102. *I–V* characteristics in light and dark conditions for a photosensor with and without blocking layers: ○, with Si_3N_4 and *p*-a-Si:H blocking layers (25°C); ●, with Si_3N_4 blocking layers (50°C); △, without Si_3N_4 blocking layer; ×, without *p*-a-Si:H blocking layers. Positive bias is applied to an ITO electrode (S. Kaneko *et al.*, 1982).

3.3.3 Optical Recording

An optical recording medium should satisfy the requirements of high sensitivity and long term stability; we shall discuss these criteria in Chapter 7. The optical recording of submicrometer imprints provides a technology to store information at ultra-high densities in digital or analog formats. The inherent advantages of a laser-driven direct read after write system are as a very high data rate, rapid access time and contactless readout. Since an optical recording system can store information in excess of 10^{10} bits/disk, then erasability is of secondary importance.

The investigation of materials for optical recording includes a-Si:H as a useful archival medium. Although it is known that the writing energy requirement is high, the mechanical and chemical stability has been viewed as an asset. The mechanism for optical storage in a-Si alloys depends on:

(a) optically induced amorphous to polycrystalline transitions (Brodsky, 1983); (b) the release of hydrogen from a-Si:H to produce either bubbling (local deformation) or ablative hole formation (Bosh, 1982).

The laser induced crystalline-to-amorphous phase transition in Si films was not accomplished because of the very high quench rates required for Si. Crystalline Si can be made amorphous only by very short light pulses (Liv et al., 1979). In a-Si:H it has been demonstrated that hydrogen inhibits pulsed-laser-induced crystallization. (Peery and Stein, 1978). Because of these aspects, a-Si:H is not considered a likely candidate for erasable optical recording.

A-Si alloys appear to satisfy some of the requirements of high density optical storage media: negligible degradation on repeated readout; low cost of fabrication and long term stability. One way of recording utilizes the effusion of hydrogen. Upon laser heating, hydrogen evolves from within the hydrogenated a-Si layer (see Section 2.2); the gas pressure deforms the films into bubbles that are smooth and symmetric and less than 1 μm in diameter. The reflectivity of the irradiated area changes from 40% in the as deposited film to a few percent from the bubble area; this provides a good signal to noise ratio.

Mcleod et al. (1984) report that hydrogen evolution can be accomplished using an Ar laser (514 nm) in a cw scanning mode over a large area (100 μm $\times$ 2 cm) for times typically of 5 min with a laser power in the range 100 mW-1 W.

A novel method of recording and storing information relies on the formation of defects that can act as recombination centers induced by electron-beam-induced currents. By operating a SEM in the charge collection mode the areas that experience bombardment become dark in comparison with the unbombarded regions; hence the realization of contrast. This writing could be performed in SB or in p^+-i-n^+-type devices. Since the SEM results did not indicate any damage, then this effect was attributed to a bulk phenomenon, i.e., a creation of defects of the dangling bond type that result from the breakage of Si-Si bonds and/or to a loss of collection, resulting in dark area images, as discussed in Section 3.1.4.2 (Yacobi et al., 1984).

3.3.4 Charge Coupled Devices (CCD)

A CCD is a shift register with a charge generator at the input stage and a charge detector at the output state. A large number of functions such as shifting of information, scanning and logic can be performed by CCDs and have achieved great importance in image sensing.

In a typical CCD operation in a single crystal semiconductor, the signal is stored as minority carriers and kept separate from the majority carriers by a space charge region. The transient state must be achieved for a sufficiently long time in order to attain CCD operation. In a-Si:H type devices CCD operation is attained by operating in the accumulation mode. Figure 3.103 shows a schematic view of the structure. The a-Si:H film is sandwiched between two insulators; the contacts are staggered as shown (Kishida *et al.*, 1982). Figure 3.104 shows the motion of charge (electrons) in CCD operation. During an interval when only the transfer electrode, A, is connected to the clock pulse, ϕ, excess electrons are created at the a-Si:H insulator interface underneath the electrode, A. When the next transfer electrode B is connected to the clock pulse, ϕ_2 following ϕ_1, these electrons move to the a-Si:H region sandwiched between A and B. When ϕ_1 falls to a low level, electrons move to and are stopped at the opposite a-Si:H insulator interface just above electrode B; thus, one transfer is accomplished.

Since a-Si:H layers possess nonuniform DOS spectra, this should result in a limitation of device performance. Detailed analysis indicated that more than 95% of the initial stored electrons could be transferred within 250 μs, a sufficient criterion for CCD operation.

Figure 3.105 shows a cross sectional view of an a-Si:H CCD structure (Matsumura, 1984). The insulators used were Si_3N_4 and the active layer was

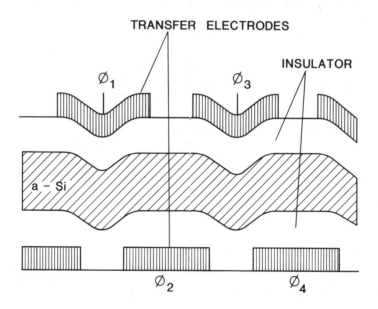

Fig. 3.103. Cross sectional view of a-Si CCDs with four-phase clock pulse (Kishida *et al.*, 1982).

MISCELLANEOUS APPLICATIONS

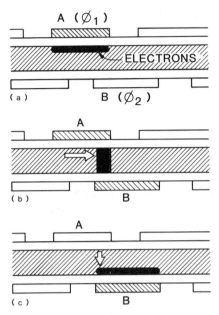

Fig. 3.104. Schematic motion of electrons in the a-Si:H CCDs. (a) Only transfer electrode A is at high level. (b) Both transfer electrodes A and B are at a high level. (c) Transfer electrode A falls to a low level and transfer electrode B remains at a high level (Kishida et al., 1982).

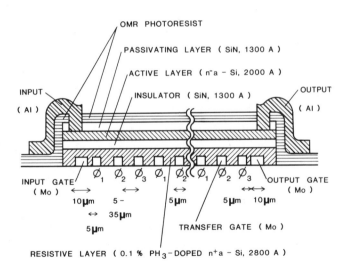

Fig. 3.105. Cross-sectional view of a-Si:H CCDs for high-speed operation (Matsumura, 1984).

an undoped a-Si:H layer. The length and width of the transfer electrodes were 5 μm and 1 mm, respectively. The device had 24 transfer electrodes and were driven by a 3 phase clock pulse with constant high voltage applied to the drain and output gate electrodes. Operating the device from 1 to 10 kHz achieved a transfer efficiency in excess of 99%.

3.3.5 Other Applications

Amorphous Si:H layers have also been used as passivation, when directly deposited onto crystalline Silicon p-n junctions, in order to reduce the leakage current caused by dangling bond surface states. The deposition of an a-Si:H layer results in a reduction of these centers, as well as acting as a protective layer that is nearly insulating (Pankove and Trong, 1979).

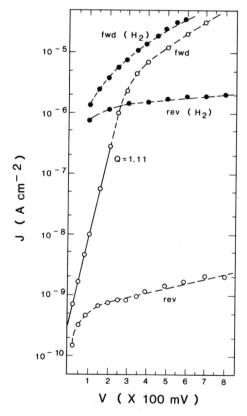

Fig. 3.106. $I(V)$ curves of the MIS structure exposed to air (○) and to 280 ppm of H_2 in an $N_2 + H_2$ mixture (●) at 295 K (D'Amico and Fortunato, 1984).

MISCELLANEOUS APPLICATIONS

Device structures such as MIS or MOS TFTs have been proposed as chemical sensors relying on surface or interface effects that can be altered by the ambient gas conditions (see Section 2.9). In particular, Pd MIS devices have been used as H_2 sensors. The H_2 sensitivity relies on the dissociation of H_2 molecules on the catylytic metal surface and the subsequent adsorption and diffusion of H atoms. This process is reversible in the presence of O_2. Typical $I(V)$ data for a Pd MIS junction exposed to air at room temperature and at 280 ppm of H_2 is shown in Fig, 3.106; note the large change in the characteristics (D'Amico and Fortunato, 1984), but degradation in the device performance is to be expected when the insulator thickness exceeds an optimum value and decouples the metal from the semiconductor (see Section 3.1.3.7).

Amorphous Silicon photovoltaic detectors have been proposed for guided wave optics (Yumoto *et al.*, 1982), and oxidized amorphous Silicon superconductors as tunnel junctions barriers (Rudman *et al.*, 1980). Piezoresistive effects in a-Si:H:F were demonstrated and have applications in strain gauges and pressure sensors (Kodata, *et al.*, 1982).

Other applications exist that couple amorphous films to crystalline materials. For example, amorphous Si alloy films have been used as emitters in bipolar transistor structures where the base and collector are crystalline Si (see, e.g., Symons *et al.*, 1987). [Devices of this type were first created by Paterson *et al.* (1976) using amorphous chalcogenide emitters.]

4

Characterization and Properties of Amorphous Chalcogenide Alloys

4.1 Introduction	318
4.2 Optical Properties of Amorphous Chalcogenide Films	322
4.2.1 Introduction	322
4.2.2 Vibrational Spectra	322
4.2.3 The Absorption Edge	324
4.2.4 Other Optical Properties: Absorption Band; Dielectric Constant; Luminescence; Photo-Induced Phenomena	327
4.3 Electrical Transport Properties	331
4.3.1 DC Conductivity and Photoconductivity	331
4.3.2 AC Conductivity and Transient Photoconductivity	337
4.3.3 Thermoelectric Power	344
4.3.4 The Hall Effect	346
4.3.5 Drift Mobility	348
4.4 Summary and Conclusions	351

4.1 Introduction

As pointed out in Chapter 1, except for states close to the band edges, states that lie well within the mobility gap are due to either impurities or defects. The origin of these states in amorphous chalcogenide alloys such as As_2Se_3 and As_2Te_3 is a difficult problem; we have discussed the models proposed by Street and Mott (1975) and Kastner et al. (1976) in Chapter 1. The experimental situation, however, is that the observation of specific defect states has not been well categorized. Different researchers emphasize diffrent interpretations, and the reports of crucial experiments that clearly distinguish between specific models are rarely found in the literature (Tauc, 1976). As an example, we might cite the discrete valence alternation pair defects predicted by Kastner et al. (1976) and their apparent absence in the transient

INTRODUCTION

photoconductivity experiments reported by Orenstein and Kastner (1981), although a spread in pair separation could reconcile this difficulty (Adler, 1981). Because of these deficiencies in our present understanding, in this chapter we will tend to avoid *detailed* comparisons between theory and experiment as much as possible.

Amorphous chalcogenide films, which are sulphur, selenium or tellurium based materials, have applications in electrophotography (see, e.g., Kawamura and Yamamoto, 1982), optical recording (see, e.g., Hecht, 1982), lithography (see, e.g., Jamai, 1981), and bias-induced switching and memory devices (see, e.g., Fritzsche and Ovshinsky, 1970). All of these device structures are in thin film geometries and are generally prepared by sputtering processes essentially identical to those outlined in Section 2.3. The properties of these films are generally different from those of the tetrahedrally bonded amorphous semiconductors such as silicon.

The usefulness of the chalcogenide glasses is often due to their transparency in the infrared. Since they are composed of atoms heavier than the oxides, their vibrational frequencies are shifted deeper into the infrared. Further, the band edge shifts towards lower energies, providing a deeper position for the first transparency edge. The free carrier absorption is also absent or, when present, very weak. Typical properties of commercially available infrared glasses are given in Table 4.1.

The chalcogenide glasses are mechanically weaker than oxide glasses, having lower softening points and hardnesses. Further, the chalcogenides

Table 4.1. Properties of Commercially Produced Chalcogenide Glasses

	Arsenic Sulphide*	Glass No. 20**	Glass No. 1173**
Composition	As_2S_3	$Ge_{33}As_{12}Se_{55}$	$Ge_{28}Sb_{12}Se_{60}$
Transmission range (μ)	0.7 to 11	1 to 15	1 to 15
Refractive index at 10μ	2.3726††	2.4919	2.6002
Δ index/°C	-8.6×10^{-6}††	—	79.0×10^{-6}
Density (g/cm³)	3.20	4.40	4.67
Expansion coefficient (per °C)	26×10^{-6}	13.3×10^{-6}	15.0×10^{-6}
Thermal conductivity cal/s cm °C	4×10^4	6.1×10^4	7.2×10^4
Strain point $10^{14.6}$	160°C	354°C	240°C
Anneal point $10^{13.4}$ Poise	177°C	364°C	259°C
Softening point $10^{7.6}$ Poise	267°C	474°C	370°C

* Produced by Servo Corporation of America under the name Servofrax, and by American Optical Company.
** Produced by Texas Instruments.
† After Rodney *et al.* (1958).
†† At 5μ (Hilton and Jones, 1967).

have much higher indices of refraction. Hence, when they are used as optical elements they are often coated with antireflective coatings. Other important applications of these glasses are as acousto-optic modulators, ultrasonic delay lines, multilayer infrared filters and as antireflective coatings themselves. One of the major features of amorphous chalcogenide films prepared by sputtering or evaporation onto substrates kept at room temperature is their lack of a measurable density of unpaired spins (Agarwal, 1973; Hauser *et al.*, 1977). Anderson (1975) explained this by invoking a negative correlation energy for these materials. Street and Mott (1975) then showed that a single defect center with a negative correlation energy would explain not only the absence of unpaired spins, but a host of other physical properties of the chalcogenides, such as the absence of variable range hopping, the large Stokes shift in the photoluminescence, and the induced EPR signal at low temperatures (Bishop *et al.*, 1975), and Adler and Yoffa (1976) showed that a negative correlation energy strongly pins the Fermi level. These models, however, failed to distinguish between the chalcogenides and the tetrahedrally bonded amorphous solids because they did not focus on the chemical nature of the component atoms. In trying to understand the nature of the defect states in the chalcogenides that produce such effects, we must realize that chalcogen atoms have nonbonding lone pair electrons that form the valence band (Kastner, 1972; Adler, 1977; Ngai *et al.*, 1978). Ovshinsky and Sapru (1974) suggested that interactions between the lone-pair electrons on different atoms and interactions with their local environment can create a spectrum of localized gap states, some of which could be charged defects (Orenstein *et al.*, 1982). Kastner *et al.* (1976) elucidated the nature of the charged defects and supported the view of Street and Mott (1975) that it is their presence that dominate the electronic properties of these materials. However, Kastner *et al.* (1976) also demonstrated that these defects have a low creation energy, so that thermodynamics requires large concentrations under all preparation conditions.

Kastner *et al.* (1976) showed that the lowest energy neutral defect in an amorphous chalcogenide is probably not a dangling bond (a singly coordinated chalcogen atom, which we call C_1^0), but rather a threefold-coordinated chalcogen atom, C_3^0 [Fig. 4.1 shows the structure and energy of a variety of chalcogen bonding configurations (Adler, 1977)]. The former costs a bond energy, E_b, while the latter only costs the excess antibonding repulsive energy, Δ. However, the lowest-energy defect is *not* neutral. Two C_3^0 centers are unstable against the reaction

$$2C_3^0 \rightarrow C_3^+ + C_1^-. \qquad (4.1)$$

The reaction of Eq. (4.1) is possible because a threefold-coordinated chalcogen together with a neighboring twofold coordinated chalcogen can

INTRODUCTION

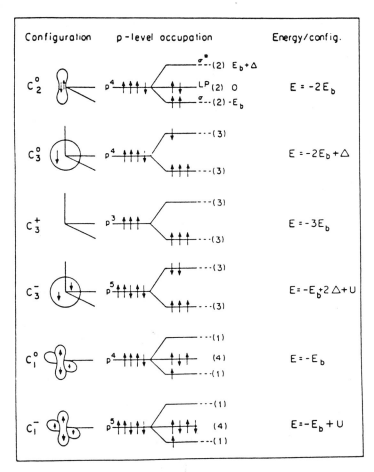

Fig. 4.1. Structure and energy of simple bonding configurations for chalcogen atoms in covalent amorphous semiconductors. In the configuration column straight lines represent bond (σ) orbitals, lobes represent lone-pair (nonbonding) orbitals and circles represent antibonding (σ^*) orbitals. Each bonding electron is paired with another from a neighboring atom. The energy of a lone-pair electron is taken as the zero of energy. U is the correlation energy and Δ is the antibonding repulsive energy (Adler and Yoffa, 1977).

spontaneously transform to neighboring singly and twofold coordinated centers by breaking one of the bonds on C_3, leaving a C_1 neighbor. This is represented by

$$C_3 + C_2 \rightarrow C_2 + C_1. \qquad (4.2)$$

The presence of an additional electron on an atomic site, as is the case with C_1^-, requires an additional energy because of the Coulomb repulsion between two electrons on the same site, which is just the correlation energy, U. The approximate energies of all C_1, C_2 and C_3 states are shown in Fig. 4.1. It is clear that the reaction of Eq. (4.1) is exothermic provided that

$$2\Delta > U, \qquad (4.3)$$

which appears to be the case (Kastner *et al.*, 1976; However, see Vanderbilt and Joannopoulos, 1983). Furthermore, since positively and negatively charged centers attract one another via the Coulomb interaction, the defect centers with two oppositely charged chalcogens as nearest neighbors have the lowest energy of all. These nearest neighboring pairs are called IVAPs (intimate valence alternation pairs). Nonintimate pairs (NVAPs) are also possible, but they will have energies greater than IVAPs (Adler and Yoffa, 1977). Bulk dipole densities in the range proposed by Street and Mott (1975), and Kastner *et al.* (1976) have been detected by an ac technique in amorphous Si and several of its alloys. But, although the VAP model has been successful in explaining many of the properties of the chalcogenides, some recent experiments have proved difficult to interpret in terms of the conventional VAP model. In particular, observations indicative of dispersive transport in these materials are consistent with an exponential, rather than discretely sharp, distribution of states in the gap (Pfister and Scher, 1978; Orenstein and Kastner, 1981). Hence, in an attempt to develop a consistent picture of what we think the basic physical properties of these films are, we will first review the extensive body of experimental data that has been garnered over the past two decades.

4.2 Optical Properties of Amorphous Chalcogenide Films

4.2.1 Introduction

Many review articles on the fundamental physical properties of amorphous materials (Zallen, 1983) have been written during the past decade. Among, them, perhaps the most general discussion of the optical properties of amorphous chalcogenides has been presented by Tauc (1974). In this chapter we will make liberal use of Tauc's review, updating with recent developments when they appear to be significant.

4.2.2 Vibrational Spectra

One view concerning the interpretation of the vibrational spectra of amorphous chalcogenide films is based on a localized model (Simon, 1960; Borelli and Su, 1968). In this view particular structural elements of the glass are isolated and are assumed to vibrate independently in a quasi-free molecular fashion. Thus, if a band is ascribable to two neighboring ions in a crystal, the band would be expected to be observed when the material is in amorphous form. Felty *et al.* (1967) interpreted their results, shown in Fig. 4.2, in this manner. The two peaks seen at wavenumbers of 300 and 215 cm^{-1} did not vary in position but only in strength when the composition of the mixed binary $As_2S_xSe_{3-x}$ was varied. These wavenumbers correspond to the restrahlen bands of pure As_2S_3 and As_2Se_3, respectively, and are ascribed to the As-S and As-Se bands. Zallen *et al.* (1971) provided a more detailed interpretation of the data for crystalline and amorphous As_2S_3 and As_2Se_3, based on Raman and infrared studies, which involves vibrational modes wherein each As atom vibrates opposite to its three neighboring S or Si atoms, and each S or Si atom vibrates opposite to its two neighboring As atoms. Further refinements were made by Lucovsky and Martin (1972); this model of the local molecular order in a-As_2Se_3 is shown in Fig. 4.3.

There have been numerous studies of the Raman and infrared spectra of a variety of amorphous chalcogenide films (Tauc, 1974). In general, a strong similarity is found between the vibrational spectra of several molecular-like complexes in both the amorphous and crystalline states. Indeed, infrared and Raman spectroscopy often provide more detailed structural information that X-ray or electron diffraction experiments.

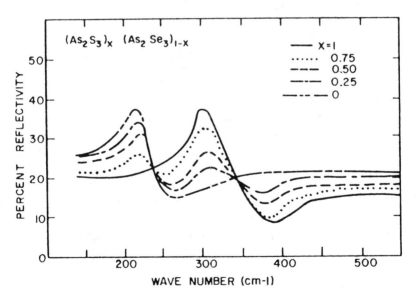

Fig. 4.2. The reststrahlen bands of the quasi-binary amorphous system $(As_2S_3)_x (As_2Se_3)_{1-x}$ (Tauc, 1974).

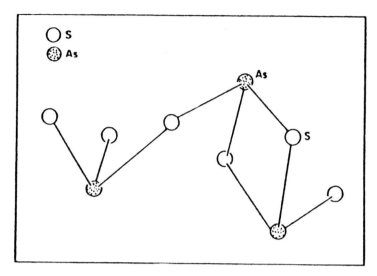

Fig. 4.3. Schematic representation of the local molecular order in a-As$_2$Se$_3$ (Tauc, 1974).

4.2.3 The Absorption Edge

As discussed in Chapters 1 and 2, perhaps the most important optical property of an amorphous semiconductor is its absorption edge; the typical shape for a chalcogenide is shown in Fig. 4.4, where crystalline (c) and amorphous (a) As$_2$S$_3$ are compared (Tauc, 1974). There is usually a high absorption region (A), an exponential part extending over about four orders of magnitude (B) and a weak absorption tail (C).

Figure 4.5 shows the absorption edges for crystalline and amorphous As$_2$Se$_3$. In general, for sufficiently high absorption levels ($\alpha \gtrsim 10^4$ cm^{-1}; region A), it is found that

$$\hbar\omega\alpha(\omega) = A(\hbar\omega - E_g^{\text{opt}})^r, \qquad (4.4)$$

where A is a constant and r is a constant of order unity. Figure 4.5 shows that for a-As$_2$Se$_3$, $r = 2$. A common view is that region A of the absorption curve is most probably associated with transitions from localized valence band states below E_v to conduction band states above E_c (delocalized), or vice-versa. Assuming that the densities of states, g, near the band edges vary as

$$g_v \sim (E_v - E)^{r_1}, \qquad (4.5a)$$

$$g_c \sim (E - E_c)^{r_1}, \qquad (4.5b)$$

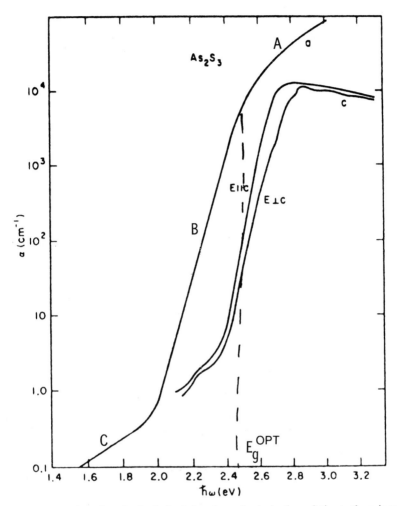

Fig. 4.4. Absorption edge of c-As$_2$S$_3$ for 2 directions of polarization relative to the axis compared with the absorption edge of a-As$_2$S$_3$ (Tauc, 1974).

then Eq. (4.4) suggests that

$$\omega\alpha \sim (\hbar\omega - E_g^{\text{opt}})^{r_1+r_2+1}, \tag{4.6}$$

where $E_g^{\text{opt}} = E_c - E_v$.

For crystals $r_1 = r_2 = 1/2$, whereas here $r = 2$. Experimental verification of this behavior has been made in many chalcogenide crystals. Chalcogenide glasses, however, have values of r different from 2, such as 1 or 3 (Tauc, 1974).

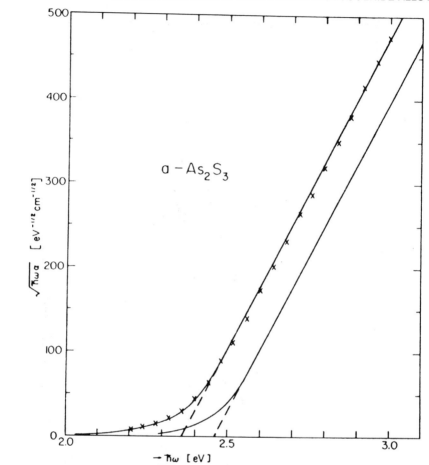

Fig. 4.5. The dependence of $\sqrt{\hbar\omega\alpha}$ on photon energy in a-As$_2$S$_3$ from which the optical gap E_g^{opt} is determined (Tauc, 1974).

Part B of the absorption edge shown in Fig 4.4, the exponential region, has the following general properties: (1) for $1 \text{ cm}^{-1} \leq \alpha \leq 10^4 \text{ cm}^{-1}$,

$$\alpha(\omega) \sim \exp\left(\frac{\hbar\omega}{E_e}\right); \tag{4.7}$$

(2) the energy E_e characterizing the slope in region B is almost independent of temperature for temperatures below room temperature and $0.05 \text{ eV} \leq E_e \leq 0.08 \text{ eV}$; (3) as the temperature increases above room temperature the slope decreases; (4) $E_e \sim T$ for sufficiently high temperature; (5) in most amorphous semiconductors parts A and B of the absorption edge move together.

Absorption in the exponential region is commonly thought to be due to the presence of band tail states that extend into the gap (Tauc, 1974). Parts B and C of the absorption edge are ascribed to absorption of this type. Wood and Tauc (1972) analyzed data for a-As_2S_3, $Ge_{28}Sb_{12}Se_{60}$ and $Ge_{33}Ae_{12}Se_{55}$ in this manner, obtaining for the total concentration of band tail states about 10^{20} cm^{-3} for part B and about 10^{17} cm^{-3} for part C. An excellent discussion of the problems associated with interpreting part B of the absorption edge can be found in Tauc (1974). The primary mechanism is involved with the presence of *local* electric fields, and thus the name "Urbach Tail" is ascribed to part B. (The effect of external applied fields on the optical properties of glasses is in general much smaller than it is in crystals, primarily because of the low mobilities of carriers in glasses.) Part C of the curve is found to be related to the preparation, purity and thermal history of the material. Experiments indicate that this weak absorption tail is often due to optical transitions rather than to scattering.

4.2.4 Other Optical Properties: Absorption Band; Dielectric Constant; Luminescence; Photo-Induced Phenomena

So far we have just considered the absorption edge. We also should point out again that because of the loss of long range order the entire fundamental absorption band is drastically changed in making the transition from the crystalline to amorphous state. Figures 4.6, 4.7 and 4.8 show this clearly for a-Se, Te, As_2S_3 and As_2Se_3.

Measurements of the indices of refraction (dielectric constant) of these types of glasses are also important. Large changes of the index of refraction (reflectivity) with additives in the system $Te_{81}Ge_{15}X_4$, where X were group V and VI additives, were observed by Feinleib *et al.* (1971). Amorphous chalcogenide films were exposed to short, intense laser pulses, which caused local crystallization. The crystalline spots could be erased by the application of a subsequent pulse of appropriate magnitude and duration. In Chapter 7 we will discuss in more detail such an "optical memory" system.

Large changes in the index of refraction indicate changes in bonding and hence the polarizability. Crystalline GeTe has a coordination of 6; a-GeTe has an average coordination of 3. A model of a-GeTe involving 2 and 4 fold coordinated Te and Ge, respectively seems consistent with infrared and photoemission data.

There are a variety of other interesting and important optical properties that are exhibited by specific chalcogenide glasses. We shall discuss several of them now. First, it is important to note that in many chalcogenide glasses such as As_2Se_3 and As_2S_3, the luminescence spectra are similar to those obtained in the corresponding crystals (Averianov *et al.*, 1980). From these data

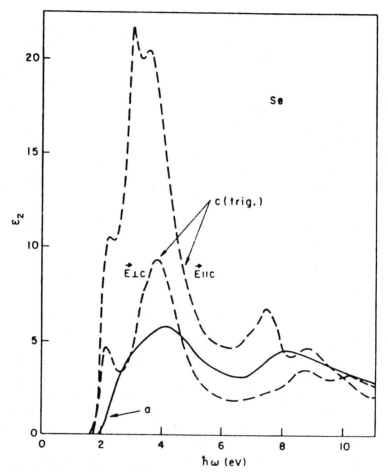

Fig. 4.6. Fundamental absorption band of a-Se and c-Se for 2 directions of light polarization (Tauc, 1974).

it has been inferred that similar recombination mechanisms and centers exist in both the crystalline and amorphous phases. Recently, however, Bishop *et al.* (1981) have shown that impurities can play a vital role in determining the optically induced properties of *crystalline* arsenic chalcogenides. Indeed, although discussions of luminescence are often made with regard to amorphous materials only, it is important to compare and contrast the crystalline and amorphous state of the same material. For example, Murayama and Bosch (1981) also have found that the two PL bands characterized by their excitation in c-As_2S_3 are quite similar to those found in a-As_2S_3. Further, one band in c-As_2S_3 is interpretable as due to the recombination of localized

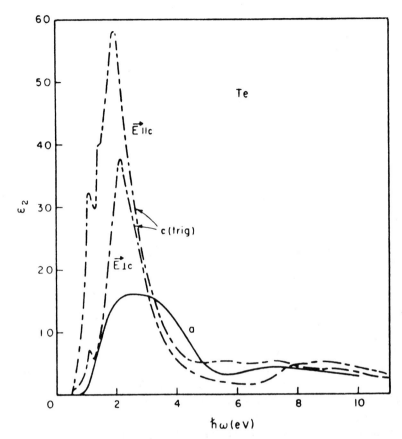

Fig. 4.7. Fundamental absorption band of a-Te and c-Te for 2 directions of light polarization (Tauc, 1974).

atoms. A photoluminescence (PL) band in a-As_2S_3 can also be interpreted in the same manner, suggesting that the microstructure of a-As_2S_3 has many of the features of c-As_2S_3. One major optical phenomenon is the observation of optically induced metastable paramagnetic states in a variety of chalcogenide glasses by Bishop et al. (1977). Here, the mid-gap optical absorption and ESR signals were attributed to optically induced localized paramagnetic states. In these low temperature experiments samples (such as As_2Se_3, $GeSe_2$ and As_2S_3 are irradiated with light of energy in the range of the Urbach tail ($\alpha \simeq 100$ cm^{-1}), which excites PL that decays with continuing excitation. As the PL fatigues, a growing ESR appears, which was not present prior to illumination. Simultaneously, an induced optical absorption occurs, demonstrating that the paramagnetic states are located in the gap of amorphous

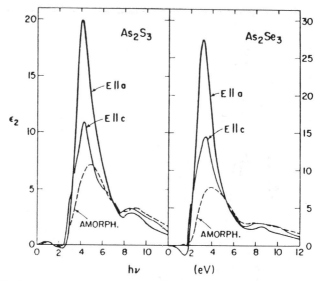

Fig. 4.8. Fundamental absorption bands of crystalline and amorphous As_2S_3 and As_2Se (Tauc, 1974).

materials. Subsequent exposure to infrared light near the mid-gap induced absorption band bleaches out both the optically induced ESR signal and the induced absorption, restoring the fatigued PL to its initial value.

The above results suggest that the radiative recombination centers that are activated during PL decay are associated with the production of metastable paramagnetic centers. The optically induced ESR centers seem unique to the amorphous phase, saturating at or below 10^{17} cm^{-3} in all the materials studied, independent of the intensity of the incident light. Bishop et al. (1977) have identified the paramagnetic centers as: a hole that consists of an electron missing from a nonbonding lone pair chalcogen orbital; a center in glasses containing As that is an electron in a p-orbital localized on an As atom.

Recently, optically induced ESR signals have also been observed in glassy sulfur (Elliott, 1981). By illuminating at 77 K with light of energy 3.1 eV, an ESR signal is observed characteristic of a completely anisotropic g tensor; the first observation of a chalcogen center in a pure chalcogenide glass. The center is most likely a simple dangling bond, C_1^0.

One of the more common optical phenomena observed in chalcogenides such as a-As_xSe is reversible photodarkening (deNeufville et al., 1974; Grigorovici and Vancu, 1981). Under optical excitation the position of the absorption edge changes with time, eventually saturating at a value that, for a given wavelength of incident light, depends upon the level of irradiation and on temperature. If the level of irradiation is reduced, photobleaching occurs. If the temperature is raised above a critical value, complete thermal bleaching occurs.

The photodarkening has little effect on the observed X-ray pattern, but all models proposed to explain the effect suppose that the photodarkening is due to some form of metastable structural modification. One popular model involves a photopolymerization mechanism (Giogorovici and Vancu, 1981) whereby the light temporarily ionizes a long pair orbital and creates a VAP of reduced stability. The simultaneous presence of a second photon in the vicinity of the center can form a dimer.

Hamanaka *et al.* (1981) have studied the photostructural changes that occur in As_2S_3 upon photodarkening. They find that the saturated values of E_g^{opt} become constant after illumination at different temperatures, independent of the previous thermal or optical history of the sample. Also, the change in the X-ray pattern observed when the sample is photoexcited is different than that observed when the temperature is changed.

Other unique optical properties have been observed in chalcogenides. For example, laser induced oscillatory phenomena have been studied by Hajto and Fustoss-Wegner (1981). When films of a-$GeSe_2$ are subjected to continuous excitation by a focussed He–Ne laser (just below E_g^{opt}) having a moderate intensity level, very low frequency oscillations in both the absorption coefficient and photocurrent occur.

Those optical properties that are most important for device applications will be emphasized in Chapters 6 and 7, where we shall discuss electrophotographic devices and optical memories.

4.3 Electrical Transport Properties

A comprehensive review of the transport properties of amorphous materials was written by Fritzsche (1974). In this section we will elborate on that review, adding significant recent advances when they are called for.

4.3.1 DC Conducitivity and Photoconductivity

As discussed in Chapter 1, over a wide temperature range most noncrystalline materials exhibit a temperature dependence of the dc conductivity of the form

$$\sigma = \sigma_0 \exp(-\Delta E/kT) \tag{4.8}$$

where $100 \leq \sigma_0 \leq 5 \times 10^3 \, (\Omega \, cm)^{-1}$. Figures 4.9 and 4.10 show data indicating that this law is valid over a wide range of temperatures for a

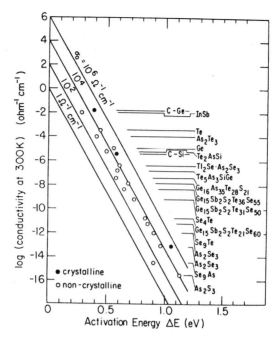

Fig. 4.9. Relation between conductivity at 300 K and activation energy ΔE as defined by Eq. (5.8) of the text for several amorphous and three crystalline semiconductors. In several cases the conductivity curve was extrapolated to 300 K in order to obtain these values. The values of σ and of ΔE are not characteristic of a given material but may depend on the method of preparation (Fritzsche, 1974).

variety of amorphous chalcogenide films. At lower temperatures the Mott $T^{-1/4}$ law associated with variable range hopping is often observed (Mott and Davis, 1979).

Many of the materials shown in Fig. 4.9 are multicomponent chalcogenide films. If the composition of these films changes, this causes a change in the physical properties. As an example of this, Fig. 4.10 shows how σ and ΔE change for the system $Tl_2SeAs_2Se_3$ as Te gradually replaces the Se. σ_0 remains constant during the process. Studies of the thermopower in this system show it to be positive, suggesting that the conduction mechanism is predominantly by holes in the valence band, a common occurrence in some chalcogenides. Others, however, exhibit negative thermopowers. Sometimes, the thermopower changes sign upon illumination (Rockstad, 1973). These results are not unexpected, since it is found that in most of these materials E_F is near the gap center; the dominant carrier species will be delicately controlled by band mobilities and defects. The position of E_F is determined by the general result that the thermal activation energy, ΔE, is about $\frac{1}{2}E_g^{opt}$ in most cases.

ELECTRICAL TRANSPORT PROPERTIES 333

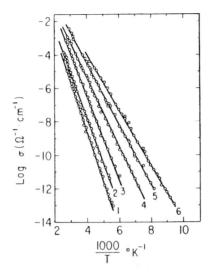

Fig. 4.10 Temperature dependence of the conductivity of several compositions of the system Tl$_2$SeAs$_2$ (Te$_x$Se$_{1-x}$) according to Andriesh and Kolomiets (1956). The composition of the samples labelled 1 through 6 correspond to $x = 0, 0.2, 0.4, 0.6, 0.8, 1.0$ respectively (Fritzsche, 1974).

The fact that E_F lies near the center of the gap does not mean that the material is intrinsic. Rather, E_F is apparently pinned there by the specific distribution of gap VAP defect states peculiar to the amorphous chalcogenide (Kastner et al., 1976; Adler and Yoffa, 1976). These states probably play a major role in determining the transport properties, which is best described as conventional dispersive transport (Schmidlin, 1977; Noolandi, 1977; Scher and Montroll, 1975; Tiedje and Rose, 1981). In dispersive transport injected carriers continuously slow down as they traverse the sample because they become trapped in ever deeper localized states that have a broad distribution of release times. Because of this the spatial distribution of the drifting carriers is not Gaussian. Rather, they are spread out more than they would be simply by Gaussian diffusion.

The local order clearly plays a major role in determining the semiconducting properties of materials. Ferrier et al. (1972) have shown this convincingly for the Tl–Te alloy systems; these results are shown in Figs. 4.11 and 4.12. The lowest values of σ and highest values of ΔE are found at impurities where a high degree of chemical order is expected. Annealing apparently improves the ordering further. It appears that compound formation occurs during film deposition and at annealing temperatures less than those required to crystallize the films; the amorphous state can occur even with a high degree of local order.

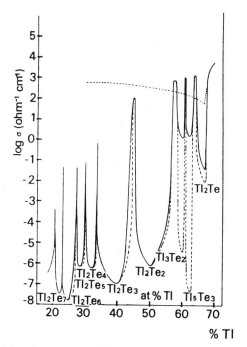

Fig. 4.11. Conductivity of co-evaporated films of Tl_xTe_{100-x} as a function of composition x. Solid curve is at 78 K as deposited. Dash-dotted curve: at 78 K after annealing. Dashed curve: conductivity of liquid at 425°C (Fritzsche, 1974).

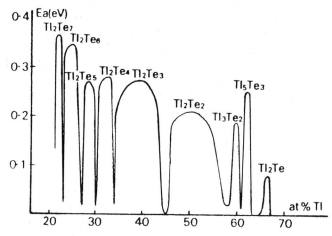

Fig. 4.12. Activation energy ΔE of conductivity obtained from the high temperature part of $\ln \sigma$ vs. $1/T$ curves (Fritzsche, 1974).

ELECTRICAL TRANSPORT PROPERTIES

Amorphous chalcogenides are often referred to as "long pair semiconductors," and these usually show annealing effects when sputtered or flash evaporated. Both ΔE and E_g^{opt} usually increase with annealing, along with the resistivity.

Equation (4.8) is valid for a wide variety of chalcogenides, but sometimes departures from it are observed, particularly in small gap materials.

In general, the chalcogenides have a high density of states in the gap due to defects and, therefore, exhibit only weak photoconductivity. However, many detailed photoconductive studies of a variety of materials have been made, particularly with regard to the spectral dependence of the dc photoconductivity (see, e.g., Mott and Davis, 1979). For example, most recently, Chamberlain and Moseley (1981) have studied the Ge_xSe_{1-x} system ($x \leq 0.4$). The photocurrent spectra obtained between 1.2 and 3.0 eV at 295° are shown in Fig. 4.13. We see that the photoconductive efficiency decreases as the Ge content is increased. Further, when the incident photon energy is low, the photocurrent follows the shape of the absorption edge. At high photon energies the photocurrent starts to rise with increasing energy. Such behavior is typical of a large number of chalcogenides, and has been attributed to bulk recombination. When the glass is doped it exhibits similar spectra.

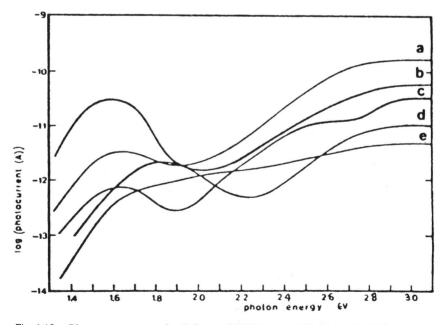

Fig. 4.13. Photocurrent spectra for $GeSe_{1-y}$ at 295 K. (a) $y = 0.05$, (b) $y = 0.15$, (d) $y = 0.33$, (e) $y = 3.8$ (Chamberlain and Moseley, 1981).

The temperature dependence of the photoconductivity is shown in Fig. 4.14, and the influence of metallic dopants on the photoconductivity is shown in Fig. 4.15. Note that at a given photon energy the inclusion of In enhances the photoconductivity but the presence of Cu or Fe decreases the photoconductivity. The dark conductivity also increases with increasing In concentration, without changing ΔE, whereas both quantities remain unchanged with Cu or Fe inclusion. Since both the dark conductivity and photocurrent increase when In is added to the glass, we expect that the carrier mobility is enhanced by some form of modification of the conducting channel by the In. The reduction of the photoconductive efficiency with Cu or Fe is probably due to the development of additional mid-gap recombination centers. Similar conclusions have been deduced for other chalcogenide glasses.

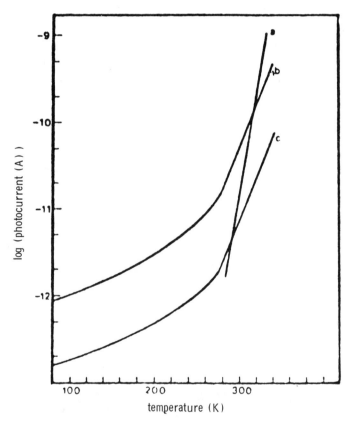

Fig. 4.14. Temperature dependence of the dark current (a) and J_{ph} (b and c) in $Ge_{0.15}Se_{0.85}$ at two different intensities for $E = 2.6$ eV (Chamberlain and Moseley, 1984).

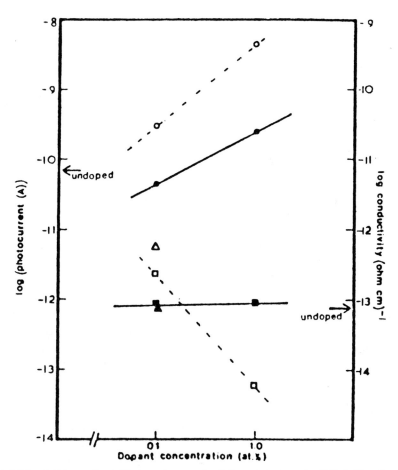

Fig. 4.15. Influence of metallic dopants on conductivity and photoconductivity in $Ge_{0.15}Se_{0.85}$. Conductivity: ●—In, ■—Cu, ▲—Pe. Photoconductivity: ○—In, □—Cu, △—Pe (Chamberlain and Mossley, 1981).

4.3.2 AC Conductivity and Transient Photoconductivity

Many chalcogenide films show an ac conductivity, σ_{ac}, whose frequency dependence varies as

$$\sigma_{ac}(\omega) = \text{const } \omega^s, \tag{4.9}$$

with $0.7 \lesssim s \lesssim 1.0$, over a wide range of frequencies (see, e.g., Fritzsche, 1974). This, and the fact that σ_{ac} is only a weakly increasing function of T suggests that the transport in these films is dominated by some form of hopping mechanism.

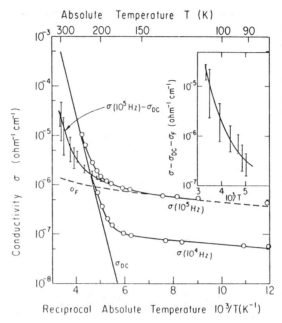

Fig. 4.16. Temperature dependence of the conductivity of evaporated a-As$_2$Te$_3$ for dc and for two frequencies (Rockstad, 1972). The dotted curve marked σ_F shows the linear T-dependence predicted by Eq. (5.33). Since $\sigma_{DC} + \sigma_F < \sigma_{ac}$, there appears to exist the additional contribution shown in the insert (Fritzsche, 1974).

The temperature dependence of the total conductivity of As$_2$Te$_3$ films at two frequencies is shown in Fig. 4.16. These results show that σ_{ac}, at least for small E_g materials, may have three contributing causes. At low T, $\sigma_{ac} \sim T$. If this portion is extrapolated to higher T and added to the dc conductivity, less than the total σ_{ac} is obtained. The additional component, σ_s, is also proportional to ω^s, but rises faster than linearly with T. Rockstad (1972) attributes σ_2 to hopping conduction in localized tail states. The insert in Fig. 4.16 shows that σ_2 is thermally activated, with a slope that is always less than the dc conductivity curve.

Pollak and Geballe (1961) developed a hopping theory which was applied by Austin and Mott (1969) for the case where tunneling conduction takes place near E_F. They obtained

$$\sigma_{ac}(\omega) = \frac{\pi}{3}[g(E_F)]^2 kTe^2 \alpha^{-5} \omega \left[\ln\left(\frac{v_p}{\omega}\right)\right]^4 \quad (4.10)$$

where v_p is a phonon frequency ($\sim 10^{13}$ s^{-1}) and α describes the spatial decay of the localized state wavefunction $\exp(-\alpha r)$.

Equation (4.10) allows for the possibility of obtaining the density of states $g(E_F)$ at E_F. Experiments have been performed in a variety of chalcogenides and the results for $g(E_F)$ have been tabulated (see, e.g., Fritzsche, 1974). The values of $g(E_F)$ obtained from Eq. (4.10) depend upon the decay parameter of the wave function that describes the localized states. It is found that $g(E_F)$ is too large to be describable simply by exponential tails of localized states extending into the gap from the band edges. In fact, the magnitude, frequency, and temperature dependence of σ_{ac} in chalcogenide glasses is very similar to that found in a wide variety of other noncrystalline semiconductors such as fused quartz and the alkali glasses. Indeed, any loss mechanism that has a wide, flat distribution of relaxation times yields the nearly linear frequency dependence given by Eq. (4.9). This behavior suggests that the behavior of σ_{ac} is dominated primarly by the lack of long range order rather than by any specific property of a particular group of materials. Other loss mechanisms can also be involved. Further, different values for σ_{ac} obtained in the same material indicate that internal inhomogeneities can play a major role. Results for Ae_2Se_3 are shown in Fig. 4.17. Recent results on the As–Se–Ag system have been reported by Kitao *et al.* (1981).

Other interesting and useful properties of the amorphous chalcogenides can be obtained from photocurrent-transient spectroscopy studies (Orenstein *et al.*, 1982; Bosch, 1982) which provide strong support for the multiple trapping model of transport in these materials. When carriers are photoinjected in a material that is dominated by dispersive transport the fraction of carriers is extended conducting states above E_0 after time t is given by (Orenstein *et al.*, 1982)

$$\theta(t) = \frac{\Delta n}{N_0} \cong \frac{g(E_0)}{g(E_0 - kT \ln v_0 t)} \gamma(v_0 t)^{-1} \qquad (4.10)$$

where Δn is the number of excess carriers at $E > E_0$, N_0 is the total expected carrier density, γ is the fraction of trapped carriers in thermal equilibrium with those at E_0, v_0 is the release rate from traps at E_0 and $g(E)$ is the density of states. Since the photocurrent is proportional to Δn, then $\theta(t)$ probes the density of states at the position of the injected carrier packet as it thermalizes.

For most chalcogenides it is found that the current decays with time in a power law fashion, which implies that the density of states decreases exponentially from E_0:

$$g(E) = \frac{N_L}{kT_0} \exp\left[-\frac{E_0 - E}{kT_0}\right], \qquad (4.11)$$

where kT_0 provides the rate of change of $g(E)$ and N_L/kT is the extrapolation

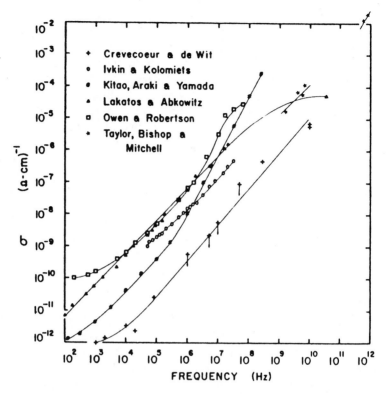

Fig. 4.17. The ac conductivity of vitreous As$_2$Se$_3$ at 300 K over a wide frequency range. References for the data can be found in Fritzsche (1974).

to E_0 of the exponential. For an exponential $g(E)$ we then have

$$\theta(t) \cong \frac{N_C}{N_L} (v_0 t)^{-1+\alpha} \tag{4.12}$$

where N_C is the number of states within kT above E_0 and $\alpha = T/T_0$, where we have taken $\gamma \cong T/T_0$ too.

The above model is correct for an infinite sample into which we inject excess carriers. The situation changes when either recombination starts in a photoconductivity experiment or when carriers reach the drain electrode in a drift mobility experiment. Because of these effects the energy distribution of the carrier packet is altered. Orenstein et al. (1982) obtain for monomolecular recombination

$$\Delta n = \frac{N_0^\alpha}{b_r n_t \tau_{\text{MR}}} \cdot \frac{[1 + (t_i/\tau_{\text{MR}})^\alpha](t/\tau_{\text{MR}})^{\alpha-1}}{[1 + (t/\tau_{\text{MR}})^\alpha]^2}, \tag{4.13}$$

where τ_{MR} is the lifetime for monomolecular recombination, n_t is the number of trapped electrons, b_r is the recombination coefficient and t_i is the time when the exciting pulse is turned off. For bimolecular recombination at short times it is found that

$$\Delta n \sim t^{-1+\alpha}, \tag{4.14}$$

and at long times

$$\Delta n = \left\{1 - \left[1 - \frac{N_0}{N_L}(v_0 t_i)^\alpha\right]\left(\frac{n_t}{N_0 + n_t}\right)^{b_c/b_r}\right\}\left(\frac{\alpha}{b_t \tau_{MR}}\right)^{-\alpha-1}, \tag{4.15}$$

where b_t is the trapping coefficient.

Both the transient photo-induced optical absorption (PA) and transient photocurrent (PC) were measured by Orenstein et al. (1982) under the same excitation conditions. Excitation was via a pulsed dye laser. Figure 4.18 shows the PA spectrum obtained for a-As_2Se_3 in the recombination-free limit. Note that Fig. 4.18a shows that the threshold of induced absorption moves to higher energy with increasing time delay, and that the magnitude of the shift is about equal during each succeeding decade of time. These data clearly show that a distribution of localized states exists within the band gap, and that the occupancy of these levels changes throughout the time of the experiment. These results are consistent with the prediction of a multiple trapping model.

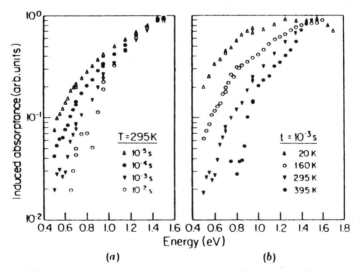

Fig. 4.18. The PA spectrum (a) for several times delays following pulsed excitation, (b) at $t = 10^{-3}$ s for various temperatures. In both cases the shift to higher energies is in qualitative agreement with the *MT* process for which the demarcation energy shifts as $KT \ln v_0 t$ (Orenstein et al., 1982).

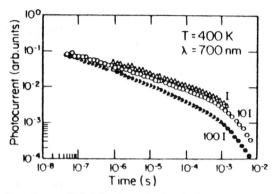

Fig. 4.19. Time dependence of PC following pulsed excitation at 400K, for three excitation intensities which differ by factors 10. The data are normalized at the earliest time to show the change in shape with intensity. Initially, the decay becomes slower as the excitation intensity is reduced. Finally, the shape becomes intensity independent indicating the absence of bimolecular recombination. More rapid decay at 10^{-3} s is due to monomolecular recombination. The excitation wavelength is $\lambda = 700$ nm (Orenstein et al., 1982).

In the recombination free limit the photocurrent should be proportional to the average value of the mobility, which is time-dependent during dispersive transport. Figure 4.19 shows PC transients in a-As_2Se_3 at 400°C. The data show both bimolecular and monomolecular recombination mechanisms. The intensity-independent cut-off at 1 ms is identified as monomolecular recombination. The bimolecular recombination slows down with decreasing laser-pulse energy until the transient eventually becomes independent of intensity. PC curves in this low intensity limit are shown in Fig. 4.20. In principle, the data of Fig. 4.20 contain the same information as is obtainable by time of flight measurements (Pfister and Scher, 1978), but more reliable values of parameters descriptive of recombination-free transport can be obtained from the latter experiments.

Orenstein et al. (1982) also provide a lucid description of experiments on recombination in a-As_2Se_3. We need not elaborate on these results in this chapter, but rather will summarize what the major results of this study teach us. These results are summarized in Table 4.2, and they tell us that multiple trapping is responsible for dispersive transport. Knowing this, we can use PC as a probe of the shape of the density of gap states, $g(E)$. Once this is known, the parameters involved with the current magnitude (mobility and infinite temperature transit time) can be obtained from time-of-flight data. After this, we can look into the effects of recombination. The effects of bimolecular recombination on PA and PC support the views that: (1) free-to-bound transitions limit the process; (2) PA measures the trapped carrier density Δn_t; (3) PC measures the conducting carrier density Δn.

Table 4.2

Experiment	General Result	Results for a-As$_2$Se$_3$
Recombination free		
PA	Direct evidence for MT	MT responsible for dispersive transport
PC	Shape of $g(E)$	$g(E) = \dfrac{N_L}{kT_0} \exp[-(E_0 - E)kT_0]$
		$T_0 = 550$ K
TOF	Parameters related to microscopic mobility	$v_0 = 10^{12}$ s^{-1}
		$\mu_0 N_c/N_L = 0.2$ cm^2 V^{-1} s^{-1}
Bimolecular recombination		
PA/PC	Free-to bound recombination	$b_r/\mu_0 = 3.7 \times 10^{-8}$ cm V
	PA measures Δn_t	Recombination radius = 10 Å
	PC measures Δn	
	Consistency of $g(E)$	$\tau_{BR} \propto N_0^{T_0/T}$
		b_t/b_r is independent of T
		$b_t N_L/b_r = 2 \times 10^{20}$ cm^{-3}
Monomolecular recombination	Find $N_{th}(T)$	$N_{th} = (8 \times 10^{19}$ cm$^{-3}) \exp\left[\dfrac{-0.5 \text{ eV}}{kT}\right]$
	Saturation behavior can provide b_t/b_r	Not measured

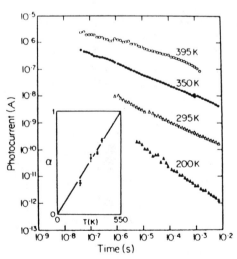

Fig. 4.20. Time decay of PC after pulsed excitation with band-gap light at several temperatures. At each temperature, the decay of the photocurrent accurately follows $t^{-1+\alpha}$. The temperature dependence of α is quite linear, as shown by the insert (Orenstein et al., 1982).

The data of a-As_2Se_3 (Orenstein et al., 1982) show that its density of states is exponential from 0.25 to 0.60 eV above the valence band, an important result because of the energy scale of the distribution, $kT_0 = 0.05$ eV. This means that the energy range investigated corresponds to seven decay constants of the exponential.

It is surprising that the measured $g(E)$ is so accurately exponential. But noting that $T_0 = 550$ K, close to the glass transition temperature T_g, Orenstein et al. (1982) argue that perhaps the exponential $g(E)$ arises from the Boltzmann factor of T_g. But they point out that this factor only determines how many centers of a given free energy of creation G_{cr} are present at T_g. On the other hand, $g(E)$ represents the number of centers that bind a carrier by the energy $E_0 - E$, so that an exponential density of states can be explained only if

$$G_{cr} = -E + \text{const}, \tag{4.16}$$

which is not true in general. It can be the case, however, if the localized states come from a high density of charged defects that freeze in at T_g, perhaps VAP's or IVAP's (Kastner et al., 1976). In this case

$$G_{cr} = G_{cr}(\infty) - C, \tag{4.17}$$

where C is the attractive Coulombic interaction between nearest defects. The binding energy at the specific defect is

$$E_0 - E = E_0 - E(\infty) - C. \tag{4.18}$$

It is interesting to note that the original suggestion of Kastner et al. (1976) was that VAP type defects produced sharp peaks in $g(E)$. This does not seem to be borne out by the PA and PC experiments, but the presence of VAP's goes a long way towards exlaining a broad body of other data, as we have discussed previously. Thus, questions of their existence and distribution still remain open (Adler, 1981).

4.3.3 Thermoelectric Power

In Chapter 2 we reviewed the subject of thermoelectric power. Briefly, the thermopower S is related to the Peltier coefficient by

$$S = \frac{\pi}{T}; \tag{4.19}$$

π is the energy carried by the electrons per unit charge, measured relative to E_F,

$$\pi = -\frac{1}{e} \int (E - E_F) \frac{\sigma(E)}{\sigma} dE, \tag{4.20}$$

where σ is the total conductivity and $\sigma(E)$ is the conductivity at energy E. Hence

$$S = -\frac{k}{e} \int \frac{E - E_F}{kT} \frac{\sigma(E)}{\sigma} dE; \qquad (4.21)$$

$S < 0$ for electrons at energies $E > E_F$.

For conduction in one band only (Fritzsche, 1974)

$$S = -\frac{k}{e}\left[\frac{E_c - E_F}{kT} + A_c\right], \quad \text{for } E > E_c; \qquad (4.22a)$$

$$S = -\frac{k}{e}\left[\frac{E_F - E_v}{kT} + A_v\right], \quad \text{for } E < E_v, \qquad (4.22b)$$

where

$$A_c = \frac{\int_0^\infty (\varepsilon/kT)\delta(\varepsilon)\,d\varepsilon}{\int_0^\infty \delta(\varepsilon)\,d\varepsilon}, \quad \text{with } \varepsilon = E - E_c; \qquad (4.23a)$$

$$A_v = \frac{\int_{-\infty}^0 (\varepsilon/kT)\delta(\varepsilon)\,d\varepsilon}{\int_{-\infty}^0 \delta(\varepsilon)\,d\varepsilon}, \quad \text{with } \varepsilon = E_v - E \qquad (4.23b)$$

are weighted averages over the carriers above the mobility edge at E_c and below the mobility edge at E_v.

Assuming that

$$E_c - E_F = \Delta E - \gamma T, \qquad (4.24)$$

where γ is a constant, then substitution of Eq. 4.24 into Eq. 4.22 yields

$$S = \pm \frac{k}{e}\left[\frac{\Delta E}{kt} - \frac{\gamma}{k} + A\right]. \qquad (4.25)$$

Since for band conduction we have

$$\sigma = \sigma_0 \exp\left(-\frac{E_c - E_F}{kT}\right) \qquad (4.26)$$

then comparison of Eqs. (4.22) and (4.26) shows that a plot of S and $\ln \sigma$ vs. T^{-1} should have the same slope for conduction in a single band. In general, however, $\ln \sigma$ vs. T^{-1} is observed to have a greater slope than S vs. T^{-1}.

For combined conduction when E_F is fixed by localized states, we obtain

$$S = \frac{k}{e}\left[\frac{1-b}{1+b} \cdot \frac{E_c - E_v}{2kT} + \frac{E_F - E_M}{kT} + \frac{1}{1+b}A_v - \frac{b}{1+b}A_c\right], \qquad (5.27)$$

where $b = \sigma_c/\sigma_v$ and $E_M = \tfrac{1}{2}(E_c + E_v)$.

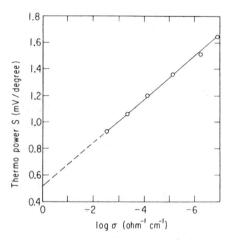

Fig. 4.21. Plot of the thermopower S against $-\ln \sigma$ at 300 K for the different compositions of the system shown in Fig. 5.10 (Fritzsche, 1974).

Fritzsche (1974) discusses the case when only states near E_F contribute to the current.

Figure 4.21 shows thermopower data for the chalcogenide system shown in Fig. 4.10. There is a clear linear dependence of S on $-\ln \sigma$.

4.3.4 The Hall Effect

The Hall coefficient in the amorphous chalcogenides was studied extensively during the late 1960's and early 1970's. It was found that it was negative in most cases (Fritzsche, 1974). The fact that it was negative regardless of the sign of the thermopower indicated that a major gap existed in our understanding of the transport process in these materials. Friedman (1971) made major inroads into our understanding of the problem by calculating the Hall mobility, μ_H, using a random phase model. He assumed that the applied magnetic field modifies the phase of the transfer integral between sites. When 3-site interactions dominate, a negative Hall coefficient is predicted. Four-site interactions yield a positive Hall coefficient, but these are apparently negligible in the chalcogenides. The Hall mobility is found to be

$$\mu_H = 4\pi \frac{ea^2}{h}[a^3 Jg(E_c)]\frac{\eta \bar{z}}{z} \qquad (4.28)$$

where J is the two site transfer integral, z is the coordination number, $\bar{z}$ the average number of interacting sites (here $\bar{z} = 3$) and the factor η is approximately $\frac{1}{3}$ for the problem at hand.

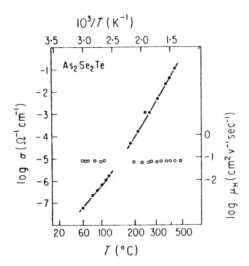

Fig. 4.22. Conductivity and Hall mobility of liquid and vitreous As$_2$Se$_2$Te as a function of temperature (Fritzsche, 1974).

The ratio of the Hall to conductivity mobility is found to be

$$\frac{\mu_H}{\mu} = \frac{6kT}{J}\eta\frac{\bar{z}}{\overline{z^2}}. \qquad (4.29)$$

Reasonable estimates of the important parameters yield $\mu_H \sim 10^{-1}$ cm^2/V s and $\mu_H/\mu \sim 10^{-2}$, results which compare favorably with experiment. Figure 4.22 shows data for As$_2$Se$_2$Te. μ_H is 10^{-1} cm^2/V s for both the liquid and solid and is independent of temperature over the range investigated. Other studies have found thermally activated behavior for μ_H, with activation energies between 0.03 and 0.05 eV. These results are shown in Fig. 4.23. Table 4.3 presents typical values of μ_H except those shown in Fig. 4.23.

The Hall mobilities of all of the above systems are quantitatively the same to within an order of magnitude, probably because of the similar prefactor values in the conductivity expressions [Eq. (4.26)]. Friedman (1971) suggests that it is basically because of the fact that $a^3Jg(E_c)$ is insensitive to structural and/or compositional variations in the material. Further, Eq. (4.28) suggests a temperature independent μ_H, whereas activated behavior is sometimes observed. Emin *et al.* (1972) suggest that polaron effects provide the activation energy for the mobility.

Recently, Friedman and Pollak (1981) calculated μ_H for systems with both positional and site-energy disorder using a percolation theoretic approach. Upper limits on μ_H for different systems were established. Butcher and McInnes (1981) have also made numerical calculations of μ_H for a Holstein

Table 4.3

Material	μ_H (cm^2/V s)	T (K)	Source
Tl$_2$SeAs$_2$(Se, Te)$_3$	−0.02		Kolomiets and Nazarova (1960)
Tl$_2$Se–As$_2$Te$_3$ at 37.5% Te	−0.09 −0.36	240–340 300	Ivkin, Kolomiets and Lebedev (1964)
As–Te–I As–Te–Br	−0.08 −0.12 −0.01 −0.10	300 365 300 365	Peck and Dewald (1964)
AsTe AsGe$_{0.1}$Te AsGe$_{0.2}$Te AsGe$_{0.3}$Te AsGe$_{0.4}$Te AsGe$_{0.3}$Te$_{0.8}$ AsGe$_{0.3}$Te$_{1.5}$	−0.065 −0.053 −0.050 −0.049 −0.043 −0.045 −0.056	300	Panus, Ksendov and Borisova (1968)
As$_2$Se$_2$Te AsSeTe$_2$ As$_2$SeTl$_2$Se As$_2$SeTl$_2$Se	−0.09 −0.10 −0.10 −0.01	300–800	Male (1967)
Tl$_2$Te.As$_2$Te$_3$ Tl$_2$Te.As$_2$Te$_3$	+0.10 +0.16 +0.028	300 325 125	Kornfeld and Sochava (1959) Nagels et al. (1970)
Se Se a-Ge	+0.37 ± +0.03 −7 ± 1.5 −0.32 ± 0.1 −(0.02 ± 0.01)	293 295 300	Juska et al. (1969a) Juska et al. (1969b) Dresner (1964) Clark (1967)

3-site hopping process. They find that their symmetrized ansatz and the Greens function formalism both agree with the numerical results. The situation at present seems to be that $\mu_H \gtrsim 10^{-4}$ cm^2/V s with a temperature dependence that depends on the specific situation.

4.3.5 Drift Mobility

In section 4.3.2 we discussed transient photoconductivity experiments (Orenstein et al., 1982) and how, when coupled with time-of-flight (TOF) measurements, we can obtain valuable information about the transport

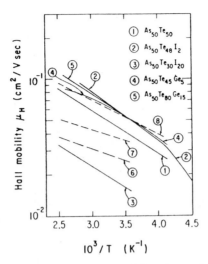

Fig. 4.23. Temperature dependence of the Hall mobility of several As–Te–I and As–Te–Ge alloys and for vitreous $(As_2Se)_x$. Curves 6, 7, 8 correspond to $x = 0.5, 1, 3$ respectively (Fritzsche, 1974).

properties of amorphous semiconductors. In this section we discuss the TOF measurement, which can reveal the drift mobility; the physics associated with this phenomenon was outlined in Section 2.2.6. Briefly, the drift mobility, μ_d, is obtained by measuring the time (of transit), t_t, for carriers injected at one electrode to reach the other. For a sample of length L in a field F

$$\mu_d = \frac{L}{t_t F} \tag{4.30}$$

The transit time should be well defined and scale with sample thickness (Enck and Pfister, 1976). Often μ_d is less than the band mobility, μ_b, or conductivity mobility, μ_0, because shallow traps tend to slow the motion of an injected pulse of carriers. It is usually the case that

$$\mu_d = \mu_0 \frac{\tau_0}{\tau_0 + \tau_t} \tag{4.31}$$

where τ_0 is the mean free time spent in the band (extended states) and τ_t is the average time spent in shallow traps during a transient. When there is only one trapping level of density N_t at E_t

$$\mu_d \cong \mu_0 \frac{N_c}{N_t} \exp\left(-\frac{E_c}{kT}\right) \tag{4.32}$$

for $\tau_0 \ll \tau_t$. N_c is the effective density of states at the band edge. When the traps are uniformly distributed from the band edge to E_t

$$\mu_d = \mu_0 \frac{N_c}{N_t} \frac{E_t}{kT} \exp\left(-\frac{E_c}{kT}\right), \tag{4.33}$$

and, when N_c falls off linearly with energy over a range ΔE,

$$\mu_d = \mu_0 \frac{\Delta E}{kT} \exp\left(-\frac{\Delta E}{kT}\right) \tag{4.33}$$

When carriers are photoinjected at one electrode of a typical chalcogenide, such as a-As$_2$Se$_3$ (Sharp *et al.*, 1981), the common current-time, $I(t)$, profiles obtained are shown in Fig. 4.24. Note that although linear plots of I is t (Figs. 4.24a and b) are relatively featureless, a plot of log I vs. log t

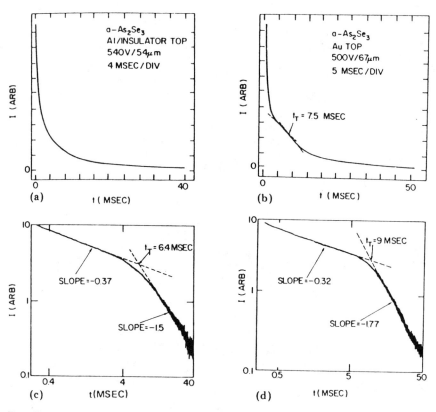

Fig. 4.24. Hole transient current signal in α-As$_2$Se$_3$ at room temperature in linear scales (*a*, *c*) and logarithmic scales (*b*, *d*) respectively (Enck and Pfister, 1976).

shows extremely interesting details. We note that there are clearly two power law regions, and it is generally agreed that the time denoted t_T in the figures corresponds to the time it takes the fastest of the carriers to reach the collecting electrode. The peculiar, yet universal, shape of the log I–log t curve is characteristic of what is now termed dispersive transport (Tiedje and Rose, 1981), caused by either a stochastic hopping mechanism (Scher and Montroll, 1975) or multiple trapping (Silver and Cohen, 1977). The transient photoconductivity experiments of Orenstein *et al.* (1982) have shown that in the chalcogenides it is most likely that multiple trapping is the origin of the dispersive transport. Both the hopping and multiple trapping theories predict that the current trace should appear as two straight lines on a logarithmic scale, intersecting at t_T, with an initial slope of $-(1 - \alpha)$ and a final slope of $-(1 + \alpha)$. When α is constant, universality is expected, and this is shown in Fig. 4.25. Note that the sum of the power exponents should be -2, independent of experimental parameters.

The most useful of the amorphous chalcogenides has been elemental selenium, which enjoys widespread use as a photoreceptor in the electrophotographic process. As opposed to a-As_2Se_3, a-Se is ambipolar; above about 200 K both carriers have well defined thermally activated drift mobilities as given by Eq. (4.30). Figure 4.26 shows representative hole drift mobility vs. T^{-1} plots, and Table 4.4 is a compilation of both the hole and electron transport parameters at room temperature. The hole transport is most readily described via shallow-trap controlled transport (Enck and Pfister, 1976), a mechanism different than the dispersive transport observed in a-As_2Se_3. In a-Se the trap release times are short compared to the transit time. The electron drift mobility is two orders of magnitude smaller than that for holes, and a trap-controlled transport model is also employed to explain this data.

4.4 Summary and Conclusions

In concluding this chapter we would like to point out that there are still no clear cut interpretations of optical and transport data that provide for a unique description of the electronic properties of a large number of amorphous semiconductors. What we have attempted to do here for the chalcogenides is to present a relevant, albeit incomplete, cross-section of the situation, while trying to avoid detailed comparisons between theory and experiment. We began by introducing a model of chalcogenide glasses based on specific defects (Adler, 1981), the VAP's, and then showed that their presence has only been inferred and never directly verified. A discussion of the optical properties followed, where interesting conclusions have been

Table 4.4. Representative hole (electron) transport parameters at room temperature for a-Se and, for comparison, α-monoclinic Se. μ_0 = microscopic mobility, μ_d = drift mobility, E = activation energy, τ = lifetime with respect to deep traps, N = density of states at the valence (conduction) band edge, N_t = density of shallow traps, S = capture cross-section of shallow traps.

	μ_0 (cm^2/V s)	μ_d (cm^2/V s)	E (eV)	τ (μs)	N (cm^{-3})	$\dfrac{N}{N_t}$	S (cm^2)
Amorphous Se holes	0.3–0.4	0.11–0.19	0.22–0.3 <250 K	<50	10^{20}	10^3–10^5	3×10^{-16}
α-Monoclinic Se holes		0.2	0.23			10^3	
Amorphous Se electrons	0.32	$(5–8) \times 10^{-3}$	0.28–0.33	<50	10^{20}	$(1–9) \times 10^3$	
α-Monoclinic Se electrons		2 lattice controlled	0.25 <200 K	10^{20}	10^6	10^{-13}	

SUMMARY AND CONCLUSIONS

extracted from photo-induced phenomena as much as from more conventional measurements. The electrical transport properties were then reviewed, and special emphasis was given to recent experiments on the photoconductivity under transient conditions. Thermopower and Hall effect experiments were then reviewed, and finally time-of-flight determinations of the drift mobility in these dispersive systems were discussed.

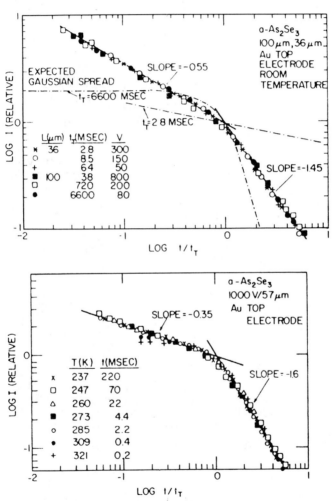

Fig. 4.25. Universality plots of current vs. time in logarithmic scales for hole transients in α-As$_2$Se$_3$. The plots were obtained by parallel shifting of the current traces along the time and current axes. (a) Room temperature, variation of applied field and sample thickness. Expected spread of Gaussian dispersion is schematically indicated. (b) Fixed applied field and sample thickness, variation of temperature (Enck and Pfister, 1976).

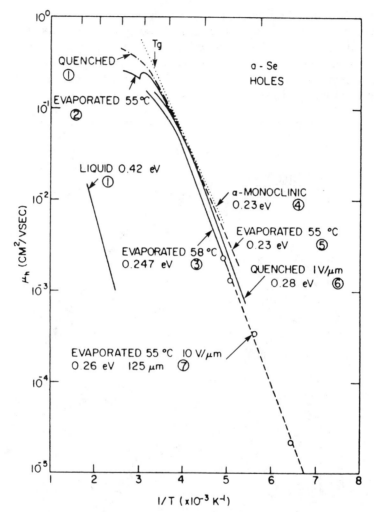

Fig. 4.26. Representative summary of hole drift mobility activation plots of hole transport in α-Se. Included are results on α-monoclinic and liquid Se. References for the data are in Enck and Pfister (1976).

5

Electrical Switching and Memory Devices Employing Films of Amorphous Chalcogenide Alloys

5.1 Introduction	356
5.2 The Thermistor	359
5.2.1 Introduction	359
5.2.2 Heat Flow in Semiconductors	362
5.2.3 An RC Network Analog of the Heating Process	366
5.3 Thermally Induced NDC	370
5.3.1 Thermal Boundary Conditions	370
5.3.2 The Effect of Inhomogeneities	375
5.3.3 Critical Electric Field Induced Switching Effects	376
5.4 Thin Chalcogenide Films	382
5.4.1 Introduction	382
5.4.2 Numerical Calculations of the First-Fire Event in Homogeneous Thin Films	391
5.4.3 Electrothermal Switching Mechanisms in Formed Chalcogenide Films	404
5.5 Thermophonic Studies of Thick Chalcogenide Films	409
5.5.1 Introduction	409
5.5.2 Experimental Results	413
5.5.3 Theory and Analysis	418
5.5.4 Discussion	428
5.6 Electronic Models for Threshold Switching	433
5.6.1 Introduction	433
5.6.2 Current Filamentation	434
5.6.3 Isothermal Analysis	438
5.6.4 A Model for the ON-State	448
5.7 Summary	454
Appendix	456

5.1 Introduction

Many important features of a wide variety of solid state devices result from situations where a sufficiently high bias is applied so that the device either switches from one conductive state to another or oscillates between two different conductive states. Many of these phenomena are classified as electronic instabilities. Among devices that behave this way are the: Tunnel diode (Esaki, 1958); Gunn diode (Gunn, 1964); Avalanche diode (Shockley, 1954); Josephson junction device (Josephson, 1962); Thyristor (Fulop, 1963); PIN diode (Prince, 1956). Others are classified as electrothermal instabilities, such as chalcogenide thin film switching devices (DeWald *et al.*, 1962; Ovshinsky, 1968), or thermal instabilities, such as Vanadium dioxide, VO_2 (Anderson, 1959), and many cases of second breakdown in transistors (Thornton and Simmons, 1958). Although substantial evidence exists for the fundamentally electronic nature of threshold switching in chalcogenide films (see, e.g., Adler *et al.*, 1978; Kotz and Shaw, 1984), the intrinsically thermal nature of memory and "forming" type processes, and the "delay time", make it important to emphasize both the thermal and electronic aspects of the problem. To this end, we will discuss in this chapter the important features of heat flow in solids that are pertinent to the development of thermal effects. Further, we will also review a reasonable cross-section of experiments and models associated with both thermal and electrothermal breakdown (Klein, 1971; Berglund and Klein, 1971), switching and memory effects.

In order to analytically understand the electrical and thermal behavior of the above structures, it is most often, but not always, convenient to characterize them in terms of an effective region of negative differential conductivity (NDC) in their current density, **J**, electric field, **F** characteristic. In this region $dJ/dF < 0$ holds. Figure 5.1 shows the two major classes of NDC characteristics: SNDC; NNDC. As can readily be seen from the curves, the S and N stand for the shape of the characteristic; for NNDC we have a $J(F)$ curve, for SNDC an $F(J)$ curve. In this chapter we concentrate on SNDC devices, with the realization that all NDC characteristics are functions of frequency and this dependence must always be considered in determining the behavior of a particular device (Shaw *et al.*, 1979).

The simplest circuit in which any of these devices operates is shown in the inset of Fig. 5.2. Here the NDC element is in series with a load resistor R_L and bias battery that provides a voltage V_B. If I is the current in the circuit and V the voltage drop across the NDC element, then

$$V_B = IR_L + V; \tag{5.1}$$

the equation of the dc load line. This line is plotted in Fig. 5.2; its slope is $-1/R_L$ and its intersection with the device characteristic $I(V)$ defines the

INTRODUCTION

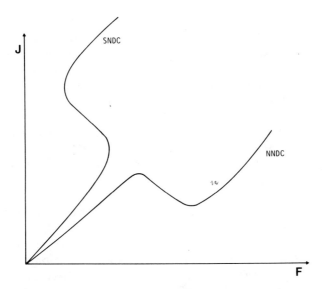

Fig. 5.1. Example SNDC (current controlled) and NNDC (voltage controlled) $j(F)$ curves.

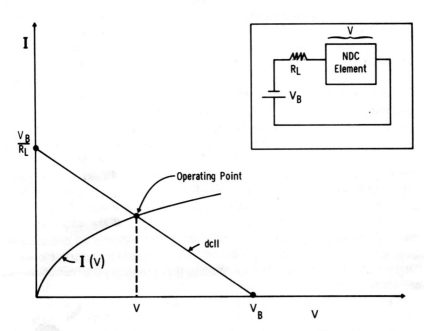

Fig. 5.2. Load line (dcll) and current-voltage ($I(V)$) curve for a nonlinear element placed in the circuit shown in the inset.

steady state operating point. [Strictly speaking, the SNDC element has a $V(I)$ characteristic, but the common terminology is $I(V)$. In what follows we will use the two interchangeably, but realize that the proper form is $V(I)$.] Intersections of the load line with the $I(V)$ characteristic are stable as long as $dI/dV > 0$, which is the case depicted in Fig. 5.2. However, for situations where either dI/dV or dJ/dF have regions of NDC, operating points at intersections in these regions are often unstable both against the formation of inhomogeneous field and/or current density distributions (space charge nonuniformities, see, e.g., Knight and Peterson, 1966, 1967) and/or circuit controlled oscillatory effects (see, e.g., Shaw et al., 1973 and the appendix of this chapter). In order to understand the detailed nature of these instabilities, we must ask two basic questions. First, what is the mechanism responsible for the NDC region? Second, how do we analyze the resultant phenomena that often occur? In this chapter we treat the case where the SNDC characteristic in amorphous films is produced by electronic means and limit ourselves to situations where the circuit plays no major role in controlling the behavior of the resultant instability. Circuit effects are discussed in the appendix; it is primarily the mechanism and steady state solutions we are going to concern ourselves with now. But in order to separate the ubiquitous heating effects from the major electronic features of the SNDC switching problem, we must appreciate how thermal effects alone can produce a form of SNDC (Shaw and Yildirim, 1983), and then understand the general problem of heat flow in semiconductors. We treat these aspects of the problem first in Section 5.2. Before we do, however, it is important to outline some of the major phenomenological aspects of bias-induced switching and memory events in thin amorphous chalcogenide films. This is best done by considering Fig. 5.3. Here the OFF and ON states are noted, along with the dc load line. V_t, I_t is the threshold point. The Lissajous type figure (thin solid line) represents the circuit response (see the appendix and Shaw et al., 1973), here seen as a damped oscillation as the energy is exchanged during the switching event between the local reactive components. V_h, I_h is the holding point, which is circuit dependent. There is a minimum holding current I_{hm}, below which the sample always switches back to the OFF state. There is also a current I_f, above which memory can occur. That is, if the sample is kept ON at currents above I_f for a sufficiently long time, rather than switch back to the OFF state when the bias is lowered, it remains ON and follows back along the dot-dashed characteristic. A crystalline filament path then connects the electrodes (Fritzsche and Ovshinsky, 1980). As we shall see later, a memory switch can readily be reset to a threshold switch by the application of a sufficiently intense and long current pulse. Figure 5.3 also shows a line of arrows. These represent the trajectory that is followed by the transport component of the current during the switching transition. The Lissajous

THE THERMISTOR

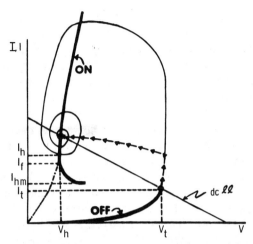

Fig. 5.3. Estimated $V(I)$ and $V(i)$ curves for the case where the minimum current does not fall low enough to quench the filamentary inhomogeneity. (All parameters are defined in the text.) The circuit oscillations damp, a filament forms, and switching occurs to the ON state. The ON state is the heavy dot. As V_B is varied the dots fall on a "filament characteristic," which is sketched (dark). The switch will take place in a damped oscillatory fashion. A "spiraling in" to the ON state is fundamental to the switching process in general. In the current-time profile this appears as a damped "ringing".

figure is shown in the plane of conduction current, i, vs. voltage, whereas the dc load line correspond to the plane of total current I, which includes the displacement current vs. voltage. We shall see that the displacement current (capacitive discharge) plays a major role in controlling the thermal aspects of the problem, especially with regard to forming and memory effects (Kotz and Shaw, 1984).

5.2 The Thermistor

5.2.1 Introduction

Thermal effects in solids have been treated in great detail over the past fifty years (see, e.g., Carslaw and Jaeger, 1959). Of particular interest has been the variety of phenomena associated with thermal runaway induced by Joule heating and the associated breakdown or switching processes often observed (see, e.g., Fock, 1927; Becker, 1936; Franz, 1956; Skanavi, 1958; Böer *et al.*, 1961; Stocker *et al.*, 1970; Shousha, 1971; Altcheh *et al.*, 1972; Thoma, 1976). These instabilities often result in regions of NDC appearing in the $I(V)$ characteristics of a variety of materials. It is now well known that NDC can appear in the static and dynamic characteristics of common materials and devices in which the current level is determined not only by the applied voltage, but also by the temperature. One reason for this is that the Joule heating of the sample often causes the average temperature to rise above that of the ambient temperatures, T_a. Figure 5.4 shows how this might arise. Linear $I(V)$ characteristics are sketched for isothermal cases where the ambient temperature $T_{a4} > T_{a3} > T_{a2} > T_{a1}$. These are the characteristics that would result were the heat sinking sufficient to maintain the system at the ambient levels shown. However, when the heat sinking is insufficient to remove heat fast enough, then, e.g., if the ambient is T_{a1}, it is possible that the steady state average T can correspond to a point on the T_{a4} line. The actual $I(V)$ characteristic might then appear as the thick solid line; a region of SNDC could occur. A device in which NDC is induced in this manner is called a thermistor. Note that every nonlinear point on the thermistor characteristic corresponds to a different average steady state T distribution. The slope of the NDC characteristic will depend primarily upon the heat sinking, heating rate, pulse repetition frequency and pulse width. Hence, its detailed form is a variable that depends on the way in which the measurement is performed, and a major feature of its shape is the question of the existence and position of a "turnover" or threshold voltage, V_T (Hayes, 1974; Hayes and Thornberg, 1973).

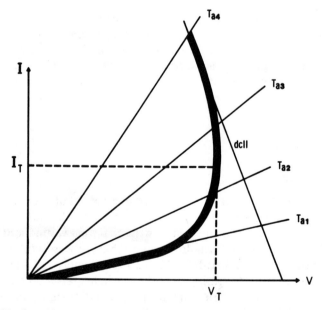

Fig. 5.4. Nonlinear thermistor characteristics (heavy line) that might arise in a material where linear $I(V)$ curves at different ambient temperatures T_a are shown at $T_4 > T_3 > T_2 > T_1$.

THE THERMISTOR

Burgess (1955a,b,c, 1960) has made an extensive study of thermistor behavior in materials having a conductance of the form $G = G_0 \exp(-b/T)$, where G_0 and b are constant, which encompasses a large number of important semiconductor materials and devices. Independent of the form of the conductance, however, if $I = I(V, T)$, then

$$\frac{dV}{dI} = R\left(\frac{1-y}{y+x}\right), \tag{5.2}$$

where $R \equiv V/I$, $X \equiv R(\partial I/\partial V)_T$, $y \equiv (V/B)(\partial I/\partial T)_V$ and $B \equiv (\partial/\partial T)(IV)$. To obtain Eq. (5.2) we write the total differential of $I = I(V, T)$, divide by dV and multiply both sides of the resulting equation by V/I. We then use the linear relationship

$$P \equiv IV = B(T - T_a), \tag{5.3}$$

where P is the power, B a constant and T_a the ambient temperature, to obtain an expression for dV/dT, which leads directly to Eq. (5.2).

It is clear from Eq. (5.2) that the turnover point occurs at $y = 1$, with NDC setting in for larger values of y. Thus, at turnover

$$1 = V_T\left(\frac{\partial I}{\partial T}\right)_{V,T}\left(\frac{dT}{dP}\right), \tag{5.4}$$

where T_T is the temperature of the contact at turnover. V_T for a given thermistor is a function only of T_a.

In order to achieve the condition $y = 1$ the conductivity of the semiconductor must have the proper temperature dependence. In particular, the common form

$$I = AV \exp\left(-\frac{b}{T}\right), \tag{5.5}$$

where A and b are constants, is sufficient to achieve the turnover condition. In order to simultaneously solve both the heat flow and current flow equations, we require knowledge of the boundary conditions. In Shaw *et al.* (1979) we emphasized the electrical boundary conditions. We must now concentrate on the thermal boundary conditions. A common thermal boundary condition is given by Eq. (5.3), rewritten here as

$$T - T_a = aP, \tag{5.3}$$

where a is a constant relating the excess temperature to the power supplied to the material or device.

Equations (5.3) through (5.5) yield for the temperature of the element at turnover

$$T_T = \tfrac{1}{2}b - [\tfrac{1}{4}b^2 - bT_a]^{1/2}, \quad (5.6)$$

which shows that a requirement for turnover is $b > 4T_a$. The power at turnover is

$$P_T = a^{-1}[\tfrac{1}{2}b - T_a - (\tfrac{1}{4}b^2 - bT_a)^{1/2}]; \quad (5.7)$$

the power at turnover increases with increasing ambient temperature. This feature is characteristic of a semiconducting thermistor having the property shown in Eq. (5.4). For $b \gg T_a$, a good approximation of Eq. (5.7) is

$$P_T \cong \frac{T_a^2}{ab} \quad (5.8)$$

and the conductance at turnover is

$$\frac{I_T}{V_T} = G_T \cong G_0 \exp\left(1 + \frac{T_a}{b}\right); \quad (5.9)$$

i.e., the conductance at turnover is enhanced by about a factor of e from its isothermal value, G_0, at the same T_a (Burgess, 1955a).

5.2.2 Heat Flow in Semiconductors

The well-studied problem of heat conduction in solids (Carslaw and Jaeger, 1959; Kittel, 1976) will be touched upon shortly. Rather than be very detailed, we will emphasize the specific problem at hand with reference to Fig. 5.5.

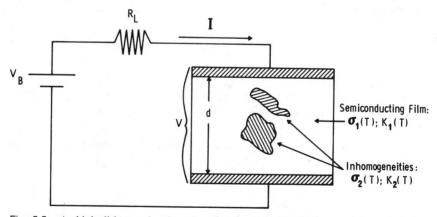

Fig. 5.5. A think disk sample of semiconductor containing inhomogeneities placed in a resistive current. The symbols are defined in the text.

THE THERMISTOR

The semiconductor has an electrical conductivity $\sigma_1(T) = \sigma_0 e^{-\Delta E/kT}$, where ΔE is the thermal activation energy and k the Boltzmann constant. σ_0 is the conductivity in the limit $T \to \infty$. For later use we also include the presence of inhomogeneities. For metallic inhomogeneities we take $\sigma_2 \gg \sigma_1$ and assume for simplicity that σ_2 is either constant or a slowly decreasing function of T. We also take the thermal conductivities as $K_1(T) = \alpha_1 T + \beta_1$, with $\alpha_1 \equiv (dK_1/dT) > 0$ and $K_2(T) = \alpha_2 T + \beta_2$, with $\alpha_2 \equiv (dK_2/dT) < 0$. The object of the excercise is to determine the $I(V)$ characteristics of the device and study the affect of the: (1) temperature dependent electrical conductivity; (2) temperature dependent thermal conductivity; (3) thermal boundary conditions; (4) presence and morphology of inhomogeneities; (5) critical electric fields for the onset of impact ionization and carrier multiplication; (6) thermally induced phase changes and latent heats (Shaw and Yildirim, 1983). To do the general analysis we must solve equations for the flow of both electric and thermal currents. Since we are discussing thermal effects now, we will first concentrate primarily on the heat flow equation. Later in the chapter we will discuss the coupled thermal and electronic equations for a general semiconductor device containing both holes and electrons.

We will first consider the simplest system and search for general criteria for: (1) thermistor behavior; (2) switching effects. To do this we remove the inhomogeneities, provide a constant current source by setting $R_L = \infty$ and treat the general case where $d\sigma/dT > 0$.

For the total electronic current $\mathbf{J}_e$ with no sources or sinks we have

$$\text{div } \mathbf{J}_e + \frac{\partial}{\partial t} \sum_i n_i q_i = 0, \tag{5.10a}$$

where $\sum_i n_i q_i$ is the total mobile charge density. In the steady state

$$\text{div } \mathbf{J}_e = 0. \tag{5.10b}$$

For a uniform homogeneous medium

$$\mathbf{J}_e = -\sigma \text{ grad } V; \tag{5.11}$$

we have for the stable steady state

$$\text{div}(\sigma \text{ grad } V) = 0 \tag{5.12}$$

or

$$\nabla^2 V + \frac{\text{grad } \sigma \text{ grad } V}{\sigma} = 0 \tag{5.13}$$

If J_h is the heat current density, then the conversion of electrical energy into heat is given by

$$-\text{div}[VJ_e + J_h] = \frac{\partial w}{\partial t}, \tag{5.14}$$

where w is the energy density and $\partial w/\partial t$ the power density. Here we have assumed that the product of the electronic current and enthalpy per unit charge carrier per unit length is independent of position and that thermoelectric power and Peltier heating effects are negligibly small (Carslaw and Jaeger, 1959).

Applying the divergence theorem to Eq. (5.14) yields

$$-\oint_s (VJ_e)\,ds - \oint_s J_h\,ds = \frac{\partial W}{\partial t}, \tag{5.15}$$

where $W = \int_v w\,dv$ (Joules) is the energy stored in the volume v enclosed by the surface s.

The first term on the LHS of Eq. (5.15) describes the electrical power flowing into the closed surface s:

$$P_e \equiv -\oint_s (VJ_e)\,ds. \tag{5.16}$$

The second term on the LHS (without the minus sign) is the heat power flowing out of s:

$$P_h \equiv \oint J_h\,ds. \tag{5.17}$$

The RHS is equal to the difference $P_e - P_h$, i.e., dW/dt denotes the rate of increase of the energy stored in v.

$$P_e - P_h = \frac{dW}{dt}. \tag{5.18}$$

The energy stored in v is related to the heat capacity of v,

$$C = \frac{dW}{dT}, \tag{5.19}$$

or

$$dW = C\,dT.$$

Thus, Eq. (5.18) becomes

$$P_e - P_h = C\frac{dT}{dt}. \tag{5.20}$$

THE THERMISTOR

Equation (5.20) states that the electrical input power P_e is used in two ways: Part of the input power flows out of S in the form of heat current with density $\mathbf{J}_h$; the rest is used to increase the temperature of the system.

The terms in Eq. (5.14) can also be interpreted in a similar way. Defining

$$p_e = \text{div}[V\mathbf{J}_e] = (\text{grad } V)\mathbf{J}_e + V(\text{div } \mathbf{J}_e) = \mathbf{J}_e \text{ grad } V$$
$$= -\sigma|\nabla V|^2 = -\frac{|\mathbf{J}_e|^2}{\sigma}, \tag{5.21}$$

which is the electrical input power density at a point, and

$$p_h = \text{div } \mathbf{J}_h = -\text{div}(K \text{ grad } T) = -K\nabla^2 T - \text{grad } K \text{ grad } T, \tag{5.22}$$

which is the heat power density flowing out of that point, Eq. (5.14) becomes

$$p_e - p_h = \frac{dw}{dt}. \tag{5.23}$$

The RHS of Eq. (5.23) is the power density required to alter the local temperature of a point. It can be rewritten as:

$$w = \frac{dW}{dv} = \frac{C\,dT}{dv}$$

$$= \frac{C\,dT}{dv}\frac{m_g}{m_g}, \tag{5.24}$$

where m_g is the mass in the volume v. Using the definitions:

$$\rho_m = \frac{m_g}{dv} = \text{mass density} = \text{specific mass} \tag{5.25}$$

and

$$c = \frac{C}{m_g} = \text{heat capacity per unit mass (specific heat)}, \tag{5.26}$$

Eq. (5.23) becomes

$$p_e - p_h = \rho_m c \frac{dT}{dt}. \tag{5.27}$$

Thus, we have two power continuity equations. One is in integral form, which can be used to relate the total input power, total heat power (efflux) and the rate of change of T:

$$C\frac{dT}{dt} + P_h = P_e. \tag{5.28}$$

This equation can be used to study the system as a whole.

The other equation is a point relation which can be used to study local regions of the material:

$$\rho_m c \frac{dT}{dt} + p_h = p_e. \qquad (5.29)$$

Both of these equations reemphasize the simple fact that the difference in input power and heat efflux goes into increasing the temperature of the system.

When the external source is switched on, dT/dt will initially be greater than zero. That is, the temperature of the material will start to rise. The rise will continue until a sufficient temperature gradient is reached whereby, neglecting radiative losses, all the incoming electrical power will flow out as heat. When steady state is reached $dT/dt = 0$, $P_e = P_h$ and $p_e = p_h$. The differential Eq. (5.29) is a forced diffusion equation, an inhomogeneous, parabolic partial differential equation which can be rewritten in terms of V, T, K, σ and $\mathbf{J}_e$ as

$$\rho_m c \frac{dT}{dt} + \operatorname{div}[K(T) \operatorname{grad} T] = -\frac{|J_e|^2}{\sigma}. \qquad (5.30)$$

For a temperature-independent K it becomes

$$\frac{dT}{dt} + \frac{K}{\rho_m c}\nabla^2 T = -\frac{|J_e|^2}{\sigma \rho_m c}, \qquad (5.31)$$

which has a diffusion constant $D_h = K/\rho_m c$.

Since the heat flow process is diffusive, let us first discuss the transient state (where $dT/dt \neq 0$) in a qualitative manner. The most important feature is that the RHS of Eq. (5.31) is the heat power generated at a point. Considering that point alone (we use the superposition principle to study the other points in a similar way), it is seen that if D_h is large (good thermal conductivity, low ρ_m and c) the temperature wave will move rapidly (higher diffusion velocity) and a local thermal pulse or disturbance will propagate over a relatively long distance before being attenuated. A simple way of visualizing this process is offered next via the behavior of an analogous distributed RC network (Sousha, 1971).

5.2.3 An RC Network Analog of the Heating Process

Heat flow in a solid is a diffusive phenomenon. The differential equation and solutions for $T(t, \mathbf{r})$ are very similar to the solution describing current and voltage waves in a distributed RC network. Figure (5.6) shows a simple one-dimensional analog distributed RC circuit in which the voltage is analogous

THE THERMISTOR 367

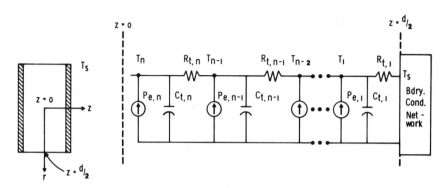

Fig. 5.6. Left-thin disk sample of thickness d. Right-distributed RC analog network. The symbols are defined in the text.

to temperature and current is analogous to heat current. For simplicity and compactness we have considered heat flow only in one-dimension (z). A similar model readily accounts for heat flow in the radial direction.

The R_t's are thermal resistances per unit length, which are inversely proportional to $K(T)$. The C_t's are thermal capacitances per unit length, which are proportional to ρ_m, c and m_g. The $p_{e,n}$'s are the "heat current sources", the Joule heating at each point. The T_n's are the node temperatures (voltages developed across the capacitors). The energy stored in a capacitor is analogous to the thermal energy stored in the thermal capacity of the system. During the transient state part of the current in R_t is used to charge C_t and the rest flows to the load network, which simulates the thermal boundary conditions. At steady state all the capacitors are charged to their final values (determined by R_t and p_e) and the input power flows to the load network.

The temperature at the zeroth node represents the surface temperature, T_s. A constant temperature boundary condition (infinite heat sink) can be simulated by connecting an ideal voltage (temperature) source with a voltage (temperature) equal to the ambient temperature T_a.

A boundary condition of the type

$$\mathbf{I}_h \cdot \hat{n}\big|_{\text{at interface}} = G(T_s - T_a), \tag{5.32}$$

Newton's law of cooling, can be represented by an ideal current-dependent voltage source,

$$T_s - T_a = \alpha I_{h1}, \tag{5.33}$$

Let's first consider an infinite heat sink and suddenly impose an electrical power source on the system. Heat will be generated everywhere in accordance with

$$p_e = \frac{|J_e|^2}{\sigma(T)}. \tag{5.34}$$

For a temperature-independent σ, p_e will be highest at points where the electric current density is highest. However, since $\sigma(T)$ also increases with temperature, the variation of p_e with T must be inspected further, and we shall do this shortly.

If we treat one current source at a time, we see that the heat power will diffuse towards the short circuit load (electrodes) with a diffusion constant

$$D_h = \frac{K}{\rho_m c} \tag{5.35}$$

and with a diffusion velocity proportional to K. Considering just the source $p_{e,n}$ for the moment, at $t = 0$, $C_{t,n}$ acts as a short circuit and all the heat current flows through it. As $C_{t,n}$ is charged, the temperature $T_n(t)$ will start to increase, which will in turn cause part of the heat current to flow to the neighboring circuit ($R_{t,n-1}C_{t,n-1}$). This transient process will stop when a sufficient temperature gradient ($T_n, T_{n-1}, T_{n-2}, \ldots, T_s$) is developed such that all the heat power will flow through the R_t's towards the boundary. That is, the C_t's are charged to their limits and draw no more current. Thus, the T_n's will not increase further. The limiting temperature (voltage) for each C_t is determined by the source strength (p_e), R_t and the boundary conditions.

The heating process, which we just discussed for a single heat source, occurs for all sources simultaneously. Since there are sources at each point in the material, the steady state may be established in a time shorter than for the single-source case. Because of the symmetry, at steady state the heat (current) flows only to the right towards the load. Therefore,

$$T_{\max} > T_n > T_{n-1} > T_{n-2} > \cdots > T_s \text{ at steady state.}$$

It is also interesting to note that during the transient states that may arise due to changes in some heat sources, the direction of heat flow may reverse. For example, let's assume that all the sources except one are dead at an instant $t = 0$. Also assume that the capacitors have initial temperatures $T_n(0) > T_{n-1}(0) > \cdots > T_0$. Let the strength of $p_{e,h}$ suddenly increase, and let's kill all the other sources at $t = 0$. The source $p_{e,h}$ will start to charge the nearest capacitors towards the new steady state value, and all the other capacitors start to discharge (or charge) towards their new steady state values as determined by the new steady state network.

THE THERMISTOR

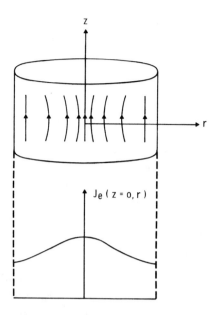

Fig. 5.7. Lines of electric current (top) and curent profile (bottom) sketched for a sample with a "hot spot" near its geometrical center.

In the above we treated the case of an infinite-heat-sink boundary condition. Similar behavior will occur for a Newton type boundary condition, which we will discuss further in the next section. First, however, we must examine the role of a temperature-dependent electrical conductivity.

Since σ increases with T, then the inner region of the example, which has the highest T, will have the highest σ. Hence, this region will draw the most current; a schematic is sketched in Fig. 5.7. The moderate "current-crowding" process shown in the figure is not self-accelerating or divergent. That is, a larger $\mathbf{J}_e$ does not necessarily mean more heat generation in this region. In fact, inspection shows that *less* heat is generated in the crowded region because $p_e = |J_e|^2/\sigma(T)$. Here an increase in $\mathbf{J}_e$ at a point is the result of an increase in $\sigma(T)$. Currents prefer to flow through the high conductivity region to dissipate less power; the crowding process is a self-stabilizing one.

The underlying process can be explained by considering two conducting plates connected to each other by two parallel resistors R and $r(T)$, where $r(T)$ is a decreasing function of T. Initially we let $R = r(T_0)$; equal currents will pass through R and $r(T_0)$. The total power drawn will be

$$P_0 = I^2(r \parallel R) = \frac{I^2 R}{2}. \tag{5.36}$$

As T increases r will decrease and draw more of the current. Let $r(T_1) \ll R$; almost all of I will pass through $r(T_1)$ and the power drawn from a constant current source will decrease to

$$P_1 = I^2 r(T_1) \ll P_0. \tag{5.37}$$

Therefore, it is not correct to assume that an increase in local temperature, which increases the local conductivity, will increase the local current density and lead to a further increase in temperature. That is, thermal runaway will not occur because of this process alone.

It is important to note one further aspect of the effect of current crowding induced by the locally heated region. The channeled current will flow through the region in the z direction producing higher p_e's there. Hence, there will be a tendency for the hot region to expand in the direction of the electrodes. This expansion will stop when a sufficient temperature gradient is established that will allow all the heat power to flow out of the film, and this brings us back to the questions of "turnover", the possible presence of NDC, and thermal runaway. All of these effects are related to the ability of the electrical resistance of the system to be reduced to a sufficiently low value by Joule heating. Ultimately, we must therefore find a mechanism by which the layers of the material adjacent to the electrodes can have their resistance lowered to sufficiently small values. The infinite-heat-sink boundary condition used above is unable to account for such effects since the material adjacent to the electrodes will remain cool and at a relatively high value of resistance. If turnover and NDC is to occur, we must have access to a mechanism that, e.g., will account for the fact that in the NDC region an increase in the current could lead to a decrease in the input power density, even though the temperature increases. To achieve this we can let the electrodes be heated (Newton's law of cooling) or realize that the electric fileds adjacent to the electrodes might be raised to sufficiently large values to induce the field stripping of carriers, carrier multiplication, or avalanche breakdown effects (Shaw et al., 1973b; Adler et al., 1978; Kotz and Shaw, 1984).

5.3 Thermally Induced NDC

5.3.1 Thermal Boundary Conditions

The development of an NDC region will lead to switching and oscillatory effects for a sufficiently lightly loaded system. It could also result in suspected thermally induced current filamentation (Stocker et al., 1970; Altcheh et al., 1972) thermal runaway and permanent or alterable memory type phenomena (memory, Ovshinsky, 1968). In order to understand how

THERMALLY INDUCED NDC

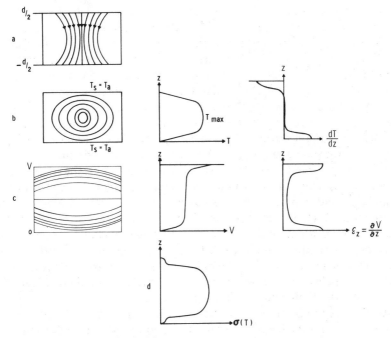

Fig. 5.8. (a) Electric current lines. (b) Isothermal lines (left), temperature profile (center), temperature gradient profile (right). (c) Equipotential lines (left), potential distribution (center), electric field distribution (right). (d) Electrical conductivity profile.

the above possibilities might occur, we return to the basic diffusion-free relation $\mathbf{J}_e = -\sigma(T)\,\mathrm{grad}\,V$. We see that $\mathbf{J}_e$ can grow when either σ, grad V or both increase. Let us imagine that we have established conditions for an infinite-heat-sink boundary condition, where substantial heating of the bulk material has occurred. The resulting current (a), temperature (b), potential (c) and electrical conductivity (d) profiles are shown in Fig. (5.8). With regard to Fig. (5.8), the following important points should be noted:

(1) There are large temperature gradients near the electrodes;
(2) There are large electric fields near the electrodes caused by the low temperature of the electrodes (ambient) and nearby regions, and these produce "low conductivity" layers.
(3) The major voltage drop across the film is caused by the narrow, low temperature, low conductivity layers.
(4) If no other factor is included, as the current increases further these narrow regions near the contacts continue to shrink, but they never vanish. Therefore, although the resistance of the structure continues to decrease, the voltage across the device continues to increase (PDC).

The situation outlined above cannot continue indefinitely. Eventually, at sufficiently high bias, the fields near the electrodes will cause a change in the transport properties of the semiconductor. Indeed, these crucial field effects will be treated shortly. Now, however, we will consider the effect that a more realistic boundary condition has on the profiles sketched in Fig. 5.8.

Instead of keeping the electrodes at a constant ambient temperature, we employ a more realistic Newton-type thermal boundary condition, which allows the electrodes to heat up and thereby decreases the resistance of the thin layer near the electrodes. The appearance of the isothermal lines will now have the form shown in Fig. 5.9. We see that the electrode surface near the z-axis will be heated first. This process can lead to current filamentation because the resistive barrier will be lowered in this region. The affect of this boundary condition is similar to the affect of a temperature-dependent electrical conductivity. That is, the electronic and heat current densities are redistributed in accordance with the new boundary condition. The development of current filamentation will be a smooth, continuous process, with filamentation growing as the current increases. (This is to be contrasted with the critical field case, where the transition might be sudden). Figure 5.10 shows how the filamentation develops.

As discussed in Section 5.2.1, the presence of a conductivity of the form $\sigma = \sigma_0 e^{-b/T}$ plus the realistic Newton boundary condition leads to turnover and NDC. However, for a sufficiently large load and proper circuit conditions (see the appendix and Shaw *et al.*, 1973), these thermally induced NDC points are stable, and the entire $I(V)$ curve can in principle be mapped out (Jackson and Shaw, 1974). Of course, if the load is made sufficiently light,

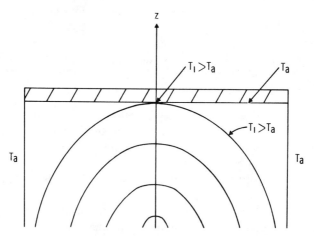

Fig. 5.9. Schematic drawing of isothermal lines near the electrodes using Newton's law of cooling boundary conditions.

THERMALLY INDUCED NDC

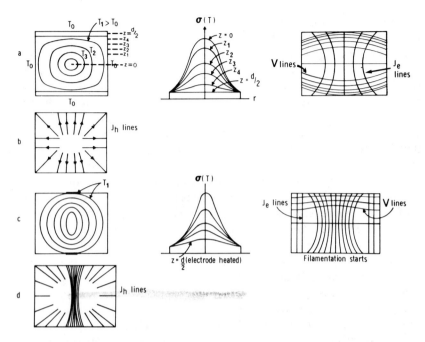

Fig. 5.10. (a) Temperature (left), conductivity (center) and electric current and potential lines (right) for a constant temperature boundary condition. (b) Heat current lines for a constant temperature boundary condition. (c) The same as (a) for a Newton's law of cooling boundary condition. (d) The same as (b) for a Newton's law of cooling boundary condition.

switching will occur from the point I_T, V_T (see Fig. 5.3) to another point at higher current and lower voltage, as determined by the load line. For very lightly loaded systems the final state may not be stable because of thermally induced phase changes in the material. Here melting and shorting or opens may occur, and we describe this event as a thermally induced breakdown or memory phenomenon (Kotz and Shaw, 1984).

A word about stability is in order here. It is readily shown via Maxwell's equations (Shaw *et al.*, 1973; 1979) that bulk NDC points produced electronically are intrinsically unstable against the formation of both inhomogeneous field and current density (Adler *et al.*, 1980; Schöll, 1984) distributions. When the system evolves into these inhomogeneous states, the domain or filamentary characteristics so produced often exhibit regions of NDC (*conductance*) themselves. These NDC points *can be stabilized*. The same situation holds for thermally induced NDC points: any uniform, homogeneous NDC region is unstable; filamentary NDC regions can be stabilized. The latter comprise essentially all the known thermally controlled conditionally unstable systems.

Let's now review some major features of the development so far. The main principle that governs the behavior of electric current lines is that these lines prefer to traverse the easiest path from source to sink. An increase in T is the cause of current concentration; not the result. Therefore, this process is self-stabilizing, not self-accelerating. The major role of the Newton boundary condition is basically that, since the electrode is heated at its center ($r = 0, z = d$), the electrical conductivity here will be increased and lead to current concentration. One effect of the increased conductivity is that the field here will not be as high as it was when the electrode was kept at ambient temperature. Therefore, it we combine a critical electric field concept (in the next section) with a Newton type boundary condition, the boundary condition will inhibit the field at the electrodes from reaching its critical value. Of further consequence is that if the thermal conductivity is temperature-dependent and if $dK/dT > 0$, then the thermal resistance to the flow of heat (at the electrodes) will be reduced as the electrode is heated. This will cause a reduction in the overall temperature level of the system, making local heating more difficult to achieve. The thermal conductivity of the inner regions will increase and lead to a thermally short circuited region. Thus, the heat generated in the inner regions will flow to the boundaries quite readily and the thermal resistances at the boundaries will be more effective in shaping the thermal behavior of the system. In the equivalent RC network model the situation can be described by reducing the R_t's in the middle of the system compared to the R_t's near the electrodes (see Fig. 5.5). At steady state, $R_{t,n} < R_{t,n-1} < R_{t,n-2}$ will be very small compared to $R_{t,1} > R_{t,2} > R_{t,3}$. We can exaggerate the situation by making $R_{t,n} = R_{t,n-1} = R_{t,n-2} = 0$. However, the $R_{t,n}$'s cannot decrease too much, otherwise the temperatures T_n, T_{n-1}, T_{n-2} will decrease because of the loss of heat. This in turn decreases the thermal conductivity, thus increasing $R_{t,n}, R_{t,n-1}$, That is, there is a lower limit for the $R_{t,n}$'s which provides a sufficient T_n level to keep $K(T_n)$ high. The property $dK/dT > 0$ has a stabilizing effect on T. It prevents an excessive increase of T by allowing easier heat flow; when T tends to decrease $K(T)$ will decrease and prevent excessive heat loss.

Since we have been emphasizing thin film configurations, the most important boundary conditions are those at the metal electrodes; the axial boundaries have been fixed at the ambient temperature. It is important to realize, however, that when the geometry becomes that of, e.g., a cylinder, the axial boundary condition also becomes important. An excellent example of its importance is the problem of the Ballast Resistor (Bedeaux et al., 1977a,b; Landauer, 1978). This device is simply a long metallic wire immersed in a gas kept at an externally controlled temperature T_G, and is a useful example of a quasi-one-dimensional system exhibiting a variety of spatial and temporal thermal instabilities. The state of the wire at time t is described by only one variable, $T(x, t)$, the temperature field along the length x. Under the proper set of conditions the system exhibits NNDC type instabilities. The reader is referred to the articles by Bedeaux *et al.* (1977) for further details.

5.3.2 The Effect of Inhomogeneities

The presence of a metallic-like inhomogeneity inside the amorphous semiconducting film (see Fig. 5.4) will influence the thermal and electrical transport processes and alter the conditions required for either turnover or switching. First, the inhomogeneity will cause current crowding. The highly conductive region draws the most current, therefore heat will be generated around the periphery of the inhomogeneity. Further, any sharp corners on the inhomogeneity will create large **F** fields leading to higher current densities via carrier generation and enhanced heating at these points (microplasmas). Although the heating process and concentration of current in the middle part of the film will also continue in the manner discussed above, the existence of a metallic inhomogeneity can affect the course of events in a number of ways. For example:

(1) local melting may occur at points surrounding the inhomogeneities, especially at sharp corners where the current density is high;
(2) the points where the critical field, $\mathbf{F}_c$, is exceeded may differ from the normal configuration at the electrodes. That is, if there were no inhomogeneity, $\mathbf{F}_c$ would be exceeded at $r = 0$, $z = 0, d$; on the electrode-film boundaries. The existence of an inhomogeneity will cause distortion of both the shape of the equipotential lines and their densities.
(3) the thermal conductivity of the inhomogeneity may be important. If we denote the thermal conductivity of the metallic inhomogeneity $K_2(T)$, then a wide range of temperatures will exist for which $dK_2(T)/dT < 0$, whereas for the semiconductor $dK_1(T)/dT > 0$ is typical. Since K_1 increases with T, the generated heat will flow out of the homogeneous material readily and inhibit excessive increases in temperature. With an inhomogeneity present, however, one must consider the system in somewhat more detail.

Let's consider the region in the vicinity of the inhomogeneity. As the input power increases, the local T will increase for two reasons: (1) heat will be produced inside the inhomogeniety; (2) heat will be produced in the region surrounding the inhomogeneity. This will act to raise the T of the inhomogeneity via the thermal boundary conditions. Consider the situation sketched in Fig. 5.11. As the temperature inside the inhomogeneity increases $K_2(T)$

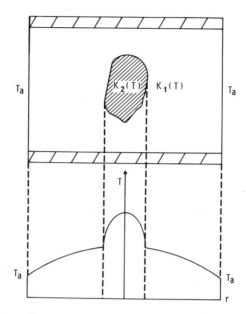

Fig. 5.11. The disk sample (top) containing an inhomogeneity and the associated temperature distribution (bottom) induced in the steady state by Joule heating.

will decrease. The heat produced inside the inhomogeneity will see an increased thermal resistance to its flow out of this region. That is, a larger portion of the heat produced here will stay inside the region, which increases T further. This in turn leads to a further growth in T, which will reach a steady state at a higher level and in a longer time than if the thermal conductivity were independent of T. Further discussion of the role of inhomogeneities, including the case where several are present, is deferred to the next section, after we discuss the concept of a critical electric field.

5.3.3 Critical Electric Field Induced Switching Effects

The presence of a critical electric field, $\mathbf{F}_c$, at which a precipitous increase in conductivity occurs, will have a profound influence on the ability of a specific system to undergo a switching transition (Shaw *et al.*, 1973b). $\mathbf{F}_c$ will play a role for either one or both of the following reasons: (1) it can act to short out the coolest region near the electrodes; (2) it can cause large current densities to flow which produce local heating and melting. As we have discussed, a sudden change in σ due either to $\mathbf{F}_c$ or a critical temperature, T_c (Jackson and Shaw, 1974), will produce filamentary NDC points that can be stabilized by a proper choice of circuit parameters. Systems that exhibit NDC due solely to thermistor type effects can also be stabilized in their NDC regions by an appropriate choice of intrinsic inductance L, load resistance R_L and package capacitance C (Shaw et al., 1973). In the appendix we discuss these important circuit parameters in detail.

When a critical electric field is present we may write for the conductivity

$$\sigma(T) = \begin{cases} \sigma_0 e^{-\Delta E/kT}, & \mathbf{F} < \mathbf{F}_c \\ \sigma_h (\gg \sigma_0 e^{-\Delta E/kT}), & \mathbf{F} \geq \mathbf{F}_c. \end{cases}$$

In any thermistor-type system, as the turnover condition is neared, the field near the electrode becomes large, with the largest field occurring just outside the metallic electrodes near the center of the sample. If $\mathbf{F}$ is made to exceed $\mathbf{F}_c$ at that point, then the region adjacent the electrodes will become highly conducting. In essence this is similar to the sudden motion of a virtual electrode into the material over the warmest region, a shorting out of the highly resistive contact region (Kaplan and Adler, 1972) in this vicinity. This is a self-accelerating process. The expansion of the virtual electrode into the film increases the field just in front of it, which causes further penetration of the virtual electrode (under infinite heat sink boundary conditions, for example). Thus, the device resistance will decrease and the current will crowd into this region and nucleate a filament. When filamentation starts, *less* power will be drawn from the source. But since this power is dissipated in a narrower region, higher local temperatures will readily be produced. (Effects due to the variation of the thermal conductivity in this region will be discussed shortly.) We see that $\mathbf{F}_c$ can lead to the occurrence of a switching event prior to the turnover condition being achieved.

The above critical field effect is electronic in nature, and is invoked as a mechanism by which a switching event can be electronically initiated. When metallic inhomogeneities are present, however, fields can be produced near sharp points that can lead to high local current densities and initiate local melting. Indeed, the two types of effects might work together to produce a switching event. A substantial increase in temperature (or melting) might occur near an inhomogeneity, which would lead to a field redistribution and the subsequent attainment of $\mathbf{F}_c$ near an electrode; switching or breakdown would then occur. Other possible sequences and/or simultaneities are easy to visualize and categorize. The order of occurrence is not critical; a self-accelerating switching event will be induced.

The above considerations will be modified somewhat when the variation of the local power density, $p_e(T)$, with temperature is considered. As T increases, $\sigma(T)$ will increase and $p_e(T)$ could decrease. If $p_e(T)$ decreases at a sufficiently rapid rate as T increases, the steady state condition may be reached in a shorter time period and be at a lower T level. However, we have seen previously that $p_e(T)$ can be affected by things other than the presence of an inhomogeneity ($\mathbf{F}_c$, T_c, $dK_2/dT < 0$, etc.). These features can make $p_e(T)$ *increase* with T. Because of this the projected steady state T distribution may not be reached; a switching event will occur first. For example, a switching event can be launched prior to turnover in a system where $dK_2/dT < 0$ and an $\mathbf{F}_c$ is present.

Many common systems will have not only an $\mathbf{F}_c$, but will also contain many inhomogeneities. It is important, therefore, to consider the affect that multiple metallic inhomogeneities might have on the above conclusions. First, it is clear that current crowding effects will be more pronounced in regions that contain more inhomogeneities per unit volume. Therefore, local heating effects will be prevalent here. Further, at the sharpest corners of the inhomogeneities the fields and current densities will be largest. If critical values are exceeded and melting occurs, the presence of a nearby inhomogeneity will aid in the development of the potential instability. This case can be seen simply by realizing that the presence of nearby inhomogeneities will lead to more current crowding about the specific inhomogeniety we are heating. Once the major hot spot is nucleated, the high conductance region will expand towards neighboring inhomogeneities and raise the local fields, for example, above $\mathbf{F}_c$, thus reducing the time it takes to switch the sample, along with the voltage at switching.

Next, it is very possible that the switching event may also start at regions near the electrode interface where the inhomogeneities are often densest. Further, since $dK_2/dT < 0$, the power generated inside the inhomogeneities will tend to remain there. Although negligible power should develop there because of the high values of $\sigma(T)$, $p_e(T)$ $[=|J_e|^2/\sigma(T)]$ may become large due to substantial increases in current density. The existence of regions with high thermal resistances (metallic inhomogeneities at elevated temperatures) will also narrow the path of the flow of heat current, which will then act to raise the average temperature of the inhomogeneous region.

Finally, the heating of the electrodes (Newton's law of cooling) may be enhanced in regions where $\mathbf{F}_c$ is reached, and this will also tend to accelerate the switching process.

It is useful to conclude this section with a general description of the possible types of thermal behavior that might occur in a system where the electrical conductivity is a function of both temperature and electric field. A qualitative-analytical approach, with support from numerical calculations, leads to the following observations and conclusions.

Consider the nonlinear eigenvalue problem

$$\Delta u = -\lambda f(x, u) \qquad x \in D, \qquad (5.38a)$$

$$\alpha u + \beta \, \partial u/\partial n = 0 \qquad x \in S, \qquad (5.38b)$$

where $D + S$ is the active region, S the boundary surface and Eq. (5.38b) a general boundary condition. λf is the source of u. Some well-known results on the existence, uniqueness and stability of the solutions of Eq. 6.38 are (Joseph, 1965; Joseph and Sparrow, 1970; Keller and Cohen, 1967; Simpson and Cohen, 1970):

(A) If $\partial f/\partial u \equiv f_u \geqq 0$ and $f(u) < F(\equiv$ a function independent of $u)$ for $0 \leqq u \leqq \infty$, then a finite unique and stable solution $u > 0$ exists for all λ in the interval $0 \leqq \lambda \leqq \infty$.

(B) If $f_u > 0$ and unbounded as $u \to \infty$, then a solution $u > 0$ exists only in a range $0 < \lambda < \lambda^*$, where λ^* is a limit determined by $f(u)$, the boundary conditions and geometry of the system. This solution is unique if $f(u)$ is concave $[f_u(u_2) < f_u(u_1)$ for $u_2 > u_1]$ and non-unique if $f(u)$ is convex $[f_u(u_2) > f_u(u_1)]$.

Equation (5.38) is in a form similar to (but more general than) Eq. (5.30) under steady-state conditions:

$$\text{div}[K(T) \text{ grad } T] = -F^2 \sigma(T, F), \qquad (5.39)$$

where T is analogous to u, $\sigma(T, F)$ to $f(u)$ and F^2 to λ. Therefore, the well-known features of Eq. (5.38) may be used to study the solutions of Eq. (5.39). Examples are:

(a) If the conductivity σ has no F dependence, then Eq. (5.39) has the same form as Eq. 5.38; for a conductivity of the form

$$\sigma(T) = \sigma_0 e^{-a/T}$$

(or any other form which remains finite for all T), Eq. (5.39) has a unique solution because $\sigma(T)$ has the property A stated above. Under these conditions a SNDC region can be generated by using Newton's law of cooling as the boundary condition for T. The filamentary solutions can be stabilized in the NDC region using a source with a sufficiently high resistance and the proper local circuit environment (Shaw et al., 1973).

(b) For a conductivity of the form

$$\lim_{T \to \infty} \sigma(T) \sim T^n \qquad (n > 0),$$

which goes to infinity as $T \to \infty$, property B above holds and solutions do not exist beyond a certain local electric field $\mathbf{F}_c$ (analogous to λ_c). Also, for $0 < n < 1$, the existing T solutions are unique and stable; for $n > 1$, the solutions are non-unique and only one (the lowest) is stable.

(c) For an electric field-dependent conductivity $\sigma(T, F)$, the problem is slightly more complicated because two coupled equations must be considered together in considering the existence, uniqueness and stability of the solutions;

$$\text{div}[K(T) \text{ grad } T] = -\sigma(T, F)F^2; \quad (5.39)$$

$$\text{div}[\sigma(T, F) \text{ grad } V] = 0. \quad (5.12)$$

Experimentally, it is known that in materials such as the chalcogenides the electrical conductivity increases with electric field,

$$\frac{d\sigma}{dF} > 0.$$

This property has a limiting effect on the maximum field value, F_{max}, which occurs at the electrode-film interface. As the material is heated, F_{max} steadily increases. However, this increase is decelerated by the increase in conductivity, which tends not to support high F fields. Thus, the right-hand side of Eq. (5.39) remains finite, and property A will hold again, yielding a unique T solution for all possible values of field.

The best fits between numerical and experimental results are obtained with such conductivities. For example, Kaplan and Adler (1972) have obtained a switching phenomenon with a conductivity of the form $\sigma(T, F) = \sigma_0 e^{-a/T} e^{F/F_0}$. Shaw and Subhani (1981) have obtained a better fit by assuming a discontinuous $\sigma(F)$ variation given by

$$\sigma(T, \varepsilon) = \begin{cases} \sigma_a = \sigma_0 \exp\{-(\Delta E - \beta E)/kT - F_0/F\}, & F < F_c \\ \sigma_h \gg \sigma_a, & F > F_c \end{cases} \quad (5.40)$$

Both conductivity functions satisfy the condition $d\sigma/dF > 0$, lead to current filamentation and cause a discontinuous jump from a low conductance to a high conductance state under sufficiently high bias independent of the external circuit. With such conductivity functions, no stabilizable uniform NDC region apparently exists. We have an inherently bistable system independent of the external circuit.

Considering all of these features and the numerical results, electrothermal switching phenomena in semiconductors can be described, in general, as follows.

As the applied voltage is increased, the temperature of the central region grows faster than the regions close to the boundaries. At low heating levels the electric field is not very effective in controlling the conductivity, hence the conductivity can be approximated by $\sigma_0 e^{-a/T}$. Therefore, the conductivity of the central region will be larger than the regions close to the electrodes. This produces a low F in the central region and a high F field near the electrodes. As the voltage is increased further, if σ were not a function of F, very high fields would be produced near the electrodes. Since the heat generation rate is σF^2, increasingly large power will be generated near the electrodes. To keep the power density at the electrode-material interface finite, the conductivity here will increase by some mechanism, which in turn will decrease the local field. This can be accomplished in a number of ways. For example, Newton's law of cooling is a realistic thermal boundary condition. It allows a certain amount of heating at the interface, thus increasing the conductivity and lowering the field. Newton's law of cooling introduces a higher thermal resistance at the electrodes, which elevates the temperature level of the material, thus increasing the conductivity everywhere. However, the electrodes are still cooler than the central region, resulting in higher fields at the electrodes. Therefore, most of the heat generation again occurs near the electrodes. The temperature distribution over the electrode surface will be such that it will be maximum at the center of the electrode if there is no inhomogeneity. However, if a metallic inhomogeneity exists, T will be maximum at the point nearest to the inhomogeneity. At that point, where the low conductivity barrier is reduced, current filamentation starts. This confines the dominant heat generation near the electrodes to these weak points, which reduces the electrical resistance further.

The dependence of σ on F with $d\sigma/dF > 0$ can also lead to a limitation on the power generation near the electrodes. For example, we have pointed out that as the Joule heating increases, the $e^{-a/T}$ dependence produces a higher conductivity in the central regions compared to the regions near the electrodes, leading to low F's in the central region and high F's near the electrodes. The e^{F/F_0} or F_c type dependence starts to be effective after a certain field level, thus increasing the conductivity more at a point on the electrode where F is maximum. In the presence of a metallic inhomogeneity, this point will be nearest the inhomogeneity. Here is where the filamentation begins, leading to higher F fields, which in turn produces a higher local conductance. This process is equivalent to the expansion of the electrode into the material at the point of interest. This is a self-accelerating event, because as the virtual electrode pushes into the material (Kaplan and Adler, 1972) the entire voltage drops across a shorter distance, which means a higher F field, and further expansion of the virtual electrode into the material.

It is clear from what we have discussed so far that the switching process in thin films is a very rich and complex subject, and the presence of inhomogeneities can play a very important role. Indeed, in a recent experimental study, Thoma (1976) provided evidence that bias-induced reversible switching transitions in a wide variety of thin insulating and semi-conducting films between 2 μm and 100 μm thick occurred when a critical amount of power per unit volume was dissipated in the samples. The materials investigated—crystalline and polymeric, as well as amorphous—included ZnS, mica, Al_2O_3, anthracene crystals, Mylar, polystyrole foils, crystalline LiF,

ZnO, CdS, Si and GeAsTl glasses. He concluded that in order to explain this ubiquitous phenomenon, one must assume that many real insulating materials contain defects or inhomogeneities arranged in chain-like patterns which give rise to higher mobility and/or higher carrier concentration paths through the films, a view taken and exploited analytically for inhomogeneous multicomponent chalcogenide films by Popeseu (1975). Under bias, these inhomogeneities lead to very narrow current filaments that extend throughout the thickness of the films. Since the switching effect in *thin* amorphous chalcogenide films has been explored in great detail, we treat it first. We shall see that here a critical electric field is fundamental to the switching process. Further, we shall find that the switching process is *electronically initiated* and *sustained* (Ovshinsky, 1968; Mott, 1969) and that thermal effects are important with regard to: (1) the "delay" time; (2) the capacitive discharge that occurs upon switching; (3) the "forming" and memory process.

5.4 Thin Chalcogenide Films

5.4.1 Introduction

5.4.1.1 Scope of the Problem

As we have already pointed out, the application of sufficiently high electric fields to any material sandwiched between metal contacts almost always results in departures from linearity in the observed current-voltage characteristics (Shaw, 1981). With further increases in bias in the nonlinear regime either a breakdown, switching or oscillatory event eventually occurs (Shaw and Subhani, 1981; Thoma, 1976). Breakdown usually results in local "opens" or "shorts", whereas switching is often involved with local changes in morphology that are not as catastrophic as those that result from a breakdown event; here reversible changes in conductance are induced (Kotz and Shaw, 1984). In many thin films, after a switching event from an "OFF" to an "ON" state occurs, when the ON-state is maintained for a sufficiently long time, a "setting" or memory can occur such that when the bias is reduced the sample will not switch back to the OFF-state until it is subjected to further treatment such as the application of high current pulses.

There are two classes of explanations for the above array of complex phenomena: thermal and electronic. In general, we have stressed that both effects must be considered in any quantitative analysis, and the two can produce a coupled response called "electrothermal". In a discussion of the physical mechanisms involved with a particular specimen, the major parameters controlling its operation must be identified and separated out from the less significant features. In this section we do this for bias-induced switching effects in amorphous chalcogenide films (De Wald et al., 1962; Ovshinsky, 1968; Pearson and Miller, 1969) typically 0.5–10.0 μm thick. It is the purpose of this section to: review the major experimental features of these phenomena; present the results of numerical calculations that model the first-fire event in homogeneous films and compare favorably with experiment; discuss switching in inhomogeneous films that have become so because of the morphological changes induced by prior switching events—a process known as "forming". Here, we suggest that specific inhomogeneous films can show a switching transition initiated by an instability that nucleates at a critical local power density (Thoma, 1976; Shaw, 1979). On the other hand, specific homogeneous films can be induced into a switching event at sufficiently high electric fields (Shaw et al., 1973b), but rather than resulting in an open or short, intermediate (inhomogeneous) states are formed which serve as basis states for subsequent switching events (Kotz and Shaw, 1984). The differences and similarities between virgin and formed films and their electronic behavior will be emphasized throughout the section.

It has been common practice among some investigators to separate switching in thin amorphous chalcogenide films into two classes, threshold and memory (Adler et al., 1978; 1980), according to whether the OFF-state can be resuscitated after the ON-state has been maintained for a given length of time. In fact, in most of these materials a memory effect will occur when the ON-state is held by direct current for sufficiently long times. Specimens called memory switches usually are of relatively high conductivity [e.g., $Ge_{17}Te_{79}Sb_2S_2$ (Buckley and Holmberg, 1975; Kotz and Shaw, 1982), where the room temperature conductivity is about $10^{-5}\ \Omega^{-1}\ cm^{-1}$] and "set" in a matter of milliseconds. Specimens called threshold switches usually are of relatively low conductivity (e.g., $Te_{39}As_{36}Si_{17}Ge_7P_1$ (Petersen and Adler, 1976), where the room temperature conductivity is about $10^{-7}\ \Omega^{-1}\ cm^{-1}$) and require times much longer than a millisecond to set. On some occasions relatively low conductivity films can still be returned to the OFF-state (without additional treatment) by reducing the applied bias after being kept ON for times on the order of 10 hours.

As we shall see in Section (5.5), it is most probably the case that the major difference between memory and threshold switches is the time required to set when excited by pulsed or continuous direct current. However, in what follows we will delineate between the two in deference to common practice. Since our emphasis here is on threshold switching, the memory effect will only be discussed when its understanding will help elucidate the threshold effect. The mechanism for the initiation of the switching event in both cases is the same (Kotz and Shaw, 1982).

5.4.1.2 Switching Parameters

The $I(V)$ characteristics of a typical threshold switch are shown in Fig. 5.12. At low currents a high resistance is observed (approx. $10^7 \, \Omega$); this is called the OFF-state regime. When a threshold voltage, V_T, typically 10–1000 V, is exceeded, the sample switches to a low-resistance operating point on the load line, with the dynamic resistance falling to about 1–100 Ω; this regime is known as the ON-state. As long as a minimum current, I_h, called the holding current, is maintained, the sample remains in the ON-state. However, if the current falls below I_h, the sample either switches back to an OFF-state operating point on the load line or, as shown in the appendix, undergoes relaxation oscillations between the ON and OFF-states, depending on the value of the load resistance (Shaw *et al.*, 1973). Since I_h depends on the circuit conditions, it is better to treat an essentially circuit-independent parameter, the holding voltage, V_h as more fundamental; V_h is usually about 1–2 V.

Voltage-pulse experiments (Buckley and Holmberg, 1975; Petersen and Adler, 1976; Pryor and Henisch, 1972; Shaw *et al.*, 1973b; Kotz and Shaw, 1982) provide a major source of information about threshold switching and

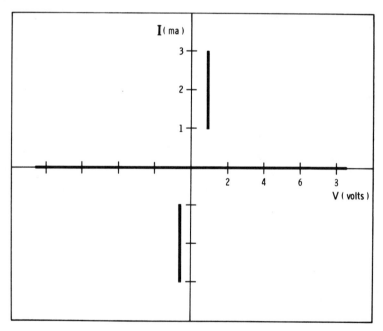

Fig. 5.12. Current as a function of voltage for a 1-μm thick film of amorphous Te$_{39}$As$_{36}$Si$_{17}$Ge$_7$P sandwiched between Mo electrodes. This is a trace from a Tektronix curve-tracer oscilloscope, which implies a 60 hz ac signal (Petersen and Adler, 1976).

THIN CHALCOGENIDE FILMS

lead to the introduction of other parameters of interest. After a voltage pulse is applied, a delay time, t_d, typically less than 10 μs, elapses before the onset of switching. The switching time, t_0, has proven to be faster than any means found of measuring it, but is known to be less than 1.5×10^{-10} s. It has also been convenient to define a pulse interruption time (Pryor and Henisch, 1972), t_s, as the time between removal of V_h and application of an ensuing pulse with $V \geq V_h$. After V_h is removed for a time t_{sm}, the maximum benign interruption time, only V_h is required to restore the ON-state (t_{sm} is typically about 250 ns, but varies with the original ON-state operating point). For longer values of t_s, the voltage required to reswitch the sample approaches the original threshold, V_T; the latter being completely restored in a recovery time, t_r. Figure 5.13 shows some of these parameters for two pulses, each of width t_p.

5.4.1.3 OFF-State Characteristics of a Homogeneous Film

For conciseness, we will emphasize a typical switching material, $Te_{39}As_{36}$-$Si_{17}Ge_7P_1$, which has been perhaps the most thoroughly studied sample (Petersen and Adler, 1976). At low fields, less than about 10^3 V/cm, the $I(V)$ characteristics are linear and the resistivity varies with temperature as $\rho(T) = 5 \times 10^3 \exp(0.5[eV]/kT) \, \Omega \, \text{cm}$. This yields a room temperature resistivity of the order of $10^7 \, \Omega$ cm. The optical energy gap is approximately 1.1 eV, or about twice the thermal activation energy, a result typical of amorphous chalcogenide semiconductors, which are usually p-type in nature (see, e.g., Tauc, 1974).

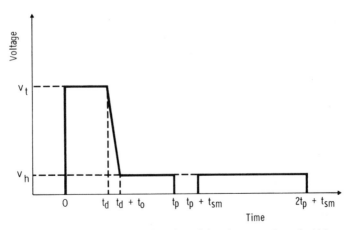

Fig. 5.13. Voltage across a sample as a function of time for two pulses of width t_p separated in time by t_{sm}. All parameters are defined in the text.

When Mo electrodes are put in contact with the chalcogenide, the bands in the latter bend upward by approximately 0.15 eV. Under extremely low applied bias the characteristics are linear, but as the applied bias is increased into the field range 10^3–10^5 V cm, the Schottky barrier manifests itself and the current becomes contact limited; it is controlled by various tunneling contributions from field and thermionic-field emission (Shaw, 1981). However, in the 10^5 V cm range a "high field" characteristic appears in which the conduction is bulk-limited and of the form $\sigma = A \exp(F/F_a)$, where F is the electric field and A and F_a are constants. In fact, in the field region above about 10^4 V cm, we shall show that the OFF-state $I(V)$ characteristics of a virgin, homogeneous film can be fit rather well by using an expression for the conductivity given by $\sigma = \sigma_0 \exp[-(\Delta E - \beta F)/kT - F_0/F]$ where β is a constant representing a field-dependent decrease in activation energy, and F_0 is a constant associated with carrier multiplication. σ_0 is the conductivity as $T \to \infty$ in the absence of a field effect (Reinhard et al., 1973).

5.4.1.4 The Switching Transition

When thin amorphous chalcogenide films containing tellurium are homogeneous, uniform and virgin, under sufficiently short pulse conditions, the initial (first-fire) switching event is a completely electronic event (Shaw et al., 1973b; Buckley and Holmberg, 1975). Although the first-fire event is classified as a switching phenomenon, it is sometimes also useful to treat it as a breakdown-type process, since it is often the case that the voltage at threshold is substantially lower after the first few fire events. Furthermore, after further firings the voltage at threshold often continues to drop slowly until, after a sufficient number of firings, in many case it stabilizes at a "running" value. (We will discuss this in more detail in Section (5.5).

The drop in threshold voltage upon firing is associated in most cases with a forming process (Shaw et al., 1973c; Kotz and Shaw, 1984) wherein either crystalline (Bosnell and Thomas, 1972), morphological (Allinson et al., 1979) or amorphous imperfections are produced locally. In a formed or inhomogeneous sample the instability that develops at a critical value of local power density has some features that are somewhat different from the first-fire event in homogeneous films. In the latter case, for short enough pulses a critical electric field is reached isothermally over the entire sample, independent of the thickness (Buckley and Holmberg, 1975). As we shall discuss, we expect that this field strips trapped carriers off local defects, and then the significantly increased Joule heating often dominated by the subsequent capacitive discharge (Kotz and Shaw, 1982) can cause morphological changes at the weakest point in the film. Further, in order to obtain a switchback effect (voltage drop), we must do more than just create excess carriers.

We must also have their presence alter some transport property of the film, such as the mobility. We shall discuss possible models for these in Section (5.6) (Adler *et al.*, 1978 and 1980).

For pulse widths greater than about 10^{-9} s the common delay-time mode is observed in formed samples of all thicknesses. (To our knowledge, single-shot data showing the existence of a delay-time event occurring during first-fire in a thin homogeneous sample is not available.) There is abundant experimental evidence (Thoma, 1976; Balberg, 1970; Reinhard, 1977) that the delay-time mode produces a switching event at a critical local power input. It was first shown (Balberg, 1970) that the intimate double inverted pulses produced identical delay times and later (Reinhard, 1977) that a critical rms voltage switched samples after identical times in a study of their response to pulse burst waveforms. These results are evidence that t_d is associated with the time it takes for a local hot spot to grow and rearrange the field in the sample (Shaw *et al.*, 1973c; Homma, 1971).

As we have emphasized above, voltage pulse measurements have been very useful in elucidating several important aspects of t_d. In fact, for sufficiently short pulses (Shaw *et al.*, 1973b) t_d can be made comparable to the time it takes for the voltage across the sample to collapse from V_T to V_h, the switching time, t_0. Whereas t_d is thermal in origin, t_0 is due to an electronic process in both virgin and formed films. Models for both will be discussed shortly.

5.4.1.5 ON-State Characteristics

The forming process produces a local inhomogeneous region typically 1–5 μm in diameter. It is through this relatively high conductance region that the major portion of the current flows when the sample is in the ON-state. This filamentary current-carrying path has a radius r_f, and its major features have recently been described for samples that did not exhibit forming (Petersen and Adler, 1976) (the first-fire and running threshold voltage were the same). In the studies of Petersen and Adler the current dependence of r_f was experimentally exposed by several independent methods. The results are shown in Fig. 5.14. First, a study of velocity saturation in crystalline-Si/amorphous-chalcogenide heterojunctions provided a means of determining the current density in the chalcogenide in the 29 mA range (prior to avalanche breakdown in the Si depletion region). It was found that the area of the current filament, A_f, increased more or less proportionally to the increase in current, indicating that the current density remains constant in the filament over a wide range of current. Second, the transient-ON-$I(V)$ characteristic (TONC) technique was used to analyze the ON-state behavior (see Fig. 5.15). It was found that the TONC was stable for only about 50 ns,

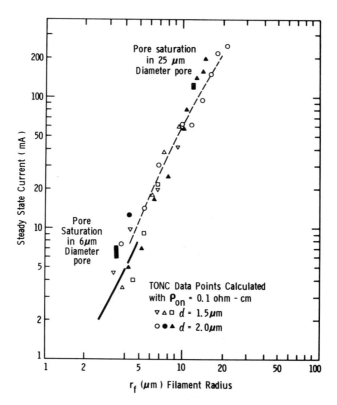

Fig. 5.14. Filament radius as a function of steady state current determined by four methods. The solid line represents the results of the velocity-saturation analysis, the data points are the TONC results, and the dashed line is calculated from Shank's (1970) carbon/chalcogenide/carbon results. Two pore saturation points are also indicated (Petersen and Adler, 1976).

after which the response gradually relaxed to the steady ON-state $I(V)$ characteristic. Therefore, for TONC pulses less than 50 ns we expect that the area of the current filament remains the same as in the steady state; the shape of the TONC should then depend upon the value of the operating steady ON-state current for which a particular TONC is taken. This is in fact the case.

In general we expect three contributions to the voltage drop across the sample in the ON-state: the resistance of the ON-state material; the contact resistance, R_c; the interfacial barrier; V_B. The TONC curves should obey

$$V_{\text{TONC}} = V_B + I\left[R_c + \frac{\rho_{\text{on}}d}{A_f(I_{\text{dc}})}\right] \tag{5.41}$$

THIN CHALCOGENIDE FILMS

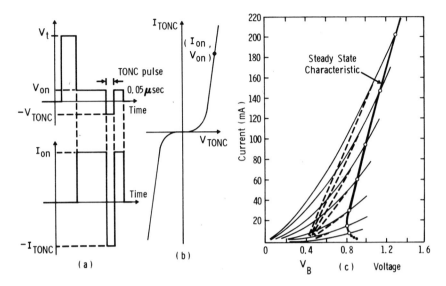

Fig. 5.15. (a) Transient ON-state characteristic (TONC) measurements. (b) A typical result. (c) Different Mo/amorphous/Mo sample TONC curves taken from different I_{on}, V_{on} points. The sample is 50 μm in diameter (Petersen and Adler, 1976).

where ρ_{on} is the ON-state resistivity and $A_f(I_{dc})$ the area of the current filament at the steady-state operating point. Extrapolation of the sub-50 ns TONC curves should yield the same value for V_B, and this value should be the same as the metal/amorphous-chalcogenide barrier measured by other means. The agreement is good. The TONC slopes then determine the variation of A_f with steady-state current.

Further, the steady ON-state voltage is

$$V_{dc} = V_B + IR_c + \rho_{on} J(I) d. \tag{5.42}$$

If, as expected, J is independent of I, extrapolation of the steady ON-state characteristics to $I = 0$ will yield an effective barrier voltage

$$V_{B(eff)} = V_B + \rho_{on} J d \tag{5.43}$$

and this should vary linearly with the thickness of the chalcogenide film. Experiments show this correlation rather well. Furthermore, extrapolation of $V_{B(eff)}$ to $I = 0$ should yield V_B. Again, this is the experimentally observed situation. These results yield $\rho_{on} \cong 0.07 \, \Omega$ cm.

Shanks' results (1970) for the ON-state $I(V)$ characteristics of a formed chalcogenide film having pyrolytic graphite electrodes were also used to determine the area of the current filament as a function of current. The results are shown as the dotted line in Fig. (5.14).

Analysis of the gain observed in an N(ON-state) PN amorphous-crystalline heterojunction transistor (Petersen et al., 1976) as a function of the crystalline-Si base doping concentration showed that the free-carrier concentration in the ON-state is of the order of 10^{19} cm^{-3}. This implies that the ON-state carrier mobility is about 10 cm^2/V s.

The total of the cathode and anode interface barriers for MO/amorphous/MO samples is 0.4 eV. If this is distributed evenly between cathode and anode, and if in the ON-state fields above 10^5 V cm must be maintained near both electrodes, the band bending will then extend about 30–70 Å into the amorphous material. This is sufficiently narrow such that it is possible that the ON-state can be maintained by either strong-field emission or thermionic-field-emission tunneling through the electrode barriers. However, if the barriers are asymmetric the depletion regions can be larger in extent, and tunneling processes become less likely. In either event, the switching process can be sustained by a double injection process, as pointed out by Mott (1969). Alternatively, the ON-state can be maintained from carrier generation in the high-field regions themselves. Since the potential drop in these regions is less than E_g, such generation would have to be from localized states rather than from across-the-gap excitation. In either event, it is likely that both electrons and holes contribute to the ON-state current, just as they do in the OFF-state. However, just as holes predominate in the OFF-state, there is evidence that electrons predominate in the ON-state (Petersen et al., 1976).

5.4.1.6 Recovery Properties

When the current is reduced below I_h, the sample switches back to the OFF-state. One possible mechanism for the initiation of this transition suggests that there might exist a minimum r_f for which radial diffusion would break the filament. This would set an absolute minimum value for the current that can be maintained in the ON-state, I_{hm}. However, observation of I_{hm} is normally difficult to achieve because of the reactive components in the circuit. If we define I_h as that current below which circuit controlled relaxation oscillations occur (Shaw et al., 1973), then for most sample configurations there will always be a range of currents between I_h and I_{hm} that are unstable against relaxation oscillations. The package capacitance, C, and intrinsic plus package inductance, L, will always produce $I_h > I_{hm}$. On the other hand, if an intimate double-pulse technique is employed (Hughes et al., 1975), where the sample is forced to remain in the ON-state after switching by first rapidly reducing the applied bias, then by minimizing C and maximizing the load resistor R_L, values of I_h as low as 10 μÅ can be observed. (These are currents that would produce relaxation oscillations were the ON-state not "held" by the second pulse.) For current densities in the filament in the range of 10^3 Å/cm², such low values of current imply that filament radii in the 0.5 μm regime can be stabilized. In this case, V_h is rather high since the ON-state (filamentary) characteristic itself exhibits a long, stabilizable NDC region for these currents. The fields are therefore sufficiently high so that the ON-state is maintained in an extremely narrow filament; as $V_h \rightarrow V_T$, r_f approaches its minimum value. There is evidence that the minimum radius of the current filament may, in fact, be in the fractional-micrometer regime; thus there is a possibility that I_{hm} exists. However, for essentially all circumstances where the battery voltage is kept constant after switching occurs, I_h should be treated as a completely circuit-controlled parameter.

Once the voltage across the sample is removed, the recovery curve can be studied. As shown in Fig. 5.16, the recovery process depends upon the steady-state operating point. One explanation of the data is that after the voltage is removed, the field at the anode adjusts almost instantaneously but the cathode field decays slowly (Frye et al., 1980), maintaining carrier generation or tunneling near the contact. Since the applied voltage is now zero, a

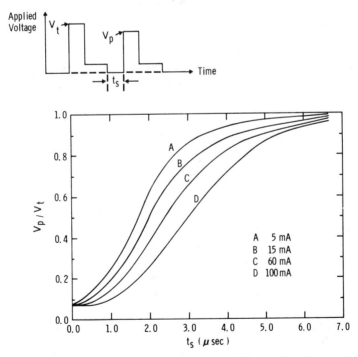

Fig. 5.16. Recovery of threshold voltage as a function of interruption time, t_s, for several values of ON-state current. The double pulse sketched on top is as shown in Fig. 5.13 but with $t_s > t_{sm}$ so that $V_p > V_h$ (Petersen and Adler, 1976).

counterfield will be built up near the anode within a dielectric relaxation time; this explains the symmetry of the TONC results shown in Fig. 5.15b. The limiting feature of the recovery process is then the ambipolar diffusion of carriers radially out of the conducting filament. As the diffusion proceeds, the radius of the filament decreases. As long as any filament remains, only V_h is required to resuscitate the ON-state. However, after a time which depends on the original r_f (and thus I_{on}), the filament shrinks to zero radius and the contact barriers begin to decay (this is the origin of the parameter t_{sm} discussed in Section 5.4.1.2. Once the equilibrium contact barriers are re-established, the original V_T is completely restored.

An alternative explanation of the recovery data is that t_{sm} is the time it takes for the contact barriers to widen to lengths insufficient to sustain the motion of large of carriers through them. At this point a sharp increase in resistance occurs as the contact-to-contact path is broken; the remainder of the recovery process involves diffusion of heat out of the filament.

5.4.2 Numerical Calculations of the First-Fire Event in Homogeneous Thin Films

5.4.2.1 Introduction

Shaw et al. (1973b) and Buckley and Holmberg (1975) have presented experimental data on the first-fire event in both virgin threshold and memory material, with the latter work being more extensive in that, among other things, a range of samples thicknesses were explored. Both sets of experiments showed that for short enough pulses a critical electric field exists that initiates a breakdown-like switching process which, in these experiments, leads to forming and a substantial drop in threshold voltage upon subsequent firings. Although the workers cited above did not explore the formed filamentary region via scanning electron microscopy, others (see, e.g., Bosnell and Thomas, 1972) have done so in great detail, and different types of inhomogeneities have been shown to be present. We will discuss one type shortly.

In this section we present the results of extensive numerical electrothermal calculations for both threshold and memory type material, and compare our results with the experiments discussed above. The details of the calculations can be found in Subhani (1977). We study both time-independent and time-dependent processes, incorporating a critical electric field into the model in order to obtain agreement with experiment. It is important to note in what follows that the only difference between the two types of samples involves the setting or lock-on in the memory ON-state. The mechanism for the initiation of the ON-state (prior to memory lock-on if it occurs) is the same for both.

5.4.2.2 Memory-Type Samples-Calculations of the Steady State

We first solve an electrothermal model for the steady state $I(V)$ and $T(r)$ characteristics of a virgin memory sample, and then compare our calculated $I(V)$ characteristics with the experimental results of Buckley and Holmberg (1975). By electrothermal we mean solutions of the heat balance equation explicitly including nonohmic contributions such as a field-dependent conductivity and/or a critical electric field.

The geometry of the sample under analysis is that of a homogeneous thin circular disc of radius R and thickness d sandwiched between metallic electrodes (see inset, Fig. 5.17). Because switching occurs primarily along a central axial path, the temperature far from the center of the sample remains at ambient. Hence, rather than apply the boundary condition $(\partial T/\partial r)|_{r=R} = 0$, the radial surface of the sample is kept at a temperature $T(r = R) = T_a$, where T_a is the ambient temperature. However, the axial surfaces of the sample (the amorphous/electrode interfaces) have finite heat losses and are modeled using the electrode boundary condition previously described,

$$K_\alpha \operatorname{grad} T \cdot \hat{n} = -G_\alpha(T - T_a), \qquad (5.32)$$

Newton's law of cooling, where $\alpha = e$ for the electrodes and $\alpha = a$ for the chalcogenide material; $\hat{n}$ is a unit vector normal to the boundary. Here both G_e and K_e are taken as finite and independent of T (G is the Newton coefficient). For simplicity K_a is also taken as a constant, although it is a slowly increasing function of T. The electrical conductivity of the amorphous material is taken as thermally activated and of the form

$$\sigma = \sigma_0 \exp\{-(\Delta E - \beta F)/kT - F_0/F\},$$

where the term in the parenthesis represents a field-induced decrease of thermal activation energy and the last term in the bracket represents carrier multiplication effects (in the actual numerical calculation, this term is written as $F_0/(F + F')$ in order to yield a finite conductivity at zero field).

For the steady-state calculations the inhomogeneous elliptic partial differential equation of heat conduction [(Eqs. 5.31) with $\partial T/\partial t = 0$],

$$K\nabla^2 T + J_e^2/\sigma = 0, \qquad (5.39)$$

together with the coupled electrical and thermal processes in the sample are solved on a grid in the finite-difference approximation. It is important to appreciate that for well heat sunk electrodes the solution of this equation for a temperature-independent K cannot successfully account for the observed virgin $I(V)$ characteristics of either memory or threshold-type material unless a critical electric field, F_c, at which a precipitous increase in conductivity

Table 5.1. Parameters Used for Memory-Type Samples

$F_0 = 7 \times 10^3$ V/cm
$F_c = 3.1 \times 10^5$ V/cm
$\Delta E = 0.43$ eV
$\beta = 1.5 \times 10^{-7}$ e cm
$\sigma(T_a, =0) = 1.1 \times 10^{-5}$ $(\Omega\text{ cm})^{-1}$
$T_a = 297$ K
$K_{a,\text{eff}} = 3.0$ mW/°C cm
$G_e/K_e = 3.5 \times 10^4$ cm^{-1}

occurs, is included in the calculations. The OFF-state characteristics used for memory material are given in Table 5.1. To compare with experiment, calculations are performed for four cases: 20 μm and 40 μm pore diameter; 1.40 μm and 2.55 μm thickness.

Figure 5.17 shows a comparison of the numerical results with the experimental values reported by Buckley and Holmberg (1975). The agreement is quite good. We determined steady state values using a constant-current

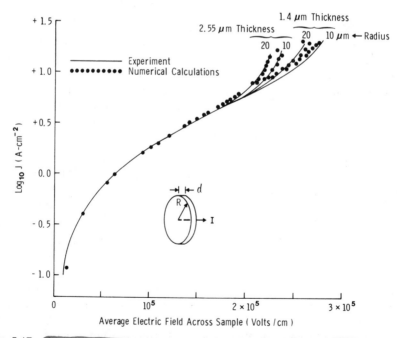

Fig. 5.17. Comparison of the experimental and calculated dc. OFF-state characteristics for a memory type sample whose parameters are given in Table 5.1. Four different sizes are modeled. The insert shows the geometry of the sample (Shaw and Subhani, 1981).

source. Starting at low current density in the OFF-state, the current was incremented slowly until the sample underwent a large change in conductance. The steady-state voltage was recorded for each current density. Figure 5.18 shows the temperature as a function of radial position on a plane through the center of the memory sample just after breakdown, where the steady ON-state current is equivalent to that produced with a load resistor, $R_L = 10$ kΩ. Note that: (1) in Fig. 5.17, the $J(\bar{F})$ characteristics exhibit a slight region of thermistor type SNDC before the onset of switching; (2) in Fig. 5.18 the temperature distribution and current density define a sharp filamentary conducting path after switching.

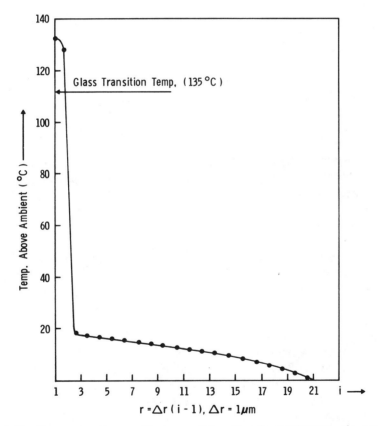

Fig. 5.18. Temperature above ambient vs. radial position for the ON-state of a memory switch. Using the experimental fact that the microcrystalline ON-state filament is about 2–4 μm in diameter, and the glass transition temperature in the material is about 135°C, we inserted the filament diameter and found that the best fit to the $I(V)$ characteristic was obtained for $K_{on}/K_{off} = 10^2$. We also used $\sigma_{on} = 10\ \Omega^{-1}\,\text{cm}^{-1}$. The sample is 2.55 μm thick and 40 μm in diameter. $R_L = 10$ K Ω (Shaw and Subhani, 1981).

Prior to switching, the maximum temperature in the OFF-state is calculated to be about 20°C above ambient. Just after switching, the maximum temperature at the center of the filament is 126°C above ambient, which is above the glass transition temperature for this material. However, to develop a crystalline filament substantially higher temperatures must be reached. In the next section we show that much of the energy required to do this comes from the capacitive discharge induced during the switching transition (Shaw et al., 1973; Kotz and Shaw, 1984). The microcrystalline filament eventually formed is approximately 2–4 μm diameter, in agreement with experimental observations. This demonstrates clearly that electrothermal numerical calculations, modified with a critical electric field, can quantitatively mimic the breakdown-type switching characteristics observed in *memory-type* chalcogenide films.

In general, our calculations yield for virgin memory-type samples:

(1) The calculated and experimental $I(V)$ values are coincident at low fields for the range of film thicknesses and radii investigated. This indicates that a uniform field and current density distribution is present at low fields and thermal effects are unimportant here.

(2) The $I(V)$ characteristics diverge from a common line near the threshold voltage, indicating that the electric field or the current density, or both, become nonuniform within the sample because of local heating.

(3) As the film thickness increases, the characteristics diverge from the common curve at lower fields and current densities, again because of local heating.

(4) As the sample diameter increases, the breakdown voltage decreases.

(5) The current density at breakdown is unaffected by variations in diameter for a given sample thickness (in the range investigated).

(6) As the film thickness decreases, the effects of the diameter variations are sharply diminished, thus making the $I(V)$ characteristics less sensitive to diameter for very thin samples.

(7) For all the memory samples under investigation there is a tendency for the OFF-state characteristic to bend back upon itself near the onset of switching (thermistor-type behavior). The resulting effect is to decrease the average applied field because of the NDC region.

(8) Complete thermistor-type behavior (see Fig. 5.19) can be observed by decreasing the thermal conductivity of the electrodes or eliminating F_c. When G_e/K_e is increased, thermistor behavior with a turnover voltage either above or below V_T can be induced. Here, in many cases F_c is never reached in the sample. For the case where F_c is removed from the calculations, a turnover voltage without breakdown occurs;

THIN CHALCOGENIDE FILMS

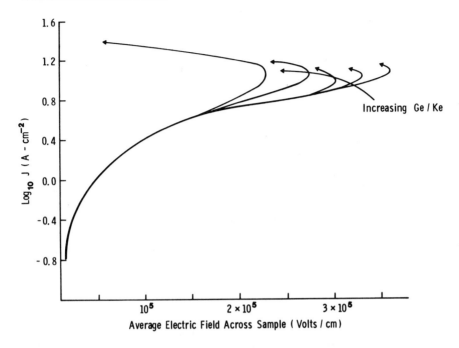

Fig. 5.19. Calculated prethreshold $I(V)$ characteristics of a typical memory sample having a 20 μm radius and 1.40 μm thickness as a function of G_e/K_e, the heat conducting properties of the electrodes (Shaw and Subhani, 1981).

the turnover voltage is much higher then V_T. (It is important to re-emphasize here that for a sufficiently light load thermistor characteristics produce a thermal runaway event.)

(9) The current density and temperature distributions define sharp filamentary conducting paths connecting the electrodes for currents above threshold.

Our calculations reveal the fact that for any one sample thickness, the switching current density is independent of radius. This is also evident from a current-density vs. radius plot in the immediate pre-switching region, where prethreshold heating is observed. In other words, any divergence from the common curve at high fields (near switching) is due to a nonuniform field distribution along the axis of the sample.

We can also expect a somewhat radially nonuniform current distribution as a result of non-metallic inhomogeneities (film imperfections). We have verified this expectation by simulating an inhomogeneous conductivity model wherein we allow for small conductivity perturbations that model imperfections having the same value of thermal conductivity as the amorphous material. Relatively small changes in the virgin $I(V)$ characteristics are observed for conductivity variations across the sample of up to 30%. The general shape of the $I(V)$ curve shown in Fig. 5.17 is maintained.

5.4.2.3 Threshold-Type Samples–Calculations of the Steady State

Now that it is clear that the OFF-state conditions of a memory sample can be modeled, we turn to the threshold case (Shaw *et al.*, 1973b), making use of those ON-state parameters which have emerged from the best fit for the memory behavior. We assume that the thermal properties of the memory and the threshold material are the same, and in our calculations use the same ratio of ON to OFF-state thermal conductivity for the threshold sample as for the memory sample. All other parameters are obtained directly from observations on the threshold material. The parameters are given in Table 5.2.

The major features of the $I(V)$ and $T(r)$ calculations shown in Figs. 5.20 and 5.21 are similar to that of memory-type samples. However, note that in Fig. 5.20 the departure from the common curve at high fields is very small compared to that of the memory samples. Thus, according to the calculation, there should be no phase change induced in threshold-type samples. Forming should be absent and the switching initiation and maintenance should be electronic processes with only minor thermal overtones. Although this is sometimes the case, many experiments, however, produce different results. First, scanning electron microscopy reveals the presence of both crystalline and morphological imperfections (Bosnell and Thomas, 1977; Allinson *et al.*, 1979) in formed films and these are the most common films encountered in practice. Hence we suggest that F_c causes a switching event that produces high temperatures often because of the capacitive discharge in both memory and threshold samples, and also because of operation at high ON-state currents (Kotz and Shaw, 1984).

Table 5.2. Parameters Used for Threshold-Type Samples

$F_0 = 7 \times 10^3$ V/cm
$F_c = 7.0 \times 10^5$ V/cm
$\Delta E = 0.55$ eV
$\beta = 1.5 \times 10^{-7}$ e cm
$\sigma(T_a, =0) = 2 \times 10^{-8}$ $(\Omega$ cm$)^{-1}$
$T_a = 297$ K
$K_{a,\text{eff}} = 3.0$ mW/°C cm
$G_e/K_e = 3.5 \times 10^4$ cm^{-1}

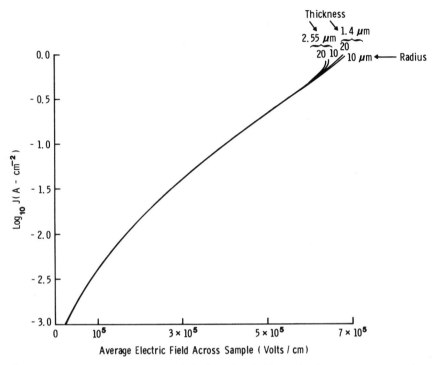

Fig. 5.20. OFF-state $J(V)$, characteristics for a threshold sample whose parameters are given in Table 5.2. Four different sizes are modeled (Shaw and Subhani, 1981).

In order to explain the divergence from the common curve at high fields for a given (memory or threshold) film thickness with different diameters, we study the temperature profile of the sample in the OFF-state. Investigation reveals that for low fields the power input is small, causing negligible heating effects, and the curves are coincident for all geometries. As the current density increases, heating effects are observed if the power input approaches the power dissipation capacity of the sample. Generally, the diameter of the sample is much larger than its thickness and heat is dissipated primarily along the axial direction. Thicker films will develop higher internal temperatures than thinner films under the same conditions because the heat transfer is limited primarily by the low thermal conductivity of the material. Furthermore, the conductivity expression descriptive of the material is a temperature-activated type; a small change in temperature will result in a comparatively large change in conductivity. Hence, the temperature gradient will redistribute the applied voltage across the colder regions of the film, causing an axially nonuniform field distribution. Therefore, it is reemphasized that aside from electronic contact effects, the highest fields will occur

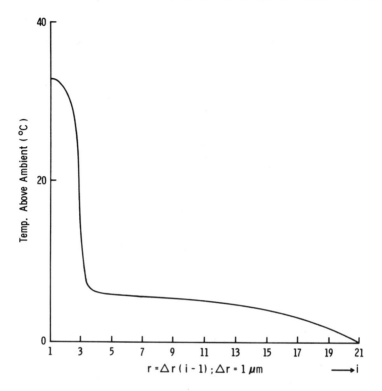

Fig. 5.21. Temperature above ambient vs. radial position for an electronic model of the ON-state of a threshold switch. The sample is 2.55 μm thick and 40 μm in diameter. (For a 1.40 μm thick sample the maximum temperature is 18°C above ambient.) $R_L = 10$ kΩ. No overvoltage is applied. These results assume that the carriers induced by F_c produce no additional joule heating (Shaw and Subhani, 1981).

next to the electrodes, where the film is coolest. When the local field exceeds F_c, the sample will switch. Our calculations indicate that for $R \gg d$, radial heat transfer is small. However, it is not negligible for $R \cong d$ (thick films or small diameters) because the ratio of diameter to thickness is reduced and the relative contribution of the radial heat transfer is increased. This can result in a lower internal temperature rise and higher calculated average breakdown field.

The experimental results as well as the calculations provide good evidence of internal heating in the immediate preswitching region for memory material. A sample will usually undergo breakdown at higher average fields and higher average current densities if heating effects are reduced by changing the sample geometry or material composition (conductivity) for a given set of thermal boundary conditions.

THIN CHALCOGENIDE FILMS   401

In comparing the memory-type virgin $I(V)$ characteristics with the threshold-type characteristics, we see that the threshold sample has: (1) a lower OFF-state conductance; (2) a higher breakdown voltage; (3) almost no departure from the "common" curve; (4) essentially the same breakdown voltage for different diameters. We therefore reemphasize that prethreshold heating in virgin threshold-type samples is not important in producing the breakdown-type switching event. Rather, it is the critical field that initiates the switching event and the ensuing capacitive discharge produces changes in the nature of the material and a formed filamentary region suitable for reversible switching events upon subsequent firings. We suggest how this might happen in Section 5.4.3.

5.4.2.4 Calculations of Time-Dependent Processes

Although to our knowledge no direct data exists on delay-time effects involved in switching events in virgin samples, we calculate the average field at breakdown as a function of pulse width, t_p, which can be compared with experiments of this type (Shaw et al., 1973b; Buckley and Holmberg, 1975). To do this, we solve the time-dependent heat equation [Eq. (5.31)] with $\mathbf{J}_e = \sigma \mathbf{F}$

$$c\rho_a \frac{\partial T}{\partial t} = K\nabla^2 T + \sigma F^2, \tag{5.44}$$

subject to the boundary conditions previously described. The results of the calculations for memory-type samples are shown in Fig. 5.22. Pulse widths in the range $2 \times 10^{-9} \leq t_p \leq 10^{-4}$ s were investigated. Comparison of Fig. 5.22 with Fig. 5.23 shows that good qualitative agreement exists between experiment and the numerical calculations. We see that the threshold voltage saturates for $t_p \leq 10^{-6}$ s and $t \geq 10^{-5}$ s, in approximate agreement with the data. However, the difference in the average fields at which the long and short pulse results saturate, which we call ΔF_T, is generally not as great in the numerical calculations as it is in the experimental data. Furthermore, the experimental value of F_c is about 20% higher than the value required to obtain the precise $I(V)$ fit in the memory material shown in Fig. 5.17. The closeness of F_c in these two cases is, in fact, evidence that our model applies rather well to this memory-type virgin material. Finally, the typical t_d's of about 5 μs predicted from the calculations for virgin samples are sufficiently close to those observed experimentally in inhomogeneous "running" samples to support a model where t_d is thermally induced in formed memory-type samples for sufficiently long pulses. In this standard model, a "hot spot" nucleates in the center or high conductance region of the sample and

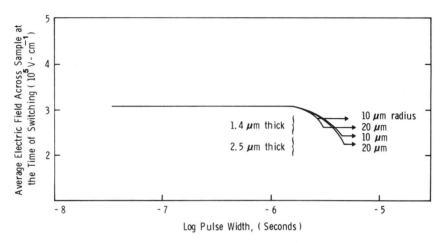

Fig. 5.22. Calculated variation of the average switching field vs. long pulse width (seconds) of a memory-type sample modeled after that producing the data shown in Fig. 5.17. The arrowheads denote the V_t was independent of t_p in both directions for all larger and smaller values of t_p (Shaw and Subhani, 1981).

the conductivity there increases, thereby reducing the voltage in the central region and increasing it near the electrodes where F_c is eventually reached and switching occurs. This model is basically the one we outlined in Section 5.3.3.

The agreement between experiment and numerical calculations is not nearly as good for the threshold-type virgin material (Shaw and Subbani, 1981) [here the data are sparse since only a single thickness was studied (Shaw et al., 1973b)]. The experiments revealed no clear cut saturation for $t_p \leq 10^{-7}$ s. Furthermore, the experimental value of ΔF_T is substantially larger than the predicted value. Finally, the predicted t_d is less than 10^6 s, a value much below those experimentally observed in inhomogeneous samples, which is of the same order of magnitude as for formed memory-type samples, less than about 10^5 s. This is an important point. Experimental values of t_d are typically the same for both formed (inhomogeneous) threshold and memory-type material. The numerical calculations for virgin samples, however, show that t_d should be about an order of magnitude longer in memory-type material. This result is in harmony with the switching model we shall discuss in the next section, and is based on the precept that F_c initiates a switching event in both types of materials. This event causes forming in both, and the mechanism for the switching effect observed in subsequent firings is the same in both—an electronic instability that is thermally modified and electronically sustained.

One other important point, most clearly seen with reference to Fig. 5.23, is that as t_p decreases below about 10^{-7} s, the rate of rise of F_T first tends to

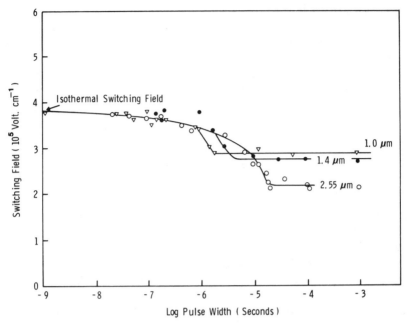

Fig. 5.23. Variation of the average switching field with voltage pulse width for virgin samples of $Ge_{17}Te_{79}Sb_2S_2$ having a 20 μm pore diameter and three different thicknesses. Note that for the shortest pulse the switching field asymptotically approaches the same value independent of sample thickness (Buckley and Holmberg, 1975).

diminish and then eventually saturates. These data are contrary to the behavior expected from a model of the first-fire event based solely on heating with weakly heat sunk electrodes. For long values of t_p ($\geq 10^{-5}$ s), F_T decreases with increasing film thickness and is almost independent of t_p for any one thickness. Furthermore, as previously discussed, the axial nonuniformity that causes F_T to diverge with thickness is consistent with our model over the range of studied t_p's. However, F_T is independent of film thickness for short t_p's. Calculated values deviate slightly in the sense that the increase of F_T with thickness disappears as t_p decreases, whereas a small slope is seen in the experimental data. This may be due to the fact that in our model the voltage drop near the electrodes is symmetric, which is probably not the case for actual samples. However, the fact that F_T is independent of film thickness in this time regime suggests that the field is only slightly nonuniform along the axial direction and the breakdown event is a bulk effect; i.e., the threshold voltage equals $F_c d$ for sufficiently short pulses.

We have seen that the first-fire event in homogeneous films can be interpreted as a switching event induced by a critical electric field. In the next section we discuss how this event might lead to forming and how formed samples might act as reversible switches. In what follows we make use of the several experimental observations showing that differences exist between virgin and formed films; e.g., virgin films show a short-pulse critical field and a dc critical voltage—formed films seem to show a critical local power density; virgin films show large prethreshold currents for short pulses—formed films only do so when the temperature is lowered substantially.

5.4.3 Electrothermal Switching Mechanisms in Formed Chalcogenide Films

5.4.3.1 Introduction

In Jackson and Shaw (1974) we discussed switching effects in vanadium dioxide (VO_2). In this material a structural phase change occurs at a critical temperature, T_c; at T_c the conductivity rises precipitously by several orders of magnitude. By modeling this phenomenon with a step function change in conductivity at T_c, we can treat the problem analytically and predict the observed $I(V)$ curves successfully. A similar phenomenon occurs for inhomogeneous chalcogenide films. Here, however, it is F_c which causes σ to rise precipitously, and it is reached locally in many cases because of thermal effects.

In the last few sections we have discussed switching effects in uniform homogeneous chalcogenide films. Thermal theories attempting to explain this phenomenon have been presented by many authors (see, e.g., Stocker *et al.*, 1970; Croitoru and Popescu, 1970; Shousha, 1971; Duchene *et al.*, 1971; Altcheh *et al.*, 1972; Kaplan and Adler, 1972; Warren, 1972; Kroll, 1974; Owen *et al.*, 1979; Shaw and Subhani, 1981). From these studies, it has been made clear that for non-thermistor type switching to occur an electronic mechanism must also be operative in order to short out the low-conductance regions adjacent to the cool electrodes. A critical electric field will certainly suffice, and it is this assumption that we have used to provide good agreement with the first-fire event data for memory samples discussed in the last section. As we shall see in Section 5.5, during the first few firing events a breakdown-type process often occurs that is driven by the capacitive discharge and/or high operating currents (Kotz and Shaw, 1984). An open, and most often, forming or a short (memory) occurs if the sample is kept ON for a sufficiently long time. In general, the first-fire event produces an intermediate state that is a narrow ($\leq 5\,\mu m$ dia.) filamentary region containing crystalline or morphological imperfections. [Sometimes several firings are required to develop a formed state that is amenable to easy observation by scanning microscopy, but it is the first-fire event that often results in the largest change in threshold voltage (Allinson *et al.*, 1979).]

For a threshold switch the formed region is of higher conductance than the surrounding homogeneous film (Coward, 1971), but still of substantially lower conductance than the ON-state. We can imagine the intermediate state as being formed in the following manner. Consider, for example, the case where sufficiently short pulses are applied such that F_c is reached isothermally (Walsh et al., 1981) over the entire sample, independent of its thickness (Buckley and Holmberg, 1975). In this region of pulse width ($\leq 10^{-8}$ s) the power, P, dissipated in the sample because of Joule heating is given by $P \cong \sigma_{RT} F^2$, where σ_{RT} denotes the room-temperature (ambient) conductivity. When F_c is reached the current increases by orders of magnitude (Buckley and Holmberg, 1975) at constant voltage. (This large increase in current at constant voltage prior to breakdown or switching has only been observed at room temperature in virgin samples for sufficiently short pulses. Formed samples show this effect at low temperatures.) In this region $P \cong \sigma_h F_c^2$, where $\sigma_h \gg \sigma_{RT}$. The large increase in conductivity induced by F_c can be due to either the field-stripping of trapped carriers and/or avalanching. Once these carriers are generated, because of the ensuing capacitive discharge, the significantly increased Joule heating causes a breakdown at the weakest point in the film. The sequence leading to forming in virgin samples subjected to short pulses is first electronic, then thermal. As previously stated, the outcome of the switching event can be: (1) an open; (2) a short (e.g., the memory state); (3) an inhomogeneous formed region (threshold switch). In what follows we support the view that a formed or intrinsically inhomogeneous region is common in conventional threshold switches (Popescu, 1975) made from thin amorphous chalcogenide films. We also suggest that the mechanism for the switching event has features that are somewhat different from that of the first-fire switching event in a homogeneous film. In the latter case F_c is either reached isothermally over the entire sample for short pulses or, for longer pulses, some thermal modification allows for switching to occur when F_c is reached only over part of the sample. In either case the switch occurs very rapidly when a critical field or voltage is reached. Formed samples, however, show a switching transition, after a delay-time t_d, when a critical local power density is reached (Balberg, 1970; Thoma, 1976; Reinhard, 1977; Shaw and Subhani, 1981). In the following section a model for these effects is presented.

5.4.3.2 An Electrothermal Model for Threshold Switching in Inhomogeneous Films

Popescu (1975) has provided a detailed analytical model of how switching can occur in inhomogeneous chalcogenide films. It is our view that his arguments center correctly on the properties of the formed region and the

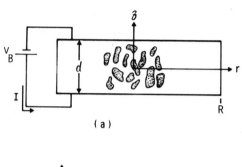

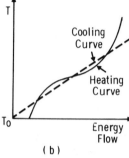

Fig. 5.24. (a) Geometry under analysis. All parameters are defined in the text. (b) Heating and cooling curves as a function of temperature. T_0 is the ambient temperature.

nature of the current instabilities possible in the vicinity of such paths. In what follows we offer a simple supplement to Popescu's work by suggesting possible means by which an electronic instability can be encouraged in such systems.

Figure 5.24a shows the geometry under analysis. As in Fig. 5.17, the sample is a thin cylindrical disk of radius R composed of material having a thermal conductivity, $K_a(T)$, and electrical conductivity, $\sigma_a(T)$, that increase with increasing temperature. These are the conductivities associated with the homogeneous parts of the film. Now, however, we have imbedded in the material an array of inhomogeneities (shaded) confined to the region $r < R_I$. (In general, we expect these inhomogeneities to be near or attached to one of the electrodes.) For the specific, but common case where the inhomogeneities are Te-rich crystallites, they have a thermal conductivity $K_i(T)$ that decreases with increasing temperature (over the temperature range of interest) and an electrical conductivity, $\sigma_i(T)$, substantially higher than $\sigma_a(T)$ and weakly dependent on temperature. A bias voltage, V, is applied across its thickness, d, and current, I, flows in the external circuit. Because of the properties of the system outlined above, the current density in the region $r < R_I$ is greater than in the surrounding homogeneous medium.

The conventional thermal instability (Landauer, 1978) that can occur at a critical value of local power density in such systems has been outlined by Landauer and Woo (1972), and treated in detail by Popescu (1975). It can be understood most simply by considering Fig. 5.24b. Here the cooling curve represents the rate at which heat can be taken away from the region $r < R_I$, $-(d/2) < z < (d/2)$, when it is excited by Joule heating. The heating curve is sketched for the case $\sigma(T) = \sigma_0 \exp(-\Delta E/kT)$. A stable solution exists at the lower intersection of the heating and cooling curves. As the input power is increased the heating curve shifts to the right and an instability results when no lower intersection point between the two curves is possible. The upper intersection point represents another stable state of the system and switching occurs between these two stable states; a sudden increase in local temperature can occur.

This type of thermal instability does not seem to be operative in amorphous chalcogenide films. Rather, the dominant thermal aspect of the initiation of the event is simply the delay-time, t_d, which is the time it takes a hot spot to spread through the formed region, approach both electrodes (Shaw et al., 1973c; Homma, 1971; Newland, 1975) and cause F_c to be reached near an electrode. (The delay time is a consequence of a thermal process, in good agreement with experiment.) When the field near an electrode reaches a magnitude sufficient to sustain field-stripping within, or tunneling through the amorphous regions adjacent the contacts (Mott, 1969), switching occurs, along with the concomitant capacitive discharge. If the entire formed region melts upon overdriving the sample, this picture would be in harmony with the observations of Pearson and Miller (1969). Here, upon turning the switch off, the molten region could revitrify. The subsequent switching event could then initiate at a different spot; the conducting path could "jump-around" from cycle to cycle. However, if partial crystallization occurred, then the same spot could initiate the switch upon consecutive firings. We shall discuss these possibilities in Section 5.5.

The switching transition electronically stabilizes the filamentary region, which can sustain relatively high temperatures in its center. The electrode temperature, however, is cooler. The regions near the electrodes are amorphous, maintain a large temperature gradient and have an average energy band gap that depends upon the temperature gradient. In the narrow amorphous regions carriers are being supplied by fields on the order of 10^5 V cm. Recombination radiation is being emitted near 0.5 eV (Walsh et al., 1978, 1979); it could be originating from either (1) defect transitions in the amorphous layers, or (2) band-to-band transitions in the core of the filament. A black-body spectrum has not been observed in these recent studies, although it has been in others.

In the above switching model the width of the current filament in the ON-state is largely constrained to the width of the formed region. Typical formed regions in threshold and memory-type samples have been measured by scanning microscopy and found to be 1–5 μm in diameter. We expect that formed threshold-type samples will generally have highly conducting ON-state paths of this size. Thus, for a given load line that produces an ON-state below current saturation of the formed region, the current filament will be smaller than the formed region. As the load is lightened or the current increased at fixed load, the current filament will widen until it fills the formed region. Further increases in current will result primarily in heating of the current-carrying path rather than its continued spatial expansion (Kotz and Shaw, 1982).

The model presented here satisfactorily explains the phenomenology of threshold switching. It is consistent with the experimental observation that the instability initiates at approximately zero-time for any overvoltage (Shaw *et al.*, 1973c) and the inference, taken from the data, that is like a convective instability. Furthermore, it explains the behavior of t_d in the "statistical" regime just at threshold. Here, very long t_d's can be observed, where the current is not observed to rise until within a microsecond or two prior to the switching event. We suggest that the instability is triggered by a fluctuation associated with the injection of carriers. Slightly past threshold, t_d is usually 1–2 μs in a 1 μm thick film. This is the time it takes for the hot spot to grow. As this occurs, the current increases with time. In the statistical regime we must wait for the fluctuation that will trigger an event in a material that will be slightly different each cycle. There will be no rise in current while we wait.

The model also explains the results of Henisch *et al.* (1974) and Rodgers *et al.* (1976). The former group found that the voltage at threshold was insensitive to light intensity at low excitation levels, even though the current at threshold increased due to the enhanced conductivity of the material. The latter group showed that the voltage at threshold decreased with intensity at high levels where the material is heated by the optical pulse. A straightforward explanation can now be given for these effects. The Te-rich crystallites are essentially unaffected by the light. At low intensities the conductivity of the region surrounding the crystallites is increased, but the local field is thereby decreased, and the local power density remains essentially unaltered. The critical condition is local, and if the temperature of the surrounding medium is unchanged, the instability will occur at the same value of local power density. Once the temperature increases locally, however, the threshold power density will drop. The excess currents observed in these experiments with increasing light intensity are due primarily to the enhanced conductance in those (major) parts of the films that remain homogeneous.

5.5 Thermophonic Studies of Thick Chalcogenide Films

5.5.1 Introduction

The thermophonic experiment described in this section (Kotz and Shaw, 1984) directly addresses the question of whether or not thermally induced structural changes occur prior to the initiation of the switching transition for a wide co-planar sample in a circuit with relatively high (external) circuit capacitance. In the section we will first briefly outline some thermal properties of the memory material used in the study ($Ga_{15}Te_{81}S_2Sb_2$). Next, we will describe the sample design and then the experimental arrangement. After displaying the results, we will devote a section to theory and analysis.

The study was originally initiated because at the time: (1) there was doubt as to whether the mechanism for the initiation of the switching event from a high resistance OFF-state to a low resistance ON-state is predominantly electronic or thermal; (2) the technique provides a novel way of studying the various phenomena and estimating the temperature-time profile during the three phases of the switching cycle. The results of the study support the view that: (1) an electronic, rather than thermal, mechanism governs the switching transition, surprisingly even in thick films; (2) the capacitive discharge provides much of the energy by which phase changes occur (Jones et al., 1978; (3) cooling occurs immediately after the capacitive discharge; (4) a memory event occurs when enough "forming" has taken place to allow melting of the partially crystallized formed filamentary region during the capacitive discharge, but with low enough ON-state power to prevent the temperature from exceeding the melting temperature at the end of the pulse; (5) there are *two* different ranges of power input where threshold switching events occur, and these ranges bracket the power conditions where memory events result.

The memory material $Ge_{15}Te_{81}S_2Sb_2$ is similar in properties to $Ge_{15}Te_{85}$, which is the eutectic composition for the Te–GeTe binary system (eutectic temperature, 375°C) (Moss and De Neufville, 1972). The addition of S and Sb is believed to affect the rate of crystallization when heating the material in glassy form. Assuming that we start with the glassy material, heating to the temperature range $140 \leq T \leq 240°C$ allows Te crystallites to form. Heating the material to $240 \leq T \leq 375°C$ leads to the formation of Te and GeTe crystallites. Further heating supplies the heat of transformation for the dissolution of the crystallites, and produces the liquid phase.

If the liquid is cooled slowly enough, a polycrystalline, low resistivity material results, whereas if it is cooled rapidly enough, a high resistivity glass results. Cooling to room temperature from the crystallization temperature

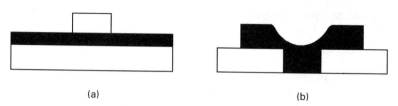

Fig. 5.25. (a) Sandwich structure. (b) Coplanar structure. (Dark = chalcogenide, light = electrode.)

region, $140°C \leq T \leq 375°C$, however, preserves whatever crystallites are present no matter what the rate of cooling (Fritzsche and Ovshinsky, 1970).

There are usually two basic choices for the geometry of the sample: "sandwich" and "coplanar" (see Fig. 5.25). The coplanar geometry was emphasized in this study because it is easy to achieve large electrode spacings, which reduces the sample capacitance and, perhaps more importantly, reduces the effect of the electrodes as heat sinks. (This is important from the standpoint of using a one dimensional model for heat flow in the system.)

Because of the large gap between the electrodes it was desired to use a chalcogenide of low threshold field, hence, relatively low threshold voltage. (In practice we are limited by the voltage output of any pulse generator used to switch our devices. Since memory materials generally have a lower threshold voltage, we chose a memory material, $Ge_{15}Te_{81}S_2Sb_2$, for this purpose.)

The coplanar sample consists of a one square inch microscope slide glass base upon which molybdenum electrodes, and then the chalcogenide were deposited via rf sputtering (see Fig. 5.26). The electrode spacing ($\geq 100\,\mu m$) was determined by a mask of either 28 or 30 gauge copper wire held firmly against the central portion of the base. The chalcogenide was deposited in a one-half inch circle on top of the electrodes and filled the space between the electrodes. Many samples were studied; two representative cases were chosen for detailed study.

Samples were tested for threshold switching, for which typical parameters were : threshold voltage, $V_T \geq 250\,V$; ON-state current, $I_{ON} \geq 0.4\,mA$; delay time, $t_d \geq 0.2\,ms$. As expected, t_d decreased with increasing bias voltage.

A chamber containing a sinusoidally electrically excited gold sheet which caused acoustic vibrations in the enclosed air was first used by Wente (1922) for the purpose of calibrating a microphone (also enclosed in the chamber). Wente modeled the acoustic waves in the chamber, or thermophone, as plane waves and solved a thermal transport equation in the enclosed (ideal) gas, thus finding the pressure as a function of time.

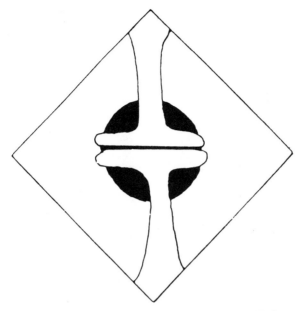

Fig. 5.26. View of coplanar sample, looking through glass substrate. (Twice actual size; before area reduction.)

Recently, Aamodt and Murphy (1978) modeled the acoustic response of a thermophonic cell to an electrically pulsed, platinum-black thin film. Using a Laplace transform technique, they solved a set of one dimensional thermal transport equations to describe the temperature in the various cell regions as a function of the input power density, physical properties of the cell, and (indirectly) time. Then, relating the temperature variation to the pressure variation in the gas region of the cell, an expression for the microphone signal was obtained. The agreement between theory and experiment was good.

The thermophonic cell used in our study encloses a coplanar electrode, memory-type chalcogenide switch. By measuring the microphone signal as a function of time during a switching event and using a model based on the work of Aamodt and Murphy, the average temperature as a function of time was estimated for the chalcogenide switch.

The major components of the thermophonic experiment are shown schematically in Fig. 5.27. The pulse generator supplies a "square" electrical pulse input for the sample. The voltage across the sample is monitored by VP1 and, in order to obtain the current, VP2 measures the voltage across the resistor R2, which is very small in comparison to the other resistances in the circuit. The heating of the sample, at one end of the thermophonic cell,

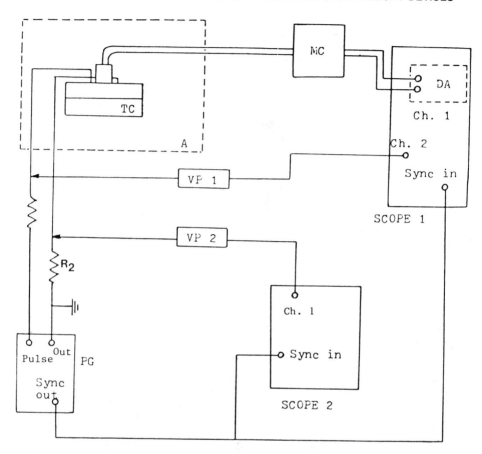

Fig. 5.27. Experimental arrangement for thermophonic experiments. Grounds are made at SCOPE 1 which is earthed. (Ground connections and shields not shown.) TC: Thermophonic Cell. A: Anechoic room. MC: Microphone Circuit. DA: Differential Amplifier (Tektronix 7A22). PG: Pulse Generator (Cober Model 606P). SCOPE 1: Tektronix Model 7403N. SCOPE 2: Tektronix Model 564 (Storage).

results in the generation of acoustic waves in the gas-space of the cell. These are detected by the miniature microphone at the other end of the cell. (The cell length is variable; see Fig. 5.28.)

The microphone output is measured by a differential amplifier whose output is displayed on an oscilloscope (SCOPE 1 in Fig. 5.27). The differential amplifier is desirable because of its low noise and its compatability with a microphone output not directly referenced to ground. The microphone circuit is designed to give an optimum low frequency response (Kotz and Shaw, 1982).

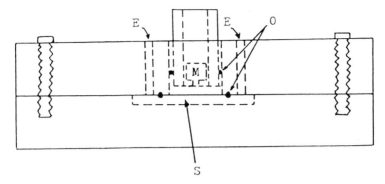

Fig. 5.28. Thermophonic Cell. S: Sample. E: Electrode entrance. O: O-ring. M: "Potted" microphone (Knowles miniature, model BT-1785, condenser).

5.5.2 Experimental Results

5.5.2.1 Non-switching behavior

A typical non-switching output display is shown in Fig. 5.29. (This applies to the case when the applied voltage is too low to switch the sample.) The curvature at the beginning of the voltage pulse is a result of capacitance (parallel to the sample) due primarily to the long lines from the pulse generator to the sample. The initial slow rise in the microphone signal is a consequence of this, but it also, perhaps, caused by the inertia in the diaphragm of the microphone. The narrow spike early in the microphone signal is apparently due to a capacitive pick-up between the microphone signal lines and the pulse generator-sample lines. Otherwise, the microphone signal is very nearly linear in rise. Any significant departure from linearity in the non-switching signal is a result of cell leaks and/or a thermal wave contribution (Aamodt and Murphy, 1978).

As will be seen, the linear behavior is consistent with a one-dimensional model of the thermophonic cell. In this model under these conditions the microphone signal is proportional to H, the power per unit volume dissipated in the sample, given by

$$H = \frac{V^2}{R} \cdot \frac{I}{Ad},$$

where V, R and d are the sample voltage, resistance and electrode spacing, respectively, and A is the electrode-sample interfacial area. For non-switching pulses with V appreciably less than V_t, it is reasonable that the electric field is uniform across the switching material (Petersen and Adler, 1976). The contact resistance is ignored because it is very small in chalcogenide films

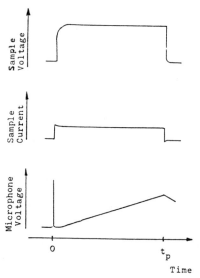

Fig. 5.29. Typical non-switching output display (single pulse excitation). Arbitrary Units. t_p on the order of several ms.

(Shaw, 1981). Expressing the resistance as $R = \rho d/A$, where ρ is the sample resistivity, we have

$$H = \frac{V^2}{\rho d^2},$$

which is independent of A, the interfacial area.

The area independence was checked by recording non-switching signals before and after reducing the sample area by removing some of the switching material. The microphone signal was found to be independent of interfacial area, as long as the sample width was less than about one-half centimeter. Also, the signal was found to be proportional to the input power for a fixed conducting volume. Hence, the use of the above one-dimensional (Aamodt-Murphy) model for non-switching behavior was apparently justified.

5.5.2.2 Switching Behavior

When the samples used in this study were initially switched in the thermophonic cell, it was desired to induce threshold switching (as opposed to memory switching), which required keeping the pulse power level at the low levels and employing only single pulse excitation. Under these conditions, the output display appears as shown in Fig. 5.30. The abrupt change in voltage leads the sudden rise in microphone signal by 28 μs when the cell length is

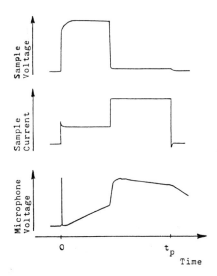

Fig. 5.30. Output display for switching with low ON-state power (single pulse excitation). Same scales as in Fig. 5.29.

approximately 0.2 cm. This is considerably longer than the acoustic transit time (~6 μs), so that it is reasonable to assume that the jump in the microphone signal is due to an effect occurring *after* the transition.

It was found that the signal increase after the switching transition depended on stored capacitive energy. The capacitance (parallel to the sample), as judged from sample voltage risetime, was on the order of 100 pf, much larger than the sample capacitance (≤ 1 pf), and depended upon the length of the power lines. (The sample chamber was placed in an anechoic room, about 15 feet from the pulse generator.) In Fig. 5.31a, only the length of the power lines was changed, making capacitance the only variable. In Fig. 5.31b the pre-switching voltage is the only variable. Thus, changes in the stored capacitive energy, $E = \frac{1}{2}CV^2$, where C is the capacitance and V is the preswitching voltage, will in turn change H, the power density, when the energy is released immediately after switching due to the abrupt change in resistance.

Referring to Fig. 5.32, on an expanded scale, the low-power threshold switching microphone signal has small oscillations superimposed on it. These oscillations are a manifestation of a resonance in the cell or in the microphone (Kotz, 1982).

It was found that the secondary slope of the threshold switching microphone signal is an increasing function of P_{ON}, the ON-state power (see Fig. 5.33). For a particular sample (CP-MO-19), for relatively high threshold

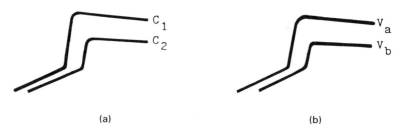

Fig. 5.31. The effect of reduction in preswitching sample voltage, V, on microphone signal ($V_a > V_b$). (Low ON-stage power.)

ON-state currents ($I_{ON} > 1$ mA, where these and the following current values apply to this particular sample), it appeared that the filamentary volume was approximately constant since a change in slope in Fig. 5.33 for high P_{ON} is proportional to the change in P_{ON} (rather than by a change in H_{ON} as predicted by our later analysis).

As P_{ON} is increased by decreasing the load resistance, there is a gradual reduction in the jump signal due to the capacitive discharge. Eventually, as we see in Fig. 5.33, no jump occurs at all. Under the appropriate conditions this occurred during a memory switching event (i.e., the sample did not revert to the OFF-state after the voltage returned to zero at the end of the pulse). Only when the capacitive jump vanished did a memory event occur.

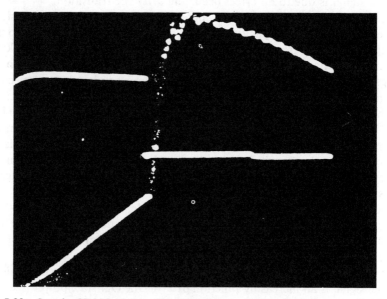

Fig. 5.32. Sample: CP-MO-13. Low ON-state power threshold switching event; $t_p \cong 1.5$ ms.

THERMOPHONIC STUDIES OF THICK CHALCOGENIDE FILMS

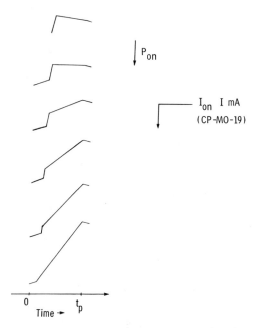

Fig. 5.33. The effect of increasing ON-state power on the secondary slope of the microphone signal.

In some samples the capacitive peak also diminished under repetitive (low P_{ON}) threshold switching (~1 pps) (see Fig. 5.34). This probably indicates that some structural change was occurring, in which case the energy was taken up as heat of transformation. After the power level was increased, a filamentary path was clearly evident.

For relatively high P_{ON} ($I_{ON} > 1$ mA, single pulse threshold and memory switching) lasting structural change was evident under 40× magnification as phase separated regions near the electrodes. Under SEM (scanning electron microscopy) after memory events had occurred, there were connected paths

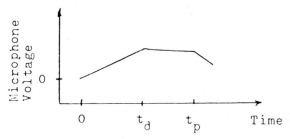

Fig. 5.34. Microphone signal for repetitive switching when the capacitive spike is not present.

present. X-ray analysis indicated a high degree of crystallinity in the paths. For high ON-state power levels, there was a noticeable reduction in pre-switching sample resistance. This reduction was between 15-30% and preceded the onset of memory switching. We refer to the process of increasing crystalline density, decreasing threshold voltage and lowering OFF-state resistance as "forming". For low values of P_{ON}, when only threshold switching occurred, there was no evidence of morphological changes or surface filamentary paths under 40× magnification. Such films maintained the appearance of unswitched samples.

5.5.3 Theory and Analysis

We now describe the thermophonic experiments via the Aamodt-Murphy model. After presenting the model, we display the calculated results of varying the heat source, H, as a function of time. We will see that, qualitatively, the calculated signals resemble the experimental switching microphone signals. However, we will find that in order to quantitatively represent the switching microphone signal for a particular sample, other physical parameters must also change as a function of time.

5.5.3.1 The Aamodt-Murphy Model

The thermophonic cell is divided into four regions: 1. substrate, 2. sample, 3. gas, 4. endplate. Considering energy conservation via thermal processes, the temperature variation (from ambient, T, in regions 1 and 4 is given by

$$\alpha_i \nabla^2 T - \frac{\partial T}{\partial t} = 0, \qquad (5.45)$$

where the subscript refers to the appropriate region and α is the diffusivity. In region 2 the sample is under electrical excitation and generates Joule heat; for this region,

$$\alpha_s \nabla^2 T - \frac{\partial T}{\partial t} = \frac{-H}{C_{ps}}, \qquad (5.46)$$

where H is the heat generated per unit volume. Here, H includes the effect of Joule heating (a positive contribution to H) as well as, when appropriate, the heat of transformation (a negative contribution to H) in the appropriate units. $C_{ps} = c_{ps}\rho$ is the thermal capacity per unit volume for the sample, where c_{ps} is the sample specific heat and ρ is the sample mass density.

In region 3 the gas temperature follows

$$\alpha_g \nabla^2 T - \frac{\partial T}{\partial t} = -\frac{1}{C_{pg}} \frac{\partial P}{\partial t}, \quad (5.47)$$

where P is the pressure variation and C_{pg} is the thermal capacity per unit volume for the gas. The effect of viscosity is neglected. Treating the gas as ideal, using the equations of state and continuity, the last equation is replaced by

$$\nabla^4 T - \nabla^2 \left(\frac{1}{\alpha_g} \frac{\partial T}{\partial t} + \gamma\tau \frac{\partial^2 T}{\partial t^2} \right) + \frac{\tau}{\alpha_g^2} \frac{\partial^3 T}{\partial t^3} = 0, \quad (5.48)$$

where, with c_0 being the speed of sound in the gas, $\tau = \alpha_g/c_0^2$ and γ is the ratio of specific heats, i.e., c_p/c_v for the gas.

In practice we want to know the temperature variation in the gas, since, as indicated in Eq. (5.47), it is related to the pressure variation, to which the microphone responds. To find the gas temperature we must partially solve the set of equations describing the four regions of the thermophonic cell, Eqs. (5.45), (5.46) and (5.48). To facilitate the solution we perform a Laplace transform, with the result that

$$\nabla^2 \hat{T}_i = \left(\frac{s}{\alpha_i}\right) \hat{T} \quad \text{for regions 1 and 4;} \quad (5.49)$$

$$\nabla^2 \hat{T} = \left(\frac{s}{\alpha_s}\right) \hat{T} - \frac{\hat{H}}{C_{ps}\alpha_s} \quad \text{for region 2;} \quad (5.50)$$

and

$$\left[\nabla^4 - \left(\frac{s}{\alpha_g}\right)(1 + \gamma_s \tau)\nabla^2 \right] \hat{T} = -\left(\frac{\tau}{\alpha_g^2}\right) s^3 \hat{T} \quad \text{for region 3.} \quad (5.51)$$

In Eqs. (5.49) through (5.51), the caret denotes the Laplace transform with variable transform pair (t, s).

As shown by Aamodt and Murphy, since $|s\tau| \ll 1$, the temperature in region 3 can be written as

$$\hat{T} = \hat{\theta} + \hat{\psi}, \quad (5.52)$$

where

$$\nabla^2 \hat{\theta} = \frac{s^2}{c_0^2} \hat{\theta} \quad (5.53)$$

and

$$\nabla^2 \hat{\psi} = \frac{s}{\alpha_g} \hat{\psi}. \quad (5.54)$$

Using the integral form of Eq. (5.47), they also show that

$$\hat{p} = C_{pg}\hat{\theta} - (\gamma - 1)s\tau C_{pg}\hat{\psi}. \tag{5.55}$$

Equation (5.53) represents the acoustic component of the temperature variation since its (transformed) solution is a wave travelling at the speed of sound in the gas. Equation (5.54) represents the "thermal" component of the temperature variation since its (transformed) solution is a dispersive, thermal-diffusion limited wave. Equation (5.55) gives the pressure dependence of these two components.

We now make the assumption that the electrical heating occurs uniformly in a "conducting channel", both prior to and after a switching transition, if one occurs. (However, there is a reduction in volume upon switching.) Further, we assume that the diffusion of heat occurs primarily along the axis of the cell. As it turns out, this is probably reasonable for non-switching behavior because the low OFF-state thermal diffusivity and geometry of the sample limit the flow of heat to the electrodes. However, as we shall see, a modification may be required in order to model the ON-state behavior properly.

The transformed Eqs. (5.49), (5.50), (5.53) and (5.54) are now each one dimensional, linear, second order differential equations whose solutions are readily shown to be:

$$\hat{T}_1 = \hat{T}_{01} \exp\left[\left(\frac{s}{\alpha_1}\right)^{1/2}(1 + x)\right], \tag{5.56}$$

$$\hat{T}_2 = M \exp\left[-\left(\frac{s}{\alpha_s}\right)^{1/2} x\right] + N \exp\left[\left(\frac{s}{\alpha_s}\right)^{1/2} x\right] + \hat{T}', \tag{5.57}$$

$$\hat{T}_3 = E \exp\left(-\frac{sx}{c_0}\right) + F \exp\left(\frac{sx}{c_0}\right) + G \exp\left[-\left(\frac{s}{\alpha_g}\right)^{1/2} x\right]$$

$$+ H \exp\left[\left(\frac{s}{\alpha_g}\right)^{1/2} x\right], \tag{5.58}$$

$$\hat{T}_4 = \hat{T}_{04} \exp\left[\left(\frac{s}{\alpha_4}\right)^{1/2}(L + x)\right]. \tag{5.59}$$

From Eqs. (5.54) and (5.58) we can identify $\hat{\theta}$ and $\hat{\psi}$:

$$\hat{\theta} = E \exp\left(-\frac{sx}{c_0}\right) + F \exp\left(\frac{sx}{c_0}\right) \tag{5.60}$$

$$\hat{\psi} = G \exp\left[-\left(\frac{s}{\alpha_g}\right)^{1/2} x\right] + H \exp\left[\left(\frac{s}{\alpha_g}\right)^{1/2} x\right]. \tag{5.61}$$

In the above $\hat{T}'$ is a particular solution of Eq. (5.50) and the other previously unidentified symbols represent constants determined by the following boundary conditions:

$$K_i \frac{\partial \hat{T}_i}{\partial x} = K_j \frac{\partial \hat{T}_j}{\partial x}, \qquad (5.62)$$

where K is the thermal conductivity,

$$\hat{T}_i = \hat{T}_j, \qquad (5.63)$$

and

$$\frac{\partial p_i}{\partial x} = 0. \qquad (5.64)$$

Equations (5.62) and (5.63) apply to all interfaces. Equation (5.64), which requires a constant gas velocity, applies at $x = 0$ and at $x = L$.

Application of the boundary conditions, Eqs. (5.62) through (5.64), to Eqs. (5.56) through (5.59) results in a linear algebraic system which is solved by standard methods for the appropriate constants. Once this is done the transformed pressure is obtained from Eq. (5.55). In the approximation $|\mu_s \tau| \ll 1$, the result is

$$\hat{p} = \hat{T}'' R_c, \qquad (5.65)$$

where

$$\hat{T}'' = \hat{T}'(0) - (1 + m) Q_0 \hat{T}'(-1)$$

$$+ \left(\frac{\alpha_s}{s}\right)^{1/2} Q_0 \left[-q_0 \left(\frac{\partial \hat{T}'}{\partial x}\right)(0) + (1 - m)\left(\frac{\partial \hat{T}'}{\partial x}\right)(-1) \right] \qquad (5.66a)$$

and where

$$R_c = C_{pg}(s)^{1/2} \{C_1 - C_2 + D[S_1(s\tau)^{1/2} - S_2]\}$$
$$\times \{2(s)^{1/2}[1 - C_1 C_2 - (b + D)][C_2 S_1' + (s)^{1/2} C_1 S_2] - (1 + Db) S_1' S_2\}^{-1}. \qquad (5.66b)$$

The unidentified terms in Eq. (5.66) are listed in Table 5.3.

Under the above conditions, $\hat{H}(s, x) = \hat{H}(s)$, i.e., $\hat{H}$ is independent of x, and the particular solution of Eq. (0.00) is $\hat{T}' = \hat{H}(s)(sC_{ps})^{-1}$. Therefore, from Eq. (5.66)

$$\hat{T}'' = [1 - (1 + m) Q_0] \hat{H}(s)(sC_{ps})^{-1}. \qquad (5.67)$$

The above expression for the transformed pressure [Eq. (5.65)], in effect, the transformed microphone signal, is not transformable to the time domain by any known "analytic" means. Therefore, an approximate numerical inversion is employed. We chose the method of Dubner and Abate (1968), which can be modified for faster convergence.

Table 5.3. (All values in c.g.s. units)

$c_0 = 34,300$

$\mu = \gamma - 1 = \left(\dfrac{c_p}{c_v}\right)_g - 1 = 0.41$

$\alpha_g = 0.187$ $\kappa_g = 5.8 \times 10^{-5}$ $C_{pg} = (c_p\rho)_g = 3.1 \times 10^{-4}$

$\alpha_s = 2.87 \times 10^{-3}$ $\kappa_s = 7.0 \times 10^{-4}$ $C_{ps} = 0.25$

$\kappa_1 = \kappa_h = 0.0024$ $C_{p1} = C_{ph} = 0.48$

$\kappa_4 = \kappa_w = 0.48$ $C_{p4} = C_{pw} = 0.5$

$l = 5.5 \times 10^{-4}$ (CP-MO-19)

$L = 0.175$

5.5.3.2 Some Features of Thermophonic Modeling. Heat Source Time Variation

We now display some results of the numerical analysis. The values used for the various physical constants are given in Table (5.4). First, treating the case of square-wave pulsed heat generation, which represents the effect of a non-switching pulse applied to the sample, the computed microphone signal appears in Fig. (5.35). All such computed curves show a linear slope to within 5%.

We now consider a heat source that varies in time as shown in Fig. (5.36a). All other physical parameters are constant. Depending on the relative magnitudes of H_1, H_2 and H_3, a wide assortment of computed signals result. However, when $H_2 \gg H_1$, $H_1 \cong H_3$ and $t_{RC} \ll T_p$, computed signals resembling threshold-switching microphone signals with increasing secondary slopes occur (Fig. 5.36b). To obtain a computed curve with a decreasing secondary slope and still keep all other parameters constant, it is necessary to take $H_1 \gg H_3$ (Fig. 5.36c).

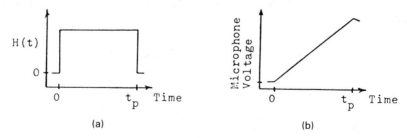

Fig. 5.35. Non-switching computer results.

Table 5.4

$$C_1 = \cosh\left(\frac{sL}{c_0}\right) \qquad S_1 = \sinh\left(\frac{sL}{c_0}\right)$$

$$C_2 = \cosh\left[L\left(\frac{s}{\alpha_g}\right)^{1/2}\right] \qquad S_2 = \sinh\left[L\left(\frac{s}{\alpha_g}\right)^{1/2}\right]$$

$$S_1' = S_1\mu^{-1}\tau^{-1/2} \qquad q = \left(\frac{s}{\alpha_s}\right)^{1/2} l$$

$$B = \left[\frac{K_g C_{pg}}{K_s C_{ps}}\right]^{1/2} \qquad D = \left[\frac{K_g C_{pg}}{K_w C_{pw}}\right]^{1/2}$$

$$\xi = \left[\frac{K_s C_{ps}}{K_h C_{ph}}\right]^{1/2} \qquad m = \left[\frac{1-\xi}{1+\xi}\right]$$

$$b = BQ_0 q_0 \qquad Q_0 = \exp(-q)[1 + m\exp(-2q)^{-1}]$$

$$q_0 = \exp(q) - m\exp(-q)$$

$$\beta_1 = \frac{sL}{c_0} \qquad \beta_2 = \left(\frac{s}{\alpha_g}\right)^{1/2L}$$

$$z = \frac{z_N}{z_D} = \frac{-2s^{1/2}C_1 S_2 - 2S_1' C_2 - 2S_1' S_2 D\mu_s\tau - 2s_1' S_2 D}{S_1'[\mu_s\tau D = D + 1]\exp\beta_2 - s^{1/2}(1 - C_1 \exp\beta_2)}$$

$$G = \hat{T}'' z_D \{4s^{1/2}(C_1 C_2 - 1) + 2(b + D)(1 + \mu_s\tau)[C_2 S_1' + S^{1/2}C_1 S_2] \\ + 2bD(1 + \mu_s\tau)^2 S_1' S_2 + 2S_1' S_2 + 2\mu_s^2\tau S_1' S_2\}^{-1}$$

$$H = z_D^{-1} G[s^{1/2}(C_1 \exp(-\beta_2) - 1 - S_1'(1 - \mu' D)\exp(-\beta_2)]$$

$$N = zB[\mu_s\tau + 1]GQ_0 \exp(q) - \tfrac{1}{2}Q_0(1 + m)\hat{T}(-l) \\ + \left(\frac{\alpha_s}{s}\right)^{1/2} Q_0\left[\left(\frac{1-m}{2}\right)\frac{\partial \hat{T}'}{\partial x}(-l) - \exp q\,\frac{\partial \hat{T}'}{\partial x}(0)\right]$$

$$M = -M\exp(-2q) - \tfrac{1}{2}(1 + m)\hat{T}'(-l)\exp(-q) + \tfrac{1}{2}\left(\frac{\alpha_s}{s}\right)^{1/2}(1 - m)\frac{\partial \hat{T}'}{\partial x}(-l)\exp(-q)$$

$$E = \frac{c_0\mu\tau}{2S_1}\left(\frac{s}{\alpha_g}\right)^{1/2}\{(G - H)\exp(\beta_1) - G\exp(-\beta_2 + H)\exp(\beta_2)\}$$

$$F = c_0\mu\tau\left(\frac{s}{\alpha_g}\right)^{1/2}(H - G) + E$$

$$\hat{T}_{01} = M\exp q + N\exp(-q) + \hat{T}'(-l)$$

$$\hat{T}_{04} = E\exp(-\beta_1) + F\exp(\beta_1) + G\exp(-\beta_2) + H\exp(\beta_2)$$

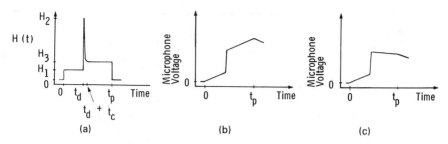

Fig. 5.36. (a) Time dependent energy density with $H_2 \cong H_1$, $t_p \gg t_c$. Large spike decays exponentially; t_c is determined from the condition $H_2 \exp[(t_d - t_c)/RC] = H_3$. (b) Computer microphone signal with $H(t)$ as in (a) with $H_3 \cong H_1$. (c) Computer microphone signal with $H(t)$ as in (a) with $H_3 \ll H_1$.

Finally, consider Fig. 5.37a, which shows another time dependent heat source, but without the intermediary "spike". This results in the computed signal shown in Fig. 5.37b; it resembles a microphone memory switching signal.

5.5.3.3 Quantitative Modeling

As guides to quantitatively modeling the switching microphone signal we use the following:

(1) the relative slopes of the pre-switching and post-switching portions of the experimental signal;
(2) the relative jump (if any) which occurs during the switching transition.

Once all the physical parameters are fixed, we can find the average temperature at the gas-sample interface, as calculated from Eqs. (5.57) or (5.58) with $x = 0$. [The same inversion technique is applied to these equations as to Eq. (5.65).]

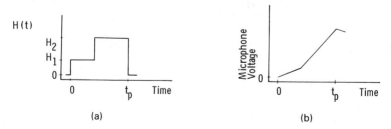

Fig. 5.37. (a) Time-dependent energy density without spike. (b) Computed microphone signal with $H(t)$ as in (a).

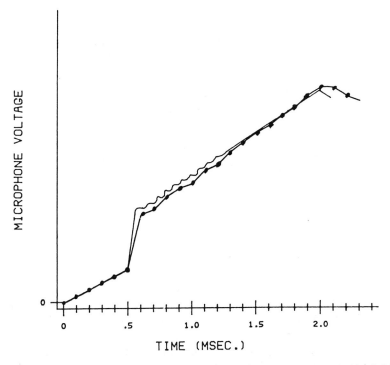

Fig. 5.38. Experimental and computed microphone voltage vs. time for a threshold switching event.

The computed curve in Fig. 5.38 models a microphone signal which occurred for a relatively high threshold ON-state power in sample CP-MO-19 and preceded the memory event for CP-MO-19 shown in Fig. 5.39.

To obtain the computed curves, the following assumptions were put into the model:

(1) A time dependent thermal conductivity, K_s, was used for the sample. It is reasonable to have this in the model as it is generally accepted that K_s increases at the switching transition.

The most convenient way to change K_s is in the form of a step function as in Fig. 5.40. For Fig. 5.38, $K_{s1} = 0.00075$ (cgs units) and $K_{s2} = 0.004$; for Fig. 5.39, $K_{s1} = 0.00081$ and $K_{s2} = 0.004$. [Both cases are assumed to occur in formed samples, hence, $K_{s1} > K_s$ ($=0.0007$), the thermal conductivity of the non-formed sample. K_{s2} is probably not the same for the two events modeled here, but at least approximately so.] Thus, we have assumed that K_s changes roughly by a factor of 5 at the transition.

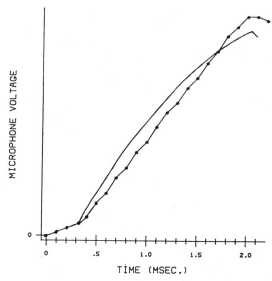

Fig. 5.39. Computed microphone voltage vs. time for a memory switching event and experimental microphone voltage vs. time for a memory switching event. (Voltage scale expanded slightly from Fig. 5.38.)

(2) A time-dependent substrate thermal conductivity, K_h. Again, a step function is used. It appears necessary to put this in the model and is reasonable on the grounds that, if indeed the sample conductivity increases significantly, the effect of the electrodes as heat sinks must be greater. Thus, we model the effect of the electrodes by an increase in K_h, and the flow of heat is effectively one dimensional. In Fig. 5.38, the substrate conductivity is increased 2.3 times and in Fig. 5.39, two times. The final values of K_h are the same in Figs. 5.38 and 5.39; $K_{h2} = 0.006$. Again, we assume slightly high initial values of K_h to represent forming.

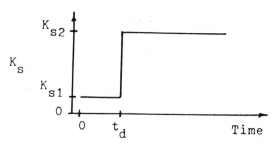

Fig. 5.40. Time-dependent thermal conductivity. $K_{s2} \cong 5K_{s1}$.

(3) The values of H are chosen to be consistent with the measured powers for the OFF- and ON-states, with the volume estimated from SEM data for the memory ON-state, and with the slopes of the microphone signals. In the case of threshold switching the relative jump in signal occurring at the switching transition is also considered.

We first estimate $H_M(t)$, the memory switching power density as a function of time. Since the ON-state power was 0.0055 cal/s and the ON-state volume (estimated from SEM data) was 8.7×10^{-9} cm^3, the ON-state power density, H_{M3}, is 6.3×10^5 cal/s · cm^3. Observing the change in the experimental signal in going from threshold to memory switching (Fig. 5.33), it is reasonable to assume that the sample acts as a heat sink for a short period of time. We take the *net* power density, H_{M2}, during this period (~ 3 μs) as zero. (We will discuss this matter in detail in Section 5.5.4.)

Finally, it is found that in order to match the slopes in Fig. 5.39, an initial power density, $H_{M1} = 1.2 \times 10^5$ cal/s · cm^3, is required. Thus, the time dependent power density appears as the inset of Fig. 5.39.

With the assumptions 1–3, the calculated microphone signal appears as in Fig. 5.39.

In order to obtain the calculated signal of Fig. 3.38, we use assumptions 1–2 and the following.

(4) We assume that it is the capacitive discharge that determines the size (width) of the filament. (Thus, both of the events discussed here have roughly the same ON-state volume.) This assumption is consistent with Fig. 5.33, where, for larger ON-state power, the filament size appeared constant. Hence, the ON-state resistance, R_{ON} ($\cong 20$ kΩ as estimated from ON-state voltage and curent) is used in the following calculations. For the threshold switching case at hand, the voltage across the capacitor (i.e., the pre-switching voltage) was 300 V. Thus, initially,

$$(H_{cap})_{max} = \frac{(300)^2}{20 \times 10^3} \cdot \frac{1}{0.87 \times 10^{-8}} \cdot \frac{1}{4.184} \cong 10^8 \frac{cal}{sec \cdot cm^3}$$

This will decay at a rate $\exp[(t_s - t)/RC]$ where $C \cong 100$ pf and thus $RC \cong 2$ μs. Therefore, the power density as a function of time, $H_{T2}(t)$, during the capacitive discharge is $H_{T2}(t) = 10^8 \exp[(t_s - t)/RC]$ cal/s · cm^3.

(5) Since the power during the ON-state of the threshold switching event was 0.0053 cal/s, the ON-state power density is taken as $H_{T3} = 6.0 \times 10^5$ cal/s · cm^3. It is found that $H_{T1} = 1.6 \times 10^5$ cal/s · cm^3 gives the best match to Fig. 5.38. The time dependent power density for the threshold event will thus appear as the inset in Fig. 5.38.

As a check on the values of H_{M1} and H_{T1}, the two preswitching slopes of Figs 5.38 and 5.39 can be compared. One might expect that since slope M1/slope T1 $\cong$ 0.6, we would have $H_{M1}/H_{T1} = 0.6$. However, because of forming, the thermal conductivities are greater for the memory case (assumptions 1 and 2). For a simple pulse excitation, the model predicts a slope ratio of slope M1/slope T1 = 0.88 due to conductivity differences alone. Thus, we should expect that the ratio of slopes is given by

$$\frac{\text{slope M1}}{\text{slope T1}} = \frac{H_{M1}}{H_{T1}} \cdot (0.88) = 0.66,$$

in good agreement with the experimental results.

As another check, we expect the preswitching volumes, v_{M1} and v_{T1}, to be about the same. Using the above estimates for H_{M1} and H_{T1} and the measured powers, P_{M1} and P_{T1}, we have,

$$v_{M1} = \frac{P_{M1}}{H_{M1}} = 0.0069 \cdot \frac{1}{1.2 \times 10^5} = 5.8 \times 10^{-8} \text{ cm}^3$$

and

$$v_{T1} = \frac{P_{T1}}{H_{T1}} = 0.0093 \cdot \frac{1}{1.6 \times 10^5} = 5.8 \times 10^{-8} \text{ cm}^3.$$

5.5.4 Discussion

In this section we will discuss and attempt to clarify several aspects of the thermophonic experiments and analysis. In particular, we will deliberate the quantitative modeling of Section 5.5.3.

First, let us review for a moment the sequence shown in Fig. 5.33. We begin by threshold switching the sample to a relatively low ON-state power. As the ON-state power is increased in subsequent switching events, the secondary microphone slope increases, eventually in a manner whereby the filamentary volume appears constant. There is a gradual decrease in the mirroring of the capacitive jump until, at sufficiently high power, there is none evident. (The capacitive discharge is always present and almost constant since both C, which is dominated by the external capacitance, and dV/dt are essentially the same even though the sample appears to undergo a continual forming process.) When this series was originally performed on sample CP-MO-19, a memory event did not occur. However, the large ON-state power threshold events probably caused significant forming to occur, as evidenced by the 28% decrease in the OFF-state resistance. Finally, a memory event occurred (Fig. 5.39) at a lower ON-state power. (In fact, the two power levels for Figs. 5.38 and 5.39 differ by only 5%.)

There are two points to be made here. First, it appears that the degree of forming (not only the ON-state power) affects the diminution of the capacitive spike in the microphone signal. Second, the ON-state power may in some cases be too large to allow the memory event to occur. (We will discuss this second, and crucial, point again later.)

In the case of large P_{ON} and significant forming, when the capacitive jump is diminished in the microphone signal, it appears that the rapid heating caused by the capacitive discharge is sufficient to cause melting. When forming has occurred (i.e., crystallites are present), the discharge energy is most probably dissipated in a heat of transformation as well as in heating of the filament. We assume that here the formed material melts quite easily and rapidly once the melting process begins. However, any ensuing crystallization occurring after the capacitive discharge and during the pulse is assumed to occur at a much slower rate and therefore the crystallization exotherm (Fritzsche and Ovshinsky, 1970) may be ignored. As far as we know, the actual heat of transformation for a thin chalcogenide film of formed $Ge_{15}Te_{81}S_2So_2$ is not known; however, for a bulk sample, a value of approximately 15 cal/gm has been estimated. Applying this value to a memory event and assuming that after the capacitive discharge occurs some part of the formed region melts in 1 μs, then the sample is a power density "sink" at a rate of 15 cal/gm $\cdot$ 5 gm/cm^3 $\cdot$ 1/1 $\times$ 10^{-6} s = 7.5 $\times$ 10^7 cal/s $\cdot$ cm^3. Thus, in terms of $H(t)$, we might expect a large positive, exponentially decaying spike, 2 μs long (i.e., the capacitive discharge) followed by an almost equally large negative, rectangular spike, 1 μs long. Since the integrated power is nearly equal for the two spikes, and the relative time span is short, it is more convenient to use $H(t) = 0$ as in Section 5.5.3.

In terms of the calculated temperature, the effect of taking $H = 0$ is that the sample-gas interface is cooler in the memory case after the transition than in the threshold case (see Figs. 5.41 and 5.42). However, the temperature (140°C) is still in the vicinity of T_g, the glass transition temperature, above which the switching material may crystallize. Physically, while melting, the sample would have to be at or near T_M, the melting temperature (≥ 375°C). Still, the switching material probably does cool immediately after melting (Fritzsche, 1974) because it is much hotter than its surroundings. It may not cool as much as indicated in the model, however, since the experimental signal shows a "bowed" shape; a higher slope immediately after the transition than just before the end of the pulse. In any event (threshold or memory), the general temporal sequence is that the sample cools immediately after switching. Cooling may continue until the voltage pulse ceases (in which case no forming occurs), or heating may resume after a short while.

Since the formation of crystallites is a relatively slow process which, we assume, takes place during the ON-state when P_{ON} is large enough, we ignore

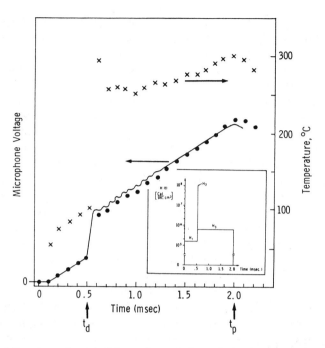

Fig. 5.41. Microphone signal (solid-experimental, dots-calculated) and calculated temperature at the sample-gas interface (crosses) as a function of time for a threshold switching event. The delay time and pulse width are noted. The oscillations in the experimental signal are due to either a resonance in the microphone or cell. Calculations suggest that they may be due to a radial mode in the cell. To obtain this fit the thermal conductivity of the substrate was increased by a factor of 2.3 at t_d; the thermal conductivity of the sample was increased by a factor of 5.3. The inset is the time-dependent power density used to fit the data.

the heat given up upon forming the crystallites. These are probably both Te and GeTe crystallites, since the calculated temperatures generally exceed 240°C for the pulses in which forming presumably occurs. It is apparent that there is a net forming from event to event in Fig. 5.33 until the highest levels of P_{ON}. This is indicated by the progressive drop in the mirroring of the capacitive discharge—more and more of the discharge energy (which is nearly constant) goes into melting at least part of the formed region, which is larger after each switching event. (Melting tends to undo the forming of the previous pulse, but there is still an apparent net forming increase from pulse to pulse.) What distinguishes a memory event from a threshold event of the same P_{ON} is then, simply, the amount of prior forming.

At the highest levels of ON-state power, melting occurs near the transition, yet P_{ON} is so large that the rate of temperature rise during the remainder of the pulse is such that $T > T_M$ at the end of the pulse also. The rapid cooling

THERMOPHONIC STUDIES OF THICK CHALCOGENIDE FILMS   431

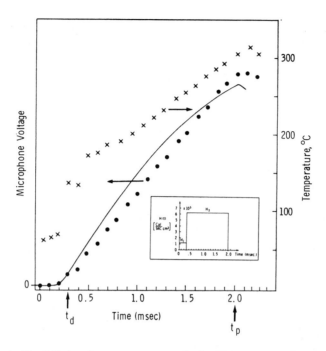

Fig. 5.42. As Fig. 5.41, but for a memory event. Setting $H_2(t) = 0$ here did not allow us to obtain the temperature maximum due to the capacitive discharge. However, we argue in the text that it must exceed the melting temperature $T_m \cong 340°C$ in order for a memory event to occur. Further, at t_p we must keep $T < T_M$, or else a threshold event will result. In this calculation the thermal conductivity of the substrate was increased by a factor of 2.0 at t_d; the thermal conductivity of the sample was increased by a factor of 4.9.

to room temperature following removal of the pulse then ensures some revitrification and development of the OFF-state; a threshold event results.

In summary, there are two factors determining whether a memory event occurs in these particular experiments:

(1) There is sufficient forming prior to the particular switching event. [A connected path of crystallites is then completed during the ON-state of the (first) memory event.]
(2) The temperature just before the end of the pulse satisfies $T \leq T_M$.

Requirement 1 is in contrast to models of memory behavior in which the external capacitance is on the order of the sample capacitance and melting is not assumed to occur (Fritzsche, 1974; Bagley, 1974). Here the sample is assumed to be Joule heated to the region $T_g \leq T \leq T_M$ and, if the pulse is sufficiently long, the crystallization of Te and/or GeTe required for lock-ON occurs during a single pulse. If we apply this model to the above experiments, we must conclude that, for the intermediate threshold case in which $T_g \leq T \leq T_M$, the pulse widths were too short to achieve lock-ON in a single shot.

So far we have discussed only the post-switching behavior. As far as the preswitching behavior is concerned, there is one particular feature which deserves attention. As forming progressed, there was a tendency for the switching to take place earlier in the pulse. (See Figs. 5.33, 5.38 and 5.39.) The resulting average temperature just before switching is smaller. This is consistent with a model of switching in which temperature-dependent electrical conductivity increases near the middle of the gap, causing field enhancement near the electrodes (Kotz and Shaw, 1982; Shaw et al., 1973). Eventually a critical electric field, F_c, is reached which causes field-stripping or avalanching; i.e., the initiation of switching (Ovshinsky, 1968; Buckley and Holmberg, 1975; Pryor and Henisch, 1972; Walsh and Vezzoli, 1974; Petersen and Adler, 1976; Adler et al., 1978; Shaw et al., 1973). As forming progresses, the effective electrode–electrode distance decreases and hence, the time it takes to reach the critical field (t_d) also decreases. In this manner, there is no difference between threshold and memory switching as far as the initiation of the event is concerned. Furthermore, when low ON-state power threshold switching was investigated, there was no visible evidence (under 40× magnification) of morphological changes, nor were the *average* temperatures calculated prior to switching sufficient to cause such changes. Indeed, no evidence for a purely thermal initiation of threshold or memory switching was obtained from the thermophonic studies.

We also note that in quantitatively modeling the microphone signal, we are limited by the fact that the expression [Eq. (5.65)] derived by Aamodt and Murphy does not give the microphone signal in absolute terms; we only know that the microphone signal is proportional to the pressure. In order to arrive at an expression in which the microphone signal is given, say, in millivolts, one would have to take into account the response of the microphone itself. This was not attempted, but the results have been of value despite this drawback.

As pointed out above, the switching and memory events observed concur with a model where the sample heats during the delay time; the heating causes the electric fields to rearrange and a critical value is reached near an electrode whereby a carrier generation process is encouraged. During t_d, voltage reversals (Balberg, 1970) or pulse bursts (Reinhard, 1977) will have no effect on t_d, since a critical local power density (Shaw and Subhani, 1981; Reinhard, 1977; Thoma, 1976) seems to be required to drive the electric fields sufficiently high near the electrodes. Once the switching event occurs, the resulting effects on the morphology of the sample are determined by the capacitive discharge energy and the ON-state power level, as discussed above. For the particular memory material discussed, the criteria for a memory event to occur in a well formed sample was simply that: (1) melting occurred at the switching transition; (2) the temperature after switching was in the crystallization range; (3) the temperature at the end of the pulse remained below the melting point. Therefore, two distinct ranges exist where threshold switching events occur: low ON-state power, where phase changes are absent or minor; high ON-state power, where T_M is exceeded at the end of the pulse. Indeed, in the high power mode of electronic switching, the sample melts twice during a switching cycle, making it understandable why it was often suggested in the past that thermal effects were fundamental to the switching process. In thin film sandwich samples, of higher glass transition temperatures and resistances, typical capacitive discharges may not provide sufficient energy for memory events to occur, and if care is taken not to overdrive these samples, the forming process is readily avoided (Petersen and Adler, 1976).

5.6 Electronic Models for Threshold Switching

5.6.1 Introduction

In prior sections we have shown that thermal effects are primarily responsible for the delay-time phenomenon, memory, forming and morphological changes in general. On the other hand, the initiation and maintenance of threshold switching events are fundamentally electronic processes. In this section we will discuss some electronic models that have been proposed to explain both the initiation and maintenance of the switched ON-state. An extensive body of literature exists on this subject (see, e.g., Adler *et al.*, 1978, 1980), with many novel ideas for the initiation (see, e.g., Haberland, 1970; Van Roosbroeck, 1972; Lee and Henisch, 1973; Zabrodshii *et al.*, 1973; Anderson, 1976; Ovshinsky, 1976; Walsh *et al.*, 1981; Walsh and Vezzoli, 1974) and maintenance (see, e.g., Adler *et al.*, 1978; Petersen and Adler, 1976; Mott, 1969, 1975) abounding in the literature. Many of these models invoke a critical electric field, F_c, which is favored because it agrees with the results of various experiments (see, e.g., Shaw *et al.*, 1973; Buckley and Holmberg, 1975). Others, such as the carrier density switching criterion proposed by Walsh and Vezzoli (1974), provide phenomenological coupled-carrier equations capable of explaining a broad range of observed switching phenomena. Since we presently favor F_c models; we will emphasize them in this section. But whatever the mechanism, electronic switching is often describable via an SNDC type model, and isothermal SNDC can lead to current filamentation without the presence of the requisite hot-spot involved in thermal models. We show this in the following section, then present a simple isothermal analysis of switching, and finally one model for the ON-state.

5.6.2 Current Filamentation

It has been thought for many years that SNDC elements are unstable against the formation of high-current-density filaments (Barnett, 1969; Schöll, 1982, 1984). A dominant mechanism by which SNDC and filamentation occurs is through the carrier concentration-dependence of the generation and/or recombination rates. Examples are: avalanche breakdown double injection (Steele *et al.*, 1962); impurity breakdown in compensated semiconductors (Melngailas and Milnes, 1962); impact ionization in Gunn diodes (Gelmont and Shur, 1973); microplasmas in PNPN structures (Shaw and Gastman, 1971; Blicher, 1976; Varlamov and Osipov, 1970). Here, the SNDC characteristics are due to the nonlinear dependence of the field-induced carrier generation or recombination on the carrier concentration; a positive feedback mechanism results and leads to filamentation. Ridley (1963) postulated the occurrence of (1) high electric field domains in NNDC elements and (2) high-current-density filaments in SNDC elements (Rockstad and Shaw, 1973) using both thermodynamic and electrical stability arguments. For NNDC elements, several subsequent detailed one-dimensional analyses (see, e.g., Shaw *et al.*, 1979) have successfully supported and augmented Ridley's principal conclusions. However, the electrical stability of SNDC elements under isothermal conditions has only recently been treated in similar detail (Schöll, 1982, 1984). Using Maxwell's equations, Shaw *et al.* (1973) demonstrated that SNDC elements are unstable against the formation of inhomogeneous current-density distributions under high-frequency excitation. In what follows, we demonstrate that filamentation is expected in general for SNDC elements under isothermal conditions.

We now demonstrate isothermal filament formation by applying the Poisson and current-continuity equations to the case of a system consisting of a semiconductor having only one type of carrier, which we choose to be n-type in this discussion. [A similar result holds for two types of carriers (Adler *et al.*, 1980).] We consider a cylindrical sample under a uniform bias applied in the z direction (using cylindrical coordinates). Mobile electrons are supplied to the conduction band, in the model presented here from shallow ionized donors for simplicity, and are trapped by deep states which, by way of example, are neutral when empty and negatively charged when filled by electrons. The population of the trapping centers depends on F_z and n. The basic equations that describe the radial component of electric field in steady state are (1) the Poisson equation

$$\frac{1}{r}\frac{\partial rF_r}{\partial r} = \frac{q}{\varepsilon}(N_D - N^- - n); \qquad (5.68)$$

(2) the current-continuity equation

$$qn\mu_n F_r + qD_n \frac{\partial n}{\partial r} = 0, \tag{5.69}$$

which ensures no radial current flow; and (3) the steady-state kinetic equation for the filled traps:

$$g(n, F_z)nN^- = \alpha(n, F_z)n(N_T - N^-). \tag{5.70}$$

Here N_D is the donor concentration, N_T is the trap concentration, N^- is the concentration of filled traps, D_n is the diffusion coefficient, and $g(n, F_z)$ and $\alpha(n, F_z)$ are the functions describing electron generation and trapping, respectively. Equations (5.68) and (5.70) yield:

$$\frac{1}{r}\frac{\partial}{\partial r}(rF_r) = \frac{q}{\varepsilon}[N_D - F(n, F_z) - n], \tag{5.71}$$

where

$$F(n, F_z) \equiv N^- = \frac{N_T}{1 + [g(n, F_z)/\alpha(n, F_z)]} \tag{5.72}$$

Equations (5.69) and (5.71) have a uniform solution F_{z0}, n_0 given by:

$$F_{r0} = 0 \tag{5.73}$$

and

$$n_0 + B(n_0, F_z) = N_D. \tag{5.74}$$

We can then solve Eq. (5.74) to find n_0 as a function of F_z and thus the current density-field characteristics of a uniform sample, $j_z = qn_0(F_z)\mu_n F_z$.

We next investigate small radial perturbations from the uniform solution: $n = n_0 + \delta n(r)$; $F_r = F_{r0} + \delta F(r)$, where $\delta n \ll n_0$. Equations (5.69) and (5.71) yield:

$$\frac{\partial^2}{\partial r^2}\delta n(r) + \frac{1}{r}\frac{\partial}{\partial r}\delta n(r) - \frac{q\gamma n_0 \mu_n}{\varepsilon D_n}\delta n(r) = 0, \tag{5.75}$$

where

$$\gamma = \frac{\partial F}{\partial n}\bigg|_{n=n_0} + 1.$$

By introducing the dimensionless variables, $N = \delta n/n_0$ and $X = r/R_D$, where $R_D = \varepsilon D_n/qn_0\mu_n|\gamma|$ is an effective Debye radius, Eq. (5.75) becomes

$$\frac{\partial^2 N}{\partial X^2} + \frac{1}{X}\frac{\partial N}{\partial X} - N = 0 \tag{5.76}$$

for $\gamma < 0$. The boundary conditions corresponding to filamentary solutions are:

$$\left.\frac{\partial N}{\partial X}\right|_{r=0} = 0 \qquad (5.77a)$$

and:

$$N_{x \to \infty} = 0. \qquad (5.77b)$$

The solution of Eq. (5.76a) is:

$$N = Z_0(ix), \qquad (5.78)$$

where

$$z_0(ix) = C_1 I_0(X) + C_2 K_0(X). \qquad (5.79)$$

I_0 is a modified Bessel function of the first kind and K_0 is a modified Bessel function of the second kind; the C's are constants. From Eq. (5.77b) we have that $C_1 = 0$, but then we cannot fulfill the other boundary condition [Eq. (5.77a)]. We thus conclude that filamentation does not occur for $\gamma > 0$.

For $\gamma < 0$, Eq. (5.76b) has a solution

$$\delta n = CJ_0(X), \qquad (5.80)$$

which satisfies the boundary conditions. Again, J_0 is a Bessel function of zeroth order. Equation (5.80) implies that a nonuniform radial carrier density distribution (current filamentation) can occur if for some value of n, a value of $\gamma < 0$ results.

We can appreciate the physical significance of the above result by considering Eqs. (5.71) and (5.73) in the steady state and graphically solving the two equations,

$$N_0^- = N_D - n_0 \qquad (5.81)$$

and

$$N_0^- = F(n_0, F_z). \qquad (5.82)$$

The solution also determines the dependence of n on F_z; both N_0^- and F_z as functions of n are shown in Fig. (5.43). Note that the condition $\gamma < 0$ (point 2 in Fig. 5.43b) implies that an SNDC curve determines the relation between n and F_z. Hence SNDC directly leads to filamentation.

It follows from the small-signal solution presented above that the characteristic size of the filament walls is approximately the Debye radius, R_D, which is characteristically quite small. We can therefore expect the following concentration profile to result (see Fig. 5.43): an extensive high-density core

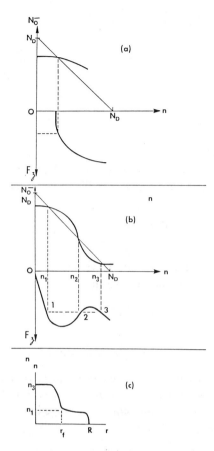

Fig. 5.43. N_0^- vs. n and n vs. F_z (see text for definition): (a) Normal case for positive differential conductance. (b) SNDC case (the number of filled traps decreases as n increases. (c) Qualitative carrier density profiles for a filament in a semiconductor having one type of carrier.

of radius $r_f \gg R_D$; a thin, charged filament wall several R_D's thick; a carrier concentration within the core of n_3; a carrier concentration far from the filament of n_1.

Although the above argument is based on small-signal considerations, we can understand large-signal solutions in the regime where the wall of the filament is small compared to the diameter of the filament. In this regime, we can approximate Eq. (5.71) as

$$\frac{\partial F_r}{\partial r} \cong \frac{q}{\varepsilon}[N_D - B(n, F_z) - n] \tag{5.83}$$

and make use of Eq. (5.69) to obtain

$$\frac{\partial}{\partial n} F_r^2 = -\frac{D_n}{\mu_n} \frac{N_D - B(n, F_z) - n}{n}, \quad (5.84)$$

or

$$F_r^2 = -\frac{D_n}{\mu_n} \int_{n_1}^{n} \frac{N_D - B(n, F_z) - n}{n} dn. \quad (5.85)$$

By considering Fig. (5.43b), we see that such a solution is only possible if the $n(F_z)$ relation exhibits a region of SNDC. Furthermore, since from Eq. (5.69) and Fig. (5.43b) we see that F_r will vanish except in the walls of the filament, then

$$\int_{n_1}^{n_3} \frac{N_D - B(n, F_z) - n}{n} dn = 0 \quad (5.86)$$

which is the analogue of the "equal areas rule" developed for NNDC systems (Butcher, 1967). It must also be true that

$$n_1 + B(n_1, F_z) = N_D \quad (5.87a)$$

and

$$n_3 + B(n_3, F_z) = N_D. \quad (5.87b)$$

By solving Eqs. (5.86) and (5.87) we can find the values n_1, n_3 and F_z that are uniquely specified for a given function $f(n_1, F_z)$. This means that an SNDC sample containing a filament is a voltage limiter. The current in the sample can increase at a fixed value of F_z simply by having r_f expand:

$$I = q\mu_n F_z \pi [n_3 r_f^2 + n_1 (R - r_f)^2]. \quad (5.88)$$

Figure 5.44 depicts the homogeneous and filamentary characteristics expected from the above considerations.

5.6.3 Isothermal Analysis

5.6.3.1 Trapping and Switching

One model of electronic switching in amorphous chalcogenide semiconductors involves the field-induced filling of positively and negatively charged traps (Adler *et al.*, 1980). Hence, we first demonstrate that the filling of traps in general can lead to an SNDC curve (Fig. 5.44, curve 2) by considering a simple idealized system where there are N_D ionized shallow donors and N_A deep traps which are negatively charged when filled by electrons.

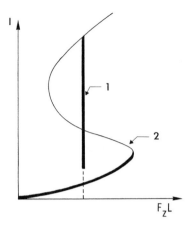

Fig. 5.44. Qualitative current-voltage characteristic for uniform applied fields for an SNDC sample containing a filament (curve 1) and in the absence of filamentation (curve 2, the homogeneous solution, which is unstable against the formation of a filament).

In the absence of field-induced carrier generation the concentration of electrons in the conduction band is

$$n = N_D - N_A. \quad (5.89)$$

When electron-hole pairs are generated, the system can be described by the well-known Shockley-Read equations:

$$G = \alpha_n n N_A (1 - f); \quad (5.90)$$

$$N_D - n - N_A f + p = 0; \quad (5.91)$$

and

$$G = \alpha_p p N_A f, \quad (5.92)$$

where G is the generation rate and $f \equiv N_A^-/N_A$, where N_A^- is the concentration of filled acceptors. For simplicity, we assume that the generation coefficients for electrons and holes are equal and that the generation rate is a function of electric field, so that

$$G = A(n + p)\lambda, \quad (5.93)$$

where A is a constant and λ is a dimensionless monotonically increasing function of the applied field. We can now rewrite Eqs. (5.90) through (5.92) in dimensionless form as:

$$\lambda(N + P) = N(1 - f); \quad (5.94)$$

$$P - N + N_0 + 1 - f = 0; \quad (5.95)$$

and
$$\xi(N + P) = Pf, \qquad (5.96)$$

where $N = n/N_A$, $N_0 = (N_D - N_A)/N_A$, $P = p/N_A$, $\lambda = A\lambda_\varepsilon/\alpha_n N_A$ and $\xi = \alpha_n/\alpha_p$. From Eq. (5.94), we have

$$f = 1 - \lambda - \frac{\lambda P}{N}. \qquad (5.97)$$

Substituting Eq. (5.97) into (5.94) and (5.96), we find

$$PN - N^2 + N_0 N + \lambda N + \lambda P = 0 \qquad (5.98)$$

and

$$\left(\frac{P}{N}\right)^2 - (\lambda^{-1} - 1 - \xi)\frac{P}{N} + \xi = 0. \qquad (5.99)$$

The solution of Eqs. (5.98) and (5.99) yields

$$\frac{P}{N_{+,-}} = \frac{\lambda^{-1} - 1 - \xi}{2} \pm \left[\left(\frac{\lambda^{-1} - 1 - \xi}{2}\right)^2 - \xi\right]^{1/2} \qquad (5.100)$$

and

$$N = \frac{N_0 + \lambda(1 + P/N)}{1 - P/N}. \qquad (5.101)$$

The existence of two possible solutions for N, P for a given value of λ indicates that an SNDC characteristic is present. Analysis of Eq. (5.100) shows that two solutions exist for

$$\frac{1}{2(1 + \xi)} \leq \lambda \leq \frac{1}{(1 + \sqrt{\xi})^2}. \qquad (5.102)$$

Equation (5.102) determines the range of voltages over which NDC exists (see Fig. 5.44, curve 2). A physical explanation of this phenomenon is the following. Under conditions of impact ionization, the generation rate is proportional to the carrier concentration [Eq. (5.93)]. For small concentrations at a given voltage, the generation rate is small, but if a larger concentration of electron-hole pairs can also exist for the same voltage, then a larger generation rate will also develop. Furthermore, the recombination mechanism depends not only on the carrier concentration but also on whether or not the traps are filled by holes. For example, we might have the following two states corresponding to one value of voltage: (1) low electron concentration, traps not filled by holes (OFF); (2) high electron concentration, traps substantially filled by holes (NDC). In essence, in this model carrier generation produces pairs; the holes neutralize the traps that were compensated by electrons and the generated electrons enhance the conductivity of the material by increasing the density of free carriers. In the model, both the generated electrons and holes fill the charged traps, thereby increasing the mobility, and also provide excess carriers. The conductivity is enhanced by both a mobility and a carrier-concentration increase.

5.6.3.2 The Phenomenological Kinetic Equations

We have now shown that carrier generation followed by a change in concentration of the filled traps can produce an SNDC characteristic which in turn leads to current filamentation. For the case of one type of carrier, it is trap emptying that produces SNDC and filamentation. For two types of carriers (Adler et al., 1980) we can have SNDC with either trap emptying or trap filling, depending on the details of the generation and recombination processes. We now use these ideas to generate and solve a set of phenomenological kinetic equations for an amorphous chalcogenide film containing equal numbers of C_3^+ and C_1^- trapping centers, specific defects discussed in Chapter 4. We first assume that in the OFF-state the following recombination processes dominate: (1) hole capture by C_1^- centers ($C_1^- + e^+ \to C_1^0 \to C_3^0$); (2) electron capture by C_3^+ centers ($C_3^+ + e^- \to C_3^0$). The second process should be faster since it does not involve a local bond rearrangement and thus has no activation barrier. This effect further enhances the p-type nature of the OFF-state (Adler and Yoffe, 1976) if we assume that emission from the charged centers is a rare process. In this model the charged centers act as recombination centers.

We further assume that in the ON-state the C_3^+ and C_1^- centers are filled, so that the following processes are important: (1) hole capture by a neutral center ($C_3^0 + e^+ \to C_3^+$); (2) electron capture by a neutral center ($C_3^0 + e^- \to C_3^- \to C_1^-$). The latter process is slower since it requires a local bond rearrangement which involves an activation barrier (Adler, 1977). However, immediately after either hole or electron capture by a neutral center, the charged center so created will rapidly capture a mobile carrier of opposite charge from the dense background of mobile carriers present in the ON-state: $C_3^0 + e^+ \to C_3^+ + e^- \to C_3^0$, a slow and then very fast process with no bond modification; $C_3^0 + e^- \to C_1^- + e^+ \to C_3^0$, a very slow and then very fast process with two bond modifications. Both processes will be much slower than the OFF-state events because the C_3^0 centers are neutral. The ON-state should therefore be intrinsic in the sense that the same number of oppositely charged mobile carriers are present. It could, however, appear as n- or p-type in nature depending upon the relative band mobilities and/or the nature of the contact conditions, which e.g., could be more blocking for one of the carrier species (Petersen et al., 1976).

Taking into account the recombination processes described above, the following set of phenomenological kinetic equations can be used to describe a thin amorphous chalcogenide film.

For holes,

$$\frac{1}{q} \text{div} \, j_p + \frac{\partial p}{\partial t} = G_t - \gamma_p N_3^0 p - \beta_p N_1^- p, \qquad (5.103)$$

where G_t is the total generation rate (thermal plus electronic) for electrons and holes, N_3^0 is the concentration of C_3^0 centers, N_1^- is the concentration of C_1^- centers and γ_p and β_p are capture coefficients. The second term on the right-hand side describes the capture of holes by neutral centers and the third term describes the capture by C_1^- centers. As stated above, we expect that $\gamma_p < \beta_p$. For electrons,

$$-\frac{1}{q} \text{div} \, j_n + \frac{\partial n}{\partial t} = G_t - \gamma_n N_3^0 n - \beta_n N_3^+ n. \qquad (5.104)$$

According to the above arguments we expect that $\beta_n > \beta_p > \gamma_p > \gamma_n$; electron capture by C_3^+ is faster than hole capture by C_1^-, which is faster than hole capture by C_3^0, which in turn is faster than electron capture by C_3^0. In general,

$$\frac{\partial N_3^+}{\partial t} = \gamma_p N_3^0 p - \beta_n N_3^+ n - A N_3^+ N_1^- + B(N_3^0)^2, \qquad (5.105)$$

where the last two terms on the right-hand side take into account the reaction,

$$2C_3^0 \rightleftharpoons C_3^+ + C_1^-, \qquad (5.106)$$

where A and B are reaction coefficients. Similarly,

$$\frac{\partial N_1^-}{\partial t} = \gamma_n N_3^0 n - \beta_p N_1^- p - A N_3^+ N_1^- + B(N_3^0)^2. \qquad (5.107)$$

Also, the total number of traps, N_T, is given by

$$N_T = N_3^0 + N_3^+ + N_1^-. \qquad (5.108)$$

Adding Eqs. (5.103) and (5.105) and then subtracting Eqs. (5.104) and (5.106) yields the continuity equation,

$$\text{div} \, j + \frac{\partial \rho}{\partial t} = 0, \qquad (5.109)$$

where

$$\rho = q(p - n + N_3^+ - N_1^-) \qquad (5.110)$$

ELECTRONIC MODELS FOR THRESHOLD SWITCHING

is the space charge, and

$$j + j_p + j_n \quad (5.111)$$

is the total current density.

In the absence of mobile charge carriers the steady state equations are:

$$AN_3^+ N_1^- = B(N_3^0)^2; \quad (5.112)$$

$$N_T = N_3^+ + N_3^0 + N_1^-; \quad (5.113)$$

$$N_3^+ = N_1^-. \quad (5.114)$$

Equation (5.114) shows that in the absence of mobile carriers the C_3^+ and C_1^- centers can be created or destroyed only pairwise [see Eq. (5.106)]; it follows from the condition of charge neutrality in the absence of mobile carriers. According to an OFF-state model of chalcogenide materials (Kastner et al., 1976) $B \gg A$, so that when p and n are small enough the first two terms on the right-hand side of Eqs. (5.105) and (5.107) are negligible and

$$N_3^+ = N_1^- \cong \frac{N_T}{2}. \quad (5.115)$$

5.6.3.3 The OFF-State Kinetic Equations

In the OFF-state, p, n and N_3^0 are all very small compared to N_3^+ and N_1^-. Under these conditions, the kinetic Eqs. (5.103), (5.104), (5.105) and (5.107) become:

$$\frac{1}{q} \operatorname{div} j_p + \frac{\partial p}{\partial t} \cong G - \frac{p - p_0}{\tau_p}; \quad (5.116)$$

$$-\frac{1}{q} \operatorname{div} j_n + \frac{\partial n}{\partial t} \cong G - \frac{n - n_0}{\tau_n}; \quad (5.117)$$

$$\frac{\partial N_1^-}{\partial t} \cong \frac{p - p_0}{\tau_p} + G; \quad (5.118)$$

$$\frac{\partial N_3^+}{\partial t} \cong -\frac{n - n_0}{\tau_n} + G; \quad (5.119)$$

where $G = G_t - G_{th}$ is the generation rate due to the electric field, which for the moment we assume is uniform across the sample, and:

$$\tau_p \cong \frac{2}{\beta_p N_T}; \tag{5.120a}$$

$$\tau_n \cong \frac{2}{\beta_n N_T}; \tag{5.120b}$$

$$n_0 = G_{th}\tau_n; \tag{5.121a}$$

$$p_0 = G_{th}\tau_p. \tag{5.121b}$$

Equations (5.116) through (5.121) represent the following model. Holes are trapped by C_1^- centers and electrons by C_3^+ centers; their concentrations diminish and the reaction of Eq. (5.115) proceeds to the right in order to replenish the C_3^+ and C_1^- centers that have been neutralized via trapping. Reemission of carriers from the C_3^0 centers is assumed negligible with respect to the reaction of Eq. (5.115). Again, in this limit, the charged centers are recombination rather than trapping centers.

Since $\beta_n > \beta_p$, then $\tau_n < \tau_p$ and $n_0 < p_0$; again, there are more free holes than electrons and the OFF-state is p-type.

Equations (5.116) through (5.121) are valid as long as

$$\beta_n \frac{nN_T}{2} \gtrsim A\left(\frac{N_T}{2}\right)^2. \tag{5.122}$$

The left-hand side of Eq. (5.122) represents the rate of production of C_3^0 centers by electron trapping and the right-hand side is the rate of production of C_3^0 centers via the reaction of Eq. (5.106). If we make the reasonable assumption that Eq. (5.112) is valid up to the relatively high field-induced generation level $G \gg G_{th}$, then the condition of validity of Eq. (5.116) through (5.121) becomes

$$G \gtrsim \frac{AN_T^2}{4}. \tag{5.123}$$

In the range of validity of Eq. (5.123), we may assume that $n \ll p$; the steady OFF-state conductivity is then:

$$\sigma \cong \sigma_p = q\mu_p p. \tag{5.124}$$

Furthermore, if we assume that G is related to an impact ionization process, then

$$G = pg(F); \tag{5.125}$$

ELECTRONIC MODELS FOR THRESHOLD SWITCHING

in steady state Eq. (5.116) yields

$$p = \frac{p_0}{1 - g(F)\tau_p}, \qquad (5.126)$$

and thus

$$\sigma = \frac{q\mu_p p_0}{1 - g(F)\tau_p}. \qquad (5.127)$$

Equation (5.127) describes avalanche breakdown in the OFF-state. The critical field for breakdown, F_c, is determined by the condition:

$$g(F_c) = \tau_p^{-1}. \qquad (5.128)$$

For voltages below V_t the conductivity will be field-dependent and, for $g(F)\tau_p \ll 1$, Eq. (5.127) yields

$$\sigma \cong q\mu_p p_0 [1 - g(F)\tau_p]. \qquad (5.129)$$

There are several forms of the conductivity expression that can be used for modeling carrier generation effects. For example, a common one is $\sigma \sim \sigma \exp\{-F_0/F\}$ (Adler et al., 1980). Another possibility is to write:

$$g(F) = A' \exp\left(\frac{\varepsilon}{\varepsilon_0'}\right), \qquad (5.130)$$

with $F_0' = CT$. Taking $T = 300$ K, $C = 200$ V/cm K ($\varepsilon_0' = 6 \times 10^4$ V/cm), the band mobility $\mu_p \cong 10$ cm^2/V s, $\tau_p \cong 10^{-8}$ s, $F_c \cong 10^5$ V/cm and $\sigma_0 \cong 1.5 \times 10^{-6}$ Ω^{-1} cm^{-1} yields $p_0 \cong 10^{12}$ cm^{-3} and $A' \cong 2 \times 10^7$ s^{-1}. (If we were to choose μ_p as the drift mobility, then p_0 would be the free plus trapped hole concentrations at states near the valence-band edge.) The resulting $I(V)$ characteristic for this *uniform field* OFF-state case is shown in Fig. 5.45. Note that the $I(V)$ characteristic is of the form obtained experimentally for short (2 ns) pulses (Buckley and Holmberg, 1975). Agreement is expected here since: (1) isothermal conditions prevail; (2) the fields across the sample will be relatively uniform; and (3) the exciting voltage pulse is turned off prior to the time it takes for the generated carriers to propagate and fill traps, the condition where all the traps are filled in the sample. A feedback loop is not yet established and the resulting characteristics are simply those representative of an avalanche diode. Were we to model an actual sample where thermal and contact effects are treated, delay-time-mode switching would then occur at points on the OFF-state $I(V)$ characteristic below V_t (see Fig. 5.45), and the avalanche characteristic would be masked.

5.6.3.4 The ON-State Kinetic Equations

In the ON-state, the electron and hole concentrations are sufficiently large so that the C_3^+ and C_1^- centers are almost completely neutralized. When p and n are sufficiently large, we assume that the rates of production and

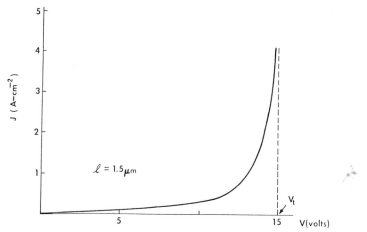

Fig. 5.45. Estimated $I(V)$ characteristic in the isothermal OFF-stage for a uniform applied electric field. $L = 1.5\,\mu m$.

decay of the charged centers due to the capture of mobile carriers are much larger than the corresponding rates due to the reaction of Eq. (5.106). This situation is opposite to that of the OFF-state, where p and n are both sufficiently small so that the reaction of Eq. (5.106) is almost completely responsible for the small value of N_3^0 and the fact that $N^+ = N^- \cong N_T/2$. Now, since p, n and N_3^0 are large compared to N_3^+ and N_1^-, and $N_3^0 \cong N_T$, Eqs. (5.103), (5.104), (5.105) and (5.107) become:

$$\frac{1}{q}\operatorname{div} j_p + \frac{\partial p}{\partial t} \cong G_t - \gamma_p N_T p - \beta_p N_1^- p; \qquad (5.131)$$

$$-\frac{1}{q}\operatorname{div} j_n + \frac{\partial n}{\partial t} \cong G_t - \gamma_n N_T n - \beta_n N_3^+ n; \qquad (5.132)$$

$$\frac{\partial N_3^+}{\partial t} = \gamma_p N_T p - \beta_n N_3^+ n; \qquad (5.133)$$

$$\frac{\partial N^-}{\partial t} = \gamma_n N_T n - \beta_p N_1^- p. \qquad (5.134)$$

In the steady state, these become:

$$N_3^+ = \frac{\gamma_p N_T p}{\beta_n n}; \qquad (5.135)$$

$$N_1^- = \frac{\gamma_n N_T n}{\beta_p p}; \qquad (5.136)$$

ELECTRONIC MODELS FOR THRESHOLD SWITCHING

$$\frac{1}{q} \text{div } j_p = G_t - \gamma_p N_T p - \gamma_n N_T n; \tag{5.137}$$

$$-\frac{1}{q} \text{div } j_n = G_t - \gamma_n N_T n - \gamma_p N_T p. \tag{5.138}$$

Furthermore, when either (1) the current is uniformly distributed in the steady ON-state (i.e., the filament fills the entire sample), or (2) we consider only the central part of the filament where the carrier density is uniform, we have:

$$G_t - \gamma_p N_T p - \gamma_n N_T n = 0 \tag{5.139}$$

and

$$p - n + \frac{\gamma_p N_T p}{\beta_n n} - \frac{\gamma_n N_T n}{\beta_p p} = 0. \tag{5.140}$$

For large generation levels, Eqs. (5.139) and (5.140) have as a solution

$$p = n = G_t \tau \tag{5.141}$$

where

$$\tau^{-1} = N_T(\gamma_p + \gamma_n) \cong N_T \gamma_p, \tag{5.142}$$

provided $\gamma_p \gg \gamma_n$. Equation (5.141) is valid if:

$$p = n \gg \frac{\gamma_p N_T}{\beta_n}, \frac{\gamma_n N_T}{\beta_p}, \tag{5.143}$$

Equations (5.141) through (5.143) then yield:

$$G \gg \frac{\gamma_p^2 N_T^2}{\beta_n}, \frac{\gamma_p \gamma_n N_T^2}{\beta_p}, \tag{5.144}$$

which determines the lower limit of the generation rate required to maintain the ON-state.

We now estimate order of magnitude values for γ_p and β_n. If we assume that the capture cross section, S_p, for hole trapping by a C_3^0 center is about 10^{-20} cm² (a reasonable value), and the thermal velocity, v_t, is about 10^7 cm s, then $\gamma_p \cong S_p v_t \cong 10^{-13}$ cm^{-3} s^{-1}. Furthermore, for $N_T \cong 10^{19}$ cm^{-3} and $\tau_n \cong 10^{-11}$ s, we find $\beta_n \cong 2/N_T \tau_n \cong 2 \times 10^{-8}$ cm^{-3} s^{-1}. With these values and Eqs. (5.141) and (5.142), we can obtain a portrait of the ON-state. First, the mobile-carrier density $p = n \gg 10^{14}$ cm^{-3}. Next, for band mobilities of the order of 10 cm²/V s and $p = n \cong 10^{19}$ cm^{-3} we obtain a minimum conductivity in the ON-state of $\sigma_{\min} = q(\mu_p + \mu_n)p \cong 20 \, \Omega^{-1}$ cm^{-1}. Finally, for holding voltages of about 1 V and bulk holding fields in the ON-state, F_h, of about 10^3 V cm, we obtain for the minimum holding current density $j_{h,\min} = \sigma_{\min} F_h \cong 2 \times 10^4$ Å cm², which is a typical value observed experimentally (Petersen and Adler, 1976).

5.6.4 A Model for the ON-State

We may now write a set of equations for the ON-state as follows:

$$p = n;$$

$$\frac{1}{q} \operatorname{div} j_p = G_t - \frac{p}{\tau};$$

$$-\frac{1}{q} \operatorname{div} j_n = G_t - \frac{n}{\tau}; \qquad (5.145)$$

$$j_p = q\mu_p p\varepsilon - qD_p \operatorname{grad} p;$$

$$j_n = q\mu_n n\varepsilon + qD_n \operatorname{grad} n.$$

Our thrust will be to solve this set of equations for a system where field-induced carrier generation occurs in a very narrow region near the electrodes and the current is nonuniform in a direction transverse to the direction of current flow. Prior to this, however, we will estimate numerical values for some important parameters that characterize the ON-state and discuss a qualitative model.

We first estimate the trapping time, $\tau \cong (\gamma_p N_T)^{-1} \cong 10^{-6}$ s, and dielectric relaxation time, $\tau_\mu \cong \varepsilon/\sigma \cong 3 \times 10^{-13}$ s, in the ON-state. Since $\tau_\mu \ll \tau$, the ON-state behaves like a "lifetime" rather than "relaxation" semiconductor (van Roosbroeck and Casey, 1972; Popescu and Henisch, 1975). The product of τ_μ and v_t yields the characteristic length of the charge separation, which in our case is about 10^{-6} cm. We have, therefore, a quasi-neutral situation since the charge separation is much less than the length of the sample.

Next, we estimate the ambipolar diffusion coefficient, $D_A \cong \mu kT/q \cong 0.3$ cm² s, and diffusion length $L_D = \sqrt{D_A \tau} \cong 6 \times 10^{-4}$ cm. Note that typical sample lengths, l, and radii, R, are such that $L_D > l$ and $R > L_D$.

Since V_h is independent of l over a wide range, we expect that the field-induced generation of carriers in the ON-state takes place in narrow regions near the electrodes, perhaps even less than 100 Å in thickness. The field at least in part of these regions is at F_c; electrons and holes created by F_c drift and diffuse out of these regions without recombining to any great extent *anywhere* in the sample (since $L_D > l$). To determine which of the two transport processes (drift or diffusion) dominate, we compare the drift current density, $j_{\text{drift}} \cong \sigma_{\text{on}} V_h/l \cong 2q\mu_p F_h$, with the diffusion current density, $j_{\text{diff}} = qD_A \partial p/\partial x \cong qD_A p/l$. Using the Einstein relation, their ratio is $2qF_h l/kT \cong 100$; the drift current is clearly dominant.

We reinforce the following basic model for the ON-state. Two very thin regions at the contacts (Mott, 1969) generate electrons (at the cathode) and holes (at the anode) at equal rates; the holes drift towards the cathode and the electrons towards the anode [double avalanche injection (Stein et al., 1962; Gunn, 1956)]. Their concentrations are equal and the electron-hole plasma [droplet? Mott (1975)] is uniform in the direction of current flow, since $L_D > l$. Because we have a homogeneous plasma, we can introduce, instead of G, an effective generation rate, G_{eff}, which does not depend on position. The simplest case is to assume that G_{eff} does not depend upon the applied voltage; it is just a constant. The conditions for the steady ON-state then yield the following estimate: $G_{\text{eff}} \cong p/\tau \cong 10^{24}\,\text{cm}^{-3}\,\text{s}^{-1}$. (We shall use this simple model in what follows. If required, however, a dependence of G_{eff} on bias or carrier density can readily be incorporated into the model.)

It is useful now to compare the inverse of the characteristic generation time at *threshold*, $g(F_c) \cong \tau_p^{-1} \cong 10^8\,\text{s}^{-1}$, with the analogous ON-state parameters at *holding*, $g_{\text{on}} \cong G_{\text{eff}}/p = \tau^{-1} \cong 10^6\,\text{s}^{-1}$. We see that due to the much longer trapping time in the ON-state, g_{on} is 2 orders of magnitude smaller than $g(F_c)$. Hence, the holding voltage can be much smaller than the threshold voltage. (Furthermore, since $\tau_p \cong 2/\beta_p N_T$, we expect that $g(F_c)$ should be relatively insensitive to the carrier concentration. The results of photo-excitation experiments (Henisch et al., 1974) suggest that this is indeed the case over a moderate range of photoinduced conductance changes; V_t is independent of light intensity in this range. However, V_t can be lowered by sufficiently intense optical radiation.) At threshold, we have seen that a current runaway (avalanche) will occur with *no* voltage switchback until the bulk traps fill. When the traps are filled (at high current levels, about the same order of magnitude as I_h) the voltage across the sample drops because a smaller inverse generation time is sufficient to compensate for recombination via the neutral traps. If F_c is reached uniformly across the sample, then excess current will flow while the traps fill. If F_c is reached in only a small part of the sample, the amount of excess current that will flow during t_d could depend upon whether one or both types of carriers are produced locally. If we are biased in the statistical regime (Shaw et al., 1973; Lee et al., 1972) just at threshold, it is possible to observe no excess current flow at all during t_d (Lee et al., 1972).

The foregoing arguments define two distinct states in which an amorphous chalcogenide film can carry current: an ON-state with a current density $j_{\text{on}} \cong 10^4\,\text{Å cm}^2$; an OFF-state where the current density is several orders of magnitude smaller. The maximum ON-state current will be given by $I_{\text{max}} = j_{\text{on}} A$, where A is the cross-sectional area of the sample.

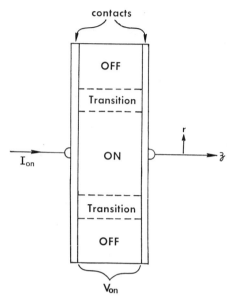

Fig. 5.46. Sketch of the ON-state current density configuration.

Suppose that we supply current to the system such that it has a value below I_{max} but greater than the critical switching current, I_t. The sample can accommodate such a current by forming a current filament, as discussed earlier. A qualitative model for the filament ("pancake") is illustrated in Fig. 5.46. In the regions of the sample which are in the OFF-state, the carrier concentration is very small; the center of the filament is in the ON-state. Carriers diffuse from the core of the filament to the transition region between the ON and OFF-states. The steady-state radially-dependent equations that describe the *transition* region (i.e., the walls of the filament) are:

$$j_{pr} = q\mu p - qD\frac{\partial p}{\partial r}; \qquad (5.146)$$

$$j_{nr} = q\mu n + qD\frac{\partial n}{\partial r}; \qquad (5.147)$$

$$\frac{1}{r}\frac{\partial}{\partial r}rj_{nr} + \frac{1}{r}\frac{\partial}{\partial r}rj_{pr} = 0; \qquad (5.148)$$

$$n = p; \qquad (5.149)$$

$$\frac{1}{q}\frac{1}{r}\frac{\partial}{\partial r}rj_{pr} = -\frac{p}{\tau_{eff}}, \qquad (5.150)$$

where τ_{eff} is an effective lifetime in the transition region comprising the walls of the filament [we estimate $\tau_{\text{eff}} \cong (\tau\tau_p)^{1/2} \cong 10^{-7}$ s]. Equations (5.146) and (5.147) describe the radial flux of holes and electrons, where we have assumed that $\mu_n = \mu_p$ and $D_n = D_p = D$. The sample is cylindrical and the ON-state plasma is homogeneous in the z direction. Equation (5.148) is the continuity equation, Eq. (5.149) represents quasineutral conditions, and Eq. (5.150) is the continuity equation for holes. Since $j_{r\text{total}} \equiv j_{nr} + j_{pr} = 0$, then $j_{nr} = -j_{pr}$. Using this fact, we obtain from Eqs. (5.146), (5.147), (5.149) and (5.150):

$$j_{pr} = -qD\frac{\partial p}{\partial r}; \tag{5.151}$$

$$\frac{1}{q}\frac{\partial}{\partial r}j_{pr} + \frac{1}{q}\frac{j_{pr}}{r} = \frac{p}{\tau_{\text{eff}}}. \tag{5.152}$$

These yield

$$D\frac{\partial^2 p}{\partial r^2} + \frac{D}{r}\frac{\partial p}{\partial r} - \frac{p}{\tau_{\text{eff}}} = 0, \tag{5.153}$$

the Bessel equation, which was analyzed earlier. As before, we employ dimensionless variables: $y = p/p_0$, where p_0 ($= G\tau$) is the concentration inside the core of the filament; $x = r/(D\tau_{\text{eff}})^{1/2}$. Equation (5.153) then becomes

$$\frac{\partial^2 y}{dx^2} + \frac{1}{x}\frac{\partial y}{\partial x} - y = 0, \tag{5.154}$$

whose solution is

$$y = Z_0(ix), \tag{5.155}$$

where

$$Z_0(ix) = C_1 I_0(x) + C_2 k_0(x). \tag{5.156}$$

I_0 is a modified Bessel function of the first kind and k_0 is a modified Bessel function of the second kind. C_1 and C_2 are constants determined from the boundary conditions:

$$y(x_f) = 1; \tag{5.157a}$$

$$y(x \to \infty) = 0. \tag{5.157b}$$

Here, $x_f = r_f/(D\tau_{\text{eff}})^{1/2}$ is the dimensionless wall width. From the boundary conditions, we have that $C_1 = 0$ because $I_0(x)$ increases exponentially for large r. The solution is

$$y(x) = \frac{k_0(x)}{k_0(x_f)}. \quad (5.158)$$

For $x \gg 1$, $k_0(x) \cong (\pi/2x)^{1/2}e^{-x}$. Thus, if the core of the filament is much larger than the diffusion length $(x, x_f \gg 1)$, then

$$y(x) \cong \left(\frac{x_f}{x}\right)^{1/2} e^{-x}. \quad (5.159)$$

This solution describes the diffusion of carriers radially outward from the filament and their subsequent recombination. The dimension of the filament walls should be of the order of a diffusion length; we therefore obtain for the minimum filament radius, $r_{f,\min} \cong (D\tau_{\text{eff}})^{1/2} \cong 10^{-4}$ cm, in reasonable agreement with experiment (Petersen and Adler, 1976). The model also predicts that the cross-sectional area of the filament increases linearly with current (Ovshinsky, 1968), as experimentally observed (Petersen and Adler, 1976). Furthermore, the maximum benign interruption time of the ON-state, t_{sm} (see Section 5.4.1.2) which is about 100–500 ns (Petersen and Adler, 1976; Pryor and Henesch, 1972), is determined by the lifetime in the transition region;

$$t_{sm} \cong \frac{\pi r_f^2}{4\pi D} = \frac{I}{4\pi D j_{\text{on}}}. \quad (5.160)$$

An approximate proportionality of t_{sm} and I has been experimentally observed (Pryor and Henesch, 1972).

The ON-state consists of about 10^{19} cm^{-3} electrons and holes moving through the conduction and valence bands, respectively (Petersen and Adler, 1976; Petersen et al., 1976). If these are excited only from the C_1^- and C_3^+ centers, rather than from interband generation, then it restricts the maximum free-carrier concentration to the density of VAPs (Kastner et al., 1976). In this model, when additional current is applied after the material is in the ON-state, the carrier concentration cannot increase; instead, the filament grows in area, maintaining a constant current density (Petersen and Adler, 1976). Such a result, however, is expected for systems that are effective voltage limiters, as the ON-state is.

Petersen and Adler (1977) have also proposed another possible, but similar, model for the ON-state, in which the conduction takes place within the C_3^0 band (centered about 0.05 eV below the conduction-band mobility edge) rather then in the valence and conduction bands. In this model, the high conductivity is initiated by a Mott transition in the C_3^0 band, which should occur at a critical carrier concentration very close to that observed in the ON-state. The most evident experimentally observable difference between that model and the one presented here is the dependence of the ON-state carrier concentration on the VAP density. In the latter, the carrier concentration in the ON-state should vary with the VAP density, while in the alternate model of Petersen and Adler it should be independent of the VAP density; however, a minimum VAP density exists in the latter, below which no switching can occur. At present, insufficient data exist on the VAP densities of different switching materials to even begin to critically test these models, but we should note that the two descriptions of the ON-state are analogous to the two possibilities which arise when a semiconductor such as crystalline silicon is heavily doped—the material becomes degenerate whenever a Mott transition takes place in the impurity band or the Fermi energy enters the conduction band. There is no *a priori* way to determine which occurs first.

The Mott-transition model has some further appealing features. First, it is consistent with having an ON-state barrier ($\cong 0.4$–0.5 eV) less than the band gap in that this configuration allows only one type of ON-state carrier (electrons) to dominate, which appears to be the case (Petersen *et al.*, 1976). Here space-charge neutrality is maintained in the C_3^0 band by electrons donated from C_3^0 sites. (Holes will be less likely to move off C_3^0 sites because a local bonding rearrangement would be required.) Second, it also provides for the second type of carrier required by the rather successful phenomenological equations of Walsh and Vezzoli (1974). (The first type is required to fill the C_3^+ and C_1^- sites; this generates the second type which moves in the C_3^0 band. Switching occurs when enough of the latter are present to cause a Mott transition.) Last, the ON-state luminescence observed by Walsh *et al.* (1978) can be explained via the $2C_3^0 \rightarrow C_3^+ + C_1^-$ transition, which produces about 0.5 eV of energy.

The above model assumes that electrons dominate; hence, the cathode barrier is assumed to drop the largest portion of the holding voltage. Another possibility for the ON-state in this case would be simply the thermal activation from the C_3^0 impurity band to the conduction band. The activation energy for this case, about 0.05 eV, is consistent with the temperature dependence of the conductivity observed in the ON-state. Finally, the ON-state luminescence (Walsh *et al.*, 1978) centered at about 0.5 eV could be associated with a collapsed band gap (Ma, 1977). If this were the case, then the ON-state barrier would be comparable with the reduced band gap and either avalanche or double injection could sustain the ON-state current. (A partially collapsed band gap could result if the filling of traps decreased the electronic disorder in the system; localized states would transform to extended states.)

5.7 Summary

In this chapter we have emphasized the major electronic and thermal features of amorphous chalcogenide switching and memory devices. The results of our analysis, coupled with experiments (Petersen and Adler, 1976) and the circuit aspects we shall discuss in the appendix are summarized in Figs. 5.46 and 5.3. Figure 5.46 depicts the filamentary current distribution in the ON-state and Fig. 5.3 (see page 359), the complete $I(V)$ characteristics of an amorphous chalcogenide switch. The thick solid line represents the steady-state characteristics of a threshold switch under normal operating conditions, i.e., as observed on a curve-tracer oscilloscope. Here we may expect some thermal contributions to the OFF-state $I(V)$ curve, depending on the sample geometery, composition, and thermal boundary conditions. The thin solid line shows the actual current and voltage during the very rapid switching transition as determined by the local lumped-element circuit parameters that are always present (see the appendix) but excluding the capacitances associated with the measuring circuit (Callarotti and Schmidt, 1977; Kotz and Shaw, 1984). At any value of current the difference between the thin and thick lines represents the inductive voltage drop in the sample. Note the damped oscillatory development of the ON-state. The near-vertical arrows (dashed) represent the critical field-induced avalanche and trap-filling processes; i.e., they represent the initial transport reponse of the carriers to the presence of a critical field, neglecting inductive and capacitive effects. In the usual delay-time mode, where heating causes the critical field to be reached in only part of the film, the avalanche process is masked by the trapping of carriers at the C_3^+ and C_1^- centers during the last part of the delay time. In this mode, Joule heating effects, usually near the center of the sample, first cause the conductivity there to rise and the field to rearrange. An enhanced concentration of thermally-generated carriers fills some additional traps, so that when the critical field is reached near one of the electrodes, fewer carriers have to be field-induced in order to fill the bulk traps. (In the electrothermal calculations we assumed that the ON-state is achieved immediately once the field in any region reached F_c; no electronic delay time mechanism was introduced. Hence, the calculated t_d's should be, and are, shorter than those observed experimentally.)

For the case where relatively uniform fields above a critical value can be established essentially isothermally over the entire sample, the avalanche region can be observed in virgin samples (Buckley and Holmberg, 1975; Allsopp et al., 1977). (The departure of the arrows from the near-vertical line in Fig. 5.3 represents the response of the carriers to the field during the very rapid switching transition. At this point, all the traps are filled.) Since carriers are generated throughout the sample in this mode, quite large prethreshold currents (Buckley and Holmberg, 1975; Allsopp *et al.*, 1977) can be attained. These, in turn, can produce local heating and thermally assisted filamentation (Allsopp *et al.*, 1977) prior to the strong thermal spike capacitively induced at the switching transition, with filamentary current densities of the same order of magnitude as those in the ON-state. However, the temperatures reached inside the preswitched filament ($\approx 100°C$) are well below those required to produce ON-state conductivity values based solely on thermal effects ($\approx 600°C$).

For a given sample and reactive-component environment, when the load resistor is increased such that $I_h < I_{on} < I_{hm}$, circuit-controlled relaxation oscillations are observed (Shaw *et al.*, 1973; Callarotti and Schmidt, 1977). The region between I_h and I_{hm} can be revealed by an intimate-double-pulse technique (Hughes *et al.*, 1975), where situations such that $I_{hm} = I_t$ have been observed.

If a sample is switched to the steady ON-state and then kept there for a sufficiently long time, either of two things may happen. For a relatively high-resistance material, where the temperature at the core of the filament is only moderately above ambient, the sample switches OFF once the bias is removed in an oscillatory manner that is essentially the same as shown in Fig. 5.3 during the OFF to ON-state transition. However, for cases where the material and operating conditions are such that a crystalline phase is induced in the material, the ON-state becomes fixed and the steady-state characteristics for $I_{on} < I_f$ are then as shown by the dashed-dot line in Fig. 5.3. A high-current reset pulse must then be established to restore the amorphous OFF-state (Ovshinsky, 1968; Ovshinsky and Fritzsche, 1973; Kotz and Shaw, 1984).

The initiation of switching in thin amorphous chalcogenide films is fundamentally an electronic process. Furthermore, the maintenance of the filamentary ON-state in threshold switches is also electronic in nature. The switching transition occurs when a critical electric field is reached somewhere in the sample, usually near an electrode. Field-induced carrier generation then causes the charged traps to fill (neutralize). When all the traps are filled, carriers can transit the sample with an enhanced mobility and the generation rate required to keep the traps filled is reduced from its threshold value; switching then occurs.

It is clear that switching in amorphous chalcogenide films, as all other phenomena involving power dissipation, exhibits thermal effects. The basic thermal aspects of the switching process appear in the delay time, which has an appreciable thermal component, the forming process, often affected most by the capacitive discharge, and in the conventional memory effect, which is explained readily via a thermally induced amorphous-to-crystalline phase transition. The thermal effects that dominate prethreshold heating and the delay time may also help determine the primary filament nucleation site. Depending upon how the sample is driven (short or long pulses), thermal effects can either lead or follow electronic effects prior to the switching transition, which always produces a capacitive circuit-induced thermal spike, unless all the capacitive energy goes into changing the phase of the material. More subtle types of electronic memory, formation and dc effects may also occur, but their origin remains somewhat speculative at this time.

Finally, we would like to report on recent progress made by Energy Conversion Devices, Inc., in the fabrication of chalcogenide switching devices using photolithographic and dry processing techniques. Figure 5.47 shows a cross-sectional view of the structure of the "Mesa" geometry device shown in Fig. 5.48. The Scanning Electron Microscope picture of the Mesa device is shown in plan view. These are dc-stable threshold switches. Figure 5.49 shows a typical stability curve as measured on one of the devices; the current density is 20,000 amperes per square centimeter in the filament during the ON-state.

Appendix

Our task is to determine the conditions necessary for either a high current density filament (switching to the ON-state) or circuit-controlled oscillations to occur. We examine the situation where oscillations occur in the absence

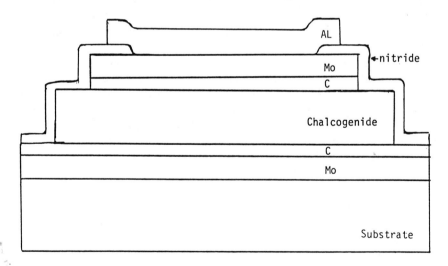

Mesa Structure

Fig. 5.47. Threshold switch mesa device structure fabricated by Energy Conversion Devices, Inc.

SUMMARY

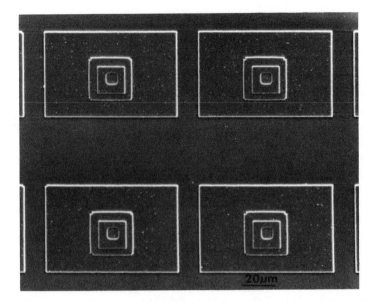

Fig. 5.48. SEM view of the threshold switch device structure shown in Fig. 5.47.

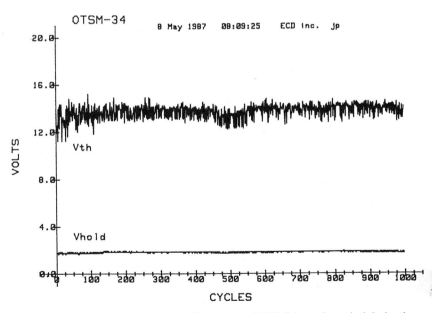

Fig. 5.49. Running threshold voltage, V_T, for each of 1000 firings of a typical device shown in Fig. 5.48.

of filament formation and then establish the additional limitation imposed by filamentation, which is demonstrated for an idealized situation. The techniques used are applicable to any SNDC element. We find that either switching or relaxation oscillations can occur for the same SNDC element in the same circuit with appropriate circuit parameters. A major aspect of the investigation is the fact that the circuit theory transforms directly between the S and N case when the important reactive components are identified for each case. The important circuit parameters in each case produce a "dual" circuit system where the voltage across the NNDC element in its "primary" circuit behaves exactly as the current through the SNDC element in its primary circuit.

A crucial aspect of the present study is the identification of the primary circuit for an SNDC element. We regard this as having been established by Shaw and Gastman (1971). The primary circuit is shown in Fig. 5.A1. Here $C(=C_p)$ is the package capacitance and $L(=L_i + L_p)$ is the sum of the intrinsic and package inductance.

For convenience we study the cylindrical SNDC element (of length l, radius a and cross-sectional area S) shown in Fig. 5.A2 in the circuit shown in Fig. 5.A3, where the SNDC element is modeled by the series combination of an appropriate L_i with an appropriate S-shaped conduction current curve of low current resistance R_0.

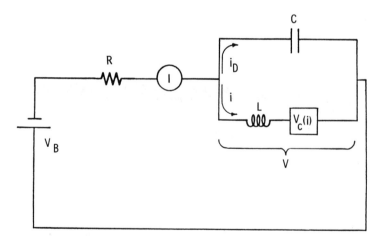

Fig. 5.A1. Circuit under analysis; the "primary" circuit for an SNDC element. i is the conduction current and i_D is the displacement current. R_L is the lead plus load resistance and V_B is the battery voltage. The inductance L is the sum of the package inductance L_p and intrinsic inductance L_i. The contact resistance R_c, which is generally nonlinear, is included in the nonlinear conduction current characteristic $V_c(i)$. The intrinsic capacitance is ignored. V is the voltage drop across C, which is the package capacitance C_p. $I = i + i_D$ (Shaw et al., 1973).

APPENDIX 459

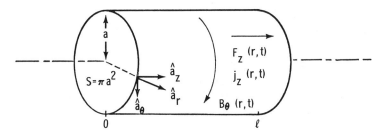

Fig. 5.A2. A cylindrically shaped inhomogeneous SNDC element. A difference of potential $V_0(t)$ is applied across the endplates. In general, for radial inhomogeneities only, the electric field $F = F_{s,r}(r, t)$ and the transport current density $j = j_{s,r}(r, t)$. $B_\theta(r, t)$ is the magnetic field produced by j_s (Shaw et al., 1973).

The electrical behavior of an SNDC element is obtained from a simultaneous solution of the circuit equations and Maxwell's equations with the appropriate boundary conditions. For the cylindrical symmetry shown in Fig. 5.A2, where all quantities are independent of z and θ, Poisson's equation and Maxwell's "curl B" equation yield

$$\frac{1}{r}\frac{\partial}{\partial r}(rF_r) = \frac{1}{\varepsilon}[\rho(r, t) - \rho^0(r, t)] \quad (5.A1)$$

and

$$j_r + \frac{\partial F_r}{\partial t} = 0 \quad (5.A2)$$

where F_r is the radial electric field, ε the permittivity, j_r the radial current density and $\rho(r, t)$ and $\rho^0(r, t)$ the mobile and background charge densities, respectively. Equation 5.A2 is an expression of zero total radial current. For longitudinal fields F_z and current densities j_z we eliminate the magnetic field from the curl B equation and Faraday's equation, yielding

$$\frac{1}{r}\frac{\partial}{\partial r}\left(r\frac{\partial F_z}{\partial r}\right) = \mu_0\frac{\partial j_z}{\partial t} + \mu_0\frac{\partial^2 F_z}{\partial t^2} \quad (5.A3)$$

where μ_0 is the permeability. The constitutive current equations are

$$j_r = \rho(r, t)\mu F_r - D\frac{\partial \rho(r, t)}{\partial r} \quad (5.A4)$$

and

$$j_z = \rho(r, t)\mu F_z \quad (5.A5)$$

where μ is the mobility and D the diffusion coefficient, both of which are chosen to be constant. Equation (5.A5) is the SNDC curve.

Equations (5.A1) through (5.A5) are the governing equations of the system. They are coupled to the external circuit through the equations for total current through the SNDC element

$$i(t) = \int \left[j_z(r, t) + \frac{\partial F_z(r, t)}{\partial t} \right] dS_z \qquad (5.A6)$$

and total voltage across the SNDC element

$$V_0(t) = F_z(r = a, t)\ell \qquad (5.A7)$$

where $F_z(r = a, t)$ is the value of the longitudinal electric field on the cylindrical surface of the SNDC element and dS_z is an element of surace area perpendicular to z. Equations (5.A1) through (5.A7), when solved simultaneously with the external circuit equations, completely specify the problem.

When current flows along a cylindrical wire, energy flows into the wire radially through the surface (Feynman et al., 1964). Conservation of energy with displacement effects neglected yields

$$-\int (F \times H) \, dS_r = \int jF \, d\tau + \frac{1}{2} \frac{d}{dt} \int \mu_0^{-1} BB \, d\tau \qquad (5.A8a)$$

where dS_r is a surface (sheath) element and $d\tau$ a volume element. $B = \mu_0 H$. The left-hand side represents the total power driving the electron stream. For the symmetry of the problem

$$-\int (F \times H) \, dS_r = F_z(r = a)\ell \int H_0 \, ds = F_z(r = a)\ell i, \qquad (5.A8b)$$

where i is the total current flowing in the SNDC element and ds is an element of circumferential length. The right side of Eq. (5.A8a) represents the Joule heating loss and the time rate of change of magnetic energy. Defining resistance and inductance as

$$R \equiv \frac{1}{i^2} \int jF \, d\tau$$
$$L_i \equiv \frac{1}{i^2} \int \mu_0^{-1} BB \, d\tau \qquad (5.A9)$$

Eq. (6.A12) becomes $[F(r = a) \equiv F_z(r = a)]$

$$F(r = a)\ell = iR + \frac{1}{2i} \frac{d}{dt} L_i i^2 \qquad (5.A10)$$

$F(r = a)\ell$ is identified as the total voltage drop $V_0(t)$ across the SNDC element and L_i depends on the distribution of current.

APPENDIX

Rather than deal with resistances, it is more convenient and instructive to define an effective conductive voltage $V_c(t)$ as

$$V_c(t) \equiv \frac{1}{i} \int jF \, d\tau \tag{5.A11}$$

Eq. (5.A10) then becomes

$$V_0(t) = V_c(t) + \frac{1}{2i} \frac{d}{dt} L_i i^2 \tag{5.A12}$$

In the absence of current density filamentation, $V_c(t)$ is a single valued function of i and can be written as $V_c(i)$, and L_i is a constant. In this case we can apply standard techniques to solve for the circuit response (Shaw and Gastman, 1971). However, when filamentation is present, $V_c(i)$ is multivalued, may not even be definable (Callarotti and Schmidt, 1977), and L_i varies over an oscillatory cycle. We will shortly show how $V_c(i)$ is obtained for this case.

We plan to represent the inhomogeneous NDC element by an appropriate SNDC conductive voltage curve $V_c(i)$ (nonlinear resistor) in series with an appropriate inductor L_i. We must therefore investigate the circuit response under these conditions. The total voltage drop across C in Fig. 5.A3 is

$$V(t) = V_0(t) + L_p \frac{di}{dt} = V_c(t) + L \frac{di}{dt} \tag{5.A13}$$

where $L = L_p + L_i$. L_p is the package inductance, which includes the geometrical inductance of the SNDC element associated with magnetic fields for $r > a$. With reference to Fig. 5.A3, we seek solutions of the

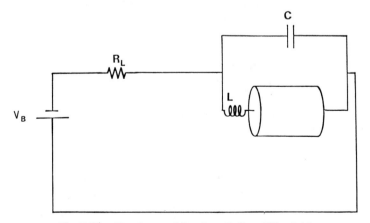

Fig. 5.A3. Circuit containing the cylinder of Fig. 5.A2.

following coupled equations:

$$L\frac{di}{dt} = V(t) - V_c(t);$$

$$C\frac{dV(t)}{dt} = \frac{V_B - V(t) - iR_L}{R_L}.$$

Solutions to the above equations come in many forms. We may obtain the current-time (or voltage-time) profiles and the current-voltage Lissajous figures. We require both in the following arguments. We therefore transform the content of the above equations into the two differential equations:

$$\frac{V_B}{R} = \frac{d^2i}{dt'^2} + \frac{1}{A}\left[\frac{L/R_L}{R_0 C} + \frac{1}{R_0}\frac{dV_c(i)}{di}\right]\frac{di}{dt'} + \left[1 + \frac{V_c(i)}{iR_L}\right]i \quad (5.A14)$$

and

$$\frac{dV}{di} = \left[\frac{(V_B/R_L) - (V/R_L) - i}{V - V_c(i)}\right](R_0 A)^2. \quad (5.A15)$$

Equation (5.A14) is the dual of the circuit equation for NNDC elements used by Shaw et al. (1979). Thus, the arguments of this section are in 1-to-1 correspondence with those. In Eqs. (5.A14) and (5.A15) $A = \sqrt{LC}/R_0 C$, $t' = t/\sqrt{LC}$, and we have written $V_c(t)$ as $V_c(i)$. The significance of the transformation on t is that in Eq. (5.A14) the bracketed non-linear damping term is usually of the order of unity or less. Thus, the strength of the damping term is determined primarily by the value of A. Equation (5.A14) is a generalization of Van der Pol's equation for a free-running oscillator. For very small A the nonlinear damping term is strong and the solutions are well-defined relaxation oscillations. For large A the damping term is small and the $i(t)$ solutions are nearly sinusoidal. Using the three-piece linear approximation for $V_c(i)$ shown in Fig. (5.A4), $i(t)$ solutions can be obtained in the three regions $i < i_p$, $i_p < i < i_v$ and $i_v < i$, which correspond to $dV/di = R_0$, $-R_n$ and 0, respectively, where R_n is the magnitude of the negative differential resistance. The individual solutions are joined smoothly from one region to the next. The current waveform begins with a slow exponential rise with time constant $\sim(R_0/L + 1/R_L C)^{-1}$ followed by a sharp spike in current when $i = i_p$. The spike is composed of a fast exponential transit through the region of negative slope ($i_p < i < i_v$) with time constant $\sim(1/R_L C - R_n/L)^{-1}$, where $|R_n/L| > |1/R_L C|$, followed by a damped sine wave for $i > i_v$, and another exponential transit for $i_v < i < i_v$. An exponential decay for $i < i_p$ completes the cycle. The time required to reach $i = i_p$ during the initial slow exponential rise depends on the applied bias ϕ_B. Thus, the frequency of the relaxation oscillations is voltage tunable.

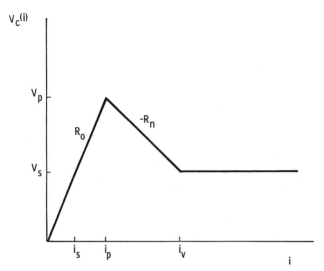

Fig. 5.A4. Three piece linear approximation for $V_c(i)$.

Returning to Eq. (5.A15), we integrate it over that portion of the cycle where $V_c(i) = V_s$, assuming $V_B/R_L \gg V/R_L$:

$$A^{-2}[V(i) - V_s]^2 + R_0^2\left(\frac{V_B}{R_L} - i\right)^2 = \text{constant}, \qquad (5.\text{A}16)$$

which is the equation of an ellipse. For a particular SNDC element the shape of the ellipse is determined by A. Plotting $V(i)$ vs. $R_0 i$, the trajectory is a circle when $A = 1$, an ellipse with the major axis along $R_0 i$ for $A < 1$, and an ellipse with the major axis along V for $A > 1$. For small A the current amplitude is large and for large A the voltage amplitude is large. If we assume the validity of extending the ellipse below i_v until it intersects the positive resistance part of $V_c(i)$, then the $V(i)$ curve is determined back to this point. At threshold $(i = i_p)$, $V(i_p) \cong V_c(i_p) = V_p$. The constant in Eq. (5.A16) may then be evaluated to give

$$A^{-2}[V(i) - V_s]^2 + R_0^2\left(\frac{V_B}{R_L} - i\right)^2 = A^{-2}(V_p - V_s)^2 + R_0^2\left(\frac{V_B}{R_L} - i_p\right)^2 \qquad (5.\text{A}17)$$

Under the assumptions leading to Eq. (5.A17), the complete $V(i)$ trajectory is obtained by joining the ellipse Eq. (5.A17) to the positive resistance segment of $V_c(i)$. This approximation is best for small A. An elliptical $V(i)$ trajectory is plotted in Fig. 5.A5. In the $V(I)$ plane [replace the abscissa in Fig. 5.A5 by $I = i + C(dV/dt)$] the trajectory collapses to the load line.

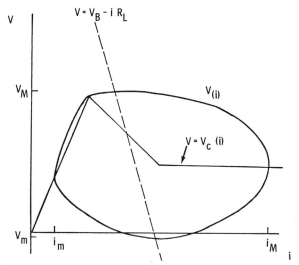

Fig. 5.A5. $V(i)$ trajectory for a relaxation oscillation. The load line in this plane, $V = \phi_B - iR_L$ is the line $dV/dt = 0$. The $V_c(i)$ curve is the curve $di/dt = 0$. The voltage and conduction current extrema on the $V(i)$ trajectory are therefore the points of intersection of $V(i)$ with $dV/dt = 0$ and $di/dt = 0$, respectively. The extrema in voltage are denoted V_M and V_m. The current extrema are i_M and i_m (Shaw et al., 1973).

Equation (5.A17) shows that besides the parameters A and V_B/R_L, the circuit response is determined by the SNDC element parameters R_0, i_p, V_p and V_s. As we shall show, the form and nature of the circuit response is a dominant factor in the formation and quenching of current density filaments. In particular, the maximum voltage V_M and minimum current i_m (Fig. 5.A5) reached during the first cycle are of major importance. From the ellipse Eq. (5.A17) it follows that for large A (sinusoidal oscillations) V_M and i_m are high, whereas for small A (relaxation oscillations) V_M and i_m are low.

We have now analyzed the response of a circuit containing an SNDC element with a uniform distribution of current. We next ask: (1) how does filamentation affect the circuit response; and (2) how does the circuit response affect filamentation? We consider the second question first. The results of the uniform current case indicated that the $V(i)$ trajectory was determined primarily by the parameter, R_0, i_p, V_p and V_s and was relatively insensitive to the slope of the NDC region. Therefore, if we assume that the major effect of filamentation when sustained circuit-controlled oscillations are present is to change the slope of the NDC region, then the circuit will simply control the extent of filamentation. As we shall see, this is in fact the case. To answer the first question we note that when uniform currents flow and circuit-controlled oscillations occur, a specific current minimum is reached once each cycle. When filamentation occurs during sustained circuit-controlled oscillations we require filament quenching once each cycle. If we assume that there is a minimum sustaining current for filamentation, then the filament quenching criterion will impose an additional limitation on the range of circuit parameters for which circuit-controlled oscillations will occur.

We illustrate both conclusions below where we draw upon an approximate scheme for computing $V_c(i)$ and obtaining a quenching criterion. The model neglects the skin effect, filament formation times, and spatial derivatives.

We divide the cylindrical SNDC element into two subelements: (1) a core cylinder of radius a_i and (2) a surrounding cylindrical shell of inner radius a_i and outer radius, a_0, as shown in Fig. 5.A6. The subelements have different SNDC curves and within each subelement j and F are uniform. For the configuration shown in Fig. 5.A6, Eqs. (5.A9) and (5.A11) yield

$$V_c(i) = \frac{1}{i}(j_1 F_1 S_1 1 + j_2 F_2 S_2 1) \tag{5.A18}$$

and

$$L_i = \frac{\mu_0 i}{2\pi i^2}\left\{\frac{i^2}{4} + \left(i_1 - \frac{a_i^2 i_2}{a_0^2 - a_i^2}\right)\left[\frac{i_2}{2} + \left(i_1 - \frac{a_i^2 i_2}{a_0^2 - a_i^2}\right)\ln\frac{a_0}{a_i}\right]\right\} \tag{5.A19}$$

where $i = i_1 + i_2$, $i_1 = j_1 S_1$ and $i_2 = j_2 S_2$. The computation of $V_c(i)$ for a

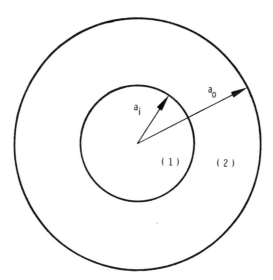

Fig. 5.A6. Two-subelement model of an inhomogeneous cylinder.

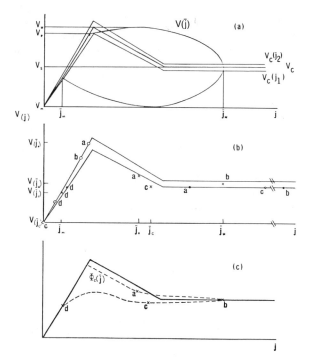

Fig. 5.A7. Illustration of the way $V_c(\bar{j})$ for a relaxation oscillation is obtained using the two subelement model. At time $a, \bar{j} = \bar{j}_a$ and $V(\bar{j}_a) = V(\bar{j}_a)$, etc. For clarity we have dropped the subscript 0 to V. In (b) the position of subelement 1 is designated by the closed circles, the position of subelement 2 by the open circles, and the values of $V_c(\bar{j})$ by the crosses. In (c) we plot a complete $V_c(\bar{j})$ curve estimated point by point throughout one cycle of $V(\bar{j})$ (Shaw et al., 1973).

relaxation oscillation with $L_p = 0$ [here $V(i) = V_0(i)$] is illustrated in Fig. 5.A7 where, for ease in plotting, we compute $V_c(\bar{j})$ rather than $V_c(i)$. Here $\bar{j} = i/S$, where $S = S_1 + S_2$. We consider the case $S_1 = S_2$, thus $a_0^2 = 2a_i^2$,

$$j = \tfrac{1}{2}(j_1 + j_2) \tag{5.A20}$$

and

$$V_c(j) = \frac{j_1}{j_1 + j_2} V_c(j_1) + \frac{j_2}{j_1 + j_2} V_c(j_2) \tag{5.A21}$$

where $V_c(j_1) = F_1 \ell$ and $V_c(j_2) = F_2 \ell$. To determined $V_c(\bar{j})$ we first obtain the $V_0(\bar{j})$ trajectory from A, V_p, i_p and V_s using the results of the circuit analysis for the uniform cylinder. Next, we consider the response of each subelement and determine $V_c(\bar{j})$ by making use of Eqs. (5.A20) and (5.A21). From Eq. (5.A12) $[V_c(\bar{j}) = V_0(\bar{j}) - (S/2\bar{j})/(d/dt)L_i\bar{j}^2]$ we see that the induced voltage drop in the cylinder is just the difference $V_0(\bar{j}) - V_c(\bar{j})$.

APPENDIX 467

In Fig. 5.A7(a) we show $V_0(\bar{j})$ for a circuit-controlled relaxation oscillation, $V_c(j_1)$, $V_c(j_2)$ and $V_c \equiv \frac{1}{2}[V_c(j_1) + V_c(j_2)]$. In Fig. 5.A7(b) we show computed values of $V_c(\bar{j})$ (crosses) at four instants of time: a, b, c and d. The conductive voltages in subelement 1 (closed circles) and subelement 2 (open circles) are also shown. At each instant of time an assumption is made about the time rate of change of current or voltage. At time a, $V_0(\bar{j})$ has reached its maximum. If at this point we make the reasonable assumption of neglecting the inductive voltage drop in subelement 2, then $V_0(\bar{j})_M = V_c(j_2)$ and j_2 is determined. From Eq. (5.A20) we obtain j_1, which in turn yields $V_c(\bar{j})$ from Eq. (5.A21). $V_c(\bar{j})$ at time a is indicated in Fig. 5.A7(b) as the cross a. At time b, $\bar{j}$ is a maximum, hence here $V_0(\bar{j}) = V_c(\bar{j})$. This point is the cross b. Similar determinations are made at c and d. A sketch of $V_c(\bar{j})$ throughout the cycle is shown in Fig. 5.A7(c).

It is of interest to examine the system in the two limiting cases: (1) $a_0 \gg a_i$ and (2) $a_0 \sim a_i$. If, for $a_0 \gg a_i$, we also consider the situation where the narrow central region has a much larger conductivity than the rest of the SNDC element, then the situation presumably corresponds quite closely to an amorphous chalcogenide threshold switch. Here we see from Eq. (5.A19) that the $\ln(a_0/a_i)$ term of L_i becomes quite important. The term has no upper limit and is produced by the flux in subelement 2 produced by current in subelement 1. For the case $a_0 \sim a_i$, as i_2 vanishes Eqs. (5.A18) and (5.A19) produce the results required of a uniform cylinder where $L_i = \mu_0/8\pi$. Performing computations on these systems similar to those discussed with regard to Fig. 5.A7 reveals several general features of the problem. We find that: (1) the $V_c(\bar{j})$ curves can be regarded as members of a family of curves characterized by similar values of i_p, V_p and V_s; (2) the $V_c(\bar{j})$ trajectory reflects the current density evolution; and (3) when inhomogeneous current density distributions occur and sustained circuit oscillations are maintained, the major affect of filament formation will be to alter the shape of the region of negative slope. Since the circuit response is relatively insensitive to this parameter, the waveforms will be essentially indistinguishable from the uniform current density case.

We have been describing the oscillation in terms of the parameter $A = \sqrt{LC}/R_0C$. When filamentation is present A may be regarded as circuit dependent insofar as L_i depends on the current distribution. As shown by Eq. (5.A19), filamentation yields values of L_i greater than its value for uniform current densities. Therefore, A is larger above than below threshold. From the arguments for the uniform current density case, a larger A would yield higher values of i_m (minimum current). The implications of filamentation are therefore clear: if the SNDC element is part of a circuit that for the uniform field case yields a value of i_m substantially below a "filament sustaining current" i_f, the effect of filamentation on the circuit oscillations will be negligible. On the other hand, if $i_m \leq i_f$, then small increases in L_i due to filamentation may well increase i_m so that it exceeds i_f. In this case the circuit oscillations will damp and switching will occur.

The above calculations illustrated filamentation when at threshold only one subelement entered its region of negative slope. The presence of relatively uniform current densities implies that at some time during the cycle both subelements are in their regions of negative slope. For the example considered above, when subelement 1 enters the NDC region its total voltage drop ceases to rise significantly and the voltage drop across subelement 2 soon begins to decrease. If we imagine a situation where $V_0(\bar{j})$ rises significantly after threshold, then subelement 2 can also be pushed into its NDC region. The current distribution will initially be relatively uniform. Therefore, an important parameter in the growth of current filaments is the maximum voltage V_M reached during the cycle. From Eq. (5.A17) we see that V_M is a strong function of A. For relaxation oscillations (small A) V_M barely exceeds the threshold voltage for the subelement with the highest carrier concentration. This leads to the highly nonuniform situation where a filament appears immediately upon reaching threshold. For near sinusoidal oscillations (large A) V_M may be much higher than the threshold voltage, producing a more uniform initial current density distribution.

In order for circuit-controlled oscillations to be sustained it is necessary that all current density nonuniformities be quenched at the end of each cycle. When this occurs, the $V_0(\bar{j})$ trajectory is closed and completed on the low-current line of positive slope of $V_c(\bar{j})$. If, however, the current density minimum $\bar{j}_m$ is not sufficiently low ($\bar{j}_m > \bar{j}_f$), the trajectory is open and the nonuniformities are enhanced. The circuit oscillation damps and a filament remains as a steady-state solution. This behavior is shown schematically in Fig. 5.A8.

It is possible to make some qualitative predictions about how small i_m must be to insure filament quenching. Certainly $V_c(i)$ will not form a closed trajectory unless i_m is sufficiently below i_p. In general we expect that i_m must typically be below i_s (see Fig. 5.A4) to insure circuit-controlled oscillations. According to Eq. (5.A17), the elliptical part of the $V_0(i)$ trajectory will pass through the line of positive slope of $V_c(i)$, when

$$A = \left[\frac{i_p - i_s}{(2V_B/R_L) - (i_p - i_s)} \right]^{1/2}$$

For A less than this critical value filaments will quench. Consider the case where $i_p \cong 2i_s$. Here, since $V_B/R_L > i_p$, a necessary condition for quenching is $A \leq 1$. For a realistic bias $V_B/R_L \sim 2i_p$ and $A \cong \frac{1}{2}$ is the critical value. Note that for a given A, as V_B increases, a transition from a relaxation

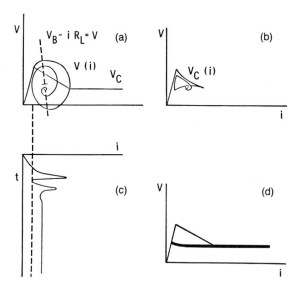

Fig. 5.A8. Estimated $V(i)$, $V_c(i)$ and $i(t)$ curves [(a), (b), (c) respectively] for the case where i_m does not fall low enough to quence the inhomogeneities. The circuit oscillations damp, a filament forms, and switching occurs to the ON state. The ON state is the heavy dot. As V_B is varied the dots fall on a "filament characteristic," which is sketched (dark) in (d). When the load lines intersects the SNDC element at points $i > i_v$ (see Fig. 5.A4), a switch will occur to that point without relaxation oscillations (unless for some reason the stable point at $i > i_v$ is never reached). The switch will also take place, however, in a damped oscillatory fashion. A "spiraling in" to the ON state is fundamental to the switching process in general. In the current-time profile thus appears as a damped "ringing" oscillation at a frequency determined in part by $\sqrt{LC}$.

oscillation to the ON-state can occur as the critical value of A is exceeded. Furthermore, if we design the SNDC element such that A is significantly greater than unity, filament quenching can be avoided. It is, therefore, possible to construct a filament-forming SNDC element that will only turn ON and not exhibit relaxation oscillations for any V_B and R_L. To achieve this we require $\sqrt{LC} > R_0 C$.

Throughout this discussion we have assumed that the package inductance L_p was zero. For a finite L_p, $L = L_i + L_p$ and some inductive voltage will be drained off the SNDC element. For a given $V(\bar{j})$ trajectory, therefore, the maximum value of $V_0(\bar{j})$ will be below its $L_p = 0$ value. Since a large V_M is necessary for relatively uniform current densities, increasing the package inductance enhances the possibility of filamentation. The two circuit factors that control filamentation are thus L_p and A, with filamentation optimized for large L_p and small A.

We conclude by considering the presence of a primary filament nucleation site. In modeling a thermal filament, the primary site is often at the center of the cylinder. However, wherever the site is, its role may be likened to that of a subelement with a high carrier concentration compared to the other subelements. Its presence will thus act to decrease $V_p - V_s$, which from Eq. (5.A17) can be seen to reduce the amplitude of $V(i)$. This will increase i_m and lead to the damping of the relaxation oscillations and the domination of a filament. Thus, if a primary site exists, a filament will not quench unless $A \geqslant 1$. Switching to the ON state will dominate. Amorphous chalcogenide threshold switches almost always exhibit relaxation oscillations for sufficiently large R and also have a high-conductivity central region (Coward, 1971; Bosnell and Thomas, 1972; Allinson et al., 1979). At first glance these two features seem incompatible. However, the high resistances of these devices ($R_0 \sim 10^6$-10^7 Ω) and values of $C_p \sim 10^{-12}$ F produce values of A much less than unity even for reasonably large estimates of the inductance. Hence, filament quenching and relaxation oscillations are still favored.

6

Electrophotography

6.1 Introduction.. 470
6.2 The Physics of Electrophotography... 476
    6.2.1 Introduction 476
    6.2.2 Emission Limited Discharge... 479
    6.2.3 Space-Charge-Perturbed and Space-Charge-Limited Discharge 482
6.3 The Physics of the Development Process...................................... 489
6.4 Charging of the Toner Particles 491
6.5 Materiasl Requirements for Photoreceptors 491
6.6 Electrophotographic Applications of Amorphous Hydrogenated Silicon 497

6.1 Introduction

The singularly most significant use of amorphous chalcogenides has been in the area of electrophotography (see, e.g., Pai and Melnyk, 1986; Burland and Schein, 1986; Schmidlin, 1976; Schaffert, 1975; Dessauer and Clark, 1965). Here the photoconductive properties of selenium or selenium based alloys are applied to the problem of the direct copying of documents, and in this chapter we will discuss the problem by making extensive use of the papers by Lucovsky and Tabak, 1974; Schmidlin, 1976; and Mort, 1973. Basically, the electrophotographic process we are primarily concerned with here, xerography (Carlson, 1942), involves spraying the exposed surface of a chalcogenide film with positively charged ions, whence an image negative charge builds up on the metallic substrates. [Now dual layers are often used in which a thin layer of a photoconductive pigment is coated onto a thicker layer capable of transporting charge (Pai, 1983)] Reflected light from the document is projected onto the charged film. The light is absorbed on the original document at places where print exists. At the places where no print exists the reflected light is absorbed in the film and produces electron-hole pairs near the free surface. The pairs are separated by the large electric field produced in the films by the charged ions on the surface. The electrons move toward the surface and neutralize the positive ions. The holes drift through the film and neutralize the induced negative charge at the substrate. The result is that the surface of the film neutralizes at places where there was no printing on the original document, but remains charged where print existed.

Subsequently, negatively charged black particles of "toner" are rolled across the film and are attracted to the positively charged regions; these particles are then transferred to a sheet of paper that is positively charged. Heating fixes the toner and the copying process is complete.

A somewhat more detailed description of the process is shown in Fig. (6.1) (Lucovsky and Tabak, 1974).

Step 1. The photoreceptor is sensitized. A film about 50 μm thick, deposited on an Al substrate, is sensitized via a corona discharge. The required fields are about 10^5 V cm.

Step 2. The film is exposed and a latent image is formed. In regions where light strikes, the surface potential is reduced. The fact that the discharge current flows normal to the surface then allows for the production of a potential distribution that replicates the image pattern.

Step 3. The image is developed. A mixture of black (or colored) toner particles, about 10 μm in diameter, and carrier beads, about 100 μm in diameter, are then spread over the film. The toner particles are triboelectrically charged and are attracted to either the surface fringe field at the light-dark boundaries or by the potential in the dark areas, depending on the system employed. The charged toner particles adhere to the film and form a visible image that corresponds to the latent electrostatic image.

Step 4. The image is transferred. The paper is charged appropriately and the toner particles are attached to the paper.

Step 5. The print is fixed. The toner particles are next fused or melted into the surface of the paper via heat, heat and pressure or solvent vapors.

Step 6. The paper is cleaned. Residual toner is removed by mechanical or electrostatic techniques.

Step 7. The image on the film is erased. The film is finally flooded with light intense enough to reduce the surface potential to a uniformly low value, perhaps 10^4 V cm. The system is then ready for the next cycle.

We see that the major function of the photoreceptive film is to convert an optical image into an electrostatic image. It must be able to be charged to high fields and the charge must decay sufficiently slowly in the dark during the processing steps so that the differences in surface potential between the light and dark areas are maintained. An ideal photoreceptor for electrophotography is shown in Fig. 6.2 (Lucovsky and Tabak, 1974).

INTRODUCTION

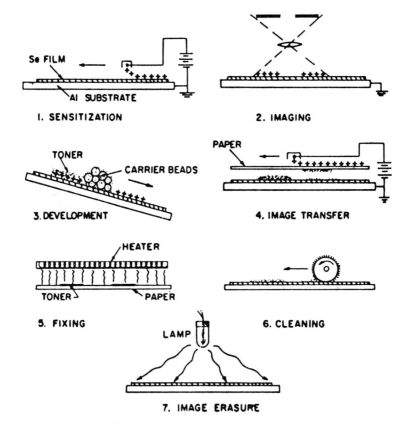

Fig. 6.1. Process steps in reuseable xerography (Lucovsky and Tabak, 1974).

This perfect system is characterized as follows:

(1) The surface charge density is proportional to the voltage.
(2) Each absorbed photon results in the motion of one unit of charge through the film.
(3) All mobile charges present are photogenerated. Hence, dark decay and surface conductivity effects, which can neutralize the image, are absent.
(4) The interface at the back conducting electrode is perfectly blocking. There is no charge injection here or a potential drop at the interface.
(5) The free surface is blocking. No charge placed on this surface can be injected into the bulk. No charge can move laterally either.
(6) The characteristics of the film do not change with cycling.
(7) The properties of the film are identical at all points on its surface.

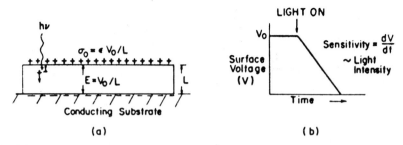

Fig. 6.2. (a) Schematic of the electrical properties of an ideal photoreceptor. Surface charge density is σ_0 and the initial voltage across the photoreceptor is V_0. Photons of energy $h\nu$ each create one hole-electron pair, the hole being transported under the action of the applied electric field F to the conducting substrate and the electron reaching the top surface. (b) Light discharge characteristic of the ideal photoreceptor. There is zero decay until the light is turned on. The sensitivity or the rate of discharge is independent of the surface potential and directly proportional to the light intensity. The total voltage discharge is directly proportional to the exposure; i.e., the light intensity multiplied by the time of exposure (Lucovsky and Tabak, 1974).

The photoreceptor is ideal; here the potential drop is proportional to the exposure. The slope of the photoreceptor voltage-exposure plot is defined as sensitivity, and is constant for the ideal case. In comparing the behavior of a-Se as a photoreceptor with the ideal case we find that it differs from ideality in three ways:

(1) A large spectral region exists where there is strong optical absorption without any observable photodischarge.
(2) The photogeneration of free carriers is a function of electric field for all wavelengths where photoconductivity is present.
(3) Every photogenerated electron and hole is not transported completely through the photoreceptive material; this leads to changes in the photoreceptive characteristics with cycling.

The dependence of the photosensitivity and optical absorption on the photon energy is shown in Fig. 6.3, where a plot of the fractional light absorption in a 40 μm thick Se film is also shown. Note that although there is strong absorption throughout the red, orange and yellow (1.7 eV to 2.2 eV), the photoconductivity or sensitivity exhibits a threshold in the green at about 2.4 eV. This means that when Si is used as photoreceptor, good copy from colored originals having black characters on a red or orange background cannot be produced.

Figure 6.4 shows the voltage sensitivity of the electrophotographic discharge for blue light. Under the conditions noted in the figure, the sensitivity, $dV/dt|_{t=0}$, measures the photogeneration rate of free carriers. For $F \leq 10^5$ V cm the discharge rate shows the same dependence for all

INTRODUCTION 475

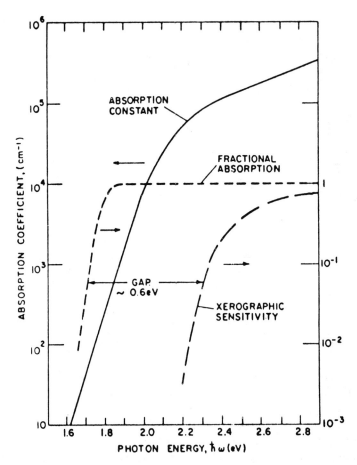

Fig. 6.3. The photon energy dependence of (1) the optical absorption constant, (2) the high field (10^5 V cm) xerographic sensitivity in a 40 μm thick sample, (3) the fractional light absorption in a 40 μm thick sample (Lucovsky and Tabak, 1974).

wavelengths. (An ideal photoreceptor would show a horizontal line at a level of dV/dt corresponding to a gain of unity.)

Both holes and electrons are trapped in the photodischarge process, with electron trapping being a greater problem since the hole mobility, and then its range, is substantially larger. In most photocopying machines the photoreceptor is positively charged and hole transport dominates. However, a negative discharge mode is also desirable since it is also useful to produce copies from white or black originals. It is easiest to do this by revising the sense of the charging step and employing all the other subsystems in the same mode.

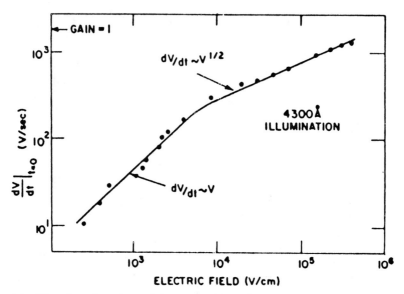

Fig. 6.4. The voltage dependence of the xerographic sensitivity for a 44 μm thick amorphous selenium sample, charged positively under 4,000 Å illumination. The xerographic sensitivity is defined in terms of the rate of change of surface potential when the light is first turned on (Lucovsky and Tabak, 1974).

In the cycled performance of a photocopier the trap release time is more important than the trapping itself. Holes release in times comparable to cycling times, on the order of seconds, so that the hole trapping will not contribute to a time-dependent build up of a residual potential, which is the potential remaining after the light erase-step. However, electron release times take minutes; the residual build up of potential here becomes cumulative. After a few cycles of negative discharge operation the discharge process ceases. Note also that trapping effects will occur only in the regions when light is incident, so the residual build up is spatially uniform and leads to ghosting type effects when the originals are changed.

6.2 The Physics of Electrophotography

6.2.1 Introduction

Although we presently have a reasonable understanding of the physics of the development process, the physics of electrostatic charging and the photoconductivity of insulators is not completely understood. Let us begin this section with a discussion of the discharge process.

Figure 6.5a shows typical curves of the photodischarge for two light intensities incident as shown in the insert. The control potential between the area where the full intensity strikes ("white" background) and the area where the light is attenuated (reflection from a dark image) as a function of time is obtained by subtracting the two curves, and this is shown in Fig. 6.5b as a function of exposure (white background flux times t). We see that there exists an optimum exposure point where the photocopying machine should be operated. Indeed, this is so because, for one, the attraction of toner particles by the fringing fields below a certain contrast potential has a strong dependence on the contrast potential. The sharpness of the curve shown in Fig. 6.5b indicates the latitude of exposure for either under or overexposure before the quality of the copy is affected. a-Se has a fairly narrow latitude in this report so that if marked, only a narrow range of object densities are reproduced well. This is due to the process of photogeneration, which we shall discuss shortly.

In the process of forming an image the way in which the surface potential falls from the initial value V_0 is crucial. As discussed above, it depends on the rate of carrier supply, mobility and lifetime. Further, the mode of exposure, i.e., bulk or surface absorption, flash or step illumination, all can determine the characteristics of the discharge.

Although the contrast potential developed because of non-uniform illumination is fundamental to the photocopying process, we treat the case of uniform illumination in order to emphasize basic processes. As outlined before, we take a photoreceptor of thickness L and permittivity ε having a capacitance per unit area $C = \varepsilon/L$ and positively charge it to an initial potential V_0 by putting a charge Q_0 on the surface. We can discharge the surface by moving charge Q_0 from the top surface to the bottom electrode. To do this we photogenerate free electron-hole pairs near the surface and drift the holes in the field across the distance L. (The electrons that drift to the surface in the region of optical excitation have only a slight effect on the characteristics.)

If the $\mu\tau F(t)$ product, the carrier Schubweg, where τ is the lifetime for deep traps, becomes less than L, the discharge ceases and a residual surface potential remains behind. Since deep trapping is undesirable, we will consider in our treatment only photoreceptors where this problem is absent or minor.

The discharge process occurs under nonconstant voltage conditions; as it proceeds the field decreases. There are two extreme cases to consider. For the case where most of the charge that moves in one transit time $t_T[=l/\mu F(t)]$ is much less than $Q(t)$, the charge remaining on the surface at time t, many transits must occur before full discharge occurs. This is called the "Emission Limited" case, since the rate of discharge, dV/dt, is controlled by the rate of emission of carriers from the region of excitation. This process will exibit

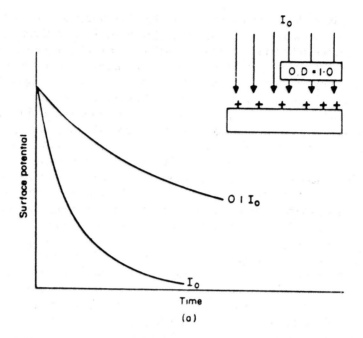

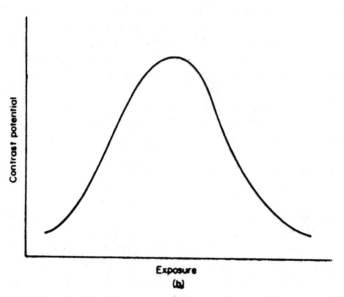

Fig. 6.5. (a) Schematic photodischarge curves vs. time for an amorphous Se photoreceptor exposed to a light pattern of two intensities (insert). (b) Schematic contrast potential curve vs. exposure.

THE PHYSICS OF ELECTROPHOTOGRAPHY   479

the light intensity dependence of the photogeneration process. For the case where an amount of charge Q_0 is generated in a time less than the initial carrier transit time, $L/\mu F_0$, the total discharge occurs within a single transit. As the sheet of charge Q_0 moves away from the surface, the field-behind it falls to zero so that additional carriers will neither drift into the bulk nor be generated. This type of discharge depends on the light intensity. It is the maximum rate of discharge for light that is highly absorbed. We term it a "trap free space-charge-limited" (TFSCL) discharge.

6.2.2 Emission Limited Discharge

Our first task is to derive an expression for the surface potential as a function of exposure (flux × time) (Chen and Mort, 1972; Chen et al., 1972) and then compare it with experiment. We assume that bulk trapping can be neglected and that all the light is absorbed in a depth small compared to the thickness of the sample. Further, we assume that the amount of charge in transit at any time is small compared to the charge remaining on the surface. With these assumptions, we have (Mort, 1972, 1973)

$$-\frac{dV}{dt} = \frac{eN_e}{C} = \frac{J_e}{C}, \qquad (6.1)$$

where N_e is the number of carriers emitted per unit area per unit time. Integration of Eq. (6.1) yields

$$t = -C \int_{V_0}^{V} J_e^{-1} \, dV, \qquad (6.2)$$

where V_0 is the initial surface potential ($t = 0$).

To integrate Eq. (6.2) we must write J_e as a function of V. N_e can be written as

$$N_e = F'\eta(F', V), \qquad (6.3)$$

where F' is the light flux and η the quantum efficiency of the carrier emission, which includes the photogeneration step and any limitations due to barriers for recombination and injection.

The quantum efficiency can be written as

$$\eta(V) = \left(\frac{V}{V_n}\right)^P, \qquad V \leq V_n \qquad (6.4)$$

$$\eta(V) = 1, \qquad V \geq V_n, \qquad (6.5)$$

where V_η is the potential at and above which $\eta = 1$. Since secondary photocurrents do not flow, the electrophotographic process has unity gain. With the above relationship Eq. (6.2) can be integrated to yield for $V_0 \leq V_\eta$

$$t = \frac{C}{eF'} \frac{V_\eta}{1-P} \left[\left(\frac{V_0}{V_\eta}\right)^{1-P} - \left(\frac{V}{V_\eta}\right)^{1-P} \right], \quad \text{if } P \neq 1$$

or

$$t = \frac{C}{eF'} V_\eta \ln\left(\frac{V_0}{V}\right), \quad \text{if } P = 1; \qquad (6.6)$$

for $V_0 > V_\eta$

$$t = -\frac{CV_\eta}{eF'} \left[\left(\frac{V_0}{V_\eta}\right) + \frac{P}{1-P} - \frac{1}{1-P}\left(\frac{V}{V_\eta}\right)^{1-P} \right], \quad \text{if } P \neq 1$$

or

$$t = \frac{CV_\eta}{eF'} \left[\left(\frac{V_0}{V_\eta}\right) - 1 + \ln\left(\frac{V_\eta}{V}\right) \right], \quad \text{if } P = 1 \qquad (6.7)$$

The above equations tell us that if carrier emission is field independent (i.e., $P = 0$), then the surface potential decays linearly with time:

$$V = V_0 - \frac{eF't}{C} \qquad (6.8)$$

When $P > 0$ the $V(t)$ curve becomes nonlinear and dispersive, with the effect more pronounced for greater P, as shown in Fig. 6.6a. Here we plot the potential coordinate normalized to $V_\eta = 1$ because photoreceptors with different P generally have different values for V_η. In the figure the case $V_0 = V_\eta = 1$ was chosen.

Figure 6.6b displays curves of contrast potential vs. exposure with $V_0 = V_\eta = 800\,\text{V}$. Note that as the field dependence of carrier emission increases from 0 to 2 the exposure latitude increases. Note also that the optimum exposure shows a corresponding increase. This is expected since as the field dependence of the supply of carriers increases, a decrease occurs in the efficient utilization of light.

Curvature in the discharge characteristic is commonly observed in amorphous Se, as shown in Fig. 6.7. The solid line is a plot of Eq. (6.6) with $P = \frac{1}{2}$ and $F = V/L = 2.8 \times 10^5\,\text{V cm}^{-1}$. Tabak and Warter (1968) pointed out that the photodischarge curve of Se can be explained fully via its optical properties. Its dispersive character is due to the field dependence of the photogeneration of carriers.

Tabak and Warter (1968) also showed that the hole lifetime for deep bulk traps is about $50\,\mu\text{s}$. Since $\mu_d \cong 0.13\,\text{cm}^2/\text{V s}$, then the Schubweg $\mu\tau F$ is about $50\,\mu\text{m}$, the typical sample thickness, only at fields near $10^3\,\text{V cm}^{-1}$. This effect produces a residual voltage near 5 V, which does not present a problem for common development systems.

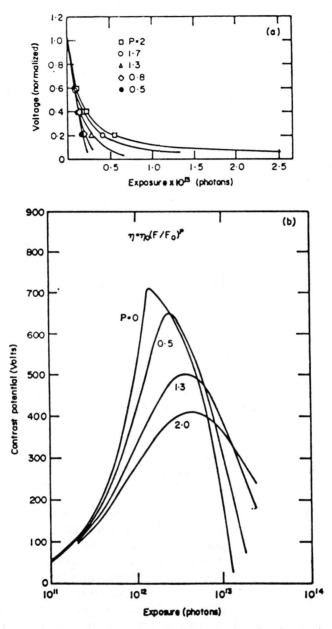

Fig. 6.6. (a) Theoretically calculated photodischarge curves vs. time for a series of model photoreceptors with different field dependences of the carrier supply P. (b) Constant potential curves obtained with differing values of P considered in (a) for an input subject density of 1.0 (Mort, 1973).

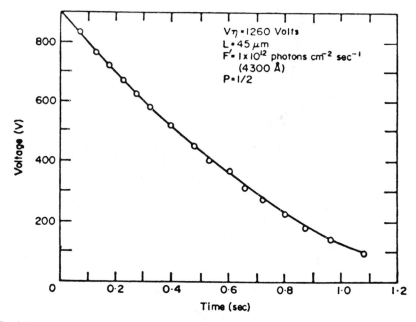

Fig. 6.7. Discharge curve (circles) of amorphous Se compared with the calculated values with $F' = 1 \times 10^{12}$ photons cm^{-2} s^{-1}, $k = 6.3$, $L = 45.0\,\mu$m; $F_\eta = V_\eta$. $L = 2.4 \times 10^5$ V cm, $P = 0.5$ (Mort, 1973).

Amorphous As$_2$Se$_3$ is also used in photocopying machines; a discussion of its features can be found in the source of our discussion, the article by Mort (1973).

6.2.3 Space-Charge-Perturbed and Space-Charge-Limited Discharge

In the prior section we discussed emission-limited discharges, where the charge in transit is less than $Q(t)$, the surface charge of the sample at any time. For a sample to discharge completely here, many transits must occur; the process can be thought of as resulting from many small equal flash exposures. Since the surface potential decreases with time, each succeeding exposure occurs at a progressively lower initial surface potential. Finally, it may be possible that during the discharge the amount produced in a flash is a substantial fraction of the charge remaining on the sample. Hence, a discharge that starts initially as emission limited can become space-charge-perturbed and ultimately space-charge-limited when the amount of free charges in the final transit equals $Q(t)$.

For the discharge to be space-charge-limited to begin with, a charge equal to Q_0 must be generated in a time much less than the carrier transit in the initial field. For a given light flux, the drift mobility of the carriers will determine whether space-charge or emission limited discharge will occur. Holes in a-Se have mobilities that are too high for the space-charge-limited regime to be reached except for very high intensity flashes. In materials with lower mobilities, the space-charge-limited case can be attained under normal excitation. Batra *et al.* (1970) solved the space-charge-limited discharge case for strongly absorbed light, flash exposure and open circuit conditions. Chen (1972) performed a more extensive study and we state some of his results now; the details can be found in his paper.

Analytic results are obtained under flash exposure conditions, where the time required to produce the surface charge is much less than the transit time, and strongly absorbed light ($\alpha L \to \infty$). For $t < t_T$, where t_T represents the transit of the leading edge of the sheet of charge; $t_T = L/\mu_0 F_0$, with μ_0 the mobility of the initial field F_0, Chen (1972) obtains

$$\tilde{V}(\tilde{t}) = 1 - \frac{1}{\omega + 2}[1 - (1 - \rho_0)^{\omega+2}]\tilde{t}; \tag{6.9}$$

$$\frac{d\tilde{V}}{d\tilde{t}}(\tilde{t}) = -\frac{1}{\omega + 2}[1 - (1 - \rho_0)^{\omega+2}], \tag{6.10}$$

where $\tilde{t} = t/t_T$ and ω represents the field dependence of the mobility via $\mu = \mu_0(F/F_0)^\omega$.

For $t > t_T$ but less than the total discharge time

$$\tilde{V}(\tilde{t}) = \frac{(1 - \rho_0)^{\omega+2}}{\omega + 2}\tilde{t} + \frac{\omega + 1}{\omega + 2}\left(\frac{1}{\tilde{t}}\right)^{1/(\omega+1)}; \tag{6.11}$$

$$\frac{d\tilde{V}(\tilde{t})}{d\tilde{t}} = -\frac{1}{\omega + 2}\left[\left(\frac{1}{\tilde{t}}\right)^{(\omega+2)/(\omega+1)} - (1 - \rho_0)^{\omega+2}\right], \tag{6.12}$$

where ρ_0 is that fraction of Q_0 that goes into the sample during the flash.

In Fig. 6.8a the crosses denote calculated values of $\tilde{V}(\tilde{t})$ for a sample where one CV_0 of charges is emitted in $t_T/100$. In Fig 6.8b we see that $d\tilde{V}/d\tilde{t}$ is independent of $\tilde{t}$ (in log scale) up to $\tilde{t} = 1$ and then it varies linearly with slope $-(\omega + 2)/(\omega + 1)$.

After $\tilde{t} = 5$ the surface potential does not change much. However, the residual potential depends on ω. So, for flash exposure, ω determines the residual potential and μ_0 determines how soon this value will occur. Therefore, a rapid discharge requires both a high mobility and a low field dependence of the mobility.

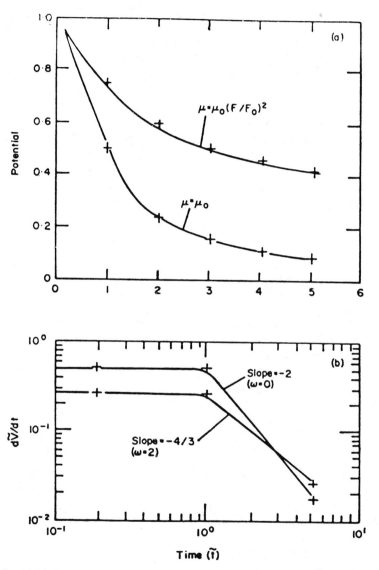

Fig. 6.8. (a) Discharge characteristics calculated (crosses) from Eqs. (20.9) and (20.11) for a photoreceptor with $\alpha L = 1000$, $P = 0.5$, $\omega = 2$ (upper curve) and $\omega = 0$ (lower curve). One CV of charge is generated in $0.01 t_T$. (b) the dV/dt vs. t curves for the discharges of (a). Crosses are calculated from the equations in the text (Mort, 1973).

The hole mobility in a-Se is independent of field ($\omega = 0$) at room temperature. From Fig. 6.8b we see that for a flash exposure $d\tilde{V}/d\tilde{t}$ is constant up to the time it takes the leading edge of the charge sheet to transit the sample. The discharge rate will be $J/C = \mu V_0^2/(2L^2)$; with $\mu = 0.13$ cm^2/V s, $L = 50\,\mu$m and $V_0 = 100$ V, dV/dt (for $t < t_T$) will be about 2.5×10^7 V/s. When $t > t_T$, dV/dt will decrease as t^{-2}. Batra et al. (1970) obtain experimental results in agreement with these results.

Heterogeneous photoreceptive samples made of about $2\,\mu$m thick Se layers evaporated onto about $20\,\mu$m of the polymer Poly N-vinylcarbazole (PVK) are also useful structures; the Se makes the structure sensitive in the visible. PVK shows unusual hole transport and hole injection into it from a-Se readily occurs (Regunsburger, 1968; Mort, 1972). Hole transport in PVK is dispersive, with the mobility of the fastest carriers about 10^{-7} cm^2/V s at $F \cong 10^5$ V/cm. Here $\mu \sim V^2$ over a range of field.

In Fig. 6.9, the discharge rate as a function of light intensity is displayed for the structure discussed above, with positive charge on the illuminated Se surface and a wavelength where all the light is absorbed in the Se. Note that at all light intensities and low fields a common value of dV/dt results that is proportional to V^4. Mort (1972) interprets this as trap-free-space-charge-limited (TFSCL) discharge of PVK. Since $dV/dt = J_s/C$ and Child's law yields $J_s = 10^{-13} K\mu\varepsilon^2/L$, then $\mu \sim F^2$ must hold. Hence, the mobility at a given field can be determined using the appropriate J_s and Child's law. For the initial potential, $\mu = 4 \times 10^{-7}$ cm^2/V s results.

In Fig. 6.10 dV/dt vs. t is displayed for four different light fluxes (Mort et al., 1972). Here we cannot apply the analytic results of Eqs. (6.10) and (6.12). Rather, an iteritive technique must be used (Chen, 1972) because one CV of free charge can be reached in the sample after a finite exposure. At higher intensities, however, Q_0 can be injected at times less than t_T, simulating flash exposure and allowing use of the analytical equations. Here dV/dt is constant until t_T and then decreases with a log scale slope of $-(\omega + 2)/(\omega + 1)$. As the flux decreases dV/dt begins at a lower value but falls less steeply after t_T. For a limited time after t_T the values of dV/dt exceed those of the highest fluxes but eventually coincide with the straight line observed at the highest fluxes. The reason for this is that at lower fluxes the samples reaches a TFSCL state at voltages well below V_0. Combining this with the fact that at lower fluxes a slower discharge time occurs indicates that the transition will take a longer time to occur.

Note in Fig. 6.10 that the linear portions of the curve for the two highest fluxes coincide. This implies that at the highest flux the sample becomes TFSCL at a time well before t_T. Since the slope of the line is $-\frac{4}{3}$, then $\omega = 2$. Using the mobility obtained from the data of Fig. 6.9, it is found that the constant dV/dt value, which equals $\mu(F_0)F_0^2/4$, equals 1.1×10^4 V s^{-1} at the

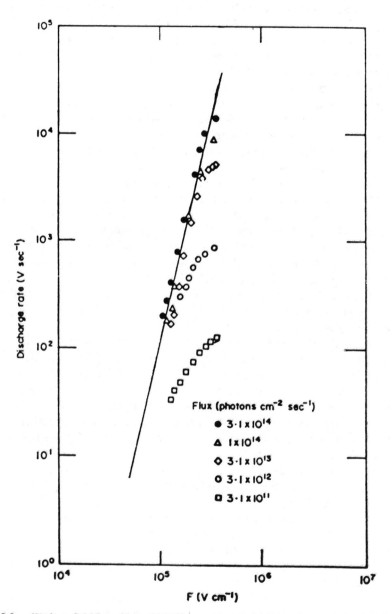

Fig. 6.9. dV/dt vs. field for a 22 μm PVK film overcoated with 2 μm of amorphous Se at five different light intensities. $F = V(t)L$ (Mort, 1973).

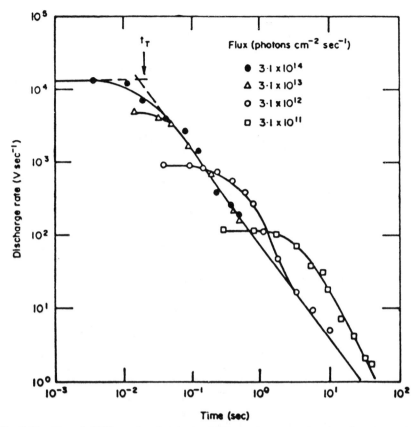

Fig. 6.10. Plot of dV/dt vs. time for the Se:PVK sample measured at four light intensities (Mort, 1973).

initial field, very close to the values observed. Extrapolating the constant dV/dt line to intercept the extended slope line yields $t_T = 2 \times 10^{-2}$ s, which gives $\mu = 3.6 \times 10^{-7}$ cm^2/V s at $F_0 = 3.3 \times 10^5$ V/cm. This agrees well with the value calculated from Fig. 6.9, and both are consistent with time of flight results (Mort, 1972).

In electrophotography the sample is usually exposed for a finite time chosen so that the white background light generates about one CV_0 of charge. This light corresponds to an object of zero optical density. The light from darker objects is attenuated by $10^{-O.D.}$, where O.D. is the optical density of the darker material. The different input densities produce different discharge rates leading to the presence of contrast potentials which are crucial to the development of the latent image. Figure 6.11 displays the consequences for the contrast potential when a TFSCL discharge occurs. Note that at an

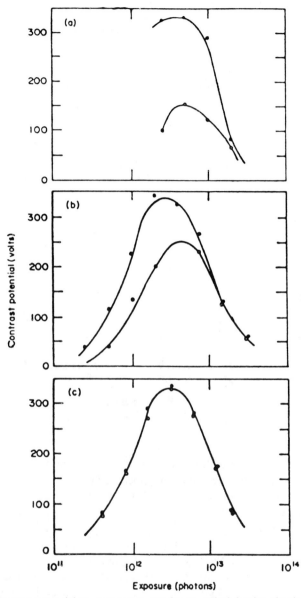

Fig. 6.11. Contrast potential vs. exposure curves for an optical density of unity at different white background fluxes. The initial surface potential was 820 V in each case. The open circles denote the instantaneous contrast potential and the black circles represent the maximum contrast potential. White background flux (a) 3.1×10^{14} photons cm^{-2} s^{-1}, (b) 3.1×10^{13} photons cm^{-2} s^{-1} and (c) 3.1×10^{12} photons cm^{-2} s^{-1} (Mort, 1973).

input density of unity the fluxes are one tenth of these values. The contrast potential was measured in two ways for each value of intensity. First, the immediate contrast potential was measured at the termination of the exposure. Next, the maximum contrast potential attainable was also measured. Sometimes this was obtained only after times longer than the exposure times. The exposure values in Fig. 6.11 are the product of the exposure time and white background flux. We see from the figure that the contrast potentials measured both ways agree at low values of flux but at higher fluxes the contrast potential measured immediately after exposure becomes increasingly smaller than the maximum attainable. This is attributed to the fact that at higher fluxes a CV of charge is already moving in the sample and continues so even after the end of the exposure. After the exposure the development of the contrast potential continues to grow. The effect results from the TFSCL discharge.

6.3 The Physics of the Development Process

Schmidlin (1976) has presented an extensive and detailed review of the electrophotographic process. In particular, he focuses on a specific view that has been useful in separating the various electrostatic imaging methods, identifying their limitations, and establishing the properties of materials needed for optimization of the process.

More precisely, the general process under investigation is termed "charged pigment electrophotography" (CPE), which encompasses xerography (Carlson, 1942) and photoelectrophoresis (Tulagin and Carreira, 1968). In both cases the development process is driven by a variation in the electrostatic force QF, where Q is the charge on a toner or pigment particle and F is the local electric field. This variation can be written $\delta(QF) = Q\,\delta F + F\,\delta Q$, whence the total force variation moving toner from an image plane can come from either a field or charge variation. If either of these occur because of a real optical image, it represents a latent image which must be developed.

It is conventional to call systems that are driven by $Q\,\delta F$ "charged pigment xerography" (CPX) and $F\,\delta Q$ driven systems "photoactive pigment electrophotography" (PAPE); the two together define CPE. The two systems are often further subdivided by considering possible non-imagewise forces that can exist which oppose the electrostatic driving force. CPX development systems can be divided into three classes (Scharfe and Schmidlin, 1975): aerosol and electrophoretic development; touchdown and magnetic brush; cascade development. The latter is the method sketched in the introduction.

Many of the limitations of practical systems can be understood from a statistical treatment of single toner development (Schmidlin, 1975). Physical processes capable of converting a real image into a latent image must first be described; CPE systems can be shown to be responsive to a single incident photon. Using a small-amplitude image, the physical nature of the latent image formation processes can be described. This is sufficient for a description of the process and also to define the quantum efficiencies η_p and η_x for PAPE and CPX respectively. It is found that both η_p and η_x depend strongly on F, and η_x even depends on the structure of the in-going image. This is explained via a solution of Poisson's equation for a periodic change pattern, which leads naturally to an extension of the analysis to large-amplitude images. Schmidlin (1976) carries through the large amplitude analysis for CPX; a periodic image is transformed through an entire system from an input density variation to an output density variation. He also provides a statistical description of the development process, using conditions of noisy adhesion between a toner and carrier. It is shown that realistic noisy adhesion lowers and bounds for the working range of the size and charge of toner, and minimum strength of the latent image (δF), which is 10^5 times greater than the idealized case. This value of δF is also found to be close to the minimum field required to drive development in practical systems.

The CPX and PAPE quantum efficiencies are developed via fundamental processes of: charge generation; displacement of these charges through the sample; transfer vs. trapping of the charges at the boundaries. We have already outlined the physics involved in this process in the last section. Schmidlin (1976) shows that the major difference between η_p and η_x lies in the transfer vs. trapping event. In PAPE, one and only one photogenerated charge must transfer off a toner. In CPX the charge approaching the surface must be trapped. This requirement, along with the fact that the photoreceptor must accept a corona charge, are sufficient conditions for showing that the surface material of a photoreceptor must be incompatible with the bulk.

More comprehensive consideration of the physics of the development process have recently been put forth (Scharfe, 1984; Williams, 1984; Schein et al., 1985) that critique the various extent models. In most development systems the mixing of the toner with the carrier beads causes them both to become oppositely charged, whence the toner adheres to the carrier. Three models have been proposed in order to explain how the toner leaves the carrier and deposits itself on the surface of the photoconductor (Scharfe, 1984; Schein et al., 1985). In the powder-cloud model the inertial forces that occur during collision free the toner from the carrier. It is then attracted to the photoconductor by electrostatic forces. In the equilibrium model the toner releases from the carrier until the electrostatic attraction of the latent images balances the attraction of the carrier beads. In the field-stripping model the electrostatic force produced by the latent image overcomes the force attracting the toner to the carrier beads. Once the model that correctly describes the development process is known, methods for improving the process suggested by the model can be implemented. In this manner the contrast can be improved and the printed areas can be made blacker.

6.4 Charging of the Toner Particles

The toner becomes charged via the electrostatic charge exchange between it and the carrier beads as they rub together before being dispersed (triboelectrification) (Harper, 1967; Lowell *et al.*, 1980). They both develop opposite charges; the smaller toner particles then adhere to the larger carrier beads. The toner particles are typically made from the blending of polymers and carbon-black pigment; the carriers have a magnetic core coated with a polymer layer. The chemical properties of the layer determine the sign of the charge acquired by the toner. Hence, to optimize this process, we must understand the physics of contact electrification; this subject is not completely understood at present (Lowell *et al.*, 1980). Our deepest understanding lies in the area of metal–metal contacts; we know the least about the subject at hand, insulator–insulator contacts. Here we lack detailed experiments and theoretical ideas. Indeed, it is difficult to determine the sign of the charge developed by one insulator when it is brought into contact with another insulator. It has been suggested (Henniker, 1962) that we arrange materials in an empirical triboelectric series (Burland and Schein, 1986). Those materials that are higher up in the series become changed positively when contacted with materials that lie below them in the series. The problem is that we cannot as yet order the materials properly; they often seem to order in rings rather than ladders.

The present situation is that it is extremely difficult to perform definitive experiments in this area. The results depend critically on the detailed microscopic properties of the surface. Roughness, chemical composition and impurities can often dominate the results.

6.5 Material Requirements for Photoreceptors

It was thought at first that the major criterion for a good photoreceptor was a high resistivity in the dark. Here it was assumed that the dielectric relaxation time, $t_d = \rho K_p \varepsilon_0$ should be larger than the development time, where K_p is the dielectric constant of the photoreceptor. Schmidlin (1973), however, showed that the t_d requirement is neither necessary nor sufficient. Rather, a high carrier range ($\mu\tau_L$, when τ_L is the time to traverse the sample) and low thermal generation rate (g_{th}) are necessary. Further, a relatively high carrier mobility ($\theta\mu$) and low equilibrium carrier concentration ($p/\theta = g_{th}t_L/\theta$) are also required. θ is the ratio of free charge to total (free plus trapped) charge that are in transport states. Since the conductivity increases the $p\mu$ product ($=g_{th}\mu\tau_L$) it can not assure that the individual factors alone will fall in the proper range.

The above considerations are by no means all that is required. It is also of fundamental importance that the photoreceptor have blocking contacts (Shaw, 1981) and that all thermal carriers must be extracted over the useful length of material. That is, the structures must be charge-depletion devices with depletion layers greater than the minimum required image dipole length or sample thickness. Further, as stated before, the material conditions in the surface region must be incompatible with the material properties of the bulk.

To determine criteria for selecting materials for efficient imaging we begin with inputs of: the extreme fields in the photoreceptor, $F_{p,\max}$ and $F_{p,\min}$; the minimum thickness or image dipole length, L, of the photoreceptor; the image process or development time, T_p. $F_{p,\max}$, $F_{p,\min}$ and L are determined by the properties of the developer and T_p is determined by the system application. The outputs are conditions on g_{th}, $\theta\mu$ and τ_L expressed in terms of the inputs. We require that the latent image be constructed prior to the beginning of the development process and that the development be completed before the decay of the latent image. Further, to obtain the special surface conditions a resolution requirement is also needed.

Every one of the processes that make up the formation of a latent image must be accomplished in a time less than T_p. These are the "photorequirements".

Generation. The time required to generate sufficient carriers to reduce the internal field from $F_{p,\max}$ to $F_{p,\min}$ is given by (Schmidlin, 1976)

$$t_g = \frac{K_p\varepsilon_0}{e[1 - \exp(\alpha L)](1 - R_v)F_0'} \int_{F_{p,\min}}^{F_{p,\max}} \frac{dF}{\gamma_{\text{eff}}(F)}$$
$$\cong \frac{K_p\varepsilon_0(F_{p,\max} - F_{p,\min})}{e[1 - \exp(-\alpha L)](1 - R_v)F_0'\gamma_{\text{eff}}(\varepsilon_{p,\min})}, \quad (6.13)$$

where R_v and α are the reflection and absorption coefficients for the incident light and F_0' is the incident light flux in photons per unit area. γ_{eff} is the effective probability that an absorbed photon is successfully converted into separable carriers. The requirement that $t_g < T_p$ is a condition on F_0'.

Displacement. Here we assume that each generation event contribute one full-length dipole. For maximum efficiency at least one member of any charge pair capable of expanding into a full-length dipole must transit the sample in less than the processing time. This must be true for all dipoles that contribute to a latent image; it must therefore be true for the slowest moving charge, that which moves in the field $F_{p,\max}$. This condition is

$$t_T = \frac{L}{\mu\theta F_{p,\min}} < T_p \qquad (6.14)$$

Further, the carrier must live long enough to accomplish this, i.e.,

$$\frac{\tau_L}{\theta} > t_T \qquad (6.15)$$

This latter condition is identical to the requirement that $\mu\tau_L F_{p,\min} > L$, which is independent of θ.

Transfer Versus Trapping. Whether or not a transfer or trapping event is desirable depends upon the specific application and interface. For example, transfer is favored over trapping at the surface of a photoactive pigment particle; $T(t = T_p) \cong 1$ for at least one of the charge carriers ("intentional transfer"). On the other hand, there is always one boundary where the length of the image dipoles terminate. Here we favor trapping over transfer; $T(t = T_p) \ll 1$ ("intentional trapping").

We must also consider the consequences of a time delay prior to either trapping or transfer. Schmidlin (1976) shows that for wavelengths less than λ_c $[= 2(\theta\mu_s\tau_s)F_{p0}(\Sigma_{10}/\Sigma_c)]$ $[1 - \exp(-t/\tau_s)]$, where $\tau_s^{-1} = \tau_{Ls}^{-1} + \tau_t^{-1}$, where τ_t^{-1} and τ_{Ls}^{-1} are the probabilities per unit time for transfer and trapping, respectively, $\Sigma_c = K_p F_0 F_{p0}$, where F_{p0} is the initial field inside the photoreceptor and Σ_c is the corona charge per unit area. $\Sigma_{10} \cong \Sigma_c - \Sigma_{\min}$, so $\Sigma_{10} = \Sigma_c$ for normal exposures cannot be resolved unless

$$2\theta_s\mu_s\tau_s F_{p0} < \lambda_c, \quad \text{if } \tau_s < T_p \qquad (6.16)$$

or

$$2\theta_s\mu_s T_p F_{p0} < \lambda_c, \quad \text{if } T_p < \tau_s \qquad (6.17)$$

The quantity $(\lambda_c/2\theta_s\mu_s F_{p0}) \equiv t_{\lambda_c}$ represents the time it takes a carrier to drift along the interface a distance $\lambda_c/2$. Hence, Eqs. (6.16) and (6.17) state that an image must be developed in times less than t_{λ_c}, unless the carrier dies by either transfer or trapping.

For intentional transfer Schmidlin (1976) shows that both $\tau_t \ll \tau_{Ls}$ and $\tau_t < T_p$ are required to insure that $T(t = T_p) \cong 1$. Hence, slow interface states must be avoided and transport states in the photoreceptor and dielectric must be fairly well matched. On the other hand, a more severe matching condition occurs for planar xerographic interfaces where the lateral motion can limit the resolution. We know that $\tau_{Ls} \gg \tau_t$ implies $\tau_s \cong \tau_t$, and $\tau_t < T_p$ means that $\tau_s < T_p$. Whence, Eq. (6.16) implies that $2\theta_s\mu_s F_{p0}\tau_t < \lambda_c$ (or $\tau_t < t_{\lambda_c}$). The significance of this requirement can be appreciated by treating the special case where the surface and bulk mobilities are equal ($\theta_s\mu_s = \theta\mu$). Here it can be seen that τ_t must be less than t_T at the highest field in order to resolve λ_c less than $2L$.

For intentional trapping Schmidlin (1976) shows that either $\tau_{Ls} < \tau_t$ or $T_p < \tau_t$ and τ_s (if $\tau_t \ll \tau_{Ls}$) must be satisfied. These requirements usually mean that transfer must be inhibited intentionally, which occurs naturally under usual xerographic conditions. However, the resolution criteria given by Eqs. (6.16) and (6.17) produce some severe restrictions. If $T_p < \tau_s$ Eq. (6.17) is directly applicable. However, if we compare this to Eq. (6.14), it is impossible to resolve $\lambda < 2L$ unless $\theta_s\mu_s F_{p,\max} < \theta\mu F_{p,\min}$, which means that the surface drift velocity at the highest field must be below the bulk drift velocity at the lowest field. For the case where $T_p > \tau_s$ we must now have $\tau_{Ls} < \tau_t$ to insure $T(t = T_p) \ll 1$. Here $\tau_s \cong \tau_{Ls}$ and Eq. (6.16) implies $2\theta_s\mu_s\tau_{Ls}F_{p0} < \tau_c$. Upon comparison with Eq. (6.15) we see that $\lambda < 2L$ will not be resolved unless $\theta_s\mu_s\tau_{Ls}F_{p,\max} < \mu\tau F_{p,\min}$. This means that the surface schubweg at the field maximum must be below the bulk schubweg at the field minimum.

The above results show that intentional trapping at the free surface of a photoreceptor requires either: (a) the surface contains many orders of magnitude more traps than the bulk; (b) the carrier that is trapped at the surface cannot traverse the bulk. To satisfy these criteria the free surface must have a sufficient number of slow surface traps to accommodate all the corona charge.

Besides the above requirements, there are also requirements on the properties of the material in the dark. It is customary to employ the initial voltage decay rate in the dark, or the initial discharge current density $[j_D = (K_p\varepsilon_0/L)\,dV/dt]$, as a measure of the discharge process. A criteria for the stability of the latent image is

$$j_D(F_{p,\max})T_p < 0.1K_p\varepsilon_0 F_{p,\max}, \qquad (6.18)$$

which means that we shall allow no more than 10% dark discharge from the initial state. To use this equation for materials selection let us consider the nature of the processes that contribute to j_D. Schmidlin (1976) discusses what is wrong with the traditional assumption that $j_D = \sigma F_{p,\max}$, which when substituted into Eq. (6.18) yields a criterion for the dielectric relaxation time. To invoke this criterion, however, requires that the contacts be linear and of low resistance ("ohmic"), or else there is no guarantee that the current will be bulk limited. But if the current is bulk limited, it must transfer to a SCLC whenever $t_T < \tau(= K_p\varepsilon_0/\sigma)$ and as Schmidlin (1976) shows, neither the bulk limited current or SCLC is compatible with the requirement that $t_T < T_p$.

In order to prevent more than 10% discharge of the photoreceptor in a transit time, less than 10% of the charge that produces $K_p\varepsilon_0 F_{p,\max}$ can be mobile. So we want the corona charge to occupy slow traps, i.e., we want

MATERIAL REQUIREMENTS FOR PHOTORECEPTORS 495

the contacts to be injection limited or "blocking" (Shaw, 1981). A blocking contact is a necessary condition for efficient latent image formation, and it is important to formulate an expression for the dark current when a photoconductor has blocking contacts. To this end we consider a unipolar material and look at the transient current at times of the order of T_p (Schmidlin, 1976).

A depletion layer always forms near a blocking contact. Sufficiently far from the contact, however, space charge neutrality is often reached. In this region detailed balance yields

$$g_{th} = \sum_{i,\,\text{slow}} \frac{p_i}{\tau_{ri}} = \sum_{i,\,\text{slow}} \frac{p}{\tau_i} = \frac{p}{\tau_L} = \frac{p}{\theta} \times \frac{\theta}{\tau_L}, \qquad (6.19)$$

where p_i is the hole concentration in the i_{th} localized state, τ_{ri} is the mean release time from this state and p is the equilibrium concentration of carriers in transport states. (These definitions of g_{th} and τ_L are unique to electrophotography and differ from the definitions used in standard semiconductor theory.)

If a corona is suddenly applied to the structure shown in Fig. 7.12, the mobile holes (p/θ) will deplete from the transport and fast traps in a distance

$$Z_d' = \frac{K_p \varepsilon_0 (F_{p,\text{max}} - F_{p,i})}{e(p/\theta)}, \qquad (6.20)$$

where $F_{p,i}$ is the field beyond the depletion layer. The current generated in the depletion region and the field required to sustain it are found from

$$j_D = e g_{th} Z_d' = e p \mu F_{p,i}. \qquad (6.21)$$

In our case we need only to examine the limiting case where the depletion region fills the photoreceptor. Here the current reaching the substrate saturates at

$$j_D = e g_{th} L. \qquad (6.22)$$

The necessary conditions for this saturation are given by

$$\mu \tau_L F_p > L, \qquad \text{if } \tau_K > \tau_L \qquad (6.23)$$

or

$$K_p \varepsilon_0 F_p > g \frac{p}{\theta} L, \qquad \text{if } \tau_K < \tau_L \qquad (6.24)$$

Schmidlin (1976) shows from these that the appearance of a saturated generation-limited current (GLC) with blocking contacts is compatible with our prior photorequirements. Equation (6.23) is an important dark requirement that must be fulfilled; it also implies a secondary photocurrent (or gain greater than unity) under sustained voltage conditions.

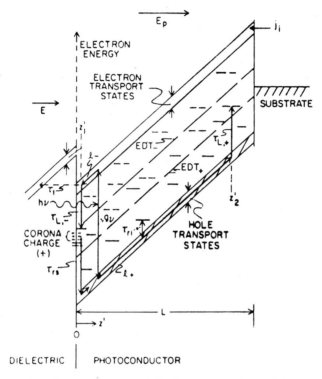

Fig. 6.12. A schematic energy diagram indicating allowed electronic energy states in a photoconductor and contiguous dielectric. g_v, l_+ and $\tau_{L\pm}$ represent the fundamental photogeneration, displacement and trapping processes of a hole and electron respectively. τ_t represents a possible transfer event competing with τ_{L-} (Schmidlin, 1976).

Substituting the dark current expression Eq. (6.22) into Eq. (6.18) yields

$$eg_{\text{th}}LT_p < 0.1 K_p \varepsilon_0 F_{p,\max}, \qquad (6.25)$$

which replaces the old dielectric relaxation criterion.

An important result of Eq. (6.25) and the photorequirements is that they become easier to satisfy as the photoreceptor becomes thinner. Indeed, the thickness of the photoreceptor is determined by the field required to drive development. Schmidlen (1976) shows that there is a minimum δF required to drive development which is proportional to the thickness of the photoreceptor. To establish the minimum thickness we assume that the photoreceptor is charged to the maximum field it can withstand. Then the minimum required δF, $[=2\sigma(F^*)]$ times about a factor of two for image neutralization, yields the minimum thickness. We must assume that this thickness is known in order to make the equations we use in this section meaningful.

When the length of the latent image dipoles is limited by the thickness of the depletion region rather than by $\mu\tau_L F_p$, any photoreceptor thickness greater than the thickness of the depletion region is ineffectual. Here we choose the maximum thickness to match the maximum depletion layer thickness. Here $\tau_K < \tau_L$ and Eq. (6.24) is the condition required to reach the saturated-current limit.

Conversely, if we regard L in Eq. (6.24) as the minimum thickness required for development and take F_p as the maximum surface field $(F_{p,\max})$, then Eq. (6.24) gives us a restriction on the equilibrium mobile carrier concentration (p/θ). Equations (6.23) and (6.24) are therefore both useful criteria for material selection; the two together cover all the possibilities for realizing a saturated GLC.

In practice, contacts are never perfectly blocking. In most cases they are so "leaky" that the bulk GLC limit is not often observed. To determine an upper bound on the allowed rate of injection we invoke Eq. (6.18) with j_D replaced by an injection current density per unit area, J_i:

$$J_i(F_p) < 0.1 K_p \varepsilon_0 F_{p,\max}/T_p \tag{6.26}$$

The injection current density at the free surface can be written as $J_i = K_p \varepsilon_0 F_{p,\max}/e\tau_{rs}$, where $K_p \varepsilon_0 F_{p,\max}/e$ is the number of corona charges per unit area and τ_{rs} their average release time from surface states. Putting this J_i into the limiting current expression gives $\tau_{rs} > 10 T_p$; the corona charge must occupy slow traps.

If the above surface trap concentration were extended into the bulk a range limitation on the discharge process ($\mu\tau F < L$) would result. Corona charged photoreceptors therefore require conditions at the surface that are incompatible with the material conditions in the bulk (Pai, 1983); dual layer systems have proven to be the best for this type of electrophotography.

Table 6.1 summarizes the above results. Here all the parameters that characterize bulk and surface or contact conditions (μ, τ_L, g_{th}, τ_{rs}, etc.) are written in terms of quantities ($F_{p,\max}$, $F_{p,\min}$, L, T_p and λ_c) that are determined by the developer, process time and resolutions required.

6.6 Electrophotographic Applications of Amorphous Hydrogenated Silicon

Although it is the purpose of this chapter to discuss the role of chalcogenides in electrophotography, recent developments (see, e.g., Kawamura and Yamamoto, 1982 and Jansen et al., 1984) have made it imperative to include a discussion of the great potential that a-Si:H has in this area (Shimizu et al., 1980; Pai and Melnyk, 1986; Hudgens, 1986). To apply a-Si:H in the

Table 6.1. Photoreceptor Material Requirements

Photorequirements	Dark Requirements
(1) Generation time: $t_g < T_p$	(1) Blocking contacts (to mobile carriers)
(2) Transit time: $t_T = \dfrac{L}{\mu F_{p,\min}} < T_p$	(2) Depletion: $\dfrac{p}{\theta} L < K_p \varepsilon_0 F_{p,\max}$
(3) Lifetime: $\tau_L > t_L$	(3) Bulk generation: $eg_{th} L T_p < 0.1 K_p \varepsilon_0 F_{p,\max}$
(4) Transfer vs. trapping; (a) for intentional transfer $\tau_t < T_p$ and τ_{LS}, and $2\theta_s \mu_s \tau_t F_{p,\max} < \lambda_c$ (b) for intentional trapping $\tau_t > T_p$; $2\theta_s \mu_s T_p F_{p,\max} < \lambda_c$ or $\tau_t > \tau_{LS}$; $2\theta_s \mu_s \tau_{LS} F_{p,\max} < \lambda_c$	(4) Contact injection: $J_i T_p < 0.1 K_p \varepsilon_0 F_{p,\max}$
	(5) Corona traps: $\tau_{rs} > 10 T_p$

area of electrophotography it is required that the resistance of the material be made to exceed 10^{13} Ω cm. Two different approaches have been used to achieve this. In one approach a blocking layer is included, which makes the structure a multilayer device (Shimizu et al., 1980). In the other approach a very high sensitivity a-Si:H film is fabricated by the inclusion of boron and oxygen (Yamamoto et al., 1981). In Chapter 2 we learned that the maximum resistivity of conventional rf glow discharge a-Si:H occurs when it is doped lightly with boron. Since most attempts to increase the resistivity above this value lead to a degradation of the photosensitivity, multilayer construction becomes one way to develop an appropriate photoreceptor. By employing a 0.3 μm thick sputtered insulating layer of SiO_x as a blocking layer, Shimizu et al. (1980) were able to obtain a sufficient amount of surface charge retentivity. The multilayer photoreceptors proposed are shown in Fig. 6.13. Here, either a phosphous doped n layer (a) or boron doped p layer (b) is deposited on a Ni–Cr conducting substrate, rather than the SiO_x layer. Next a lightly boron doped layer is applied (optimum photoconductivity). Although these structures had some utility as photoreceptors, their charge retentivity decreased by 75% within two hours. By employing an insulating layer of Si_3N_4 rather than the doped a-Si:H blocking layer and As_2S_3 top layer, Sato et al. (1981) made a bi-chargeable device with sufficient low residual surface charge potential after illumination.

Figure 6.14 shows another multilayer structure (Takagi and Ozawa, 1980), the heavily oxygen doped film is insulative whereas the lightly oxygen doped

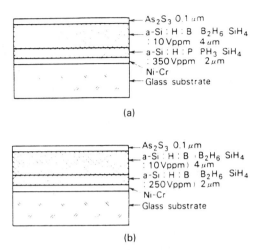

Fig. 6.13. Multilayered a-Si:H photoreceptor for the use of positive (a) and negative (b) charge polarities (Shimizu et al., 1980).

film has sufficient photoconductivity. Fukuda (1979) suggested the structure shown in Fig. 6.15. Here the a-Si:H film is deposited at low temperatures (30–80°C) and then annealed at 200°C in vacuum. The dark resistivity was 10^{13} Ω cm and the light to dark conductivity changed by two orders of magnitude.

At about the same time that the above multilayer structures were developed, Nakayama et al. (1979) developed a-Si:H films of high dark resistivity and sufficiently good photosensitivity by simultaneously doping films with boron and oxygen. Figure 6.16 shows the important film properties under a constant level of oxygen doping. (These films contained about 0.01 wt.% oxygen.) It was also found that the optimum substrate temperature for obtaining the maximum photo and dark conductivities was about 200°C. These results have allowed for the preparation of monolayered a-Si:H photoceptors with bi-chargeable properties that have great promise in the area of electrophotography.

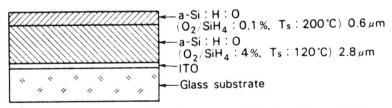

Fig. 6.14. Multilayered a-Si:H photoreceptor (Takagi and Ozawa, 1980).

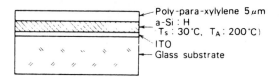

Fig. 6.15. Multilayered a-Si:H photoreceptor (Fukuda, 1979).

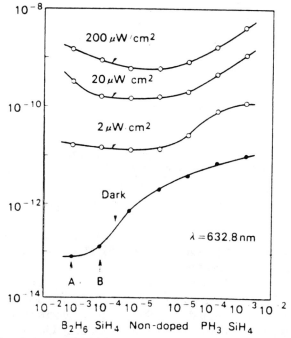

Fig. 6.16. Dark and photoconductivities at room temperature for a-Si:H doped with boron and oxygen (Yamamoto et al., 1981).

7
Optical Memories

7.1 Introduction.. 501
7.2 General Requirements for the Optical Data Disc......................... 505
7.3 Optical Massmemories Based on Crystalline/Amorphous Phase Transitions in Chalcogenide Glasses 507
7.4 Optical Massmemories Based on the Thermal Creation of Holes..................... 512
7.5 Recent Technological Developments..................................... 516

7.1 Introduction

In Chapter 5 we discussed the details of bias-induced switching and memory phenomena in thin chalcogenide films. In the memory ON-state a crystalline filamentary path forms between the electrodes; the optical properties of this region are sufficiently different from the amorphous region that the use of this phase transition for an optical memory device is evident (Adler and Feinleib, 1971). Further, since the memory filament is in most cases developed via local heating, the use of optical excitation (e.g., a focused laser) to generate the requisite phase transition is also evident. In early experiments with laser induced phase transitions it was also found that another mode of information storage was possible—the thermal creation of bubbles, or holes, in the film (Feinleib *et al.*, 1972; Terao, 1979). Both phenomena are useful for optical recording and read-out, and in this book we will discuss devices that rely on amorphous to crystalline, and the inverse, phase transitions, along with those that depend primarily on morphological changes in the material.

The key to understanding memory switching induced by either Joule or optical heating lies in thermal cycling experiments on bulk material (Fritzsche and Ovshinsky, 1970; Adler *et al.*, 1970). In Fig. 7.1 we see data representative of a differential thermal analysis experiment. Here the material under investigation is heated and cooled at a constant rate; the resulting signal is a measure of any change in the rate of heat generated or absorbed.

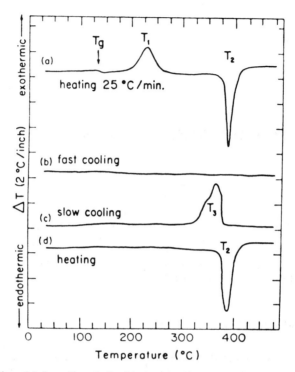

Fig. 7.1. Differential thermal analysis of a memory-type chalcogenide alloy, $Ge_{16}Te_{82}Sb_2$: (a) heating from the amorphous phase; (b) rapid cooling from the liquid phase; (c) slow cooling from the liquid phase; (d) heating from the cystallized phase (Adler and Feinleib, 1971).

These changes result from phase transitions in the material. Note the following in Fig. 7.1a. First we see that when the chalcogenide material is heated from room temperature, near 125°C the signal drops, indicating that the glass transition temperature, T_g, has been reached. Here there is a sharp drop in the viscosity of the material. When a temperature of about 225°C is reached, the signal increases because of the exothermic transition associated with the nucleation and growth of Te crystallites (Bienenstock et al., 1970); If the material is next cooled at any rate from the region between the exothermic peak near 225°C, T_1, and the large endothermic peak near 375°C, T_2, a low resistance state results because of the presence of the Te crystallites. The endothermic peak at T_2 represents the melting of the crystallized material.

Next, when the material is rapidly cooled, or quenched from the liquid phase at $T < T_2$, the result shown in Fig. 7.1b is obtained. There are no phase transitions evident; an amorphous or glassy state appears, and the material has a high resistance.

INTRODUCTION

However, if the material is cooled slowly from the melt, an exothermic transition occurs, as shown in Fig. 7.1c. Here the low-resistance crystalline state grows from the melt and, if the material is reheated an endothermic melting signal is observed (Fig. 7.1d).

The changes that occur in the optical properties of the amorphous material after transforming to the crystalline state are shown in Fig. 7.2 for pure Te. The low energy crystalline absorption edge shifts to higher energy and spreads out when the material is made amorphous. We see from Fig. 7.2 that large changes in transmission are possible for light between 0.5 and 0.7 eV when Te changes its phase. On the left in Fig. 7.2 is a plot of the imaginary part of the dielectric constant, ε_2, that mirrors the reflectivity. We see that there are large differences between the crystalline and amorphous phases, primarily because ε_2 is very sensitive to a shift in the absorption edge.

It has been suggested that the occurrence of a light-induced phase transition in certain chalcogenide materials involves both heating and direct photocrystallization (Adler and Feinleib, 1971). It is speculated that when light of the proper wavelength is absorbed a sharp increase in the free carrier concentration associated with broken covalent bonds occurs. The broken bonds weaken the metastability of the amorphous state and crystallization is enhanced at an accelerated rate; this is photocrystallization. Indeed, this phenomenon has been observed in materials like a-Se (Dresser and Strongfellow, 1968), and suggested in alloys of the type $Ge_{15}Te_{81}X_4$ in the work of Evans et al. (1970), and in Se-Te alloys by Feinleib et al. (1971).

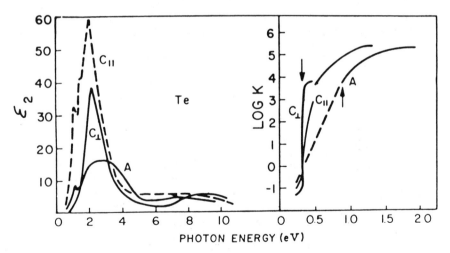

Fig. 7.2. Imaginary part of the dielectric constant and absorption as a function of energy for amorphous (A) and single crystal Te; $C_\perp$ and $C_\parallel$ indicate light with electric vector perpendicular or parallel to the crystalline c-axis (Adler and Feinleib, 1971).

We have already stated that the most probable physical mechanism for the amorphous to crystalline phase transition is the combined thermal and photocrystallization process. A focused laser should then be able to provide us with an optimum density of electron-hole pairs. In Fig. 7.3 Adler and Feinleib (1971) show that the best condition for crystallization at ambient temperature T_A occurs when the intensity and duration of the pulse bring the temperature of the material to that of the maximum crystallization rate without exceeding the melting temperature T_M (curve A). The temperature is then kept between T_G and T_M for a time t_1 during which crystallites grow. During a fraction t_2 of this time the laser is on, so the rate of crystallization is enhanced by the affect of photocrystallization. Further, Fagen and Fritzsche (1970) have shown that a non-equilibrium excited state is maintained for a time on the order of milliseconds after the laser is turned off, and this also enhances the crystallization rate.

Curve C of Fig. 7.3 shows the reverse process, where the crystallized material is heated by absorbing the laser light. Now the absorption constant of the crystal is larger than that of the amorphous material at the photon energies employed. Further, the crystal has a lower heat capacity for $T_G < T < T_M$. Hence the crystallized material will reach a higher T for the same or even less applied power than will the amorphous material. When T exceeds T_M we will get melting, as shown in Fig. 7.3. After the light is removed the liquid cools through the range between T_M and T_G in a time t_3, with the light off. Photocrystallization does not occur, and if t_3 is short enough, the material is quenched back to the amorphous state.

The above discussion emphasized the process of optical recording by an amorphous to crystalline phase transition. As we have pointed out, we can

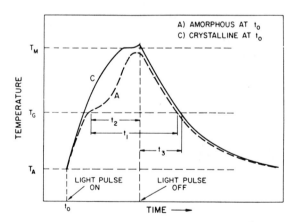

Fig. 7.3. Temperature–time profile of the active material illuminated by a square light pulse with energy above both the amorphous and crystalline band gaps (Adler and Feinleib, 1971).

also make optical recordings by the thermal creation of physical holes in the film. We will discuss some specific details of both these processes later, after we look into the requirements of the system in general.

7.2 General Requirements for the Optical Data Disc

Bell (1983) has discussed in detail the general requirements for an optical data disc. In what follows we borrow liberally from his discussion and appreciate that the materials system of the optical data disc contains three strongly interactive sub-systems: the substrate, which provides mechanical strength and stability, the storage medium, in which the data are recorded, and the protective encapsulation.

(a) *High resolution.* In principal, the storage density of optical data storage devices is limited only by the size of the diffraction limited focused light spot. This limitation yields a maximum theoretical data storage density of 10^8 bits/cm^2 for a typical system. In practice the figure is reduced by less than an order of magnitude by the additional requirements that cross talk must be limited during playback and that a constant angular velocity (CAV) mode be used for the drive. Further, because of the relationship between the playback bit-error-rate (BER) and the timing (position) of the signal transitions on the track, the material must be able to form an optical mark whose perimeter can be reproducibly set to within a few hundred angstroms.

(b) *No intermediate processing.* We desire direct-read-after-write (DRAW) operation. The absence of processing is required not only for the "instantaneous" requirement, but also for obtaining a low BER. Further, it provides flexibility in the rate of disc usage with respect to add-on data and is also important when it is required to utilize a trailing read spot after the write spot for data verification.

(c) *High recording sensitivity.* The semiconductor diode laser is the preferred read/write source for optical storage since it is of compact size and develops the power necessary to record (Ettenberg and Botez, 1982), in the range of 25 mW CW. Further, the output beam intensity can be modulated directly by the input data signal. By carefully designing the optical system an overall optical efficiency of 50% can be obtained between the diode laser and the focused read/write spot. This means that no more than 125 mW CW or 25 mW peak pulsed power at a 50% duty cycle is available for recording. To obtain a practical operating margin the material should have a recording sensitivity at a wavelength of 8200 Å better than 10 mW incident pulsed power for 50 ns pulses, which corresponds to a recording rate of 10 Mb/s. Under the same conditions this implies a threshold sensitivity of only a few milliwatts.

(d) *Recording Threshold.* In order not to be limited by shot noise and obtain an acceptable signal-to-noise rates (SNR) during playback, it is required that a minimum read power level be available. This level is determined by the playback rate, optical playback contrast and efficiency of the playback optical system. Since we must not damage the data during playback, we must have the threshold power for recording greater than the minimum required read beam power. In practice the difference is usually above a factor of three.

(e) *Sustained low BER and high SNR.* If the BER after correction during playback exceeds the drive specification, the useful life of the optical data disc will end. The BER is probably the most complex and system-dependent parameter in the problem. The intrinsic defect level and statistics of the substrate, storage layer and encapsulation system, along with the data storage density and modulation scheme, all determine the initial BER characteristic of the disc. DRAW techniques can reduce the initial BER to a tolerable level (about 1 in 10^{12} bits) and use only a few percent of the total disc capacity. How the BER evolves with time is determined by the stability, on a microscopic scale, of the material components, along with the effectiveness of the encapsulation and other environmental protection measures taken to avoid the build up of airborne particulates between the playback optics and the storage layer.

The SNR is in general not limited by photon shot noise in the playback beam. Rather, it is often determined by the statistical fluctuations of the optical mark geometry, which can produce timing errors. For digital data a wideband SNR of 30 dB is sufficient to reduce the BER due to the source well below that due to intrinsic material defects. For analog signal recording, the SNR requirement can be much more restrictive.

(f) *Data formatting.* It is generally essential for the optical disc to offer the ability for formating and random access during recording. One way to achieve this is the pregrove concept (Bulthuis *et al.*, 1979), wherein the substrate is formed with a microscopic pattern of data grooves, headers and timing marks molded into its surface (Carasso and Huijser, 1982). Another is to coat a flat, featureless substrate blank with the storage medium and process to a finished disc. A servo-writer can then be used to record heading, tracking and timing information directly. The first approach seems the most promising at the present time.

Rather than discuss the details of further specific requirements and how they might be met, we will leap directly to those materials that have the greatest potential for solving the general problem of optical storage today, the chalcogenides.

7.3 Optical Massmemories Based on Crystalline/ Amorphous Phase Transitions in Chalcogenide Glasses

Chalcogenide films for optical recording have been developed over the past 15 years (see, e.g., Feinleib *et al.*, 1971 and 1972; Watanabe *et al.*, 1983). Sb_2Se_3 and Sb_2Te_3 films have recently been reemphasized (Watanabe *et al.*, 1983) and, because of their great promise for reversible video disc applications, will be stressed in this section. These films are generally vapor deposited into an amorphous phase at about 10^{-6} torr. They are then annealed into a crystalline phase. Large changes in transmissivity and absorption coefficient occur after the films are annealed above the crystallization temperature; these are shown in Figs. 7.4, 7.5 and 7.6.

We next consider the affect of a diode laser on these films when rotating at a speed of 1800 rpm. Since, as we shall discuss shortly, the output of the laser will be as small as 5 mW at the surface of the disc-shaped film in a DRAW system, and the absorption of a 1200 Å thick film of Sb_2Se_3 is about 10%, we see that the film absorbs very little energy. Were the thickness of the film increased, the thermal capacity would also increase, whence the temperature of the film would not exceed the threshold temperature of 160°C. We see, therefore, that it is difficult to record on a single 1200 Å thick film of Sb_2Se_3 with a diode laser.

A way to increase the absorption of the film is to employ a double layer; a transparent layer plus a reflective layer. With this structure we can reduce both the transmissivity and reflectivity, and also enhance the difference in reflectivity before and after reading (Watanabe *et al.*, 1983). Figure 7.7

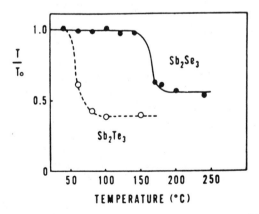

Fig. 7.4. Dependence of transmissivity T on annealing temperature of Sb_2Se_3 and Sb_2Te_3 films. T_0 is the transmissivity of As-deposited film (Watanabe *et al.*, 1983).

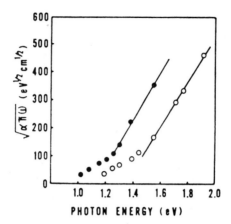

Fig. 7.5. $\sqrt{\alpha h \omega}$ vs. photon energy of Sb_2Se_3 before (open circles) and after 200°C annealing (closed circles) (Watanabe *et al.*, 1983).

shows the thickness dependence of the reflectivity of single layer films and Fig. 7.8 shows the thickness dependence of the reflectivity of a double-layer film. Here the transparent layer is Sb_2Se_3, which can be changed by the laser, and the reflective layer is a 400 Å thick Te film. (Te is used because its thermal conductivity and diffusivity is low compared to other metallic films, and these properties result in a small energy loss due to thermal diffusion.)

For a high recording efficiency we want as low a reflectivity as possible. For high playback sensitivity, the difference in reflectance before and after recording must be greater than 20%. For Sb_2Se_3 these conditions are satisfied

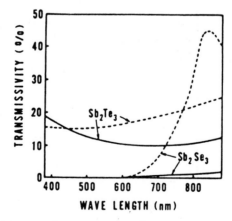

Fig. 7.6. Transmissivity of 150 Å Sb_2Te_3 and 3500 Å Sb_2Se_3 films with a glass substrate before (dotted lines) and after thermal annealing (solid lines) (Watanabe *et al.*, 1983).

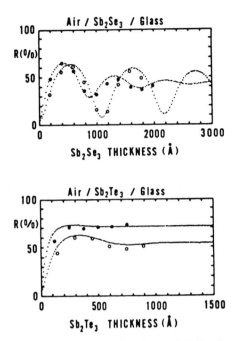

Fig. 7.7. Thickness dependence of reflectivity of Sb_2Se_3 and Sb_2Te_3 films. The open and closed circles are the experimental results of the film before and after annealing, respectively. The dotted lines are the theoretical curves of reflectivity using the values of the refractive index of Sb_2Se_3 and Sb_2Te_3 given in the text (Watanabe et al., 1983).

by the thicknesses X and Y shown in Fig. 7.8. But, for high recording sensitivity we must have a small thermal capacity, and the film thickness in region X satisfies this requirement.

From the standpoint of stability, Watanabe et al. (1983) chose Bi_2Te_3 rather than Te for the reflective layer. Further, the refractive index of Bi_2Te_3 changes upon annealing to 200°C whereas that of Te does not. Also a PMMA (polymethyl-methacrylate substrate)/Sb_2Se_3/Bi_2Te_3 structure provides a better playback signal than the same structure with a Te rather than a Bi_2Te_3 layer. Figure 7.9 shows the theoretical reflectivity vs. thickness curve for the above structure.

Figure 7.10 is a schematic of a high density DRAW optical recording system. The diode emits at 8300 Å with a peak output power near 15 mW. The collimated lens has a numerical aperture (NA) of 0.3. The asymmetric beam is rounded by two prisms which expand the beam twice in one direction. The beam is also split by a grating in order to use the threespot method for the tracking servo. The beam focuses onto a spot less than 2 μm in diameter on the sample and an overall transmission efficiency of 30% was obtained

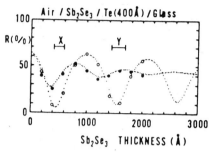

Fig. 7.8. Thickness dependence of the reflectivity of Sb_2Se_3 on a 400 Å Te layer. The open and closed circles are experimental results of the films before and after annealing, respectively. The dotted lines are theoretical curves of reflectivity (Watanabe et al., 1983).

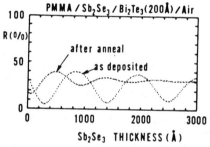

Fig. 7.9. Theoretical curves of reflectivity of PMMA/Sb_2Se_3/Bi_2Te_3 before and after thermal annealing (Watanabe et al., 1983)

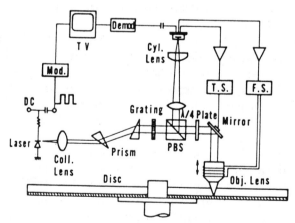

Fig. 7.10. Optical DRAW system for video signals. T.S. and F.S. mean tracking servo and focusing servo (Watanabe et al., 1983).

Disc	Structure	Reflective change after irradiation	Optically changed layer
A	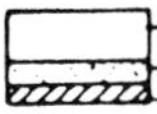 PMMA / Sb_2Se_3 / Bi_2Te_3	10%→30%	Sb_2Se_3
B	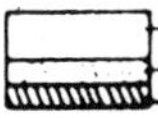 PMMA / Sb_2Se_3 / Sb_2Te_3	25%→10%	Sb_2Te_3
C	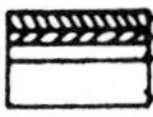 Sb_2Te_3 / Bi_2Te_3 / Se / Glass	45%→60%	Sb_2Te_3

Fig. 7.11. Three types of recordable discs (Watanabe *et al.*, 1983).

by Watanabe *et al.* (1983). The beam is then reflected by the disc, collected by the objective lens and sent back towards the polarizing beam splitter (PBS).

Watanabe *et al.* (1983) recorded FM video signals on three types of Sb_2Se_3 and Sb_2Te_3 discs, as shown in Fig. 7.11. In the type A disc the Sb_2Se_3 layer is optically changeable and the Bi_2Te_3 is the reflective layer. In the type B disc the Sb_2Te_3 is optically changeable and Sb_2Se_3 is the anti-reflection layer. In both A and B the beam is incident through the PMMA substrate. In the type C disc the Se is a thermal barrier, the Bi_2Te_3 is the reflective layer and the Sb_2Te_3 is optically changeable. Here the beam is incident through the Sb_2Te_3.

Figure 7.12 shows the optimum recording power of the three types of discs. The type B disc can be recorded with the lowest laser power because the

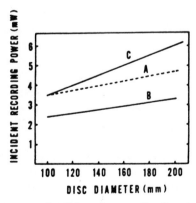

Fig. 7.12. Recording power on the disk surface vs. disc diameter for three types of discs (Watanabe *et al.*, 1983).

Fig. 7.13. The shape of the dots on the type B disc: 400 Å Sb_2Te_3/400 Å Sb_2Se_3/PMMA substrate (Watanabe *et al.*, 1983).

Sb_2Te_3 can be changed at a lower temperature than the Sb_2Se_3 layer (Fig. 7.13) shows the shape of the dots recorded in a type B disc). The type C disc requires the highest power because its reflectivity is highest due to the position of the Sb_2Te_3 layer.

The type A disc is preferred over the others because of the following reasons. First, the playback signal is highest in constant reading power. Second, since the Sb_2Te_3 layer in the type B and C discs undergoes optical changes at temperature as low as 60°C, discs employing this material cannot be recommended for commercial applications. The Sb_2Se_3 layer, on the other hand, doesn't exhibit changes until 160°C. Further, it is desirable for the laser beam to be incident through the substrate since here dust on the surface of the disc will not affect the beam.

7.4 Optical Massmemories Based on the Thermal Creation of Holes

Tellurium based chalcogenide films have been shown to be excellent for optical video recording (Feinleib *et al.*, 1972; Terao *et al.*, 1979; Zembutsu, 1982) via the thermal creation of cleanly shaped holes. A typical technique is first to deposit about 400 Å of an $As_{20}Te_{80}$ film on a glass or PMMA disc substrate. The disc is rotated at 1800 rpm and a pulsed agron–ion laser beam irradiates the film in order to record the desired information. The shape of the holes, which depends upon film thickness, is shown in Fig. 7.14 (Terao *et al.*, 1979). Similar nicely shaped holes have also been fashioned in $Ge_{10}Te_{10}$, $As_{15}Se_{85}$, $Ge_{10}Se_{90}$ and $Sb_{20}S_{80}$ films. The mechanism for the creation of the holes is thought to be as follows. Initially, the laser partially evaporates the film and the material is moved from the center to the edge of the exposed area. After the substrate is exposed at the center of the irradiated region, the softened film is pulled towards the end of the exposed area by surace tension. A rim is formed around the hole. The chalcogenides are useful for this purpose because their chainlike structures produce high

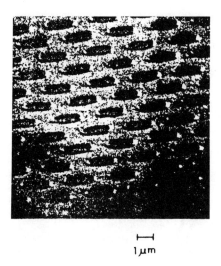

Fig. 7.14. SEM photograph of examples of holes created in an As–Te–Se amorphous thin film (Terao et al., 1979).

viscosities in their liquid or softened states. The addition of As, Ge or Sb cross links the chains and further increases the viscosity. (The production of clean holes depends upon the viscosity of the material.)

When Se is added to As–Te films the stability of the films is enhanced, especially when the surface contains high concentrations of Se and Te. The key here is that the formation of crystalline grains of As_2O_3 and a surface layer of TeO_x ($0 < x < 2$) are inhibited by the addition of Se, and the overall properties of the film are changed only in a minor way. The As_2O_3 grains form during light irradiation when As atoms are liberated on the surface. These grains make it difficult to record and produce noise in the reproduced signals. The presence of a TeO_x layer makes it difficult to record because it does not absorb light well and has a high melting point (Zembutsu, 1982).

It is also useful to coat the surface of the substrate with a spin-coated organic film. This decreases the surface roughness and enhances the signal-to-noise ratio of the reproduced signal. For example, a 2000 Å thick layer of cellulose acetate (CA) decreases the surface noise by 10 dB at 7.5 Mhz. Further, the presence of the CA film reduces the recording power required by nearly 40%, and its presence also increases the adhesion of the As–Te–Se film.

Terao et al. (1979) have developed a high speed random access picture file with a capacity of 20,000 still color pictures per disc employing an As–Te–Se film. FM video signals were recorded and read out via reflected light with a high signal level because of the high reflectivity of the recording film. A reproduced picture is shown in Fig. 7.15.

514 OPTICAL MEMORIES

Fig. 7.15. Examples of reproduced color pictures displayed on a color picture tube (Terao *et al.*, 1979).

We have seen that, in general, amorphous chalcogenide films have excellent properties for optical memory applications and will probably outperform magnetic discs. Their high resolving power, uniformity and ability to undergo structural and morphological phase transitions are the important points here. Zembutsu (1982) also discusses some other major optical device applications of these materials. A holographic supermicrofiche application employing (Se, S)-based films allows for rewritability. Further, these films are also useful in optical IC element applications since they have a high degree of transparency in the near infrared. Also, the As–Se–S–Ge films are attractive for use in integrated optics because they exhibit photoinduced and electron-beam induced changes in their refractive indices, have high resolving power and are easily fabricated and modified. It is also important to realize that the chalcogenides have great promise in the area of photoprocessing and lithography (Mizushima and Yoshikawa, 1982), since photoirradiation can result in major changes in the chemical properties of these films. The major applications here are an inorganic photoresist and lithographic printing plate.

Litwak and Siryj (1983) have provided a comprehensive evaluation of the properties of a variety of materials employed as optical massmemories. Some of the more promising recent developments in this field include the following. Heyman (1983) has studied Te, Te:Bi and Sb_2Se_3 in capped bilayer and trilayer structures, using Al_2O_3 as the phase and capping layers. He has obtained fully erasable discs having a 49 dB SNR in a 4.2 MHz bandwidth that can be recycled at least 500 times. The data retention times were about 4 to 6 hours. Takenaga *et al.* (1983) have developed a new optically erasable medium using TeO_x thin films, with x less than the stoichiometric value, containing small amounts of metal additives. Over one million record/erase cycles have been demonstrated with no major degradation in performance noted. The SNR was more than 44 dB at 5 MHz with a 30 kHz bandwidth. However, Chen *et al.* (1984) argue that no Te based alloy has yet been reported that possesses sufficient reversibility, retention time or SNR for use as an optical massmemory. Further, they indicate that the mechanisms responsible for the failure of reversibility have neither been understood nor identified.

A more recent review of the problem of reversible optical recording has been given by van Uijen (1985), who compares the present status with the corresponding features of magneto-optical recording.

Koshino *et al.* (1985) have presented results on Se films alloyed with In and Sb. The films are chemically stable and durable; writing and erasing can be performed at speeds at least as high as 600 rpm. Terao *et al.* (1985) have developed potentially useful Sn–Te–Se films that are sandwiched between two SiO_2 layers. They can be cycled over one million times. Chen *et al.* (1985)

have presented some new ideas toward the achievement of sub-microsecond erasure with data stability. For example, in a GeTe based film they achieved full erasure with a single pulse less than 700 ns long. The written spots were stable for over 7 months, and the SNR was 47 dB (30 kHz bandwidth). Sb_2Se also gave good results. Rubin *et al.* (1985), of the same group, have also presented results for the crystallization and critical quench rate of the Te–Ge system. They find that the Te–Ge system is a good one for understanding the role that composition plays in controlling phase transitions in phase-change optical recording materials. Lastly, Rose *et al.* (1986) have recently studied the crystallization rate and thermal stability of Te-based films. They find that the crystallization speed is determined by three factors: the intrinsic maximum crystallization velocity of the pure crystal; the diffusion rate of crystal-insoluble cross-linking elements away from the crystallization front; the geometry of the amorphous and peripheral crystalline zones.

7.5 Recent Technological Developments

Write-once optical memories have been on the market since 1983 as video products, and since 1985 as computer peripherals. IBM introduced a phase change (write-once) product in 1987, and this has legitimized the approach of using optical techniques for information storage. Erasable (vcad-write) memories are not on the market yet. However, some magneto-optical drives are available on a sample only basis.

As of this writing (April, 1988) amorphous semiconductor devices have become practical and useful as solar cells, optical memories, electrophotographic components, and as drivers for flat panel displays.

References

Aamodt, L. C., and Murphy, J. C. (1978). *J. Appl. Phys.* **49**, 3036.
Abeles, B., Wronski, C. R., Goldstein, Y., Stasrewski, H., Gutkowicz-Krusin, D., Tiedje, T., and Cody, C. D. (1981). *Proc. APS Topical Conference on Tetrahedrally Bonded Amorphous Semiconductors*, Carefree, p. 298.
Abeles, B., Wronski, C. R., Tiedje, T., and Cody, G. D. (1980). *Solid State Commun.* **36**, 537.
Abkowitz, M., and Pai, D. M. (1977). *Phys. Rev. Lett.* **38**, 1412.
Abou-Chacra, R., and Thouless, D. J. (1974). *J. Phys. C*, **7**, 65.
Ackley, D. E., Tauc, J., and Paul, W. (1979). *Phys. Rev. Lett.* **43**, 715.
Adler, D., Franz, J. M., Hewes, C. R., Kraemer, B. P., Sellmyer, D. J., and Senturia, S. (1970). *J. Non Cryst. Solids* **4**, 330.
Adler, D. (1971). *Amorphous Semiconductors*, CRC, Cleveland, Ohio.
Adler, D., and Feinleib, J. (1971). *In* "The Physics of Opto-Electronic Materials" (W. A. Albers, Jr., ed.), p. 233. Plenum, New York.
Adler, D., and Yoffa, E. J. (1976). *Phys. Rev. Lett.* **36**, 1197.
Adler, D. (1977). *Sci. Amer.* **236**, 36 (May).
Adler, D., and Yoffa, E. (1977). *Cand. J. Chem.* **55**, 1920.
Adler, D. (1978). *Phys. Rev. Lett.* **41**, 1755.
Adler, D., Henisch, H. K., and Mott, N. (1978). *Rev. Mod. Phys.* **50**, 209.
Adler, D., Silver, M., Madan, A., and Czubatyj, W. (1980). *J. Appl. Phys.* **51**, 6429
Adler, D., Shur, M. S., Silver, M., and Ovshinsky, S. R. (1980). *J. Appl. Phys.* **51**, 3289.
Adler, D. (1981). *J. de Physique*, Coll. C4, Supp. 10, Vol. **42**, C4-3.
Adler, D. (1982). *Naturwissenschaften* **69**, 574.
Adler, D. (1983). *Solar Cells* **9**, 133.
Adler, D. (1984). *In* "Semiconductors and Semimetals" (J. I. Pankove, ed.), Vol. 21, Part A, Chapter 14. Academic Press, New York.
Agarwal, S. C. (1973). *Phys. Rev.* **B7**, 685.
Al Jalali, S., and Weiser, G. (1980). *J. Non Cryst. Solids* **41**, 1.
Alben, R., Goldstein, S., Thorpe, M. F., and Weaire, D. (1972). *Phys. Stat. Solidi* (b) **53**, 545.
Allen, D. C., and Joananopolous, J. D. (1980). *Phys. Rev. Lett.* **44**, 43.
Allinson, D. L., Barry, T. I., Clinton, D. J., Hughes, A. J., Lettington, A. H., and Savage, J. A. (1979). *J. Non Cryst. Solids* **31**, 307.
Allsopp, D., Thomson, M. J., and Allison, J. (1977). *In* "Amorphous and Liquid Semiconductors" (W. Spear, ed.), p. 732. Edinburgh, Scotland.
Altcheh, L., Klein, N., and Katz, I. N. (1972). *J. Appl. Phys.* **43**, 3258.
Alvarez, F., and Chamouleyron, I. (1984). *Solar Energy Mats.* **10**, 151.

Alwin, V. C., Navon, D. H., and Turgeon, L. J. (1977). *IEEE Trans.* **ED-24**, 1297.
Anderson, D. A., and Spear, W. E. (1977). *Phil. Mag.* **36**, 695.
Anderson, D. A. (1978). *Bull. Am. Phys. Soc.* **23**, 249.
Anderson, D. A., and Paul, W. (1981). *Phil. Mag.* **44**, 187.
Anderson, F. (1959), *Acta Chem. Scand.* **8**, 1599.
Anderson, P. W. (1976). Inter Colloq. on Metal–Nonmetal Transitions, Autrans, France. *J. de Physique* **37**, Colloq. 4, Suppl. 0–10, p. C4-339.
Anderson, P. W. (1975). *Phys. Rev. Lett.* **34**, 953.
Anderson, P. W. (1985). *Phys. Rev.* **109**, 1492.
Anderson, V., Dilecce, G., Murri, R., and Schiavulli, L. (1985). *J. Non Cryst. Solids* **77–78**, 303.
Andriesch, A. M., and Kolomiets, B. T. (1965). *Sov. Phys. Sol. State* **6**, 2652.
Armstrong-Russel, M. K., Freedman, W., and Spitler, E. E. (1983). *Proc. SPIE Conf. on Photovoltaics* (D. Adler, ed.), Vol. 407, p. 132.
Arya, R. R., Catalano, A., and Dowd, J. (1986). *MRS Proc.* (D. Adler, Y. Hamakawa, and A. Madan, eds.), Vol. 70, p. 517.
Ashok, S., Lester, A., and Fonash, S. J. (1980). *IEEE Trans. Electron Devices Lett.* **EDL-1**, 200.
Ast, D. G., and Brodsky, M. H. (1980). *Phil. Mag.* **B41**, 273, 1980.
Ast, D. G. (1983). *IEEE Trans. Electron Devices* **ED-30**, 532.
Attala, M. M., Tannenbaum, E., and Scheibner, J. E. (1959). *Bell Systems Tech. J.* **38**, 749.
Augelli, V., Murri, R., and Schiavulli, L. (1982). *Thin Solid Films* **90**, 153.
Augelli, V., Dileece, G., Murri, R., and Schiavulli, L. (1985). *J. Non Cryst. Solids* **77–78**, 303.
Austin, I. G., and Mott, N. F. (1969). *Adv. in Phys.* **18**, 41.
Austin, I. G., Nashashibi, T. S., Searle, T. M., LeComber, P. G., and Spear, W. E. (1979). *J. Non Cryst. Solids* **32**, 373
Averianov, V. L., Kolobov, A. V., Kolomiets, B. T., and Lyubin, V. M. (1980). *Phys. Stat. Sol.* **57**, 81.
Bagley, B. G. (1974). In "Amorphous and Liquid Semiconductors" (J. Tauc, ed.), Chapter 1. Plenum, New York.
Bahl, S. K., and Bhagat, S. M. (1975). *J. Non Cryst. Solids* **17**, 409.
Balberg, I. (1970). *Appl. Phys. Lett.* **16**, 491.
Balberg, I. (1980). *Phys. Rev.* **B22**, 3853.
Baraff, D. F., Long, J. R., MacLaurin, B. K., Miner, C. J., and Streater, R. W. (1981). *Proc. SID Dig. Tech. Papers*, Vol. 22, p. 310.
Barbe, D. F. (1971). *J. Vac. Sci. Technol.* **8**, 102.
Bardeen, J. (1947). *Phys. Rev.* **75**, 1208.
Barna, A., Barna, P. B., Radnoczi, G., Toth, L., and Thomas, P. (1977). *Phys. Stat. Solidi* **41**, 81.
Barnett, A. M. (1969). *IBM J. Res. Dev.* **13**, 522.
Basterfield, J., Shannon, J. M., and Gill, A. (1975). *Solid State Electron.* **18**, 290.
Batra, I., Kanazawa, P., Keiji, K., and Seki, H. (1970). *J. Appl. Phys.* **41**, 3416.
Bauer, G. H., Bigler, G., Mohring, H. D., Nebel, C. E., and Paasche, S. M. (1984). *Tech. Digest of PVSEC-I*, Kobe, Japan, p. 747.
Becker, L. (1936). *Arch. Elektrotech.* (Berlin) **30**, 411.
Bedeaux, D., Mazur, P., and Pasmanter, R. A. (1977a). In "Statistical Mechanics and Statistical Mathematics in Theory and Application" (U. Landman, ed.). Plenum, New York.
Bedeaux, D., Mazur, P., and Pasmanter, R. A. (1977b). *Physica* **86a**, 355.
Bell, A. (1983). *Proc. of Soc. Inform. Display* **24**, No. 1, p. 17.
Berglund, C. N. (1969). *IEEE Trans.* **ED-16**, 432.
Berglund, C. N., and Klein, N. (1971). *Proc. IEEE* **59**, 1099.
Bergmann, F., and Gerstner, D. (1963). *Archiv. Elektr. Übertr.* **17**, 467.

REFERENCES

Bergmann, F., and Gerstner, D. (1966). *IEEE Trans.* **ED-13,** 630.
Beyer, W., Medeisis, A., and Mell, H. (1977). *Comm. in Physics* **2,** 121.
Beyer, W., and Wagner, H. (1981). *J. de Physique* **10,** 42, C4-783.
Bhattacharya, E., Guha, S., Krishna, K. V., and Bapat, D. R. (1982). *J. Appl. Phys.* **53,** 6285.
Bienenstock, A., Betts, F., and Ovshinsky, S. R. (1970). *J. Non Cryst. Solids* **2,** 334.
Bishop, S. G., Strom, U., and Taylor, P. C. (1975). *Phys. Rev. Lett.* **34,** 1346.
Bishop, S. G., Strom, U., and Taylor, P. C. (1977). *Phys. Rev.* **B15,** 2278.
Bishop, S. G., Shanabrook, B. V., Strom, U., and Taylor, P. C. (1981). *J. de Physique* **42,** Suppl. 10, C4-383.
Blicher, A. (1976). "Thyristor Physics." Springer-Verlag, New York.
Boccara, A. C., Fournier, D., Jackson, W., and Amer, N. M. (1980). *Amer. Opt. Lett.* **5,** 377.
Boer, K. W., Jahne, E., and Newbauer, E. (1961). *Phys. Stat. Sol.* **1,** 231.
Booth, D. C., Allred, D. D., Seraphin, B. O. (1980). *J. Non Cryst. Solids* **35–36,** 213.
Borelli, N. F., and Su, G. J. (1968). *Mat. Res. Bull.* **3,** 181.
Bosch, M. A. (1982). *Appl. Phys. Lett.* **40,** 8.
Bosch, M. (1982). *Phys. Rev. Lett.* **48,** 1228.
Bosnell, J. R., and Thomas, C. B. (1972). *Solid State Electron.* **15,** 1261.
Brattain, W. H., and Bardeen, J. (1953). *Bell System Tech J.* **32,** 1.
Brillson, L. J. (1978). *Phys. Rev. Lett.* **40,** 260.
Brodsky, M. H., Cuomo, J. J., and Evangelisti, F. (1977). *In* "Amorphous and Liquid Semiconductors" (W. Spear, ed.). CICL, Univer. of Edinburgh, p. 397.
Brodsky, M. H., Cardona, M., and Cuomo, J. J. (1977). *Phys. Rev.* **B16,** 3556.
Brodsky, M. H. (1977). *Thin Solid Films* **40,** L23.
Brodsky, M. H. (1983). U.S. Patent 3,716,844.
Brody, T. P., Asars, J. A., Dixon, G. D. (1973). *IEEE Trans. on Electron Devices* **ED-20,** 995.
Bruno, G., Capezzuto, P., and Cramarossa, F. (1983). *Thin Solid Films* **106,** 145.
Bubenzer, A., Dischler, B., and Nyaiesh, A. (1982). *Thin Solid Films* **91,** 81.
Buckley, W. D., and Holmberg, S. H. (1975). *Sol. St. Elec.* **18,** 127.
Ballot, J., Gauthier, M., Schmidt, M., Catherien, Y., and Zamouche, A. (1984). *Phil. Mag.* **B49,** 489.
Bulthuis, K., Carasso, M. G., Heemskerk, J. P. J., Kivits, P. J., Kleuters, W. J., and Zalm, P. (1979). *IEEE Spectrum,* p. 26, August.
Burgess, R. E. (1955a). *Proc. Phys. Soc.* **B68,** 706.
Burgess, R. E. (1955b). *Proc. Phys. Soc.* **B68,** 908.
Burgess, R. E. (1955c). *J. Electronics* **297,** 459.
Burgess, R. E. (1960). *Can. J. Phys.* **38,** 369.
Burland, D. M., and Schein, L. B. (1986). *Phys. Today* **39,** No. 5, p. 46.
Butcher, R. N. (1967). *Rpts. Prog. Phys.* **30,** 97.
Butcher, P. N., and McInnes, J. A. (1981). *J. de Physique,* Coll. C4, Supp. 10, Vol. 42, C4-91.
Campbell, D. S. (1971). *In* "Handbook of Thin Film Technology" (L. I. Maissel, and R. Glang, eds.). McGraw Hill, New York.
Caplan, P. J., Poindexter, E. H., Deal, B. E., and Razouk, R. R. (1979). *J. Appl. Phys.* **50,** 5847.
Card, H. C., and Rhoderick, E. H. (1971). *J. Phys. D., Appl. Phys.* **4,** 1589.
Carlos, W. E., Greenbaum, S. G., Taylor, P. C. (1982a). *Bull. Am. Phys. Soc. (2)* **27,** 208.
Carlos, W. E., and Taylor, P. C. (1982b). *Phys. Rev.* **B25,** 1435.
Carlson, C. (1942). Electrophotography, U.S. Patent 2,297,691.
Carlson, D. E., and Wronski, C. R. (1976). *Appl. Phys. Lett.* **28,** 671.
Carlson, D. E. (1977). *IEEE Trans. on Electron Devices,* **ED-24,** 449.
Carlson, D. E. (1982). *J. Vac. Sci. Tech.* **20,** 290.

Carlson, D. E. (1982). *Solar Energy Mats.* **8**, 129.
Carlson, D. E. (1983). *Proc. of SPIE Conf.* (D. Adler, ed.), Vol. 407, p. 9. Virginia.
Carlson, D. E., Catalano, A., D'Arello, R. V., Dickson, C. R., and Oswald, R. S. (1984). *Proc. 17th IEEE PV Conf.*, p. 330.
Carlson, D. E., Catalano, A., D'Daiello, R. V., Dickson, C. R., and Oswald, R. S. (1984). *AIP Conf.*, No. 20, p. 234.
Carslaw, H. S., and Jaeger, J. C. (1959). "Conduction of Heat in Solids." Oxford, New York.
Callarotti, R., and Schmidt, P. (1977). *In* "Amorphous and Liquid Semiconductors" (W. Spear, ed.), p. 717. Edinburgh, Scotland.
Carasso, M. G., and Huijser, A. (1982). *CLEO '82 Technical Digest*, p. 30, April 14-16. Phoenix, Arizona.
Castleberry, D. E., Becker, C. A., and Levinson, L. M. (1982). *Proc. SID Symp. Dig. Tech. Papers*, p. 246.
Catalano, A., D'Arello, R. V., Dresner, J., Faugnan, B., Firester, A., Kane, J., Schade, H., Smith, Z. E., Swartz, G., and Triano, A. (1982). *Proc. 16th IEEE Photovoltaic Conf.*, San Diego, p. 1421. IEEE, New York.
Catalano, A., Newton, J. L., Arya, R. R., and Wiedman, S. (1987). *Tech. Digest of PVSEC-III* (Tokyo), 61.
Ceaser, G. P., Grimshaw, S. F., and Okumura, K. (1981). *Sol. St. Comm.* **38**, 89.
Chamberlain, J. M., and Moseley, A. J. (1981). *J. de Physique*, Vol. 42, Suppl. 10, C4-309.
Chen, I. (1972). *J. Appl. Phys.* **43**, 1137.
Chen, I., and Mort, J. (1972). *J. Appl. Phys.* **43**, 1164.
Chen, I., Mort, J., and Tabak, M. D. (1972). *IEEE Trans.* **ED.-19**. Special Issue on Electrophotography, p. 413.
Chen, M., Rubin, K., Marello, V., and Gerber, U. G. (1984). Topical Meeting on Optical Data Storage, Monterey, CA, Paper ThC-D2-1 (Opt. Soc. Amer.).
Chen, M., Rubin, K. A., and Barton, R. W. (1985). Topical Meeting on Optical Data Storage, Washington, DC, Paper PD-1 (Opt. Soc. Amer.).
Chevallier, J., Weider, H., Onton, A., Guarinreri, C. R. (1977). *Solid State Commun.* **27**, 867.
Ching, W. J., Lam, D. J., and Lin, C. C. (1979). *Phys. Rev. Lett.* **43**, 805.
Chino, K. (1973). *Solid State Electron.* **16**, 119.
Chittick, R. C., Alexander, J. H., and Sterling, H. F. (1969). *J. Electrochem. Soc.* **116**, 77.
Clark, A. H. (1967). *Phys. Rev.* **154**, 750.
Cocks, F. H., Scharman, A. J., Jones, P. L., Cogan, S. F. (1980). *Appl. Phys. Lett.* **36**, 909.
Cody, C. D., Wronski, C. R., Abeles, B., Stephens, R. B., Brooks, B. (1980). *Solar Cells* **2**, 227.
Cody, G. D., Tiedje, T., Abeles, B., Brooks, B., and Goldstein, Y. (1981a). *Phys. Rev. Lett.* **47**, 1480.
Cody, G. D., Tiedje, T., Abeles, B., Moustakas, T. D., Brooks, B., and Goldstein, Y. (1981b). *J. de Physique* **42**, C4-301.
Cohen, J. D., Lang, D. V., Harbison, J. P., and Bean, J. C. (1980). *Solar Cell* **2**, 331.
Cohen, J. D. (1984). In *Semicond. and Semimet.* "Hydrogenated Amor. Si." Vol. 21, p. 9. Academic, N.Y.
Cohen, M. H., Fritzche, H., and Ovshinsky, S. R. (1969). *Phys. Rev.* **22**, 1065.
Cohen, M. H. (1970). *J. Non Cryst. Solids* **4**, 391.
Conwell, E. M. (1970). *IEEE Trans.* **ED-17**, 262.
Coward, L. A. (1971). *J. Non Cryst. Solids* **6**, 107.
Crandall, R. S., Williams, R., and Tompkins, B. F. (1979). *J. Appl. Phys.* **50**, 5506.
Crandall, R. S. (1980). *Phys. Rev.* **44**, 749.
Crandall, R. S. (1981). *Phys. Rev.* **B-24**, 7457.
Crandall, R. S. (1982). *J. Appl. Phys.* **53**, 3350.

Crandall, R. S., Carlson, D. E., Catalono, A., and Weakleim, H. A. (1984). *Appl. Phys. Lett.* **44**, 200.
Crider, C. A., and Poate, J. M. (1980). *Appl. Phys. Lett.* **36**, 417.
Croitoru, N., and Popescu, C. (1970). *Phys. Stat. Solidi* **3a**, 1047.
Crowell, C. R., and Sze, S. M. (1966). *Solid State Electron.* **9**, 1035.
Cutler, M., and Mott, N. F. (1969). *Phys. Rev.* **181**, 1336.
Czubatyj, W., Shur, M., Ng, K., and Madan, A. (1980). *Proc. 14th IEEE Photovoltaic Conference*, San Diego, p. 1214. IEEE, New York.
Dalal, V. L. (1980). *IEEE Transactions on Electronic Devices* **ED-27**, 662.
D'Amico, and Fortunato, G. (1980). *In* "Semiconductors and Semimentals" (J. Pankove, ed.), Chap. 11, p. 209. Academic Press.
Davis, E. A., and Mott, N. F. (1970). *J. Non Cryst. Solids* **4**, 107.
Debney, B. T. (1978). *Solid State Electron.* **2**, 815.
Deckman, H. W., Wronski, C. R., Witzke, H., and Yablonovitch, E. (1983). *Appl. Phys. Lett.* **42**, 968.
Delahoy, A. E., and Griffith, R. W. (1981). *J. Appl. Phys.* **52**, 6337.
Delel, V. L., Fortmann, C. M., Eser, E. (1981). *In* "Tetrahedrally Bonded Amorphous Semiconductors" (R. A. Street, D. K. Biegelsen, J. C. Knights, eds.), p. 15. AIP, New York.
Dellafera, B. P., Labusch, R., and Roscher, H. H. (1981). *Phil. Mag.* **B43**, 169.
Dellafera, P., Labusch, R., and Roscher, H. H. (1982). *Phil. Mag.* **B45**, 607.
Demond, F. J., Muller, G., Damjantschitsch, H., Mannsperger, H., Kalbitzer, S., LeComber, P. G., and Spear, W. E. (1981). *J. de Physique* **42**, C4-779.
den Boer, W. (1981). *J. de Physique* **42**, C4-451.
den Boer, W. (1982). *Appl. Phys. Lett.* **40**, 812.
DeNeufville, J. P. (1974). *Proc. 5th Int. Conf. Amor. & Liq. Semicond.* (Stuke & Brenig, eds.), p. 1351. Taylor & Francis, London.
Deneuville, A., and Brodsky, M. H. (1979). *J. Appl. Phys.* **50**, 1414.
Dersch, H., Stuke, J., and Beichler, J. (1981). *Phys. Stat. Solidi* **B105**, 265.
Dersch, H., Stuke, J., and Beichler, J. (1981). *Appl. Phys. Lett.* **38**, 456.
Dessauer, J., and Clark, H. (eds.) (1965). *Xerography and Related Processes.* Focal Press, London.
DeWald, J. F., Pearson, A. D., Northover, W. R., and Peck, W. F. (1962). *J. Electrochem. Soc.* **109**, 243C.
Dexter, D. L. (1956). *Phys. Rev.* **101**, 48.
Dey, S. K., and Fong, W. T. J. (1979). *J. Vac. Sci. and Technol.* **16**, 240.
Dey, S. K. (1980). *J. Vac. Sci. and Technology* **17**, 445.
Dohler, G. H. (1979). *Phys. Rev.* **B19**, 2083.
Dong, N. V., Dank, T. H., Leny, J. Y. (1981). *J. Appl. Phys.* **52**, 338.
Dresner, J., and Stringfellow, G. B. (1968). *J. Phys. Chem. Sol.* **29**, p. 303.
Dresner, J., and Szustak, D. Z. (1981). *J. Appl. Phys.* **38**, 998.
Dresner, J., Goldstein, B., and Szustak, D. Z. (1981). *Appl. Phys. Lett.* **38**, 998.
Drevillion, B., Huc, J., Lloret, A., Perrin, J., deRosny, G., and Schmitt, J. P. M. (1981). *Proc. 5th Symp. on Plasma Chemistry*, Edinburgh.
Dubner, N., and Abate, J. (1968). *J. Ass. Comput. Mach.* **15**, 115.
Duchene, J., Adam, G., and Augier, D. (1971a). *Phys. Stat. Sol.* **8**, 459.
Duchene, J., Terraillon, M., Pailly, M., and Adam, G. (1971b). *Appl. Phys. Lett.* **19**, 115.
Duchene, J., Terraillon, M., Pailly, M., and Adam, G. (1971c). *IEEE Trans.* **ED-18**, 1151.
Duchene, J., Terraillon, M., and Pailly, M. (1972). *Thin Solid Films* **12**, 231.
Economou, E. N., and Cohen, M. H. (1970). *Phys. Rev. Lett.* **25**, 1145.
Egerton, R. F. (1971). *Appl. Phys. Lett.* **19**, 203.

Elliott, S. F. (1981). *J. de Physique* **42**, Supp. 10, C4-387.
Ellis, F. B., Gordon, R. G., Paul, W., and Yacobi, B. G. (1984). *J. Appl. Phys.* **55**, 4309.
Emin, D., Seager, C. H., and Quinn, R. K. (1972). *Phys. Rev. Lett.* **28**, 813.
Enck, R. G., and Pfister, G. (1976). In *"Photoconductivity and Related Phenomena"* (Mort & Pai, ed.), Chap. 7. Elsevier, New York.
Engemann, D., Fischer, R. (1976). *Proc. Of 12th Int. Conf. on the Physics of Semiconductors* (M. H. Pilkuhn, ed.), p. 1042. Teubner, Stuttgart.
Engemann, D., Fischer, R. (1973). *Proc. of 5th Int. Conf. on Amorphous and Liquid Semiconductors* (J. Stuke and W. Brenig, eds.; Taylor and Francis, pub., 1974), p. 947. Garmisch: Germany.
Esaki, L. (1958). *Phys. Rev.* **109**, 603.
Ettenberg, M., and Botez, D. (1982). *Elec. Lett.* **18**, 153.
Evans, E. J., Helbers, J. H., and Ovshinsky, S. R. (1970). *J. Non Cryst. Solids* **2**, 334.
Fagen, E. A., and Fritzsche, H. (1970). *J. Non Cryst. Solids* **2**, 180.
Fang, C. J., Ley, L., Shanks, H. R., Gruntz, K. J., and Cardona, M. (1980). *Phys. Rev.* **B22**, 6140.
Faughnan, B. W., and Crandall, R. S. (1984). *Appl. Phys. Lett.* **44**, 537.
Fedders, P. A., and Carlsson, A. E. (1987). *Phys. Rev. Lett.* **58**, 1156.
Feinleib, J., de Neufville, J. P., Moss, S. C., and Ovshinsky, S. R. (1971). *Appl. Phys. Lett.* **18**, 254.
Feinleib, J., Iwasa, S., Moss, S. C., de Neufville, J. P., and Ovshinsky, S. R. (1972). *J. Non Cryst. Solids* **8-10**, 909.
Felty, G. J., Lucovsky, G., and Myers, M. B. (1967). *Solid State Commun.* **5**, 555.
Ferraton, J. P., Donnadieu, A., Berger, J. M., De Chelle, F., Coulibaly, S. P., and Ance, C. (1983). *J. Non Cryst. Solids* **59 & 60**, 313.
Ferrier, R. P., Prado, J. M., and Anseau, M. R. (1972). *J. Non Cryst. Solids* **8-10**, 552.
Feynman, R. P., Leighton, R. B., and Sands, M. (1964). "Lectures on Physics", Vol. 2. Addison-Wesley, Reading, MA.
Fisher, B. (1975). *J. Phys. C.* **8**, 2072.
Flasck, R., and Rockstad, H. K. (1973). *J. Non Cryst. Solids* **12**, 353.
Fock, W. A. (1927). *Arch. Elektrotechn.* **19**, 71.
Fortunato, G., Evangelisti, F., Bruno, G., Capazzuto, P., Cramarossa, F., Augelli, V., and Murri, R. (1981). *J. Non Cryst. Solids* **46**, 95.
Fox, S. J., and Locklar, H. C. (1971). *J. Non Cryst. Solids* **8-10**, 552.
Franz, W. (1956). *Enc. of Physics* (S. Flügge, ed.) **18**, 166. Berlin.
Friedman, L. (1971). *J. Non Cryst. Solids* **6**, 329.
Friedman, L., and Pollak, M. (1981). *J. de Physique* Coll. C4, Supp. 10, Vol. 42, C4-87.
Fritzsche, H., and Ovshinsky, S. R. (1970). *J. Non Cryst. Solids* **2**, 148.
Fritzsche, H. (1971). *J. Non Cryst. Solids* **6**, 49.
Fritzsche, H. (1974). In *"Amorphous and Liquid Semiconductors"* (J. Tauc, ed.), p. 341. Plenum Press, New York.
Fritzsche, H., and Tsai, C. C. (1978). *Solid State Technol.* **21**, 55.
Fritzsche, H., and Tsai, C. C. (1979). *Appl. Phys. Lett.* **50**, 3366.
Fritzsche, H., Tanielen, M., Tsai, C. C., and Gaczi, P. J. (1979). *J. Appl. Phys.* **50**, 3368.
Frye, R., Adler, D., and Shaw, M. P. (1980). *J. Non Cryst. Solids* **35-36** (2), 1099.
Fuhs, W., Milleville, M., and Stoke, J. (1978). *Phys. State* **89**, 495.
Fukuda, T. (1979). 38th JSAP Fall meeting, 2a-R-2.
Fulop, W., 1963, *IEEE Trans.* **ED-10**, 120.
Gau, S. C., Weinberger, B. R., Akhtar, M., Kiss, Z., and MacDiarmid, A. G. (1981). *Appl. Phys. Lett.* **39**, 436.

REFERENCES

Gaur, S. P., and Navon, D. H. (1976). *IEEE Trans.* **ED-23**, 50.
Gelmont, B. L., and Shur, M. S. (1973). *J. Phys.* **D6**, 842.
Gibson, R. A., LeComber, P. G., and Spear, W. E. (1980). *Appl. Phys.* **21**, 307.
Goldstein, B., and Szostak, D. J. (1980). *J. Vac. Sci. and Technol.* **17**, 718.
Goodman, A. (1961). *J. Appl. Phys.* **32**, 2550.
Goodman, N. B., and Fritzsche, H. (1980). *Phil. Mag.* **B42**, 149.
Greeb, K. H., Fuhs, W., Mell, H., and Welsch, H. M. (1982). *Solar Energy Materials* **8**, 253.
Green, M. A., King, F. D., and Shewchun, J. (1974). *Solid State Electron.* **17**, 551.
Greenbaum, S. G., Carlos, W. E., and Taylor, P. C. (1982). *Solid State Commun.* **43**, 663.
Grigorovici, R., and Vancu, A. (1981). *J. de Physique* **42**, Supp. 10, C4-391.
Grove, A. S. (1967). "Physics and Technology of Semiconductor Devices." John Wiley and Sons, Inc., N.Y.
Gruntz, K. J. (1981). *Ph.D. Thesis*, University of Stuttgart.
Gruntz, K. J., Ley, L., and Johnson, R. J. (1981). *Phys. Rev.* **B24**, 2069.
Gunn, J. B. (1956). *Proc. Phys. Soc.* **69, 8-B**, 781.
Gunn, J. B. (1964). *IBM J. Res. Dev.* **8**, 141.
Gutkowicz-Krusin, B., Wronski, C. R., and Tiedje, T. (1981). *Appl. Phys. Lett.* **38**, 87.
Haberland, D. R. (1970). *Solid State Electron.* **13**, 207.
Hack, M., McGill, J., Czubatyj, W., Singh, R., Shur, M., and Madan, A. (1982). *J. Appl. Phys.* **53**, 6270.
Hack, M., and Shur, M. (1983). *IEEE Electron Device Letters* **EDL-4**, 140.
Hack, M., and Shur, M. (1984). *IEEE Trans. on Electron Devices* **ED-31**, 539.
Hack, M., and Shur, M. (1984). *J. Appl. Phys.* **55**, 2967.
Hagedus, S. S., Rocheleau, R. E., and Baron, B. N. (1984). *Proc. of 17th IEEE Photovoltaic Conf.*, Orlando, 239.
Hajto, J., and Fustoss-Wegner, M. (1981). *J. de Physique* **42**, Suppl. 10, C4-313.
Hama, S. T., Okamoto, H., Hamakawa, Y., Matsubara, T. (1983). *J. Non Cryst. Solids* **59-60**, 333.
Hamakawa, Y. (1986). *Technical Digest of PVSEC-II*, Beijing, p. 347.
Hamanaka, H., Tanaka, K., Tsuji, K., and Minomura, S. (1981). *J. de Physique* **42**, Suppl. 10, C4-399.
Hamasaki, T., Kurato, H., Hirose, M., and Osaka, Y. (1980). *Appl. Phys. Lett.* **37**, 1084.
Han, M. K., and Anderson, W. A. (1981). *Technical Digest of International Electron Device Meeting*, Washington, p. 34.
Hanak, J. J. (1979). *Solar Energy* **23**, 145.
Harbison, J. P., Williams, A. J., and Lang, D. V. (1984). *J. Appl. Phys.* **55**, 946.
Harper, W. R. (1967). "Contact and Frictional Electrification." Clarendon, Oxford.
Harris, A. J., Walker, R. S., and Sneddon, R. (1980). *J. Appl. Phys.* **51**, 4287.
Hasegawa, R. (1983). "Glassy Metals: Magnetic, Chemical and Structural Properties." CRC Press, Boca Raton, Fla.
Hasegawa, S., Kasajima, T., and Shimizu, T. (1981). *Phil. Mag.* **43**, 149.
Hata, N., Matsuda, A., Tanaka, K., Kajiyama, K., Moro, N., and Sajiki, K. (1983). *Jap. J. Appl. Phys.* **22**, L1.
Hauser, J. J., Di Salvo, F. J., and Hutton, R. B. (1977). *Phil. Mag.* **35**, 1557.
Hayes, T. M., and Thornburg, D. D. (1973). *J. Phys.* **C6**, 450.
Hayes, T. M. (1974). *J. Phys.* **C7**, 371.
Heavens, O. S. (1979). "Thin Film Physics." Methuen & Co., London.
Hecht, J. (1982). *High Technology*, **May-June**, p. 60.
Henisch, H. K., Smith, W. R., and Wohl, W. (1974). *In* "Amorphous and Liquid Semiconductors" (J. Stuke, and W. Brenig, eds.), p. 567. Taylor and Francis, London.

Henniker, J. (1962). *Nature* **196**, 474.
Henry, C. H. (1980). *J. Appl. Phys.* **51**, 4494.
Hensel, F. (1977). *In* "Amorphous and Liquid Semiconductors" (W. E. Spear, ed.), p. 815. CICL, University of Edinburgh.
Heyman, P. M. (1983). *SPIE Proc.* **420**, 162.
Higaki, R., Arga, M., Terazono, S., and Yukimoto, Y. (1984). *Tech. Digest of PVSEC-II*, p. 209. Kobe, Japan.
Hilsum, C. (1962). *Proc. IRE* **50**, 185.
Hilton, A. R., and Jones, C. E. (1967). *Appl. Optics* **6**, 1513.
Himpsel, F. J., Hollinger, G., and Pollak, R. A. (1983). *Phys. Rev.* **B28**, 7014.
Hindley, N. K. (1970). *J. Non Cryst. Solids* **5**, 17.
Hirose, M., Susuki, T., and Dohler, G. H. (1979). *Appl. Phys. Lett.* **34**, 234.
Hirose, M., Hamasaki, T., Mishima, Y., Kurata, H., and Osaka, Y. (1981). *AIP Conf. Proc. No. 73, Tetrahedrally Bonded Amorphous Semiconductors* (R. A. Street, D. K. Biegelson, and J. C. Knights, eds.), p. 10. AIP, New York.
Hirose, M. (1981). *J. Phys.* **42**, C4-705.
Ho, V. Q., and Sugano, T. (1981). *IEEE Trans.* **ED-28**, 106.
Hofstein, S. R., and Heiman, F. P. (1963). *Proc. IEEE* **51**, 1190.
Hollinger, G., and Himpsell, F. J. (1984). *Appl. Phys. Lett.* **44**, 93.
Hollingsworth, R. E., Bhat, P. K., and Madan, A. (1987). *19th IEEE Specialist Conference on Photovoltaics*, 684.
Holmes, F. E., and Salama, C. A. T. (1974). *Solid State Electron.* **17**, 791.
Homma, K. (1971). *Appl. Phys. Lett.* **18**, 198.
Hower, P. L., and Reddi, V. G. K. (1970). *IEEE Trans.* **Ed-17**, 320.
Hudgens, S. J. (1986). *SPIE Proceedings* **617**, 95.
Hughes, A. J., Holland, P. A., and Lettington, A. H. (1975). *J. Non Crystal Solids* **17**, 89.
Huyn, C. H., Shur, M. S., and Madan, A. (1982). *Appl. Phys. Lett.* **41**, 178.
Inoue, T., Konagai, M., and Takahashi, K. (1983). *Appl. Phys. Lett.* **43**, 774.
Inoue, T., Tanaka, T., Konagai, M., and Takahashi, K. (1984). *Appl. Phys. Lett.* **44**, 871.
Ioffe, A. F., and Regel, A. R. (1960). *Prog. Semicond.* **4**, 237.
Iqbal, Z., Capezzuto, P., Braun, M., Oswald, H. R., Veprek, S., Bruno, G., Caramarossa, F., Stussi, H., Brunner, J., and Scharli, M. (1982). *Thin Solid Films* **87**, 43.
Ishibashi, K., and Matsumura, M. (1982). *Appl. Phys. Lett.* **41**, 455.
Ishihara, S., Hirao, T., Mori, K., Kitagawa, M., Ohno, M., and Kohiki, S. (1982). *J. Appl. Phys.* **53**, 3909.
Itoh, H., Matsuhara, S., Muramatsu, S., Nakamura, N., Ikegaki, T., and Shimada, T. (1984). *Tech. Digest of PVSEC-I*, p. 119. Kobe, Japan.
Ivkin, E. B., Kolomiets, B. T., and Lebedev, E. A. (1964). *Bull. Acad. Sci. (USSR) Phys. Soc.* **28**, 1190.
Jackson, W. (1982). *Solid State Commun.* **44**, 477.
Jackson, J. L., and Shaw, M. P. (1974). *Appl. Phys. Lett.* **25**, 666.
Jackson, W. B., and Amer, N. M. (1982). *Phys. Rev.* **B25**, 5559.
Jan, Z., Bube, R. H., and Knights, J. C. (1980). *J. Appl. Phys.* **51**, 3278.
Janai, M. (1981). *J. de Physique*, Coll. C4, Supp. 10, Tome **42**, 1105.
Janai, M., Weil, R., Levin, H. K., Pratt, B., Kalish, R., Braunstein, G., and Teicher, M. (1981). *J. Appl. Phys.* **52**, 3622.
Jansen, F., Mort, J., Gummatica, S., and Morgan, M. (1984). *J. Appl. Phys.* **55**, 4128.
Jantsch, O. (1965). *J. Phys. Chem. Solids* **26**, 1233.
Jeffrey, F. R., Dubois-Murphy, P., and Gerstein, B. C. (1981). *Phys. Rev.* **B23**, 2099.

REFERENCES

Jeffrey, F. R., Venstrom, G. D., Weber, M., and Epstein, K. A. (1986). *MRS Proc.* (D. Adler, Y. Hamakawa, and A. Madan, eds.), Vol. 70, p. 531.
Jelks, E. C., Walser, R. M,., Bené, R. W., and Neal III, W. H. (1975). *Appl. Phys. Lett.* **26**, 355.
Joannopoulous, J. D., and Cohen, M. L. (1973). *Phys. Rev.* **B8**, 2733.
Joannopoulous, J. D. (1980). *J. Non Cryst. Solids* **35-36**, 781.
Jones, B. L. (1985). *J. Non Cryst. Solids* **77-78**, 1405.
Jones, D. I., LeComber, P. G., and Spear, W. E. (1977). *Phil. Mag.* **36**, 541.
Jones, S. R., Thompson, M. J., Allison, J., and Ormondroyd, R. F. (1978). *J. Non Cryst. Solids* **28**, 1.
Joseph, D. D. (1965). *Int. J. Heat Mass Transfer* **8**, 281-288.
Joseph, D. D., and Sparrow, E. M. (1970). *Quart. Appl. Math.* **28**, 327-342.
Josephson, B. D. (1962). *Phys. Lett.* **1**, 251.
Juska, G., Matulionis, A., and Viscakas, J. (1969). *Phys. Stat. Sol.* **33**, 533.
Juska, G., Matulionis, A., Sakalas, A., and Viscakas, J. (1969). *Phys. Stat. Sol.* **36**, 16121.
Kagawa, T., Matsumoto, N., and Kumabe, K. (1983). *Phys. Rev.* **B28**, 4570.
Kampas, F. J., and Griffith, R. W. (1981). *Appl. Phys. Lett.* **39**, 407.
Kampas, F. J. (1982). *J. Appl. Phys.* **53**, 6408.
Kaneko, S., Sakamoto, M., Okumura, F., Itano, T., Kataniwa, H., Kajiwara, Y., Kanamori, M., Yasumoto, M., Saito, T., and Okubu, T. (1982). *Tech. Digest—Int. Electron Device Meet.*, p. 328.
Kaneko, S. (1984). In "Semiconductors and Semimetals" (J. Pankove, ed.), Chap. 8, Vol. 21, p. 139. Academic Press.
Kaplan, T., and Adler, D. (1972). *J. Non Cryst. Solids* **8-10**, 522.
Karcher, R., Ley, L., and Johnson, R. L. (1984). *Phys. Rev.* **B30**, 1896.
Kastner, M. (1972). *Phys. Rev. Lett.* **28**, 355.
Kastner, M., Adler, D., and Fritzsche, H. (1976). *Phys. Rev. Lett.* **37**, 1504.
Kawamura, T., and Yamamoto, N. (1982). *Japan Ann. Rev. In* "Electronics, Computers and Telecommunications" (Y. Hamakawa, ed.), Vol. 2, p. 296. North Holland, Amsterdam.
Keller, H. B., and Cohen, D. S. (1967). *J. Math. Mech.* **16**, 1361-1376.
Kirby, P. B., Paul, W., Lee, C., Lin, S., Von Roedern, B., and Weisfield, R. L. (1983). *Phys. Rev.* **B28**, 3635.
Kirby, P. B., Paul, W., Lee, C., von Roedern, B., and Weisfield, R. L. (1983). *Phys. Rev.* **B28**, 3685.
Kishida, S., Nara, Y., Kobayashi, O., and Matsumura, M. (1982). *Appl. Phys. Lett.* **41**, 1154.
Kitagawa, M., Mori, M., Ishihara, S., Ohno, M., Hirao, T., Yoshioka, Y., and Kohiki, S. (1983). *J. Appl. Phys.* **54**, 3269.
Kitao, M., Hirata, K., and Yamada, S. (1981). *J. de Physique*, Coll. C4, Supp. 10, Tome **42**, 927.
Kittel, C. (1976). "Introduction to Solid State Physics." Wiley, New York.
Klazes, R. H., van den Broek, M. H. L. M., Mezemer, J., and Radelarr, S. (1982). *Phil. Mag.* **B45**, 377.
Klein, N. (1971). *Thin Solid Films* **1**, 149.
Knight, B. W., and Peterson, G. A. (1966). *Phys. Rev.* **147**, 617.
Knight, B. W., and Peterson, G. A. (1967). *Phys. Rev.* **155**, 393.
Knights, J. C., and Davis, E. A. (1974). *J. Phys. Chem. Solids* **35**, 543.
Knights, J. C., and Lucovsky, G. (1980). *CRC Critical Rev. Solid State Sci.* **9**, 211.
Knights, J. C., Hayes, T. M., and Mikkelsen, J. C. (1977a). *Phys. Rev. Lett.* **39**, 712.
Knights, J. C., Biegelsen, D. K., and Solomon, I. (1977b). *Solid State Commun.* **22**, 133.
Knights, J. C., Lucovsky, G., and Nemanich, R. J. (1978). *Phil. Mag.* **B37**, 467.
Knights, J. C., and Lujan, R. A. (1979). *Appl. Phys. Lett.* **35**, 244.

Knotek, M. L. (1975). *Solid State Commun.* **17,** 1431.
Kodama, T., Takagi, N., Kawai, S., Nasu, Y., Yanasigawa, S., and Asama, K. (1982). *IEEE Trans. on Electron Devices* **EDL-3,** 187.
Kodata, S., Siguira, L., Ikada, A., Otaka, S., Nishida, S., Konagai, M., and Takahashi, K. (1982). *2nd Sensor Symposium*, Japan, IEEE.
Koinuma, H., Manako, T., Natsuki, H., Fujioka, H., and Fueki, K. (1985). *J. Non Cryst. Solids* **77-78,** 801.
Kolomiets, B. T., and Nazarova, T. F. (1960). *Sov. Phys. Sol. St.* **2,** 369.
Konagai, M., Lim, K. S., Sichanugrist, P., Komori, K., and Takahashi, K. (1982). *Proc. 16th IEEE Photovoltaic Conf.*, p. 1321.
Konagai, M., Kenne, J., Ohashi, Y., Lim, K. S., and Takahashi, K. (1982). SERI-Amorphous Silicon Review Meeting, Washington.
Konagai, M., Taniguichi, H., Lim, K. S., Sichanugrist, P., and Takahashi, K. (1983). *Jap. J. Appl. Phys.* **22,** 211.
Konogai, M., Matsushita, T., Sichanugrsit, P., Takahashi, K., and Komori, K. (1984). *Proc. 17th IEEE Photovoltaic Conf.*, Kissimee.
Konagai, M. (1986). *MRS. Proc.* (D. Adler, Y. Hamakawa, and A. Madan, eds.), Vol. 70, p. 257.
Konnenkamp, R., Hermann, A. M., and Madan, A. (1985). *Appl. Phys. Lett.* **46**(4), 405.
Kornfeld, M. I., and Sochava, L. S. (1959). *Fiz. Tverd. Tela.* **1,** 1370.
Koshino, N., Maeda, M., Goto, Y., Itoh, K., and Ogawa, S. (1985). *SPIE Proc.* **529,** 40.
Kotz, J., and Shaw, M. P. (1982). *Proc. 16th Int. Conf. Phys. of Semicond.*, Montpellier, France, Physica B & C, Vol. 117-118, Part 2, p. 986 (pub. in 1983).
Kotz, J. (1982). *Ph.D. Thesis.* Wayne State Univ., Detroit, MI.
Kotz, J., and Shaw, M. P. (1984). *J. Appl. Phys.* **55,** 427.
Krishma, K. V., Guha, S., and Narasimhan, K. L. (1981). *Solar Cells* **4,** 153.
Kroemer, H. (1966). IEEE Trans. **ED-13,** 27.
Kroemer, H. (1971). *Proc. IEEE* **59,** 1844.
Kroll, D. M. (1974). *Phys. Rev.* **9,** 1669.
Kuboi, O. (1981). *Jap. J. Appl. Phys.* **20,** L783.
Kuboi, O., Hashimoto, M., and Yatsurugi, Y. (1984). *Appl. Phys. Lett.* **45,** 543.
Kuhl, C., Schlotterer, H., Schwidefsky, F. (1974). *J. Electrochem. Soc.* **121,** 1496.
Kumar, S., and Agarwal, S. C. (1984). *Appl. Phys. Lett.* **45,** 575.
Kurtin, S., McGill, T. C., and Mead, C. A. (1969). *Phys. Rev. Lett.* **22,** 1433.
Kuwano, Y., Ohnishi, M., Nishikawa, H., Tsuda, S., and Nakano, S. (1981). *Proc. 15th IEEE Photovoltaic Conf.*, Florida, p. 698.
Kuwano, Y., Ohnishi, M., Nishiwaka, H., Tsuada, S., Fukatsu, T., Enomoto, K., and Taraui, H. (1982). *Proc. 16th IEEE Photovoltaic Conf.*, San Diego, p. 1338.
Kuwano, Y., Ohnishi, M., Tsuda, S., Nakashima, Y., and Nakamura, N. (1982). *Jap. J. Appl. Phys.* **21,** 413.
Kuwano, Y. (1986). *MRS Proc.* (D. Adler, Y. Hamakawa, and A. Madan, eds.), Vol. 70, p. 455.
Lachter, A., Weisfield, R. L., and Paul, W. (1982). *Solar Energy Materials* **7,** 263.
Lamotte, B. R., and Rousseau, A. (1981). *J. Physique* **42,** C4-839.
Lampert, M. A., and Mark, P. (1970). "Current Injection in Solids." Academic Press, New York.
Landauer, R., and Woo, J. W. (1972). *Comments on Sol. St. Phys.* **4,** 139.
Landauer, R. (1978). *Physics Today* **31**(11), 23.
Landford, W. A., Trantretter, H. P., Ziegler, J. F., and Keller, J. (1976). *Appl. Phys. Lett.* **28,** 566.
Lang, D. V. (1974). *J. Appl. Phys.* **45,** 3023.
Lang, D. V., Cohen, J. D., and Harbison, J. P. (1982). *Phys. Rev.* **B25,** 5285.
Lang, D. V., Cohen, J. D., Harbison, J. P., and Sergent, A. M. (1982). *Appl. Phys. Lett.* **40,** 474.

Langevin, P. (1903). *Am. Chem. Phys.* **28**, 443.
Langon, H. P. D. (1963). *Phys. Rev.* **130**, 134.
Lax, M. (1960). *Phys. Rev.* **119**, 1502.
LeComber, P. G., and Spear, W. E. (1970). *Phys. Rev. Lett.* **25**, 509.
LeComber, P. G., Madan, A., and Spear, W. E. (1972). *J. Non Cryst. Solids* **20**, 239.
LeComber, P. G., Jones, D. I., and Spear, W. E. (1977). *Phil. Mag.* **35**, 1173.
LeComber, P. G., Snell, A. J., MacKenzie, K. D., and Spear, W. E. (1981). *J. de Physique* **C4**, 423.
LeComber, P. G. (1985). *MRS Proc.* (D. Adler, A. Madan, and M. J. Thompson, eds.), Vol. 49, p. 341.
LeComber, P. G. (1986). *J. Non Cryst. Solids* **77-78**, 1081.
Lee, C., Ohlsen, W. D., Taylor, P. C., Ullal, H., and Ceaser, G. P. (1984). *AIP Conf. Proc.* **20**, 205.
Lee, S. H., Henisch, H. K., and Burgess, W. D. (1972). *J. Non Cryst. Solids* **8-10**, 422.
Lee, S. H., and Henisch, H. K. (1973). *Appl. Phys. Lett.* **22**, 230.
Ley, L., Kowalczyk, S., Pollak, R., and Shirley, D. A. (1972). *Phys. Rev. Lett.* **29**, 1088.
Ley, L. (1984). *In* "Topics in Applied Physics" (J. Joannopoulous, and G. Lucovsky, eds.), Vol. 56, p. 61. Springer.
Lillenfield, J. E. (1930). U.S. Patent 1745175.
Lillenfield, J. E. (1935). U.S. Patent 1900018.
Litwak, A. A., and Siryj, B. W. (1983). *SPIE Proc.* **420**, 178.
Liv. P. L., Yen, R., Bloenbergen, N., and Hodgsen, R. T. (1979). *Appl. Phys. Lett.* **34**, 864.
Longeaud, Ch., Baixeras, J., Arene, E., and Mencaraglia, D. (1984). *J. Appl. Phys.* **55**, 1508.
Longeway, P. A., Wiekliem, H. A., and Estes, R. D. (1984). *J. Phys. Chem.* **88**, 73.
Loveland, R. J., Spear, W. E., and Al-Sharbaty, A. (1973). *J. Non Cryst. Solids* **13**, 55.
Lowell, J., and Rose-Innes, A. (1980). *Adv. Phys.* **29**, 947.
Luborsky, S. E. (1983). "Amorphous Metallic Alloys." Butterworth, London.
Lucovsky, G., and Martin, R. M. (1972). *J. Non Cryst. Solids* **8-10**, 185.
Lucovsky, G., and Tabak, M. D. (1974). *In* "Selenium" (R. A. Zingaro, and W. S. Cooper, eds.), p. 788. Van Nostrand-Reinhold, New York.
Lucovsky, G., Nemanich, R. J., and Knights, J. C. (1979). *Phys. Rev.* **B19**, 2064.
Luo, F. C. (1984). *MRS Proc. On Comparison of Thin Film Transistors and SOI Technology*, Elsevier, New York.
Ma, H. B. (1977). *J. Non Cryst. Solids* **24**, 345.
MacKenzie, K. D., LeComber, P. G., and Spear, W. E. (1982). *Phil. Mag.* **B46**, 377.
MacKenzie, K. D., Snell, A. J., French, I., LeComber, P. G., and Spear, W. E. (1983). *Appl. Phys.* **A31**, 87.
MacKenzie, K. D., Eggert, J. R., Leopold, D. J., Lin, Y. M., and Paul, W. (1985). *Phys. Rev.* **B31**, 2198.
Madan, A. (1973). *Ph.D. Thesis,* University of Dundee.
Madan, A., LeComber, P. G., and Spear, W. E. (1976). *J. Non Cryst. Solids* **20**, 239.
Madan, A., and LeComber, P. G. (1977). *Proc. 7th Int. Conf. on Amorphous and Liquid Semiconductors* (W. E. Spear, ed.), p. 377. CICL, University of Edinburgh.
Madan, A., Ovshinsky, S. R., and Benn, E. (1979). *Phil. Mag* **40**, 259.
Madan, A. (1980). *Solar Cells* **2**, 277.
Madan, A., Czubatyj, W., Adler, D., and Silver, M. (1980). *Phil. Mag.* **42**, 257.
Madan, A., McGill, J., Ovshinsky, S. R., Czubatyj, W., Yang, J., and Shur, M. (1980). *Proc. SPIE* **284**, 26.
Madan, A., McGill, J., Ovshinsky, S. R., Czubatyj, W., Yang, J., and Shur, M. (1980b). *SPIE Trans.* **284**, 26.

Madan, A., McGill, J., Czubatyj, W., Yang, J., and Ovshinsky, S. R. (1980b). *Appl. Phys. Lett.* **37**, 826.
Madan, A., and Ovshinsky, S. R. (1980). *J. Non Cryst. Solids* **35-36**, 731.
Madan, A., Czubatyj, W., Yang, J., McGill, J., and Ovshinsky, S. R. (1981). *J. Physique* **42**, CH-463.
Madan, A. Patent #4441 113, filed Feb. 1981, issued 1984.
Madan, A., Czubatyj, W., Yang, J., Shur, M., and Shaw, M. P. (1982). *Appl. Phys. Lett.* **40**, 234.
Madan, A. (1983). Unpublished.
Madan, A. (1984). *In* "Hydrogenated Amorphous Silicon, Topics in Applied Physics" (J. D. Joannopoulous, and G Lucovsky, eds.), Chap. 7, Vol. 5. Springer Verlag.
Madan, A. (1984). *In* "Topics in Applied Physics" (J. Joannopoulous, and G. Lucovsky, eds.), Vol. 55, p. 244. Springer.
Madan, A. (1985). *Proc. of SPIE Conf.* **503**, 83, Washington, DC.
Madan, A. (1986). *Proc. of SPIE Conf.* **706**, 72.
Madan, A. (1985). Chapter 8 in "Silicon Processing for Photovoltaics" (Ravi, and Khaltak, eds.). North-Holland.
Madan, A. (1987). Amorphous Silicon Subcontractors Meeting, Palo Alto, CA.
Magarino, J., Kaplan, D., Friedrich, A., and Deneuville, A. (1982). *Phil. Mag.* **B45**, 285.
Mahan, A. H., Williamson, D. L., and Madan, A. (1984). *Appl. Phys. Lett.* **44**, 220.
Mahan, A. H., von Roedern, B., Williamson, D. L., and Madan, A. (1985). *J. Appl. Phys.* **57**, 2717.
Male, J. C. (1967). *Br. J. Appl. Phys.* **18**, 1543.
Many, A., Goldstein, Y., and Grover, N. B. (1965). "Semiconductor Surfaces." North-Holland. Amsterdam.
Mariucci, L., Gision, D., Coluzza, C., and Frova, A. (1987). *J. Appl. Phys.* (to be published).
Marshall, J. H., and Owen, A. E. (1971). *Phil. Mag.* **24**, 1281.
Matsuda, A., and Tanaka, K. (1982). *Thin Solid Films* **92**, 171.
Matsuda, A., Kaga, T., Tanaka, H., Malhotra, L., and Tanaka, K. (1983). *Jap. J. Appl. Phys.* **22**, L115.
Matsuda, A., Kaga, T., Tanaka, H., and Tanaka, K. (1984). *Jap. J. Appl.Phys.* **23**, L567.
Matsumura, M. (1984). *In* "Semiconductors and Semimetals" (J. Pankove, ed.), Chap. 9, p. 161.
Matsumura, M., and Hayama, H. (1980). *Proc. IEEE* **68**, 1349.
Matsumura, M., and Nara, Y. (1980). *J. Appl. Phys.* **51**, 6443.
Matsumura, M., and Uchida, Y. (1982). *J. de Physique* **C4**, 671.
Matsumura, H., Nakagome, Y., and Furukawa, S. (1980). *Appl. Phys. Lett.* **36**, 439.
Matsumura, H., Sakai, K., Kawakyu, Y., and Furukawa, S. (1981). *J. Appl. Phys.* **52**, 5537.
Matsumura, H., Ihara, H., Tachibana, H., and Tanaka, H. (1985). *J. Non Cryst. Solids* **77-78**, 793.
McLeod, R. D., Pries, W., Card, H. C., and Kao, K. C. (1984). *Appl. Phys. Lett.* **45**, 628.
McMahon, T. J., Yacobi, B. G., Sadlon, K., Dick, J., and Madan, A. (1984). *J. Non Cryst. Solids* **66**, 375.
McMahon, T., and Madan, A. (1985). *Appl. Phys. Lett.* **57**, 5302.
McMahon, T. J., and Madan, A. (1985). *MRS Proceedings* (D. Adler, A. Madan, and M. J. Thompson, eds.), Vol. 49, p. 287.
Melngailis, I., and Milnes, A. G. (1962). *J. Appl. Phys.* **33**, 995.
Meyerson, B. S., Scott, B. A., and Wolford, D. J. (1983). *J. Appl. Phys.* **54**, 1461.
Miller, J. N., Lindau, T., and Spicer, W. E. (1981). *Phil. Mag.* **B43**, 273.

Mizushima, Y., and Yoshikawa, A. (1982). Electronics, Computers, and Telecommunications. In *Jap. Ann. Rev.* (Y. Hamakawa, ed.), Vol. 2, p. 277. North-Holland, Amsterdam.
Moore, A. R. (1977). *Appl. Phys. Lett.* **31**, 762.
Moore, A. R. (1980). *Appl. Phys. Lett.* **37**, 327.
Moore, A. R. (1982). *Appl. Phys. Lett.* **40**, 403.
Moore, A. R. (1983). *J. Appl. Phys.* **54**, 222.
Morel, D. L., and Moustakas, T. D. (1981). *Appl. Phys. Lett.* **39**, 612.
Morimoto, A., Miura, T., Kumeda, M., and Shimizu, T. (1982). *Jap. J. App. Phys.* **21**, 2.
Morimoto, A., Miura, T., Kumeda, M., and Shimizu, T. (1982a). *J. Appl. Phys.* **53**, 7299.
Morimoto, A., Kataoka, T., and Shimizu, T. (1984). *Jap. J. Appl. Phys.* **23**, 10.
Moro Zumi, Oguchi, S. K., Misaura, T., Araki, R., and Ohohima, H. (1984). *SID Digest* **84**, 316.
Mort, J. (1972). *Phys. Rev.* **B5**, 3329.
Mort, J., Chen, I., Emerald, R. L., and Sharp, J. H. (1972). *J. Appl. Phys.* **43**, 2285.
Mort, J. (1973). In "Electronic and Structural Properties of Amorphous Semiconductors" (P. G. LeComber, and J. Mort, eds.), p. 589. Academic Press, New York.
Mort, J., Troup, A., Morgan, M., Grammatica, S., Knights, J., and Lujan, R. (1981). *Appl. Phys. Lett.* **38**, 277.
Moss, S. C., and Graczyk, J. F. (1969). *Phys. Rev. Lett.* **23**, 1167.
Moss, S. C., and DeNeuville, J. P. (1972). *Mat. Res. Bull.* **7**, 423.
Mosserei, R., Sella, C., and Dixmier, J. (1979). *Phys. Stat. Solid* **52**, 475.
Mostefaoui, R., Chevalier, J., Meichenin, S., and Auzel, F. (1985). *J. Non Cryst. Solids* **77-78**, 307.
Mott, N. F. (1938). *Proc. Cambridge Phil. Soc.* **34**, 568.
Mott, N. F. (1968). *J. Non Cryst. Solids* **1**, 1.
Mott, N. F., and Davis, E. A. (1968). *Phil. Mag.* **17**, 1269.
Mott, N. F. (1969). *Contemp. Phys.* **10**, 125.
Mott, N. F. (1970). *Phil. Mag.* **22**, 7.
Mott, N. F. (1975). *Phil. Mag.* **32**, 159.
Mott, N. F., Pepper, M., Pollitt, M., Wallis, R. H., and Adkins, C. J. (1975). *Proc. R. Soc.* **A345**, 169.
Mott, N. F., and Davis, E. A. (1979). "Electronic Processes in Non Crystalline Materials." Clarendon Press, Oxford.
Mott, N. F. (1985). *J. Non Cryst. Sol.* **77-78**, 115.
Moustakas, T. D., Anderson, D., and Paul, W. (1977). *Solid State Commun.* **23**, 155.
Moustakas, T. D., and Paul, W. (1977). *Phys. Rev.* **B16**, 1564.
Moustakas, T. D. (1979). *J. Elec. Materials* **8**, 391.
Moustakas, T. D., Wronski, C. R., and Tiedje, T. (1981). *Appl. Phys. Lett.* **39**, 721.
Moustakas, T. D., Friedman, R., and Weinberger, B. R. (1982). *Appl. Phys. Lett.* **40**, 587.
Moustakas, T. D., and Friedman, R. (1982). *Appl. Phys. Lett.* **40**, 575.
Muller, G., Kalbitzer, S., Spear, W. E., and LeComber, P. G. (1977). In "Amorphous and Liquid Semiconductors" (W. E. Spear, ed.), p. 442. CICL, University of Edinburgh.
Muller, G., Demond, F., Kalbitzer, S., Damjantschitsch, H., Mannsperger, H., Spear, W. E., LeComber, P. G., and Gibson, R. A. (1980). *Phil. Mag.* **B41**, 571.
Muller, G., Simon, M., and Winterling, G. (1982). *Proc. of 16th IEEE Photovoltaic Conf.*, San Diego, p. 1129.
Murayama, K., and Bosch, M. A. (1981). *J. de Physique* **42**, Suppl. 10, C4-343.
Nagels, P., Callaerts, R., Dunayer, M., and De Coninck, R. (1970) *J. Non Cryst. Solids* **4**, 293.
Nakamura, G., Sato, K., Kondo, H., and Yukimoto, Y. (1982). *Proc. 4th Commission of EEC, Conf. on Photovoltaics*, Stresa, Riedel, Dordrecth, p. 616.

Nakamura, G., Sato, K., and Yukimoto, Y. (1982). *Proc. 16th IEEE PV Conf.*, p. 1331.
Nakamura, G., Sato, K., Ishihara, T., Usui, M., Sasaki, H., Okonina, K., and Yukomoto, Y. (1984). *Tech. Digest of PVSEC-I*, Kobe, Japan, p. 587.
Nakano, S., Kawada, H., Matsuoka, T., Kigama, S., Sakai, S., Murata, K., Shibuya, H., Kishi, Y., Nagoaka, I., and Kuwano, Y. (1984). *Tech. Digest of Intern. PVSEC*, Kobe, Japan, p. 583.
Nakano, S., Takahama, T., Isomura, M., Nishikoni, M., Watanabe, K., Nakamura, S., Tsuda, S., Ohnishi, M., Kishi, Y., and Kuwano, Y. (1986). *Tech. Digest of PVSEC-II*, Beijing, China, p. 483.
Nakano, S., Tsuda, S., Tariu, M., Takahama, T., Haku, H., Watanbe, K., Nishikuni, M., Hishikawa, Y., and Kuwano, Y. (1986). *MRS Proc.* (D. Adler, Y. Hamakawa, and A. Madan, eds.), Vol. 70, p. 511.
Nakayama, Y., Nakano, M., Minami, T., and Kawamura, T. (1979). 38th JSAP Fall Meeting, 30-a-S-1.
Nakayama, Y., Wakimura, K., Takahashi, S., Kita, H., and Kawamura, T. (1985). *J. Non Cryst. Solids* **77–78**, 797.
Nashashibi, T. S., Austin, I. G., and Searle, T. M. (1977). *Phil. Mag.* **35**, 831.
Nespurek, S., and Sworakowski, J. (1980). *J. Appl. Phys.* **51**, 2098.
Newland, F. J. (1975). *Jap. J. Appl. Phys.* **14**, 1.
Nielsen, P., and Greden, R. (1983). *J. Vac. Sci. Technol.* **A1**, 583.
Nishimura, T., Ishizu, A., Matsumoto, T., and Akasaka, Y. (1984). *MRS Proc. on Comparison of Thin Film Transistor and SOI Technologies*, Elsevier, New York, p. 221, Vol. 333.
Ngai, K. L., Reinecke, T. L., and Economou, E. N. (1978). *Phys. Rev.* **B17**, 790.
Noolandi, J. (1977). *Phys. Rev.* **B16**, 4466.
Oda, S., Yamaguchi, Y., Hanna, J., Ishihara, S., Fujiwara, R., Kawate, S., and Schimizu, I. (1984). *Tech. D. International PVSEC-I*, Kobe, Japan, p. 429.
Ogawa, K., Shimizu, I., and Inoue, E. (1981). *Jap. J. Appl. Phys.* **20**, L639.
Oha, H., and Oshima, S. (1962). *Mitsubishi Duki Lab. Reps.* **3**, 165.
Ohnishi, M., Nishiwaki, H., Tanaka, M., Nakamura, N., Tsuda, S., Nakano, S., and Kuwano, Y. (1984). *Technical Digest of the International PVSEC-I*, Kobe, Japan, p. 419.
Okamoto, H., Kida, H., Nonomura, S., and Hamakawa, Y. (1983). *Solar Cells* **8**, 317.
Okoniewski, A. M., and Yelon, A. (1985). *J. Appl. Phys.* **58**, 414.
Okushi, H., Tokumaru, Y., Yamasaki, S., Oheda, H., and Tanaka, K. (1982). *Phys. Rev.* **B25**, 4314.
Ondris, M., and den Boer, W. (1980). *3rd EC Photovoltaic Solar Energy Conf.*, Cannes, France. p. 809. D. Riedel Pub. Co.
Onsager, L. (1938). *Phys. Rev.* **54**, 554.
Orenstein, J., and Kastner, M. (1981). *Phys. Rev. Lett.* **46**, 1421.
Orenstein, J., Kastner, M. A., and Vaninov, V. (1982). *Phil. Mag.* **B46**, 23.
Ortenbureger, I. B., Rudge, W. E., and Herman, F. (1972). *J. Non Cryst. Solids* **8–10**, 653.
Overhof, H., and Beyer, W. (1980). *J. Non Cryst. Solids* **35**, & **36**, 375.
Overhof, H., and Beyer, W. (1981). *Phil. Mag.* **B43**, 433.
Overhof, H., and Beyer, W. (1983). *Phil. Mag.* **B47**, 377.
Ovshinsky, S. R. (1968). *Phys. Rev. Lett.* **21**, 1450.
Ovshinsky, S. R., and Sapru, K. (1974). In "Amorphous and Liquid Semiconductors" (J. Stuke, and W. Brenig, eds.), p. 447. Taylor & Francis, London.
Ovshinsky, S. R. (1976). *Phy. Rev. Lett.* **36**, 1469.
Ovshinsky, S. R., and Madan, A. (1978). *Nature* **276**, 482.
Ovshinsky, S. R. (1984). *Tech. Digest of PVSEC-I*, Kobe, Japan, p. 577.
Owen, A. E., Robertson, J. M., and Main, C. (1979). *J. Non Cryst. Solids* **32**, 29.

REFERENCES

Owen, A. E., LeComber, P. G., Sarrabayrouse, G., and Spear, W. E. (1982). *IEEE Proc.* **129**, 51.
Ozawa, T., Takenouchi, M., and Tomiyama, S. (1985). *MRS Proc.* (D. Adler, A. Madan, and M. J. Thompson, eds.), Vol. 49, p. 417.
Paesler, M. A., and Paul, W. (1980). *Phil. Mag.* **B41**, 393.
Pai, D. M., and Enck, R. C. (1977). *Phys. Rev.* **B11**, 5163.
Pai, D. M. (1983). *Proc. 10th Int. Conf. Amor. and Liq. Semicond.*, Tokyo, p. 1255, North-Holland.
Pai, D. M., and Melnyk, A. R. (1986). *SPIE Proceedings* **617**, 82.
Pankove, J. I., and Carlson, D. E. (1976). *Appl. Phys. Lett.* **29**, 610.
Pankove, J. I., and Trong, M. L. (1979). *Appl. Phys. Lett.* **34**, 136.
Pankove, J. (ed.) (1984). *Semiconductor and Semimetals. Hydrogenated Amorphous Silicon—Part A Preparation and Structure*, Vol. 21, Academic Press.
Pantelides, S. T. (1986). *Phys. Rev. Lett.* **57**, 2979.
Pantelides, S. T. (1987). *Phys. Rev. Lett.* **58**, 1344.
Panus, V. R., Ksundzov, Ya. M., and Bousova, Z. V. (1968). *Inorg. Mater. (USSR)* **4**, 778.
Papaconstantapoulous, D. A., and Economou, E. N. (1981). *Phys. Rev.* **B24**, 7233.
Paul, W., Paul, D. K., von Roedern, B., Blake, J., and Oguz, S. (1981). *Phys. Rev. Lett.* **46**, 1016.
Pearson, A. D., and Miller, C. F. (1969). *Appl. Phys. Lett.* **14**, 280.
Peck, W. F., and DeWald, J. F. (1964). *J. Electrochem. Soc.* **111**, 561.
Peery, P. S., and Stein, H. J. (1978). *AIP Conf. Proc.* **50**, 331.
Perkins, G. G., Austin, E. R., and Lampe, F. W. (1979). *J. Am. Chem. Soc.* **101**, 1109.
Perrin, J., Solomon, I., Bourdon, G., Fontenille, J., and Ligeon, E. (1979). *Thin Solid Films* **62**, 327.
Petersen, K. E., and Adler, D. (1976). *J. Appl. Phys.* **47**, 256.
Petersen, K. E., Adler, D., and Shaw, M. P. (1976). *IEEE Trans.* **ED-23**, 471.
Petersen, K. E., and Adler, D. (1977). *In* "Amorphous and Liquid Semiconductors" (W. Spear, ed.), p. 707. CICL, Univ. of Edinburgh, Scotland.
Pfister, G., and Scher, H. (1978). *Adv. Phys.* **27**, 74.
Phillips, J. C. (1979a). *J. Non Cryst. Solids* **34**, 153.
Phillips, J. C. (1979b). *Phys. Rev. Lett.* **42**, 151.
Plattner, R. D., Kruhler, W. W., Rauscher, B., Stetter, W., and Grabmaier, J. B. (1979). *Proc. Photo. Conf.*, Berlin, p. 860.
Polk, D. E. (1971). *J. Non Cryst. Solids* **5**, 365.
Pollak, M., and Geballe, T. H. (1961). *Phys. Rev.* **122**, 1792.
Pollak, M. (1977). *Phil. Mag.* **36**, 1157.
Pontuschka, W. M., Carlos, W. E., Taylor, P. C., and Griffith, R. W. (1982). *Phys. Rev.* **B25**, 4362.
Popescu, C. (1975). *Solid State Electron.* **18**, 671.
Popescu, C., and Henisch, H. K. (1975). *Phys. Rev.* **B88**, 1563.
Postol, T. A., Falco, C. M., Kampwirth, R. T. Schuleer, I. K., and Yelon, W. B. (1980). *Phys. Rev. Lett.* **45**, 648.
Potts, J. E., Peterson, E. M., and McMillan, J. A. (1984). *Solar Energy Mats.* **10**, 145.
Powell, M. J., Easton, B. C., and Hill, O. F. (1981). *Appl. Phys. Lett.* **38**, 794.
Powell, M. J. (1981). *Phil. Mag.* **43**, 93.
Prince, M. B. (1956). *Bell Sys. Tech. J.* **35**, 661.
Pryor, R. W., and Henisch, H. K. (1972). *J. Non Cryst. Solids* **7**, 181.
Regensburger, P. (1968). *J. Photochem. Photobiol.* **8**, 249.
Rehm, W., Engemann, D., Fischer, R., and Stuke, J. (1976). *Proc. 13th Int. Conf. on the Physics of Semiconductors* (F. G. Fumi, ed.), p. 525, Rome.

Rehm, W., Fischer, R., Stuke, J., and Wagner, H. (1977). *Phys. Stat. Solid.* **79**, 579.
Reichmann, (1981). *Appl. Phys. Lett.* **38**, 251.
Reimer, J. A., Vaughan, R. V., and Knights, J. C. (1980). *Phys. Rev. Lett.* **44**, 193.
Reimer, J. A., Dubois-Murphy, P., Gerstein, B. C., and Knights, J. C, (1981). *J. Chem. Phys.* **74**, 1501.
Reinhard, D. K. (1977). *Appl. Phys. Lett.* **31**, 527.
Reinhard, D. K., Arntz, F. O., and Adler, D. (1973). *Appl. Phys. Lett.* **23**, 521.
Rhoderick, E. H. (1978). "Metal-Semiconductor Contacts." Oxford Press, England.
Ridley, B. K. (1963). *Proc. Phys. Soc.* **82**, 954.
Roberts, G. E., and Crowell, C. R. (1970). *J. Appl. Phys.* **41**, 1767.
Robertson, J., and Powell, M. J. (1984). *Appl. Phys. Lett.* **44**, 415.
Robertson, J. (1985). *Phys. Rev.* **B31**, 3817.
Robertson, R., Hils, D., Catham, H., and Gallagher, A. (1983). *Appl. Phys. Lett.* **43**, 544.
Rockstad, H. K. (1972). *J. Non Cryst. Solids* **8–10**, 621.
Rockstad, H. K., and Shaw, M. P. (1973). *IEEE Trans.* **ED-20**, 593.
Rockstad, H. K., and Flasck, R. (1973). *Proc. 5th Int. Conf. Amor. & Liq. Semicond.* (Stuke & Brenig, eds.), p. 1311. Taylor & Francis, London.
Rodgers, D., Thomas, C. B., and Reehal, H. S. (1976). *Phil. Mag.* **31**, 1013.
Rodney, W. S., Malitson, I. H., and King, T. A. (1958). *J. Opt. Soc. Amer.* **48**, 633.
Rose, A. (1960). "Concepts in Photoconductivity and Allied Problems." Interscience, New York.
Ross, R. C., Strand, D. A., Bjornard, E. J., and DeNeufville, J. P. (1986). *MRS Symposium*, Palo Alto, CA, April.
Rothwarf, A. (1978). *13th IEEE Photovoltaic Conf.*, Washington, p. 1312.
Rothwarf, A. (1982). *Appl. Phys. Lett.* **40**, 694.
Rubin, K. A., Barton, R., and Chen, M. (1985). Topical Meeting on Optical Data Storage, Washington, DC, Paper PD-2 (Opt. Soc. Amer.).
Rudder, R. A., Cook, J. W., and Lucovsky, G. (1984). *Appl. Phys. Lett.* **45**, 887.
Rudman, D. A., and Beasley, M. R. (1980). *Appl. Phys. Lett.* **36**, 1010.
Saito, N., Aoki, K., Sannomiya, H., and Yamaguchi, T. (1984). *Thin Solid Films* **115**, 253.
Saitoh, T., Muramatsu, S., Shimada, T., and Migitaka, M. (1983). *Appl. Phys. Lett.* **42**, 678.
Sakai, H., Maruyama, K., Yoshida, T., Ichikawa, Y., Hama, T., Veno, M., Kamiyama, M., and Uchida, Y. (1984). *Tech. Digest of Intern. PVSEC*, Kobe, Japan, p. 591.
Sakata, I., and Hayashi, Y. (1983). *Appl. Phys. Lett.* **42**, 279.
Sakata, I., Hayashi, Y., Yamanaka, M., and Karasawa, H. (1981). *J. Appl. Phys.* **52**, 4334.
Sato, K., Shimizu, I., and Inoue, E. (1981). 28th JSAP Spring Meeting 29a-T-10.
Scarlett, R. M., Shockley, W., and Haitz, R. B. (1963). *In* "Physics of Failure in Electronics" (Goldberg and Vaccaro, eds.), p. 194. Spartan Books, Baltimore, MD.
Schaffert, R. M. (1975). "Electrophotography." Halstead Press, New York.
Schafft, H. A., and French, J. C. (1962). *IEEE Trans.* **ED-9**, 129.
Schafft, H. A., and French, J. C. (1966). *IEEE Trans.* **ED-13**, 613.
Schafft, H. A., and French, J. C. (1966b). *Solid State Electron.* **9**, 681.
Scharfe, M. E. (1984). "Electrophotography Principles and Optimization." Wiley, New York.
Scharfe, M. E., and Schmidlin, F. W. (1975). *Adv. In Electronics and Electron Physics* **38**, 83.
Schein, L. B., and Fowler, K. J. (1985). *J. Imaging Tech.* **11**, 295.
Scher, H., and Montroll, E. W. (1975). *Phys. Rev.* **B12**, 2455.
Schmidlin, F. W. (1973). *26th Ann. Conf. SPSE.*
Schmidlin, F. W. (1976). *In* "Photoconductivity and Related Phenomena" (Mort and Pai, eds.), p. 476. Elsevier, Amsterdam.
Schmidlin, F. W. (1977). *Phys. Rev.* **B16**, 2362.

REFERENCES

Schmidt, M. P., Solomon, I., Tran-Quoc, H., Bullot, J., Gauthier, M., and Cordier, P. (1985). *Phil. Mag.* **B51**, 581.
Scholl, E. (1982). *Z. Phys.* **B46**, 23.
Scholl, E. (1984). *Proc. 17th Int. Conf. Phys. of Semicon.*, San Francisco, CA, p. 1353 (Springer-Verlag).
Schönhammer, K. (1971). *Phys. Lett.* **A36**, 181.
Schuller, M., and Gartner, W. W. (1961). *Proc. IEEE* **49**, 1268.
Scott, B. A., Brodsky, M. H., Green, D. C., Kirby, P. B., Placenik, R. M., and Simonyi, E. E. (1980). *Appl. Phys. Lett.* **37**. 725.
Scott, B. A., Placenik, R. M., and Simonyi, E. E. (1981). *Appl. Phys. Lett.* **39**, 73.
Scott, B. A., Reimer, J. A., Placenik, R. M., Simonyi, E. E., and Renter, W. (1982). *Appl. Phys. Lett.* **40**, 973.
Shanks, R. L. (1970). *J. Non Cryst. Solids* **2**, 504.
Sharp, A. C., Marshall, J. M., and Fortuna, H. P. (1981). *J. de Physique*, Coll. C4, Suppl. 10, Vol. 42, C4-159.
Shaw, M. P., Solomon, P. R., and Grubin, H. L. (1969). *IBM J. Res. Dev.* **13**, 587.
Shaw, M. P., and Gastman, I. J. (1971). *Appl. Phys. Lett.* **19**, 243.
Shaw, M. P., and Gastman, I. J. (1972). *Proc. 4th Int. Conf. Amorphous and Liquid Semiconductors, J. Non Cryst. Solids* **8–10**, 999.
Shaw, M. P., Grubin, H. L., and Gastman, I. J. (1973). *IEEE Trans.* **ED-20**, 169.
Shaw, M. P., Holmberg, S. H., and Kostylev, S. A. (1973b). *Phys. Rev. Lett.* **23**, 521.
Shaw, M. P., Moss, S. C., Kostylev, S. A., and Slack, L. A. (1973c). *Appl. Phys. Lett.* **22**, 114.
Shaw, M. P., Grubin, H. L., and Solomon, P. R. (1979). "The Gunn-Hilsum Effect." Academic Press, New York.
Shaw, M. P., and Subhani, K. F. (1981). *Solid State Electron.* **24**, 233.
Shaw, M. P. (1981). "Handbook on Semiconductors," Vol. 4, Chap. 1. North-Holland, Amsterdam.
Shaw, M. P., and Yildirim, N. (1983). *Advances in Electronics and Electron Physics* **60**, 307.
Shen, S. C., and Cardona, M. (1981). *Phys. Rev.* **B23**, 5322.
Shimizu, T., Nakazawa, K., Kumeda, M., and Ueda, S. (1982). *Jap. J. Appl. Phys.* **21**, L351.
Shimizu, I., Komatsu, T., Saito, K., and Inoue, E. (1980). *J. Non Cryst. Solids* **35 & 36**, 723.
Shockley, W. (1939). *Phys. Rev.* **56**, 317.
Shockley, W., and Read, W. T. (1952). *Phys. Rev.* **87**, 835.
Shockley, W. (1954). *Bell Syst. Tech. J.* **33**, 799.
Shockley, W., and Queisser, H. (1961). *J. Appl. Phys.* **32**, 510.
Shousa, A. M. (1971). *J. Appl. Phys.* **42**, 5131.
Shur, M., Czubatyj, W., and Madan, A. (1980). *J. Non Cryst. Solids* **35 & 36**, 731.
Shur, M., and Hack, M. (1984). *J. Appl. Phys.* **55**, 3831.
Sichanugrist, P., Kumada, M., Konagai, M., Takahashi, K., and Komori, K. (1983). *J. Appl. Phys.* **54**, 6705.
Sichanugrist, P., Konagai, M., and Takahashi, K. (1984). *J. Appl. Phys.* **55**, 1155.
Silver, M., Dy, K. D., and Huang, D. L. (1971). *Phys. Rev. Lett.* **27**, 21.
Silver, M., and Cohen, L. (1977). *Phys. Rev.* **B15**, 3276.
Silver, M., Madan, A., Adler, D., and Czubatyj, W. (1980). *Proc. 14th IEEE Photovoltaic Conf.*, San Diego, p. 1062.
Silver, M., Giles, N. C., Snow, E., Shaw, M. P., Cannella, V., and Adler, D. (1982). *Appl. Phys. Lett.* **41**, 935.
Silver, M., Adler, D., Shaw, M. P., Cannella, V., and McGill, J. (1986). *MRS Proc.* **70**, 119 (D. Adler, Y. Hamaka, and A. Madan, eds.).

Silver, M., Adler, D., Shaw, M. P., and Cannella, V. (1986). *Phil. Mag. Lett.* **53**, L-89.
Simmons, J. G., and Taylor, G. W. (1971). *Phys. Rev.* **B4**, 502.
Simon, I. (1960). *In* "Modern Aspects of the Vitronic State" (J. D. Mackenzie, ed.), Vol. 1, p. 120. Butterworth, London.
Simpson, R. B., and Cohen, D. S. (1970). *J. Math. Mech.* **19**, 895-910.
Sinencio, F. R., and Williams, R. (1971). *J. Appl. Phys.* **4**, 502.
Singh, J., and Cohen, M. H. (1980). *J. Appl. Phys.* **51**, 431.
Singh, J., Budhani, R. C., and Chopra, K. L. (1984). *J. Appl. Phys.* **56**, 109.
Skanavi, G. I. (1958). *Fizila Dielektrihov* (Gosudarstuennyi Izdatelstvo Fiziko-Mathematicesko Literatury, Moscow, USSR).
Smith, B. L. (1969). *Ph.D. Thesis*, Manchester University, England.
Snell, A. J., MacKenzie, K. D., Spear, W. E., and LeComber, P. G. (1981). *Appl. Phys.* **24**, 357.
Snell, A. J., Spear, W. E., and LeComber, P. G. (1981). *Phil. Mag.* **B43**, 407.
Sol, N., Kaplan, D., Dieumegard, D., and Dubreuil, D. (1980). *J. Non Cryst. Solids.* **35-36**, 291.
Solomon, I. (1981). *In* "Fundamental Physics of Amorphous Semiconductors" (F. Yonezawa, ed.), p. 33. Springer, Berlin.
Solomon, I., Dietl, T., and Kaplan, D. (1978). *J. de Physique* **39**, 1241.
Solomon, I., and Brodsky, M. H. (1980). *J. Appl. Phys.* **51**, 4548.
Solomon, P. R., Shaw, M. P., and Grubin, H. L. (1972). *J. Appl. Phys.* **43**, 159.
Solomon, P. R., Shaw, M. P., Grubin, H. L., and Kaul, R. (1975). *IEEE Trans.* **ED-22**, 127.
Spear, W. E. (1969). *J. Non Cryst. Solids* **1**, 197.
Spear, W. E., and LeComber, P. G. (1972). *J. Non Cryst. Solids* **8-10**, 727.
Spear, W. E. (1973). *Proc. of 5th Int. Conf. on Amorphous and Liquid Semicond.*, p. 1, Taylor and Francis.
Spear, W. E., Loveland, R. J., and Al-Sharbatyj, A. (1974). *J. Non Cryst. Solids* **15**, 410.
Spear, W. E., LeComber, P. G., Kinmond, S., and Brodsky, M. H. (1976). *Appl. Physics Lett.* **28**, 105.
Spear, W. E., and LeComber, P. G. (1976). *Phil. Mag.* **33**, 935.
Spear, W. E. (1977). *Advances in Physics* **26**, 811.
Spear, W. E., LeComber, P. G., and Snell, A. J. (1978). *Phil. Mag.* **38**, 303.
Spear, W. E., Allan, D., LeComber, P. G., and Ghaith, A. (1980). *Phil. Mag.* **B41**, 419.
Staebler, D. L. (1980). *J. Non Cryst. Solids* **35 & 36**, 387.
Staebler, D. L., and Wronski, C. R. (1980). *J. Appl. Phys.* **51**, 3262.
Staebler, D. L., Crandall, R. S., and Williams, R. (1981). *Appl. Phys. Lett.* **39**, 733.
Steele, M. C., Ando, K., and Lampert, M. R. (1962). *J. Phys. Soc. Jap.* **17**, 1729.
Stocker, H. J., Barlow, Jr., C. A., and Weirauch, D. F. (1970). *J. Non Cryst. Solids,* **4**, 523.
Street, R. A., and Mott, N. F. (1975). *Phys. Rev. Lett.* **35**, 1293.
Street, R.A., Knights, J. C., and Biegelsen, D. K. (1978). *Phys. Rev.* **B19**, 3027.
Street, R. A., Knights, J. C., and Biegelsen, D. K. (1978b). *Phys. Rev.* **B18**, 1880.
Street, R. A. (1978). *Phil. Mag.* **B37**, 35.
Street, R. A., Biegelsen, D. K., and Knights, J. C. (1981). *Phys. Rev.* **B24**, 969.
Street, R. A., and Knights, J. C. (1981). *Phil. Mag.* **43**, 1091.
Street, R. A., Biegelsen, D. K., and Knights, J. C. (1981a). *Phys. Rev.* **B24**, 969.
Street, R. A. (1981). *Advances in Physics* **30**, 593.
Street, R. A. (1982). *Phys. Rev. Lett.* **42**, 1187.
Street, R. A., and Biegelsen, D. K. (1983). *In* "Topics in Applied Physics" (J. D. Joannopoulous and G. Lucovsky, eds.), Vol. 56, p. 195. Springer-Verlag.
Street, R. A., Zesch, J., and Thompson, M. J. (1983). *Appl. Phys. Lett.* **43**, 672.
Stuke, J. (1970). *J. Non Cryst. Solids* **4**, 1.

REFERENCES

Stutzmann, M., Jackson, W. B., and Tsai, C. C. (1985). *MRS Proc.* (D. Adler, A. Madan, and M. J. Thompson, eds.), Vol. 49, p. 301.
Stutzmann, M., Jackson, W. B., and Tsai, C. C. (1985). *AIP Conf. Proc.* No. 20, p. 213.
Subhani, K. F. (1977). *Ph.D. Dissertation,* Wayne State University, Detroit, MI.
Sukurai, T., and Sugano, T. (1980). *Proc. of Intern,. Conf. on Physics of MOS Insulators,* p. 241. Pergamon, New York.
Sussman, R. S., and Ogden, R. (1981). *Phil. Mag.* **B44,** 137.
Suzuki, T., Hirose, M., and Osaka, Y. (1980). *Jap. J. Appl. Phys.* **19-2,** 91.
Suzuki, T., Hirose, M., and Osaka, Y. (1982). *Jap. J. Appl. Phys.*, Vol. 21, No. 5, L315.
Swartz, G. A. (1982). *J. Appl. Phys.* **53,** 712.
Symons, J., Baert, K., Nijs, J., and Mertins, R. (1987) *J. Non-Cryst. Sol.* **97 & 98,** 1315.
Sze, S. M. (1969). "Physics of Semiconductor Devices." Wiley, New York.
Tabak, M. D., and Warter, Jr., P. J. (1968). *Phys. Rev.* **173.** 899.
Takagi, N., and Ozawa, K. (1980). Spring Meeting of the IEE of Japan, S-3-5.
Takenaga, M., Yamada, N., Ohara, S., Nishiuchi, K., Nagashima, M., Kashihara, T., Nakamura, S., and Yamashita, T. (1983). *SPIE Proc.* **420,** 173.
Tanielen, M. H. (1982). *Phil. Mag.* **B45,** 435.
Taniguchi, M., Hirose, M., and Osaka, Y. (1978). *J. Cryst. Growth* **45,** 126.
Taniguchi, M., Hirose, M., Hamasaki, T., and Osaka, Y. (1980). *Appl. Phys. Lett.* **37,** 787.
Tauc, J., and Abraham, A. (1957). *Phys. Rev.* **108,** 936.
Tauc, J. (1970). In "Optical Properties of Solids" (F. Abeles, ed.), p. 277. North-Holland, Amsterdam.
Tauc, J. (1974). In "Amorphous and Liquid Semiconductors" (J. Tauc, ed.), Plenum, New York
Tauc, J. (1976). *Physics Today*, October, p. 27.
Tawada, Y., Kondo, M., Okamoto, H., and Hamakawa, Y. (1981). *Proc. 15th IEEE Photo. Conf.*, Florida, p. 245.
Tawada, Y., Tsuge, K., Kondu, M., Okamoto, H., and Hamakawa, Y. (1982). *J. Appl. Phys.* **53,** 5273.
Tawada, Y., Nishimura, K., Nonomura, S., Okamoto, H., and Hamakawa, Y. (1983). *Solar Cells* **9,** 53.
Terao, M., Shigematsu, K., Ojima, M., Taniguchi, Y., Horigome, S., and Yonezawa, S. (1979). *J. Appl. Phys.* **50,** 6881.
Terao, M., Nishida, T., Miyaauchi, Y., Nakao, T., Kaku, T., Horigoms, S., Ojima, M., Tsunoda, Y., Sugita, Y., and Ohta, Y. (1985). *SPIE Proc.* **529,** 46.
Thanailakis, A., and Rasul, A. (1976). *J. Phys. C.* **9,** 337.
Thoma, P. (1976). *J. Appl. Phys.* **47,** 5304.
Thompson, J. J., and Thompson, G. P. (1933). "Conduction of Electricity Through Gases, Vol. III, Chap. VIII. Cambridge University Press.
Thompson, M. (1984). In "Topics in Applied Physics" (J. Joannopoulous, and G. Lucovsky, eds.), Vol. 56, p. 119. Springer.
Thompson, M. J., Allison, J., Al-Kaisi, M. M., and Thomas, I. P. (1978). *Rev. de Physique* **13,** 625.
Thompson, M. J., Johnson, N. M., Moyer, M. D., and Lujan, R. (1982). *IEEE Trans. on Electron Devices* **ED-29,** No. 10, p. 1643.
Thompson, M. J. (1984). *J. Vac. Sci. and Technol.* **2,** 827.
Thornton, C. G., and Simmons, C. D. (1958). *IRE Trans.* **ED-5,** 6.
Thouless, D. J. (1974). *Phys. Rev.* **C13,** 93.
Tiedje, T., Abeles, B., Morel, D. L., Moustakas, T. D., and Wronski, C. R. (1980). *Appl. Phys. Lett.* **36,** 695.

Tiedje, T., Wronski, C. R., and Cebulka, J. M. (1980). *J. Non Cryst. Solids* **30-36**, 743.
Tiedje, T., and Rose, A. W. (1981). *Solid State Commun.* **37**, 49.
Tiedje, T., Moustakas, T. D., and Cebulka, J. M. (1981). *J. Phys.* **42**, C4-155.
Tiedje, T., Cebulka, J. M., Morel, D. L., and Abeles, B. (1981). *Phys. Rev. Lett.* **46**, 1425.
Tiedje, T. (1982). *Appl. Phys. Lett.* **40**, 627.
Tomlin, S. G. (1968). *J. Phys.* **D1**, 1667.
Tong, B. Y., John, P. K., Wong, S. K., and Chik, K. P. (1981). *Appl. Phys. Lett.* **38**, 789.
Tsai, C. C. (1979). *Phys. Rev.* **B19**, 2041.
Tsai, C. C., Thompson, M. J., and Nemaich, R. J. (1981). *J. de Physique* **C4**, 1077; (1981). *AIP Proc.* **73**, 312.
Tsai, C. C., Stutzmann, M., and Jackson, W. B. (1984). *AIP Proc.* **20**, 242.
Tsang, and Street, R. (1978). *Phil. Mag* **B37**, 601.
Tsu, R., Howard, W. E., and Esaki, L. (1970). *J. Non Cryst. Solids* **4**, 322.
Tsuda, S., Nakamura, N., Watanabe, K., Nishikuni, M., Ohnishi, M., Nakano, S. Kishi, Y., Shibuya, H., and Kuwano, Y. (1984). *Tech. Digest of PVSEC 1*, Kobe, Japan.
Tsuda, S., Nakamura, N., Nishikuni, M., Watanabe, K., Takahama, T., Hishikawa, Y., Ohnishi, M., Kishi, Y., Nakano, S., Kuwano, Y. (1985). *J. Non Cryst. Solids* **77-78**, 1465.
Tuan, H. C., Thompson, M. J., and Johnson, N. H. (1982). *IEEE Trans. on Electron Devices* **EDL-3**, 357.
Tulagin, V., and Carreira, L. M. (1968). U.S. Patent 3,384,565.
Turban, G., Catherine, Y., and Grolleau, B. (1979). *Thin Solid Films* **60**, 147.
Turban, G., Catherine, Y., and Grolleau, B. (1980). *Thin Solid Films* **67**, 309.
Turner, M. J., and Rhoderick, E. H. (1968). *Solid State Electron.* **11**, 291.
Uchida, Y., Sakai, H., Nishiura, M., and Haruki, H. (1981). *Proc. 15th IEEE Photovoltaic Specialist Conf.*, p. 922.
Uchida, Y. (1982). US-Jap. Seminar "Technological Applications of Tetrahedral Amorphous Solids," Palo Alto.
Uchida, Y., Nishiura, M., Sakai, H., and Haruki, H. (1983). *Solar Cells* **9**, 3.
Uchida, Y., Nara, Y., and Matsumura, M. (1984). *IEEE Trans. on Electron Devices* **EDL-5**, 105.
Ugai, Y., Murakami, Y., Tamamura, J., and Aoki, S. (1984). *SID Symp. Digest*, 308.
Uhlmann, D. R., and Kreidl, N. J. (1980). "Glass: Science and Technology." Academic, New York.
Ullal, H. S., Morel, D. L., Willett, D. R., Kanani, D., Taylor, P. C., and Lee, C. (1984). *Proc. 17th IEEE Photovoltaic Specialist Conf.*, Florida, p. 359.
Urbach, F. (1953). *Phys. Rev.* **92**, 1324.
Usui, S., and Kukuchi, M. (1979). *J. Non Cryst. Solids* **34**, 1.
Vanderbilt, D., and Joannopoulous, J. D. (1983). *Phys. Rev.* **B27**, 6296; 6302; 6311.
Vanier, P. E., Delahoy, A. E., and Griffith, R. W. (1981). *J. Appl. Phys.* **52**, 5235.
Van Roosbroeck, W. (1972). *Phys. Rev. Lett.* **28**, 1120.
Van Roosbroeck, W., and Casey, Jr., H. C. (1972). *Phys. Rev.* **B5**, 2154.
Van Uijen (1985). *SPIE Proc.* **529**, 2.
Varlamov, I. V., and Osipov, V. V. (1970). *Soc. Phys. Semicond.* **3**, 803 & 978.
Veprek, S., and Maracek, V. (1968). *Solid State Commun.* **11**, 683.
Veprek, S., Iqbal, Z., Oswald, H. R., Sarott, F. A., and Wagner, J. J. (1981). *Proc. 9th Int. Conf. on Amorphous and Liquid Semiconductors.* In *J. Phys. Colloq. C4* **42**, 251.
Verie, C., Rochette, J. F., and Rebouillat, J. P. (1981). *J. Phys.* **42**, C4-667.
Viktorovitch, P. (1981). *J. Appl. Phys.* **52**, 1392.
Viktorovitch, P., Moddell, G., and Paul, W. (1981). *In* "Tetrahedrally Bonded Amorphous Semiconductors" (R. A. Street, D. K. Biegelsen, and J. C. Knights, eds.), p. 186.

REFERENCES

Viktorovitch, P., Moddell, G., Blake, J., and Paul, W. (1981). *J. Appl. Phys.* **52,** 6203.
von Roedern, B., Ley, L., and Cardona, M. (1977). *Phys. Rev. Lett.* **39,** 1576.
von Roedern, B., Ley, L., Cardona, M., and Smith, F. W. (1979). *Phil. Mag.* **B40,** 433.
von Roedern, B., Mahan, A. H., Williamson, D. L., and Madan, A. (1984). *Proc. of 5th Symp. on Materials and New Proc. Tech. for Photo.*, New Orleans.
von Roedern, B., and Madan, A. (1985). *In* "Tetrahedrally Bonded Amorphous Semiconductors" (D. Adler, and H. Fritzsche, ed.), p. 79. Plenum Press.
von Roedern, B., Mahan, A. H., McMahon, T., and Madan, A. (1985). *MRS Symposium on Materials Issues of Amorphous Silicon Technologies* (D. Adler, A. Madan, and M. J. Thompson, eds.), Vol. 49, p. 167.
von Roedern, B., and Madan, A. (1986). Presented at PVSEC-II, Beijing.
Wagner, J., Stasiewski, H., Abeles, B., and Landford, W. A. (1983). *Phys. Rev.* **B28,** 7080.
Walker, C., Hollingsworth, R., del Cueto, J., and Madan, A. (1986). *Proc. MRS Conference* **70,** 563.
Walsh, P. J., and Vezzoli, G. C. (1974). *Appl. Phys. Lett.* **25,** 28.
Walsh, P. J., Ishioka, S., and Adler, D. (1978). *Appl. Phys. Lett.* **33,** 593.
Walsh, P. J., Pooladdej, D., Thompson, M. S., and Allison, J. (1979). *Appl. Phys. Lett.* **34,** 835.
Walsh, P. J., Thompson, M. J., and Allison, J. (1981). *J. Non Cryst. Solids* **45,** 209.
Wang, E. Y., Bavanona, C. R., and Brandhorst, H. W. (1974). *J. Electrochem. Soc.* **121,** 973.
Warren, A. C. (1973). *IEEE Trans.* **ED-20,** 123.
Watanabe, K., Sato, N., and Miyaoka, S. (1983). *J. Appl. Phys.* **54,** 1256.
Webb, A. P., and Verprek, S. (1979). *Chem. Phys. Lett.* **62,** 173.
Weber, W. H., and Ford, G. W. (1970). *Solid State Electron.* **13,** 1333.
Weiser, K., and Brodsky, M. (1970). *Phys. Rev.* **B1,** 791.
Weisfield, R., Viktorovitch, P., Anderson, D. A., and Paul, W. (1981). *Appl. Phys. Lett.* **39,** 263.
Weisfield, R. (1985). *Ph.D. Thesis,* Harvard University.
Weisfield, R. L., and Anderson, D. A. (1971). *Phil. Mag.* **B44,** 83.
Weisfield, R. L. (1983). *Appl. Phys.* **54,** 6401.
Weisz, S. Z., Gomes, M., Moir, J. A., Resto, O., Perez, R., Goldstein, Y., and Abeles, B. (1984). *Appl. Phys. Lett.* **44,** 634.
Wente, E. C. (1922). *Phys. Rev.* **19,** 333.
Werzer, V. W., Brandhorst, H. W., Broder, J. D., Hart, R. E., and Lamneck, T. H. (1979). *J. Appl. Phys.* **50,** 4443.
Williams, E. M. (1984). "The Physics and Technology of Xerographic Processes." Wiley, New York.
Williams, R. H., Varma, R. R., Spear, and LeComber, P. G. (1979). *J. Phys.* **C122,** L209.
Wilson, J. I. B., McGill, J., and Kinmond, S. (1978). *Nature* **272,** 152.
Wood, D. L., and Tauc, J. (1972). *Phys. Rev.* **5,** 3144.
Woodyard, J. R., Bowen, D. R., Gonzalez-Hernandez, J., Lee, S., Martin, D., and Tsu, R. (1985). *J. Appl. Phys.* **57,** 2243.
Wronski, C. R., Carlson, D. E., and Daniel, R. E. (1976). *Appl. Phys. Lett.* **29,** 602.
Wronski, C. R. (1978). *Jap. J. Appl. Phys. Suppl.* **17-1,** 299.
Wronski, C. R., Abeles, B., Cody, G. D., and Tiedje, T. (1980). *Appl. Phys. Lett.* **37,** 96.
Wronski, C. R., Abeles, B., and Cody, G. D. (1980). San Diego Workshop.
Wronski, C. R., and Daniel, R. E. (1981). *Phys. Rev.* **B23,** 794.
Wronski, C. R., Goldstein, Y., Keleman, S., Abeles, B., and Witzkes, H. (1981). *J. de Physique* **C4,** 475.
Wronski, C. R., Goldstein Y., Kelemen, S., Abeles, B., and Witzke, H. (1981). *9th Int. Conf. Amor. and Liq. Semicond.*, Grenoble. *J. Phys. Colloq.* **42,** C4, Part 1, p. 475.

Wronski, C. R., Abeles, B., Tiedje, T., and Cody, G. D. (1982). *Solid State Commun.* **44**, 1423.
Wu, Zhi-Qiang, Yi, Xu-Cun, Wei-Pang, Z., Zhang-Bo, Z., and Rong-Chuan, F. (1983). *J. Non Cryst. Solids* **59-60**, 217.
Yablonovitch, E., and Cody, G. D. (1982). *IEEE Trans. Electron Devices* **ED-29**, 300.
Yacobi, B. G., McMahon, T. J., and Madan, A. (1984). *Solar Cells* **12**, 329.
Yamaguchi, T., Okamoto, H., Nonomura, S., and Hamakawa, Y. (1980). *2nd Photovoltaic Science and Eng. Conf.*, Japan.
Yamaguchi, T., Okamoto, H., Nonomura, S., and Hamakawa, Y. (1981). *Jap. J. Appl. Phys.* **20**, 191.
Yamamoto, N., Nakayama, Y., Wakita, K., Nakano, M., and Kawamura, T. (1981). *Jap. J. Appl. Phys. Suppl.* **20-1**, 305.
Yamano, M., Takesada, H., Yamasaki, B. M., Okita, Y., and Hada, H. (1984). *Consumer Electronics Conf.*, Chicago.
Yamano, M., and Takesada, H. (1985). *J. Non Cryst. Solids* **77-78**, 1383.
Yamasaki, K., Yoshida, M., and Sugano, T. (1979). *Jap. J. Appl. Phys.* **L8**, 113.
Yamasaki, S., Nakagawa, K., Yamamoto, H., Matsuda, A., Okushi, H., and Tanaka, K. (1981). *AIP Conference Proceedings* (R. A. Street, D. K. Biegelsen, and J. C. Knights, eds.), Vol. 73, p. 258.
Yang, J., Ross, R., Mohr, R., and Fournier, J. P. (1985). *Proc. of 18th Photovoltaic Specialist Conf.*, p. 1519.
Yang, J., Ross, R., Mohr, R., and Fournier, J. P. (1986). *MRS Proc.* (D. Adler, Y. Hamakawa, and A. Madan, eds.) **70**, 475.
Yaniv, Z., Hansell, G., Vijan, M., and Canella, V. (1984). *MRS Proc.* **33**, 293.
Yaniv, Z., Canella, V., Hansell, G., and Vijan, M. (1985). *MRS Proc.* **49**, 353.
Yumoto, J., Yajima, H., Seki, Y., and Shimeda, J. (1982). *Appl. Phys. Lett.* **40**, 632.
Zabrodskii, A. G., Ryvkin, S. M., and Shlimak, I. S. (1973). *JETP Lett.* **18**, 290.
Zacharisen, W. H. (1932). *J. Am. Chem. Soc.* **31**, 1164.
Zallen, R. (1983). "The Physics of Amorphous Solids." Wiley, New York.
Zallen, R., Drews, R. E., Emerald, R. L., and Slade, M. L. (1971). *Phys. Rev. Lett.* **26**, 1564.
Zembutsu, S. (1982). *Japan Ann. Rev. in Electronics, Computers and Telecommunications* (Y. Hamakawa, ed.), Vol. 2, p. 296. North-Holland.
Zanzuchi, P. J., Wronski, C. R., and Carson, D. E. (1977). *J. Appl. Phys.* **48**, 5227.

Subject Index

A

Absorption, 62ff
Activation energy,
  of conductivity, *see* conductivity
Air Mass solar spectra, 163, 165, 175, 176, 219
Amorphous chalcogenide alloys, *see* chalcogenide alloys
  Aamodt-Murphy model for a thermophonic cell, 418
  As-S based films, *see* chalcogenide alloys and 323ff
  As-Se based films, *see* chalcogenide alloys and 323ff, 340ff, 350
  As-Te based films, *see* electrical switching, threshold switches and 338
Amorphous semiconductor alloys,
  a-Si, *see* Silicon-unhydrogenated
  a-Si:C:H,
    properties of, 155ff
    influence on solar cell performance, 242ff
    solar cells of, 257
  a-Si:F, *see* Silicon-fluorinated
  a-Si:F:H, *see* Silicon-fluorinated and hydrogenated
  a-Si:H, *see* Silicon-hydrogenated
  a-Si:Ge:H,
    properties of, 149ff
    solar cells of, 256, 257
  a-Si:Ge:F:H, 152
    solar cells of, 257
  a-Si:Sn:H,
    properties of, 151, 153ff
    solar cells of, 256

a-Si:N:H,
  properties of, 159ff
  solar cells of, 257
Auger Electron Spectroscopy, 200, 201, 208

B

Brownian motion, 9

C

Capacitance–Voltage technique, 61, 203ff
Chalcogenide alloys
  absorption edges of, 324ff
  ac conductivity of, 337ff
  circuit effects in switching devices, 356ff, 457ff
  contrast potential, 477, 488
  critical electric field induced switching effects, 376
  critical field, 377, 380, 393
  critical local power density, 383
  dc conductivity of, 331ff
  density of states in, 344
  dispersive transport in, 333, 351
  drift mobility in, 348
  ESR in, 330
  electrical properties of, 331ff
  electrophotography in, 470ff
  Fermi level in, 332
  Hall effect in, 346ff
  hopping in, 338
  index of refraction of, 327
  infrared spectra of, 323
  multiple trapping in, 341, 351

optical properties of, 322ff, 470ff, 501ff
photoconductivity of, 331ff
photocurrent transient spectroscopy of, 339ff
photodarkening of, 330
photoluminescence in, 329
properties of, 318ff
Raman spectra of, 323
switching and memory in, see electrical switching and memory devices
thermal cycling in, 501, 502
thermoelectric power of, 344
vibrational spectra of, 322
Chalcogenide glasses, 36, 37, 65, 66, 173, 174, 309, 319
Charged coupled device, 313ff
Chemical bonding, 3, 36
Chemical sensors, 317
Coherent Anti-Stokes Raman Spectroscopy (CARS), 18
Cohesive energy, 3
Conduction band tail, 36, 37, 105, 110
Conductivity,
　activation energy, 77
　hopping, 9, 75ff
　in extended states, 9, 73ff
　in tail states, 74ff
　minimum metallic, 65, 73
　prefactor, 74, 78
Co-ordination number, 3, 66, 73
Correlation energy, 39, 41, 152
Critical energy, 8, 9

D

Deep Level Transient Spectroscopy, 50ff
Defect,
　capture cross sectional areas of, 54, 87, 92, 94, 98, 105, 177, 185
　capture rates, 51
　complexes of, 39
Degradation, see stability
Delayed collection field technique, 174
Density of states, 35ff, 39, 44, 46, 48, 49, 50
Depletion width, see space charge region
Dielectric relaxation time, 57, 85
Diffraction
　electron, 4, 29
　X-ray, 3, 4, 29
Diffusion length, 253ff

Diode admittance measurement, 128
Disilane, 25, 275, 276
Disorder
　compositional, 3, 36
　effects of, 4, 5, 7, 8
　hydrogen influence on, 67
　positional, 3, 36
　structural, 68
　thermal, 68
Dispersive transport, 88, 90, 91, 92, 93, 94, 96
Divacancies (a-Si:H), 39
Doping,
　in hydrogenated a-Si, see Silicon-hydrogenated
Doping efficiency, 41, 117

E

Electrical switching and memory devices in chalcogenide films, 382ff
　circuit effects in, 356ff, 457ff
　current filamentation, 358, 434ff
　critical field in, 393
　delay time in, 401ff, 407
　electroacoustic spectroscopy of, 409ff
　electronic models for switching devices, 433ff
　electronic studies of, 433ff
　electrothermal model for, 405ff
　electrothermal switching mechanism in formed films, 404ff
　filament of current in, 387, 392, 400, 434ff, 450, 464ff
　first fire effects, 386, 391ff
　forming in, 386, 397, 404ff, 408, 428ff
　general overview of, 454ff
　holding current, 358, 384
　holding voltage, 358, 384
　isothermal analysis, 438ff
　maximum benign interruption time in, 385, 391
　memory filament, 358, 409ff
　memory switch, 393, 409ff, 414ff
　Mott transition model for, 452
　OFF state, 351, 358, 384, 386, 391, 392ff, 398ff, 405ff, 443ff
　ON state, 358, 384, 387, 390, 405ff, 417, 445ff, 448ff, 452
　phenomenological kinetic equations, 441ff
　pulse interruption time in, 385, 391

SUBJECT INDEX

radius of current filament in, 388
recent developments in, 456
recovery properties of, 390ff
switching time in, 385
switching transition in, 386
thermophonic studies of, 409ff, 418, 422ff
thick films, 409ff
threshold switch, 383, 398, 405ff, 414, 416
transient ON characteristics (TONC), 387ff
trapping and switching in, 438ff
voltage pulse experiments on, 384ff, 387
Electroabsorption, 66
Electroluminescence, 112
Electron beam induced current (EBIC), 224
Electron spin resonance (ESR), 18, 24, 39, 41
Electrophotography, 470ff
   amorphous hydrogenated silicon in, 497ff
   chalcogenides in, 470ff
   charge displacement in, 492
   charge generation in, 492
   charged pigment electrophotography (CPE) in, 489ff
   charging of the toner particles in, 491
   contact effects in, 494, 497
   contrast potential in, 477, 478
   emission limited discharges in, 479ff
   ideal, 473, 474
   material requirements for photoreceptors in, 491ff
   multilayerd photoreceptors, 499
   photoactive pigment electrophotography (PAPE) in, 489ff
   photodischarge in, 477, 478
   physics of, 476
   physics of the development process in, 489ff
   process of, 472
   space charge limited discharge, 482ff
   space charge perturbed discharge, 482ff
   transfer vs. trapping in, 493ff
Energy bands,
   width of, 6, 8
   tails of, 8
Excitons, 65, 66, 174
Extended states, 9

F

Field effect technique, 44ff, 60, 61, 128, 140, 142
Field effect transistor, *see* thin film transistors

G

Ge-Se based films, 335, 336, 337
Ge-Te based films, *see* electrical switching, memory switching
Glass transition temperature, 2
Glow discharge technique, *see* Silicon-hydrogenated
Growth kinetics, 12ff
   in disilane gas, 25ff
   in silane gas, 13ff
   involving halogens, 26ff

H

Heat flow in semiconductors, 362ff
   RC analog network of, 367ff
Heterojunction transistor, 317, 390
Hopping, phonon assisted, 9, 75ff

I

Image sensing, 313
Internal photoemission, 197
Ion implantation, 117
Isothermal capacitance transient spectroscopy, 54

J

Junction recovery technique, 239, 253

L

Linear image sensors, 310ff
Liquid crystal displays, *see* thin film transistor
Localized states, *see* density of states
Lone pair bonding, 37
Lone pair electrons, 320, 335

M

Mass spectrometry,
   in silane gas, 14, 18
Mean free path, 7, 9
Memory switching
   electrical, 383, 393
   optical, 501ff
Metal-insulator-semiconductor devices, 61, 191, 205ff

Metal-insulator-semiconductor-hydrogenated a-Si, 205ff
  conversion efficiency in, 208
  current-voltage characteristics, 209, 212, 213
  ideality or diode quality factor in, 209
  insulator effect of, 205, 211, 213
  J_{sc}, increase of, 208
  MIS devices, 205ff
  V_{oc}, increase of, 208, 209
Metastable states, 1, 262, 264
Microcrystallinity, 116, 143, 219
Mobility, 5
  drift, 84ff
  edge, 9, 73
  extended state, 9
  field effect, 292
  gap, 9, 10, 39, 73
  hopping, 9, 74
Mott $T^{1/4}$ law, 75, 81, 332

N

Negative differential conductivity (NDC), 356, 439
  thermally induced, 370ff
n-i-p devices, see p-i-n devices
Nuclear magnetic resonance, see Silicon-hydrogenated
Nuclear reaction analysis, 34

O

Optical emission spectrometry,
  in halogenated gas mixtures, 26
  in silane gas, 14, 18, 24
Optical memories, 501ff
  data formatting in, 506
  general requirements for, 505ff
  phase transition devices for, 507ff
  recording threshold for, 506
  resolution of, 505
  sensitivity of, 505
  technological progress in, 516
  use of physical holes for, 512ff
Optical recording, 312ff, 501ff
Optical trapping effects, 245ff
Overlap integral, 6, 73

P

Passivation layer, 316
Phase transitions, 414ff, 510ff, 507
Phonon,
  assisted hopping, 9, 75ff
  frequency, 9, 74
Photoconductivity, 97ff
Photoemission, 4, 65, 127, 128, 132
Photoinduced effects, see stability
Photoluminescence, 106ff
Photothermal acoustic spectroscopy, 69, 72
Photothermal deflection spectroscopy, 69
p-i-n or n-i-p devices, Hydrogenated a-Si
  B profiling of, 234
  built in potential, 223
  collection efficiency of, see quantum efficiency in,
  compositional profile of, 223
  current-voltage characteristics of, 220ff
  Dember potentials in, 237, 238
  doping influence of, 226ff, 239, 240
  electron beam induced current in, 224
  fabrication of, 219ff
  impurity influence of, 241ff, 272
  injection current in, 221, 239
  macrostructural defects in, 224
  minority carrier diffusion length in, 253ff
  minority carrier lifetime in, 226
  on ITO, 219
  on polyemide material, 219
  on SnO_2, 219
  on stainless steel, 219, 272
  open circuit voltage in, 132
  optical trapping effects in, 245ff, 253
  optimization of, 242ff
  p-type SiC in, 242ff, 253
  quantum efficiency in, 171, 249ff
  recombination current in, 221, 223, 239
  reverse breakdown voltage in, 225
  self field in, 237
  shunt component in, 221
  space charge region in, 223, 249
  substrate effect of, 219
  theoretical efficiency of, 180ff
  thickness dependence of, 248
  tunneling current in, 221, 222, 223
Plasma enhanced chemical vapor deposition, see silicon-hydrogenated
Primary photocurrent technique, 67

SUBJECT INDEX

Q

Quasi Fermi levels, 52, 58, 99, 105, 106, 178, 185

R

Radial distribution function, 29
Recombination,
  bimolecular, 99, 101, 102, 104
  centers, 98, 100, 101, 102, 179, 180
  current, 197, 221
  diffusion limited kinetics, 179
  geminate, 169, 171, 172, 174
  interface velocity, 184, 255
  lifetime, 99
  monomolecular, 99, 102, 104
  non-radiative, 107, 110, 111
  radiative, 107, 110, 165
  rates of, 99, 105
  Shockley-Read, 181, 184
  velocity, 182
Reflectors, use in solar cells, 245

S

Sb alloys, 508, 509
Scanning Electron Microscopy, 32, 201
Schottky barriers, generalized,
  barrier height, 189, 191, 194
  diffusion theory of, 194, 212
  ideality factor of, 194
  interface index in, 191
  silicide formation in, 192
  thermionic emission theory of, 194
  work function of metal-dependence of, 190, 212
Schottky barriers, hydrogenated a-Si, 188ff
  annealing effects of, 199ff
  autocompensation in, 232
  barrier height of, 197ff
  C-V measurements of, 203ff
  characterization of, 193
  collection efficiency, see quantum efficiency in
  contacts, 193, 200, 201
  contamination in, 195
  current-voltage characteristics of, 193, 195, 212

  depletion region, 193
  doping effects of, 229, 230, 231
  doping of, 54, 57
  fabrication of, 189, 193
  field emission tunneling in, 193
  ideality factor in, 196, 200
  injection current in, 211
  microprobe analysis of, 195
  minority carrier diffusion length in, 229
  oxide layers on, 190, 193
  quantum efficiency in, 63, 64, 198, 249ff
  recombination current in, 196, 211
  silicide formation in, 200, 202
  work function of metal-dependence of, 212, 215ff
Schubweg, 477
Secondary ion mass spectroscopy, 34, 141, 223
Seebeck,
  activation energy, 77
  coefficient for defect states, 76ff
  coefficient for extended states, 76ff
Selenium, 328
Silicide formation, 191
Silicon (Amorphous),
  CVD of, 134ff
  evaporation of, 81, 108, 140, 173, 174
  HOMOCVD of, 134
  ion plating of, 139
  plasma enhanced CVD of, see silicon-hydrogenated
  photo CVD of, 134, 135
  preparation of, 134ff
  sputtering of, 128, 135ff, 140
Silicon, fluorinated
  CVD of, 146
  sputtering of, 143
Silicon, fluorinated and hydrogenated (a-Si:F:H)
  doping of, 116
  electrical conduction in, 82, 140
  ESR spin density in, 145
  growth of, 26
  microcrystallinity of, 116, 143
  photoconductivity in, 142
  preparation of, 26, 140, 275
  properties of 26, 140
  role of F, 146, 147
  solar cells, 208
  stability in, 140

Silicon, hydrogenated
  boron doping, effects on electrical transport, 122ff
  capture cross sectional areas of traps in, 106
  contamination in, 33, 114
  correlation energy in, 41, 43
  dangling bonds in, 24, 41, 54, 55, 67, 71, 72, 81, 111, 127
  density of states in, 14, 44ff
  dihedral angles in, 28
  dispersive transport in, 82, 88, 90, 91, 92, 93, 94, 96
  doping compensation effects of, 112, 113, 114, 115, 125
  doping of, 35, 39, 41, 46, 71, 93, 104, 115ff, 133
  drift mobility in, 84ff
  ESR in, 18, 24, 39, 41, 71, 72, 111, 129, 130, 262, 265
  electrical conduction process in, 77ff
  electroabsorption in, 66
  electroluminescence in, 112
  electron lifetime in, 106
  electrophotography with, 497ff
  energy band gap of, 68, 69
  field effect in, 44, 46, 48, 128, 129
  growth kinetics of, 21, 22, 23, 24, 25
  hole lifetime in, 106
  hydrogen concentration in, 14, 21, 22, 32, 33, 34, 35, 39, 65, 68, 80
  hydrogen effusion in, 32, 33, 112, 238, 239, 313
  infra red spectra of, 14, 15, 18, 25, 30, 33, 34, 35, 46, 123
  microcrystallization of, 219, 116, 143
  NMR of, 39, 43, 122, 263
  optical absorption in, 62ff, 110
  phosphorous doping, effect on electrical transport, 117ff
  photoconductivity of, 63, 64, 97ff
  photoluminescence in, 81, 111, 112
  Photothermal acoustic spectroscopy (PAS) in, 69, 72
  Photothermal deflection spectroscopy (PDS) in, 69
  preparation in Disilane, 25, 275, 276
  preparation in Silane, 13, 14, 16, 18, 20, 24
  primary photocurrent in, 67
  recombination lifetimes in, 78
  scanning electron microscopy of, 32
  statistical shifts in, 121, 132
  stress in, 20, 266, 307
  structure of, 27, 28, 29, 30, 31, 32, 78
  subband gap absorption in, 69, 71
  surface states in, 44, 46, 49, 50, 53, 54, 127ff, 211, 288
  thermopower of, 76ff, 79, 118, 119, 120, 121, 122
Silicon, unhydrogenated, 4, 81, 132
  crystallization of, 29
Solar cells, 163ff
  conversion efficiency (ideal), 164
  fill factor, 164
  losses in, 164
  maximum power point, 164
  open circuit voltage, 164
  recombination in, 166
  rectification in, 166
  short circuit current, 164
  structures, 166
  superposition principle in, 182
Solar cells, Hydrogenated a-Si
  collection efficiency in, *see* quantum efficiency in
  conversion efficiency, 168ff
  current-voltage characteristics of, 180
  economics of, 274ff
  geminate recombination in, 169, 171, 172, 174
  impurities in, 185, 272
  large area, 268ff
  mass production of, 272ff
  minority carrier diffusion length in, 176, 187, 253ff
  multijunction cells, *see* tandem cells
  open circuit voltage in, 176ff
  patterning of, 269, 273
  quantum efficiency or yield in, 170, 171, 174, 249ff
  recombination in, 177, 179, 187
  short circuit density in, 174ff
  space charge region in, 169, 180
  tandem cells, 186ff
Space charge limited current technique, 55ff, 129
Space charge region, 169, 180, 193, 222, 249, 254, 255
Stability of hydrogenated Si, 259ff
  changes in opto-electronic properties, 262
  impurity effects, 265, 266

SUBJECT INDEX

influence on solar cell performance, 267
metastable states, 262, 264
stress influence, 266
surface effects, 261
Staebler–Wronski effect, *see* stability
Stokes shift, 108, 110
Supercooled liquid, 2
Superlinearity, 106
Surface photovoltage technique, 253ff
Switching in amorphous chalcogenide films, *see* electrical switching
Switching in vanadium dioxide, 404
Switching in hydrogenated Si,
   delay time in, 309
   filamentary conduction in, 309
   memory, 309ff
   threshold, 309ff
   threshold voltage of, 309

T

Tail states
   conduction band, 36, 37, 110
   valence band, 36, 37, 110
Tandem cells, 186ff, 256ff
Tellurium, 329
Tellurium alloys, 331ff
Texture, 245
Thermistor, 359ff, 396, 397
Thermal behavior determined by the electrical conductivity, 378
Thermal coupling experiments, 501, 502
Thermal instability, 407
Thermally induced negative differential conductivity (NDC), 370ff
   thermal boundary conditions for, 370
   the effect of inhomogeneities on, 375
   Newtons law of cooling, 372
Thermally stimulated capacitance technique, 50
Thermopower, *see* Seebeck coefficient
Thermalization of carriers, 95, 173, 250, 251
Thin film transistor,
   accumulation case of, 46, 282, 284, 286, 291
   addressing of, 279, 298
   adsorption effect of, 295
   channel length in, 44, 280, 296, 298
   channel width in, 44, 280
   conductance in, 45, 46, 48, 280

   contact resistance, influence of, 286ff, 291, 296
   contaminants, effect of, 295
   current-voltage characteristics of, 285, 290
   depletion case of, 46, 282, 288ff
   dielectric, influence of, 278, 291, 292, 293, 294, 295
   differential resistance in, 281
   display applications, use of, 276ff, 305ff
   distribution of charge in, 286
   dual gate, 299ff
   enhancement mode of, *see* accumulation case
   field effect mobility in, 292, 295
   gate electrodes in, 44, 280
   generalized theory of, 280ff
   hysterisis in, 295
   inversion case of, 46, 282
   off-current, 295
   on-current, 291, 295
   p-channel, 303ff
   performance of, 291ff, 297
   saturation voltage in, see threshold voltage
   space charge layer in, 280, 282
   stability of, 308
   stress, effect on, 307
   surface potential in, 281, 282
   surface states—effects of, 289ff
   threshold voltage in, 286, 288
   transconductance of, 291
   transfer characteristics of, 282, 286, 292
   turn-on time, 297
   vertical stacked, 300
   yield of, 306
Threshold switch, 383, 398
Tight binding approximation, 6, 39, 65
Topology, 4
Transparent conducting oxides, 246, 247, 248, 249

U

Urbach tail, 66, 67, 69, 187, 327

V

Valence alternation pairs (VAP), 37, 320ff, 333
Valence band tails, 36, 37